ALLE ZEIT WACH
1842

Springer Series in Wood Science

Editor: T. E. Timell

M. H. Zimmermann
Xylem Structure and the Ascent of Sap (1983)

J. F. Siau
Transport Processes in Wood (1984)

R. R. Archer
Growth Stresses and Strains in Trees (1986)

W. E. Hillis
Heartwood and Tree Exudates (1987)

S. Carlquist
Comparative Wood Anatomy (1988)

L. W. Roberts / P. B. Gahan / R. Aloni
Vascular Differentiation and Plant Growth Regulators (1988)

C. Skaar
Wood-Water Relations (1988)

J. M. Harris
Spiral Grain and Wave Phenomena in Wood Formation (1989)

B. J. Zobel / J. P. van Buijtenen
Wood Variation (1989)

P. Hakkila
Utilization of Residual Forest Biomass (1989)

J. W. Rowe
Natural Products of Woody Plants (1989)

K.-E. L. Eriksson / R. A. Blanchette / P. Ander
Microbial and Enzymatic Degradation of Wood and Wood Components (1990)

K.-E. L. Eriksson   R. A. Blanchette   P. Ander

Microbial and Enzymatic Degradation of Wood and Wood Components

With 173 Figures

Springer-Verlag Berlin Heidelberg New York
London Paris Tokyo Hong Kong

KARL-ERIK L. ERIKSSON
The University of Georgia, Franklin College of Arts & Sciences,
Department of Biochemistry, Athens, GA 30602, USA

ROBERT A. BLANCHETTE
University of Minnesota, Department of Plant Pathology, 495 Borlaug Hall,
1991 Buford Circle, St. Paul, MN 55108, USA

PAUL ANDER
STFI, Swedish Pulp and Paper Research Institute, Box 5604, S-11486 Stockholm,
Sweden

Series Editor:

T. E. TIMELL
State University of New York, College of Environmental, Science and Forestry,
Syracuse, NY 13210, USA

Cover: Transverse section of *Pinus lambertiana* wood. Courtesy of Dr. Carl de Zeeuw,
SUNY College of Environmental Science and Forestry, Syracuse, New York

ISBN-13: 978-3-642-46689-2 e-ISBN-13: 978-3-642-46687-8
DOI: 10.1007/978-3-642-46687-8

Library of Congress Cataloging-in-Publication Data
Eriksson, Karl-Erik. L. Microbial and enzymatic degradation of wood and wood components / K.-E. L. Eriksson, R. A. Blanchette, P. Ander. p. cm. –
(Springer series in wood science)
ISBN-13: 978-3-642-46689-2
1. Wood – Deterioration. 2. Wood-decaying fungi. 3. Cellulose – Biodegradation. 4. Microbial metabolism. I. Blanchette, R. A. (Robert A.) II. Ander, P. (Paul), 1941– . III. Title. IV. Series.

Softcover reprint of the hardcover 1st edition 1990

Typesetting: K+V Fotosatz GmbH, Beerfelden;

2131/3020-543210 – Printed on acid-free paper

Preface

The oil crisis during the 1970s turned interest towards the utilization of renewable resources and towards lignocellulosics in particular. The 1970s were also the cradle period of biotechnology, and the years when biotechnical utilization of lignocellulosic waste from agriculture and forestry gained priority. This was a logical conclusion since one of nature's most important biological reactions is the conversion of wood and other lignocellulosic materials to carbon dioxide, water and humic substances. However, while biotechnology in other areas like medicine and pharmacology concerned production of expensive products on a small scale, biotechnical utilization and conversion of lignocellulosics meant production of inexpensive products on a large scale. Biotechnical utilization of lignocellulosic materials is therefore a very difficult task, and the commercial utilization of this technology has not progressed as rapidly as one would have desired. One reason for this was the lack of basic knowledge of enzyme mechanisms involved in the degradation and conversion of wood, other lignocellulosics and their individual components. There are also risks associated with initiating a technical development before a stable platform of knowledge is available. Several of the projects started with enthusiasm have therefore suffered some loss of interest. Also contributing to this failing interest is the fact that the oil crisis at the time was not a real one. At present, nobody predicts a rapid exhaustion of the oil resources and fuel production from lignocellulosics is no longer a high priority.

The massive amount of effort devoted over the past few decades to a better understanding of the mechanisms of the degradation of wood and wood components provides a vast literature base that is essential for successful biotechnological utilization of lignocellulosic materials. To an academic scientist involved in these investigations, it is clear that these research investments have been a dazzling success. We hope that this book will demonstrate how, from the mid-1970s, this enormous library of knowledge has accumulated. It has been a complicated and difficult task to accomplish, and many countries, laboratories and scientists have been involved. This book is meant to serve as a source where this knowledge is combined in an understandable way.

This collation will hopefully provide important basic information of value to those interested in the biotechnological utilization of lignocellulosic materials. It should also be of interest to individuals working in the fields of forestry and forest products, botany, microbiology, mycology, ecology and biochemistry. It begins with a description of wood morphology and how wood components are decayed by different microorganisms, particularly by wood-rotting fungi. Most of this knowledge has been gained through electron microscopy studies. The chapter on cellulose describes the different mechanisms by which cellulose is degraded by different fungi and bacteria. It is evident that very rapid progress has been made in this area. It also turns out that degradation of the heteropolysaccharides, the hemicelluloses, is a more complex matter than had been expected and many different enzymes are involved in their conversion. The understanding of the mechanisms of lignin degradation has been particularly rapid, although a concerted research effort in this area did not really begin until 1978. The understanding of the physiological requirements and the enzyme mechanisms for the conversion of the lignin polymer into water-soluble degradation products will ultimately contribute to better processes for delignification and bleaching. Eventually, delignification will be carried out in a biomimetic way, i.e. catalyzed by low molecular mass organic compounds mimicking the ligninolytic peroxidases and other enzymes essential for the depolymerization process.

Acknowledgements

Many people have been involved in this production and we wish to thank in particular: Tore E. Timell, Editor of this series, for his valuble help and advice, for spreading good will and for fast and efficient handling of all questions relating to this project. Barbro Almqvist, Berit Brodén and Marian Farrow for skillful and dedicated typing of the manuscript. Sven Rasmusson, The Troedsson Library, STFI for valuable help with the collection of literature information. André Abad, Todd Burnes, Kory Cease, Lewis Otjen and Mike Carlson for technical assistance and Malcolm Brown, Jr., John Buswell, Ron Crawford, Debby Delmer, Kent Kirk, Jonathan Knowles, Matti Linko and Göran Pettersson for reading certain parts of the book.

Robert A. Blanchette has been mainly responsible for Chapter 1, Karl-Erik L. Eriksson for Chapters 2 and 3, and Paul Ander for Chapter 4.

Also I would like to thank the Swedish Pulp and Paper Research Institute for financial support and facilities for the period during which parts of this book were written.

Athens, January 1990 K.-E. L. Eriksson

Contents

Chapter 1

Morphological Aspects of Wood Degradation by Fungi and Bacteria

1.1 Morphology of Wood and Wood Components

The structure and chemical composition of wood have a significant influence on its degradation by microorganisms and the resulting patterns of decay. The type of cell, chemical composition, and cell wall morphology may all govern the effects of enzymes on the woody substrate. Wood is made up largely of cellulose, lignin, and hemicellulose in various proportions (Table 1.1). The lignin content of angiosperms is generally lower than that of gymnosperms. Differences in the major hemicellulose present are reflected in the xylose and mannose content (Table 1.1), which represents glucuronoxylan in hardwoods and galactoglucomannans in softwoods. Large differences in the chemical components exist among various woods (Fengel and Wegener 1983) and the composition of different types of cells within wood may also vary (Saka and Goring 1985). The type of lignin found in hardwoods and softwoods is also different.

Table 1.1. Percent lignin and wood sugar residues in sapwood of various tree species

Wood	Percent			
	Lignin[a]	Glucose[b]	Xylose[b]	Mannose[b]
Angiosperms				
Acacia koa	27.3	45.6	18.6	1.7
Acer saccharum	24.4	46.2	17.2	2.4
Alnus rubra	24.6	47.0	17.6	0.4
Betula papyrifera	19.0	44.9	24.3	2.1
B. verrucosa	21.7	35.1	20.7	0.9
Populus tremuloides	22.0	46.2	18.9	1.6
Quercus rubra	24.5	41.6	23.5	3.0
Tilia americana	21.5	43.9	16.1	0.3
Ulmus americana	23.6	55.8	16.0	3.1
Gymnosperms				
Abies balsamea	29.1	46.6	5.6	11.7
Picea abies	27.1	41.6	5.2	13.6
P. mariana	26.6	49.0	7.3	13.8
Pinus banksiana	29.9	44.6	8.4	10.0
P. resinosa	27.9	44.9	8.4	12.3
P. strobus	28.1	48.2	6.0	15.6
P. sylvestris	30.0	38.3	6.5	11.1
Tsuga canadensis	32.3	47.9	4.2	13.8

[a] Sulfuric acid lignin by the method of Effland (1977).
[b] HPLC analysis using method of Pettersen et al. (1985).

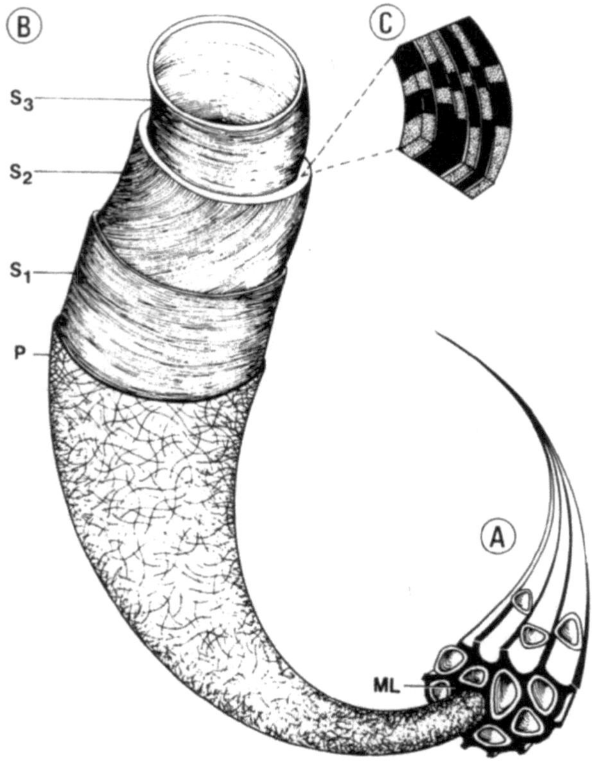

Fig. 1.1. **A** Tracheids, **B** cell wall layers; and **C** ultrastructural arrangement of lignin and carbohydrates in the secondary wall. *ML* Middle lamella; *P* primary wall; S_1, S_2, S_3 layers of the secondary wall. Each layer has a different thickness and microfibrillar orientation. Some species have an additional warty layer (not shown) over the S_3. The model depicted in **C** (redrawn from Kerr and Goring 1975 demonstrates the distribution of the lignin-hemicellulose matrix (*black*), hemicellulose (*white*), and cellulose fibrils (*stippled*)

The basic structural unit of lignin may be substituted in two or three positions; the addition of one methoxyl group to the phenol ring results in a guaiacyl unit, the addition of two methoxyl groups results in a syringyl unit. Hardwoods contain varying ratios of syringyl and guaiacyl types of lignin, whereas conifers have primarily guaicyl lignin (Fengel and Wegener 1983).

Lignin is distributed throughout the secondary wall and compound middle lamella, but the greatest concentration is in the middle lamella. The secondary wall constitutes a large proportion of the total cell wall area, and most (60% to 80%) cell wall lignin is located in this region (Fergus 1969, Fergus et al. 1969, Musha and Goring 1975, Saka and Thomas 1982a, b). The distribution of lignin in the cell wall of a birch fiber, as determined by bromination and X-ray microanalysis, is presented in Chap. 2. The distribution of hemicellulose parallels that of lignin within the wall (Parameswaran an Liese 1982). Hemicelluloses surround the cellulose microfibrils and occupy spaces between fibrils.

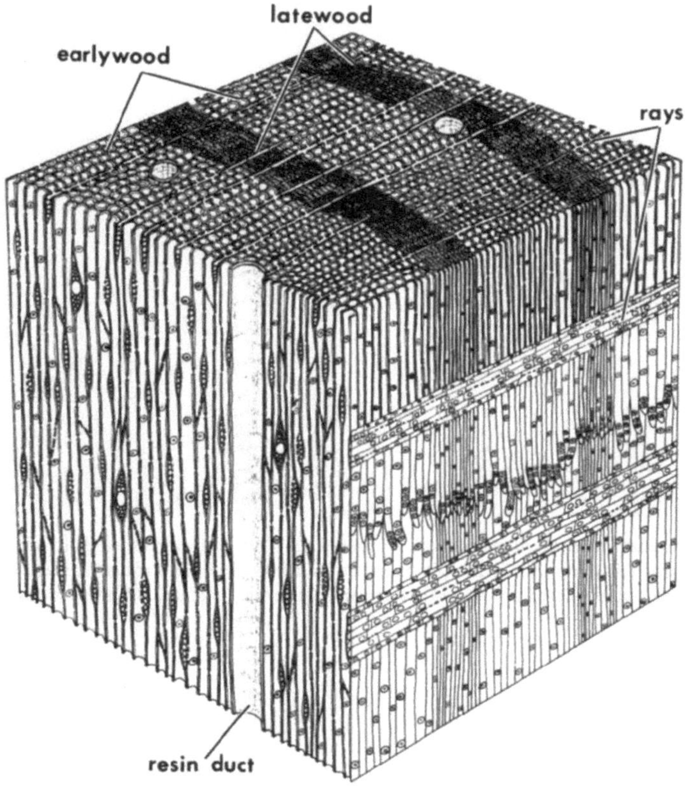

Fig. 1.2. A block of sound *Pinus strobus* wood with tracheids in earlywood and latewood regions, uniseriate and fusiform rays, and resin ducts

Crystalline and amorphous forms of cellulose occur within the cell wall. Cellulose molecules form microfibrils, and these form fibrils (see Chap. 2). A model depicting the arrangement of lignin, hemicellulose, and cellulose within the cell wall has been proposed by Kerr and Goring (1975). In this model a matrix of lignin and hemicellulose encrusts the cellulose fibrils (Fig. 1.1).

The woody cell wall is composed of various layers (Fig. 1.1). The middle lamella and primary wall make up the compound middle lamella, which is located between the secondary walls of adjacent cells (Core et al. 1979). The secondary wall has three layers designated S_1, S_2, and S_3 (Fig. 1.1). The S_3 is located nearest the cell lumen. The S_2 region, usually the largest, is the middle layer, and S_1 is the outermost layer of the secondary wall. The layers are distinct from one another because the cellulose occurs in different microfibrillar orientations around the cell axis in each layer.

Gymnosperm wood is composed of tracheids and parenchyma cells that occur in latewood or earlywood regions (Fig. 1.2). The tracheids overlap other tracheids and are connected by way of bordered pits (Fig. 1.2). The bordered

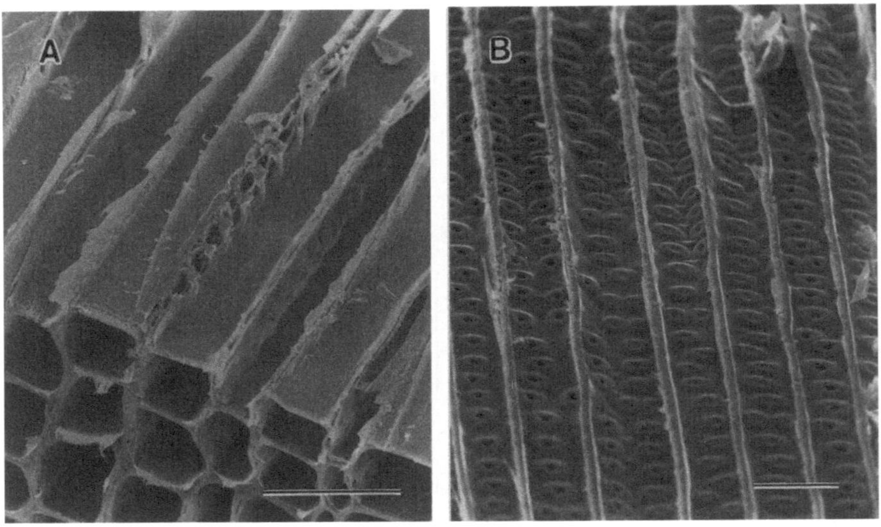

Fig. 1.3. Tracheids of sound *Pinus monticola* wood. Bordered pit pairs connect tracheids to adjacent tracheids. When tracheids join a ray, the pit pairs are half-bordered. The radial surface of tracheids may have numerous bordered pits. **A** Transverse and tangential surface. **B** Radial surface. SEM; *bar* = 50 μm

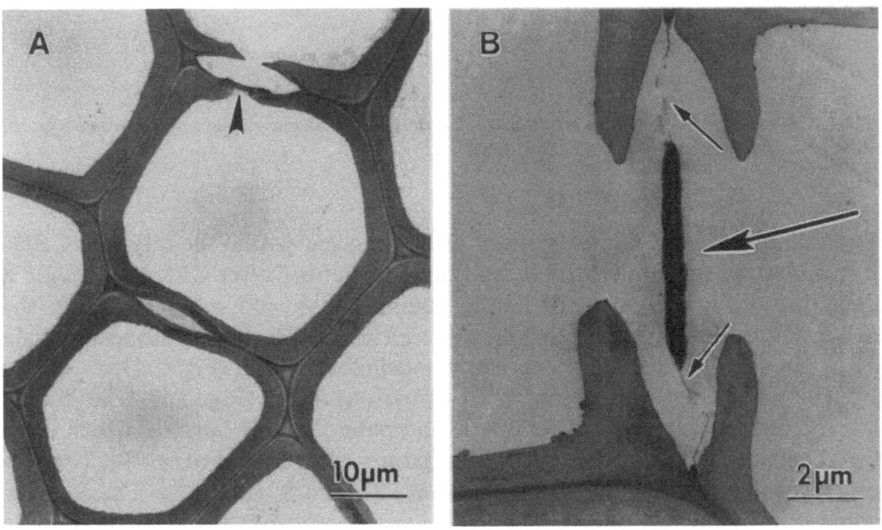

Fig. 1.4. A Tracheids of sound *Pinus strobus*, showing cell wall layers and **B** bordered pit pairs. The bordered pit pair has a membrane in the pit aperature that consists of a margo (*small arrows*) and torus (*large arrow*). When the bordered pit pair is aspirated, the torus is pressed against the pit aperature (*arrowhead* in **A**). Transverse section; TEM

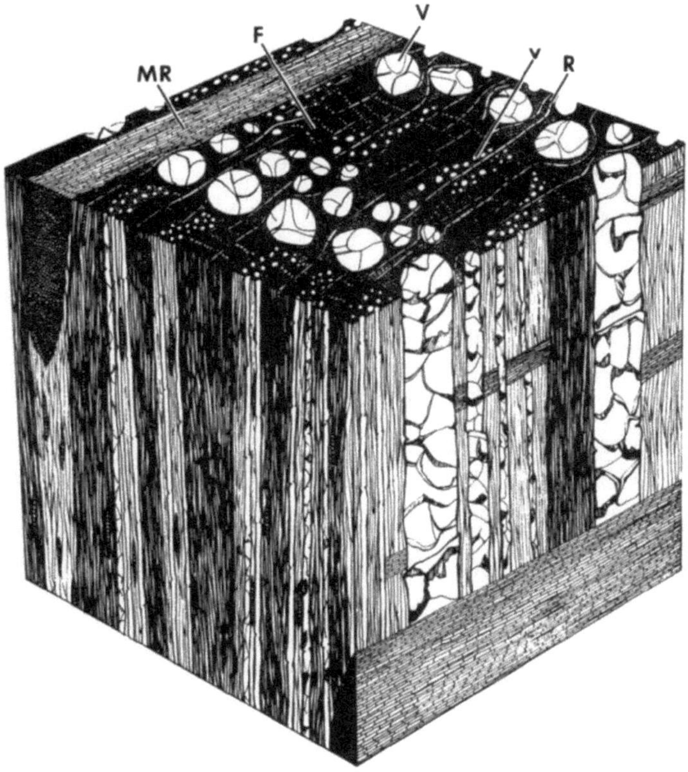

Fig. 1.5. A block of sound *Quercus alba* heartwood. This ring-porus species has large earlywood vessels (*V*) and smaller tracts of latewood vessels (*v*) surrounded by axial parenchyma cells, uniseriate rays (*R*), medullary rays (*MR*), and fibers (*F*). The vessel elements of white oak heartwood contain tyloses that occlude the vessels

pit projects out into the lumen and has a membrane in the pit chamber (Figs. 1.3 A, B; 1.4 A, B). This membrane may become aspirated when it moves to the side of the chamber to close the pit aperture. The radial wall of tracheids often has numerous bordered pits (Fig. 1.3 B). The circumference and thickness of the wall vary depending upon the location of the tracheid in the earlywood or latewood zone of the annual ring (Fig. 1.2). Compression wood tracheids, however, are very different than normal tracheids. In compression wood, the secondary wall layers are reduced to only an S_1 and S_2, and the S_2 layer is very large with helical checks that follow the microfibrillar angle (Timell 1982, 1986). Longitudinal resin ducts surrounded by thin-walled parenchyma cells are also found in many conifers (Fig. 1.2).

There is tremendous diversity in wood structure among the hundreds of hardwood species that grow in the temperate zones of the world (Panshin and de Zeeuw 1980). In general, the annual growth rings of angiosperms are composed of an array of vessels, fibers, and ray parenchyma cells in various ar-

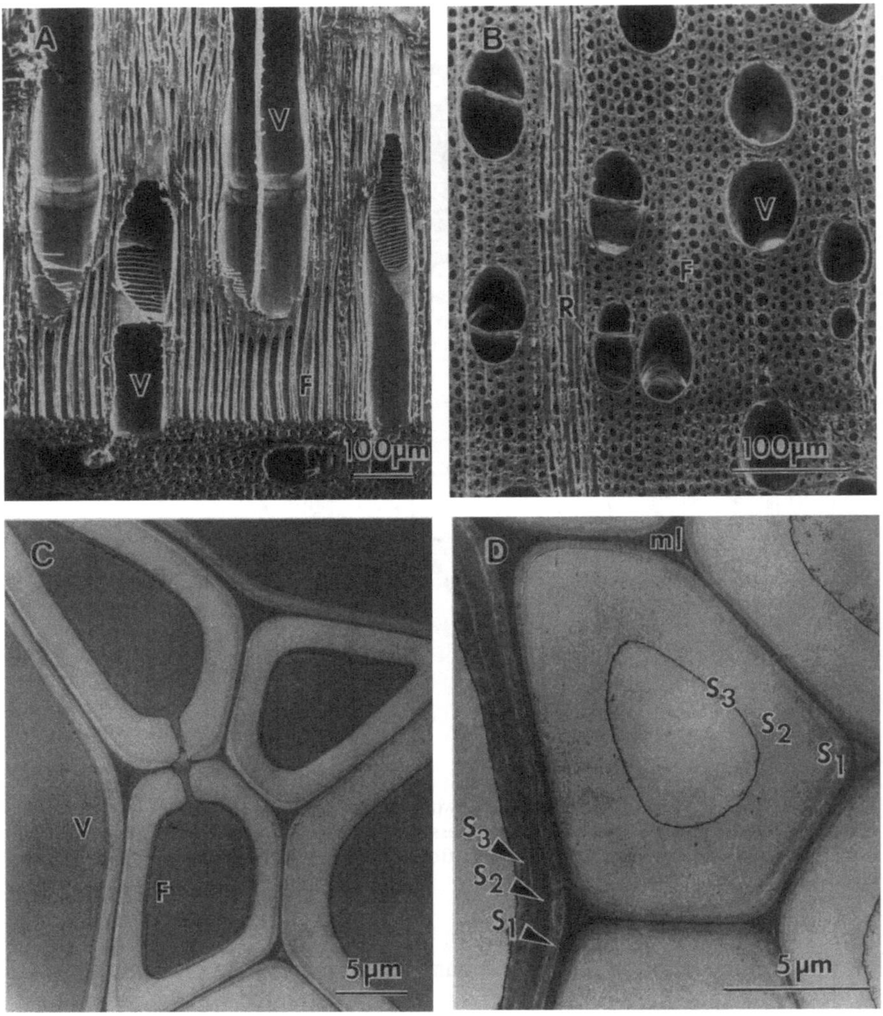

Fig. 1.6 A–D. Sound cells of *Betula papyrifera* showing vessel elements (*V*), with scaliform perforation plates, fibers (*F*) and ray parenchyma cells (*R*). The size of the cell wall and cell wall layers varies between fibers and vessels. Fibers have large S_2 layers and relatively small S_1 and S_3 layers. A fiber and an adjacent vessel share a common middle lamelle (*ml*). **A** Radial surface and **B** transverse surface; SEM. **C**, **D** Transverse sections; TEM

rangements (Fig. 1.5). These cells function for water conduction, support, and transport and storage of nutrients. The vessels may be distributed across the entire growth ring as in *Betula* and *Acer* (Fig. 1.6 A, B), or large vessels may be concentrated in the earlywood and smaller vessels across the annual ring (Fig. 1.5). The thick-walled fiber tracheids and libriform fibers are located around the vessels and are connected to other cells via bordered or simple pits

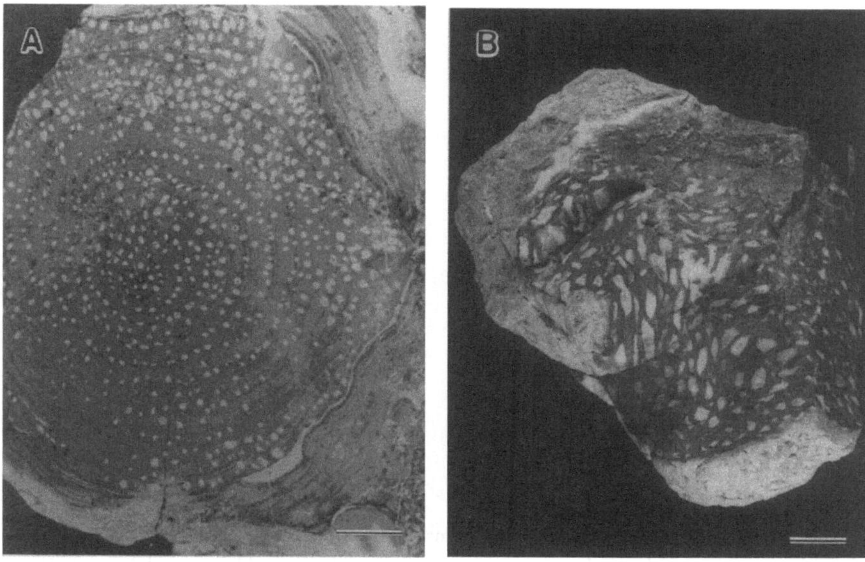

Fig. 1.7. Fossilized *Araucarioxylon* wood, a silicified gymnosperm genus from the Triassic age, found in the Fremouw Peak locality, Antarctica. The fossil wood contains a white-pocket rot with spindle-shaped pockets of delignified cells. **A** Cross-section of fossil wood with white-pockets (Stubblefield and Taylor 1986). **B** Photograph of fossil wood with multiple planes of the rot visible. *Bar* = 1 cm (Courtesy of S.P. Stubblefield and T.N. Taylor)

(Panshin and de Zeeuw 1980). The longitudinal parenchyma and ray parenchyma have relatively thin walls, and the pits range from simple to bordered and minute to very large. The composition and size of various cell wall layers may be different depending on the type of cell (Fig. 1.6C, D). Vessels have larger S_1 and S_3 layers and a smaller S_2 layer than fibers.

Wood is altered as a result of the normal aging process, and heartwood forms (Shigo and Hillis 1973, Shigo 1986, Hillis 1987). These inner zones of woody stems contain no living cells, and transport of water ceases. As heartwood is being formed, large amounts of extractives are deposited in the cells. Aliphatic compounds, terpenes, and phenolic substances accumulate in the heartwood (Hillis 1968, Sjöström 1981) and may exert a significant amount of resistance to the activities of microorganisms.

Under natural conditions organic matter is quickly decomposed by microorganisms, and complex structures are reduced to simpler forms. Environmental and substrate conditions must be conducive to microbial activity, or the degradation process will be interrupted. A good example to demonstrate this is found in wood from the St. Maries River Area, Clarkia, Idaho fossil beds (Smiley et al. 1975). Wood was covered with lake sediments, which excluded air for 15 to 25 million years. The wood structure after this prolonged period of entombment was well preserved and even the delicate pit membranes within the bordered pits were intact (Blanchette, unpublished). Usually, degradation

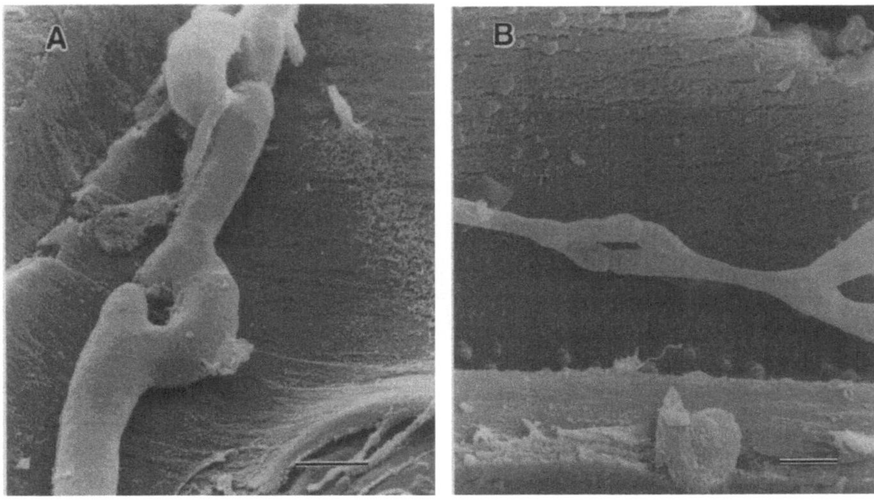

Fig. 1.8. Hyphae of Basidiomycetes with clamp connections in lumina of *Pinus monticola*. **A** Clamp connection of *Coriolus versicolor*, a white-rot fungus; **B** medallion clamp connection of *Poria placenta*, a brown-rot fungus. Radial surface; SEM; *bar* = 2 μm

is not delayed, and woody tissues are rapidly colonized and degraded as soon as the plant tissue dies. Commonly, even the heartwood of living trees is degraded. Today, fungi are the major decomposers of wood, and evidence suggests that they have been the major decomposers of higher plants since the Upper Devonian period (Stubblefield and Taylor 1986, Stubblefield et al. 1985). Evidence of wood decay has been found in fossilized gymnosperms collected in Antarctica from the Triassic and Permian periods (Fig. 1.7 A, B). The decay, found in species of *Araucarioxylon* and *Vertebraria*, was caused by a white-rot Basidiomycete. Extensive lignin degradation had occurred, and white-pockets consisting of delignified tracheids were observed. This decay was very similar to some present-day rots caused by Basidiomycetes that attack the heartwood of living conifers (Blanchette 1980a, Otjen and Blanchette 1986a, b.) Those early gymnosperms were probably highly lignified and contained large amounts of phenolic extractives in their heartwood. It is not suprising to find that the first evidence of decay resulted from fungi that tolerate high concentration of phenolic substances and have a superior capacity for lignin degradation.

The largest group of fungi to degrade wood are the Basidiomycetes. In North America alone there are 1600 to 1700 species of wood-rotting Basidiomycetes (Gilbertson 1980). Most of those species, which have dikaryotic hyphae, have clamp connections along the septate mycelium (Fig. 1.8 A, B). These structures are a unique characteristic common to the Basidiomycetes and was one of the key features observed in the fossilized decay from Antarctica (Stubblefield and Taylor 1986). Many different forms of clamp connections may be found (Fig. 1.8 A, B), and multiple whorls of clamp connections may also occur.

Hyphae in wood often have various substances deposited along their length. Fungi are known to excrete substantial quantities of organic acids as part of their normal metabolism, and substances such as oxalic acid may accumulate in large amounts as the oxalate salt (Cromack et al. 1977). These crystalline structures are often observed along the outside of basidiomycetous hyphae and may take unusual forms (Fig. 1.9 A to C). Accumulation of calcium in excess of physiological needs and release of calcium in the form of stable complexes appear to have significant nutrient recycling implications in forest ecosystems (Cromack et al. 1975). Although fungi do not appear to break down calcium oxalate, some species of bacteria can degrade it, forming calcium carbonates. These substances are subsequently utilized by soil insects and earthworms (Cromack et al. 1977). Oxalate is an extremely effective chelator of cations such as Fe and Al, and it may have important implication in the decomposition process of wood as well as in the weathering of soils.

Wood decayed by Basidiomycetes has been separated into categories based on different macroscopic characteristics. Pronounced and varied color changes occur after advanced stages of decomposition. Decayed wood can turn light to dark brown or can be bleached white (Kirk and Cowling 1984). The terms brown rot and white rot have been used to classify the two types of decay. Only about 6% of the total number of North American wood-rotting fungi cause brown rots (Gilbertson 1981). These relatively few species are primarily found associated with conifers. However, there are also some brown-rot fungi that attack hardwoods. Brown rots are also the major cause of decay to wood in service. Usually brown-rotted wood appears with a reddish-brown color, and it breaks into cube-like pieces due to the formation of shrinkage cracks (Fig. 1.10A). The brown rot may be distributed throughout the substrate, or it may occur in localized areas delimited by adjacent sound wood (Fig. 1.10B). When dry, brown-rotted wood easily breaks into a fine powder (Fig. 1.10C).

Brown-rotted wood has an important role in the functioning of forest ecosystems (Stark 1972). The total amount of brown-rotted wood in forest soils may be as high as 30% of total soil matter (McFee and Stone 1966). Brown-rotted residues are areas of high microbial activity and act as nutrient and moisture sinks. The sponge-like absorbent properties of brown-rotted wood allows it to maintain a high moisture content even during dry periods of the year. Tree feeder roots, ectomycorrhizal fungi, and nitrogen-fixing bacteria are frequently associated with it (Harvey et al. 1976, Larsen et al. 1979, Gilbertson 1981). Decayed woody residues in the soil appear to be essential to sustain good site quality and tree growth (Harvey et al. 1976). Greater consideration must be given to decaying forest residues in forest management plans to ensure the success of intensive forestry practices.

A different type of decay that is similar in macroscopic appearance to brown-rotted wood is caused by soft-rot fungi (Savory 1954). Although brown and soft rot look similar, differences in microscopic characteristics are used to differentiate the two types of decay (Chap. 4). Fungi in the *Ascomycotina* and *Deuteromycotina* are responsible for the soft rot. The decay is usually associated with wood in contact with high levels of moisture, and a brown, soft decay

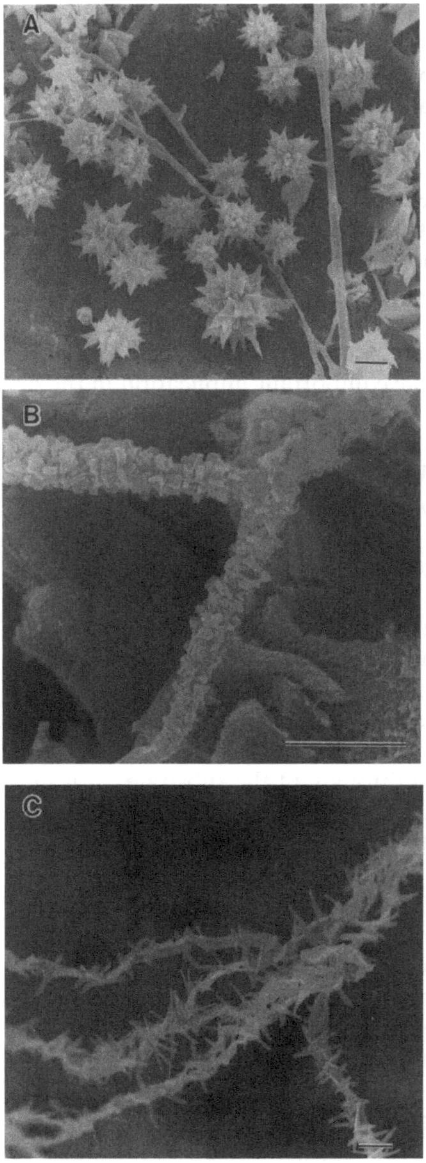

Fig. 1.9 A–C

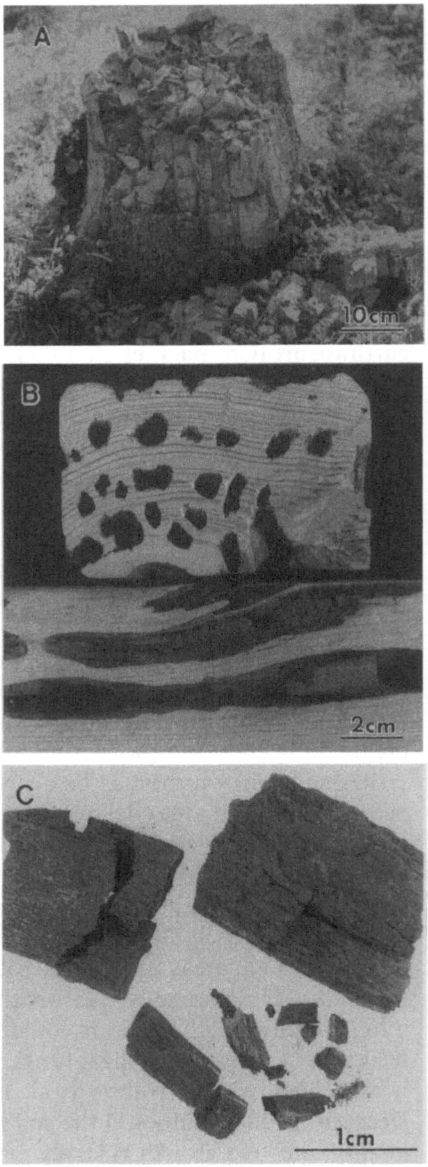

Fig. 1.10 A–C

Fig. 1.9 A–C. Different forms of calcium oxalate crystals deposited on the hyphae of *Hirschioporus abietinus* (**A**), *Poria placenta* (**B**), and an unidentified Basidiomycete isolated from forest soil (**C**). SEM; *bar* = 5 µm

Fig. 1.10 A–C. Brown-rotted wood. **A** Brown cubical decay in stump of a hardwood species. **B** Large "pockets" of brown cubical decay found in the heartwood of *Thuja*. **C** Dried wood with brown rot cracked and checked into cubical pieces. The decayed wood is crushed into a fine powder with only slight pressure

Fig. 1.11. Soft-rot decay of redwood members from cooling towers. The attack by soft-rot fungi causes a softened, brown surface of the wood that cracks and checks when dry (Duncan 1960). Courtesy (US Forest Products Laboratory, Madison WI)

occurs at the wood surface (Fig. 1.11). Degradation progresses slowly into the substrate over extended periods of time. When dry, the decayed wood cracks and checks and is easily crushed to a fine powder. These features are also characteristic of brown-rotted wood. Soft-rot decay has been found associated with wooden slats used in industrial cooling towers and wood in contact with the soil. Soft-rot fungi appear to be important decomposers of wood subject to extreme wetness or even frequent dryness (Duncan 1960), where more aggressive wood degrading fungi, such as the Basidiomycetes, cannot survive.

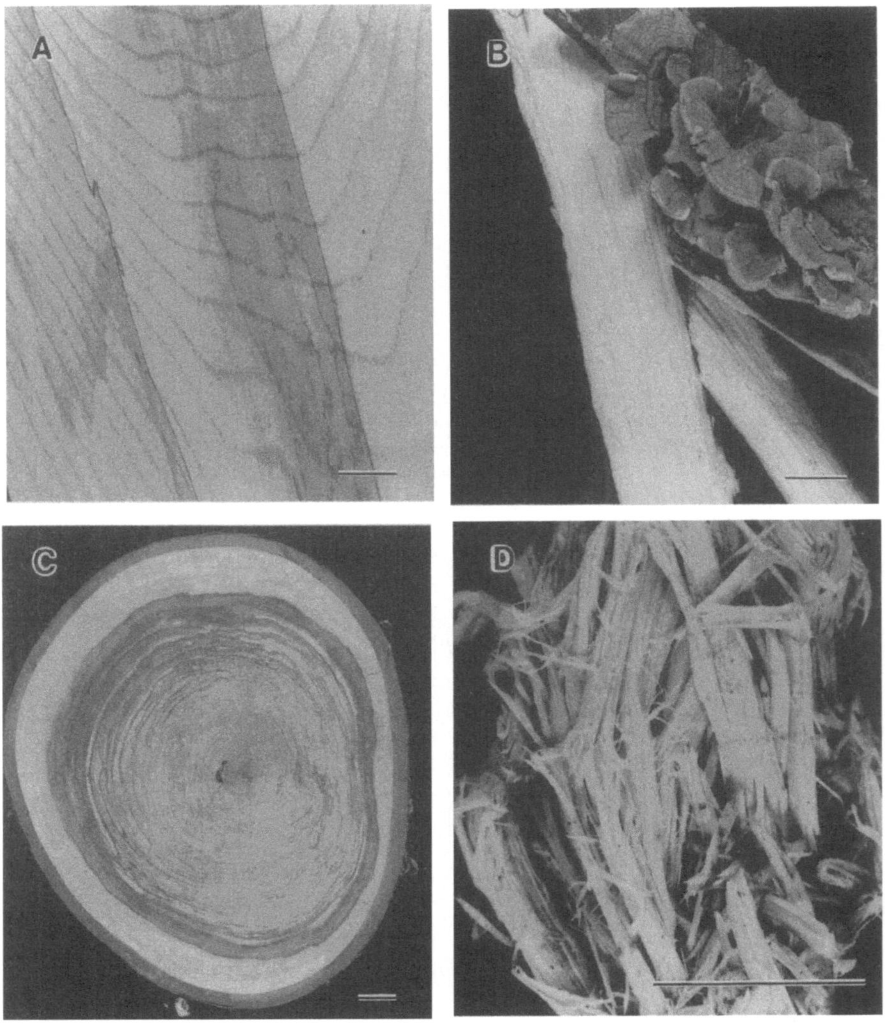

Fig. 1.12. White-rotted wood. **A** Board with incipient stages of white rot showing a "bleached" white appearance. **B** Wood with advanced stages of white rot and basidiocarps of *Coriolus versicolor*. **C** Cross-section of living *Betula papyrifera* with white-rotted wood that is surrounded by discolored wood. The decay is caused by *Inonotus obliquus*. **D** Advanced stages of stringy rot in wood of a deciduous species. This type of white rot is not characterized by a white coloration. Instead, the decayed wood is yellow to tan and can be easily separated into string-like fragments. *Bar* = 2 cm

A large amount of diversity exists among the macroscopic decay characteristics caused by white-rot fungi. In the incipient stages of decay, wood is bleached to a white coloration (Fig. 1.12A). Advanced stages of decay commonly are uniformly white throughout the substrate (Fig. 1.12B). White-rot fungi may attack wood of living trees, causing a central column of defect (Fig.

1.12C). White-rotted wood in a living tree is always preceded or surrounded by a zone of discolored wood. The white-rot fungus may move directly into the discolored wood as a pioneer organism (Blanchette 1982b) or colonize it only after a succession of microorganisms have altered the occluded xylary cells (Shigo 1967). The final coloration of the decayed wood varies from whitish-yellow to whitish-tan. The decayed wood may be brittle, soft, and spongy or separates into string-like fragments (Fig. 1.12D).

White rots can also cause localized areas of degradation. These white zones may be in the form of small spindle-shaped pockets surrounded by sound wood (Fig. 1.13A,B), long narrow pockets (Fig. 1.13C), or large patches of white tissues (Fig. 1.13D). White-pocket rot occurs frequently in the heartwood of living trees, and the white zones are found within individual annual rings, as with the decay by *Phellinus pini* (Fig. 1.13A, B), or can traverse many annual rings and cause large pockets, such as those produced by *Phellinus nigrolimitatus* (Boyce 1961). In tropical tree species, a recently identified white-pocket rot fungus, *Phellinus kawakanii* (Larsen et al. 1985), produces extensive white pockets that are not delimited by annual rings since the trees have a constant growth pattern throughout the year. In some types of decay, the white tissues within the pockets remain until very late stages of degradation, whereas in other decays the delignified cells within pockets completely deteriorate. The resulting decay has a honeycomb appearance, and only the wood that surrounds the pocket is left. The white-rot fungus first delignifies the cells, and after extensive lignin removal the remaining cellulosic white tissues are completely degraded. *Xylobolus frustulatus* (Otjen and Blanchette 1984), *Hirschioporus pargamenus* (Otjen and Blanchette 1986) and *Hirschioporus abietinus* (Blanchette 1980b) are fungi that cause white-pocket rots, but as decay continues the cellulose remaining in the pockets is also consumed. The wood between the white-pockets is not degraded.

Those white rots with a mottled appearance (Fig. 1.13D) also show morphological variation. In temperate areas, fungi such as *Ganoderma applanatum* cause a typical mottled decay composed of tan and white wood (Blanchette et al. 1985). The decay is a composite of different types of white rot, and the proportion of tan and white wood varies. In the evergreen rain forests of southern Chile the white zones produced by *Ganoderma applanatum* may be extensive and occupy the entire interior of logs (Dill and Kraepelin 1986). This white-rotted wood, called "Palo Podrido", has a high rumen digestibility and has been used as feed by foraging cattle for over a century (Gonzalez et al. 1986, Philippi 1893). The lignin content of this decayed wood may be as low as 1%–2% (Dill and Kraepelin 1986).

The patterns of white rot in wood are often governed by anatomical structures of the wood. Decay of *Quercus* species by *Inonotus dryophilus* is a good example of how various cell types affect decomposition. The fungus causes a white-pocket rot in heartwood of living *Quercus*. The pockets of white tissues are usually long, narrow regions that are separated by strips of firm, brown wood (Fig. 1.13C). The decay is commonly referred to as a white-piped rot or white-striped rot (Hedgecock and Long 1914, Boyce 1961). The fungus first

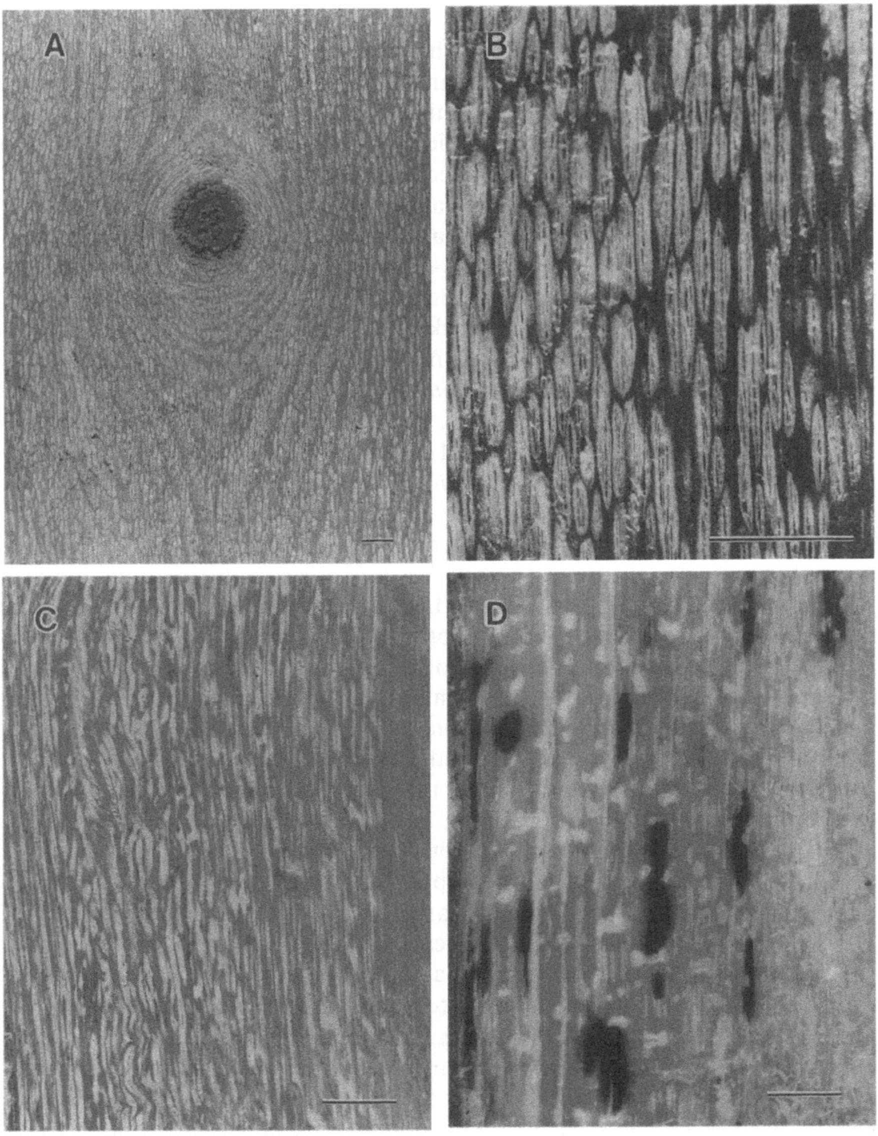

Fig. 1.13 A–C. White-rotted wood with white-pocket and **D** white-mottled appearance. **A** and **B** Decay caused by *Phellinus pini* in living *Tsuga canadensis* (**A**) and *Pinus monticola* (**B**) consisting of spindle-shaped pockets of white tissues (Blanchette 1980a, b). **C** Decay by *Inonotus dryophilus* in heartwood of living *Quercus alba*. The *white regions* are long, narrow pockets. **D** White-mottled decay in *Betula papyifera* caused by *Ganoderma applanatum*. Large white zones are dispersed throughout the tan colored wood. *Black regions* are often associated with the white zones (see Fig. 1.14). *Bar* = 1 cm

colonizes the large earlywood vessels and penetrates the numerous tyloses that occlude the vessel elements (Fig. 1.5). Vessels and adjacent axial parenchyma cells are degraded (Otjen and Blanchette 1982). The fungus selectively removes lignin from the cells without extensive cellulose depletion, and zones of white cells result. As degradation continues, the tracts of latewood vessels and parenchyma cells are also delignified to form white pockets. Latewood fibers, however, are not degraded. These cells become densely occluded with brown, pigmented substances and form the regions of sound wood between the pockets. Uniseriate rays are readily colonized and attacked by the fungus, but medullary rays are not and form barriers to degradation (Otjen and Blanchette 1982).

The anatomical structure of coniferous wood also influences the patterns of white-rot attack. Earlywood and latewood cells may affect colonization and subsequent degradation by white-rotters. *Phellinus pini*, which causes a white-pocket rot in the heartwood of living gymnosperms, shows a preference for latewood cells of *Pinus* and *Larix* species (Blanchette 1980a). In living trees, the degraded areas are first formed in the latewood, and small spindle-shaped white pockets form. After the tree has died, the white-pocket rot continues to degrade other annual rings. Long, unrestricted zones of decayed latewood cells are found, causing long, narrow white-pockets. All latewood cells are degraded, but the fungus does not decompose earlywood cells. In contrast, another white-rot fungus, *Ganoderma tsugae*, rapidly degrades earlywood cells of *Tsuga canadensis*, but not latewood cells (Blanchette 1984a). These different white-rot fungi apparently have a preference for different cells within the annual ring. Cells of compression wood in gymnosperms also are resistant to degradation by white-rot fungi (Blanchette 1982a, Timell 1986). The unusually high lignin content of compression wood and the lower methoxyl content of the lignin, resulting in a more condensed structure, are important factors that contribute to resistance of this wood toward decay (Jurasek 1964, Timell 1986).

Black spots and flecks within decaying wood are characteristically produced by many white-rot fungi. These features were first described and illustrated by Hartig (1878) for *Heterobasidion annosum* (Fig. 1.14A) and have been used to macroscopically identify certain types of decay (Hiley 1919, Hubert 1931, Boyce 1961). Until recently, the composition of these black substances was obscure. With the aid of X-ray microanalysis and atomic emission spectrometry, the black spots and flecks located in delignified cells were found to consist of extensive manganese dioxide deposits (Blanchette 1984a). In some areas, over a 100-fold increase in manganese was detected when compared to sound wood (Fig. 1.14B–D). Some fungi produce irregular patches of black zones while others produce small black flecks (Fig. 1.15A, B). The manganese deposits have been found to be located within cell lumina, covering the lumen surface and fungal hyphae (Fig. 1.15C). The deposits also may be found throughout the delignified cell walls (Fig. 1.15D). Other elements, such as phosphorus, calcium, and potassium occur in the black area of decayed wood in slightly higher concentrations than in sound wood. These concentrations were not unusually high for substrates degraded by Basidiomycetes (Cromack et al. 1975, Vogt and Edmonds 1980, Vogt et al. 1981). The accumu-

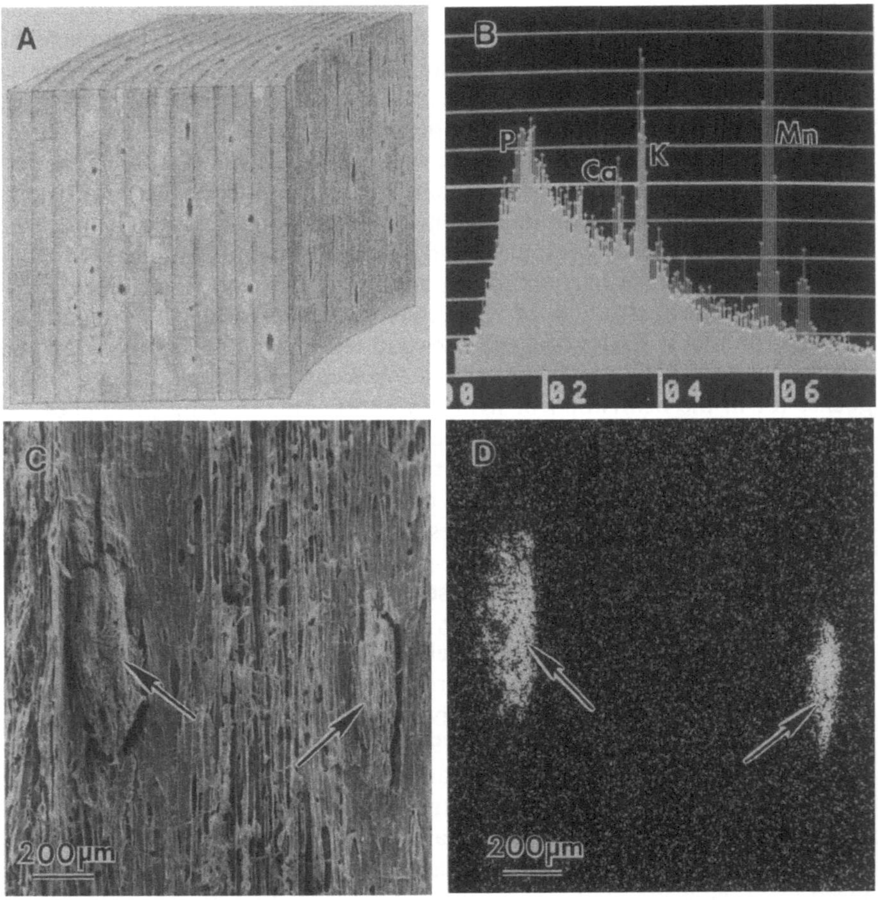

Fig. 1.14. Manganese accumulation in white-rotted wood. **A** A drawing by Robert Hartig (1878) of a block of pine wood decayed by *Heterobasidion annosum*. *Black spots* occurred within the white-pockets. **B** X-ray microanalysis demonstrated the composition of black areas in the white-pocket rot caused by *Heterobasidion annosum* (Blanchette 1984b). The graph shows two superimposed X-ray spectra of decayed wood from white-pocket (*white graph*) and black area in center of pocket (*gray graph*) collected for 100 s. The K-alpha energy peak for manganese shows very high concentrations of manganese in the black area. Phosphorus, calcium, potassium, and other elements were in equal proportions in the white-pocket and black areas. **C** Wood decayed by *Heterobasidion annosum* with two black spots (*arrows*) in the center of white pockets; SEM. **D** X-ray microanalysis dot mapping for manganese showing the same area in **C**. *Arrows* indicate location of black spots (Blanchette 1984b)

lation of manganese in excessively large amounts only occurred in the black areas of delignified wood. No manganese accumulation has been reported for basidiocarps, rhizomorphs, pseudosclerotial plates, or other structures produced by Basidiomycetes. The black areas are different from any other black stromatic structure produced by fungi.

Fig. 1.15. Manganese deposits in white-rotted wood. **A** Large black patches of manganese deposits in *Tsuga canadensis* wood decayed by *Ganoderma tsugae*. **B** Small black flecks in *Pinus resinosa* wood decayed by *Dichomitus squalens*. **C** Manganese dioxide deposits in cell lumen of *Tsuga canadensis*, covering lumen surface and hyphae of *Ganoderma tsugae*. **D** Manganese deposits within delignified tracheids after advanced stages of cell wall degradation. **A, B** Longitudinal surface of decayed log. **C** Radial surface; SEM. **D** Transverse section; TEM

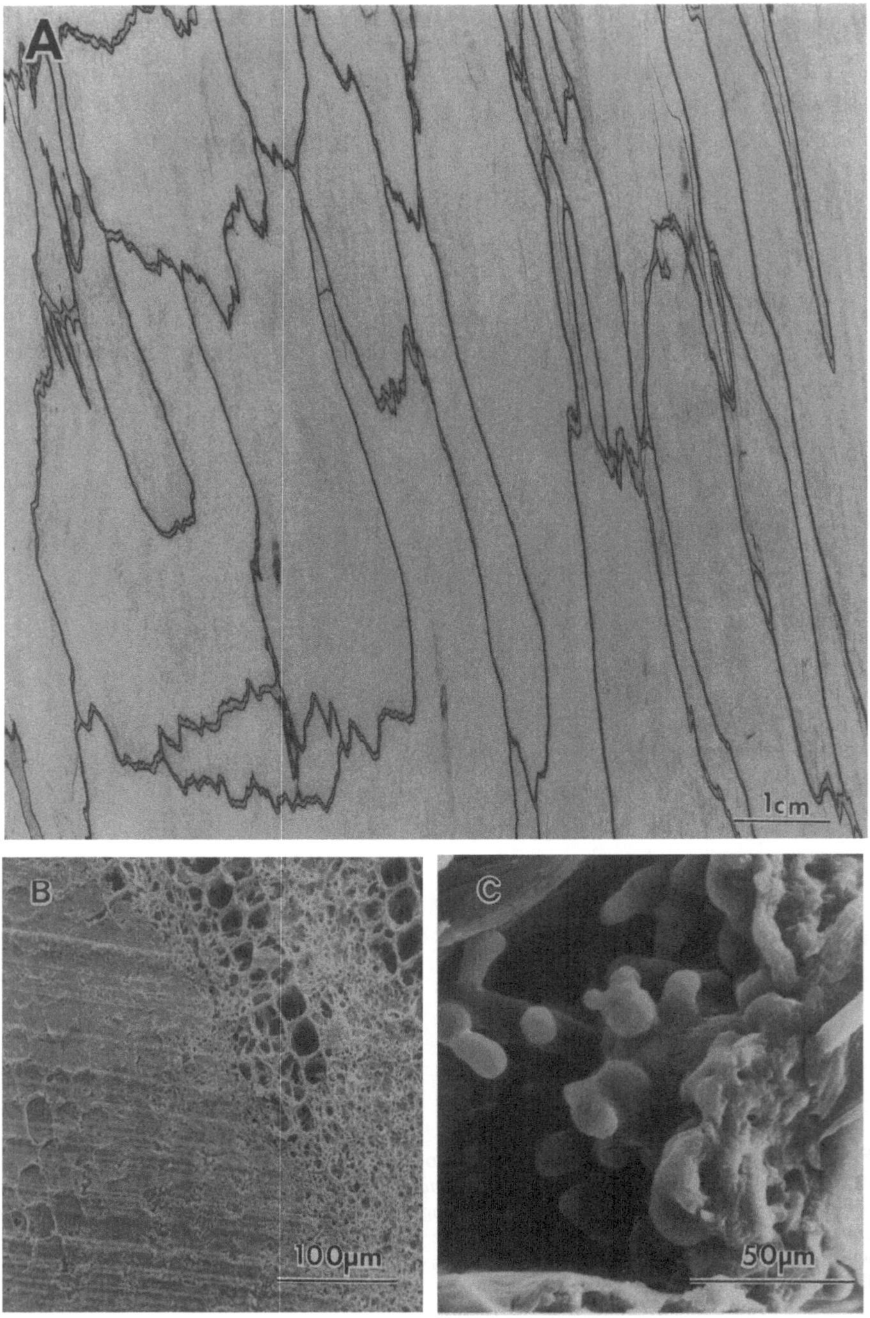

Fig. 1.16. Pseudosclerotial plates or zone lines in white-rotted wood. The pseudosclerotial plate appears as *thin black lines* on the radial surface of a decayed log (**A**). The black line is actually part of a three-dimensional plate that separates different areas of the decayed wood. The pseudosclerotial plate consists of fungal mycelia and copious amounts of pigmented substances that occlude cell lumina (**B**). The hyphae within the plate area have thick walls and appear swollen (**C**). **A** Longitudinal surface of decayed log. **B** Transverse surface; SEM. **C** Radial surface; SEM

Recently, manganese-dependent peroxidases have been shown to play a central role in the fungal decay of lignin by white rots (Paszazynski et al. 1985); (Chap. 2). The precipitation of manganese oxides appears to establish a gradient that results in additional soluble forms of manganese being transported into the area (Blanchette 1984b). The deposition of manganese is most often associated with white-pocket or white-mottled rots, but manganese can also be distributed as tiny flecks throughout the decayed wood caused by *Fomes fomentarius*, *Coriolus versicolor*, or other white rot fungi.

Pseudosclerotial plates within wood (Fig. 1.16A) are formed in relation to the moisture content of the substrate (Hopp 1938, Lopez-Real and Swift 1975), demarcation zones between different species of fungi, or between genetically distinct mycelia of the same species (Rayner and Todd 1977). The process of pseudosclerotial plate formation involves a proliferation of hyphae within the zone, hyphal swelling, causing bladder-like cell formation, and copious amounts of pigmented substances that occlude the wood (Fig. 1.16B, C). Once formed, these structures resist degradation by fungi, bacteria, and insects. Gradually, the wood below the pseudosclerotial plates, at the perimeter of a log or between plates within the substrate, may be completely degraded, leaving only the pseudosclerotia (Campbell and Davidson 1940, Manion 1981).

Many fungi do not affect the integrity of the wood but may impart a discoloration or stain in the sapwood of conifers or hardwoods (Fig. 1.17A). The stain may be superficial or penetrate deeply into the sapwood, causing a stain that is blue, brown, red, or a variety of other colors. These fungi primarily colonize ray parenchyma cells and utilize stored nutrients (Seifert 1964). Hyphae, however, may be found ramifying within lumina of any cell (Hubert 1929). Cell wall deterioration is not associated with this type of condition, and no loss of wood strength occurs (Liese 1970). Hyphae pass from one cell to another by growing through the pit membranes or directly through the cell wall (Fig. 1.17B). To penetrate the cell wall, the hyphal tip forms a specialized structure, called the transpressorium (Liese and Schmid 1964). This constricted hypha of small diameter moves into the wall and bores a hole through all cell wall layers (Fig. 1.17B). Once it has reached the lumen of the adjacent cell, the hypha resumes normal diameter. It has been suggested by Liese (1970) that bore holes from hyphae of stain fungi result from a combination of localized enzymatic action at the tip of the penetrating hyphae and mechanical pressure.

One common feature that has been observed around hyphae of fungi that stain wood or hyphae of brown, soft, and white-rot fungi is the extracellular mucilage or slime layer (Schmid and Liese 1966, Crossley 1980, Highley et al. 1983, Palmer et al. 1983a, b, Blanchette and Reid 1986). These hyphal sheaths are known to contain proteinaceous materials and various polysaccharides (Olah and Reisinger 1974, Van der Valk et al 1977). They coat the growing tip and also occur along the length of the hypha but are not found around necrotic hyphae (Palmer et al. 1983a). Although the role of this mucilage is uncertain, it does appear to be an important source of support and hyphal contact with the cell wall. Several investigations suggest that the sheath contains cellular contents that are released from the hypha (Palmer et al. 1983a, b). The muci-

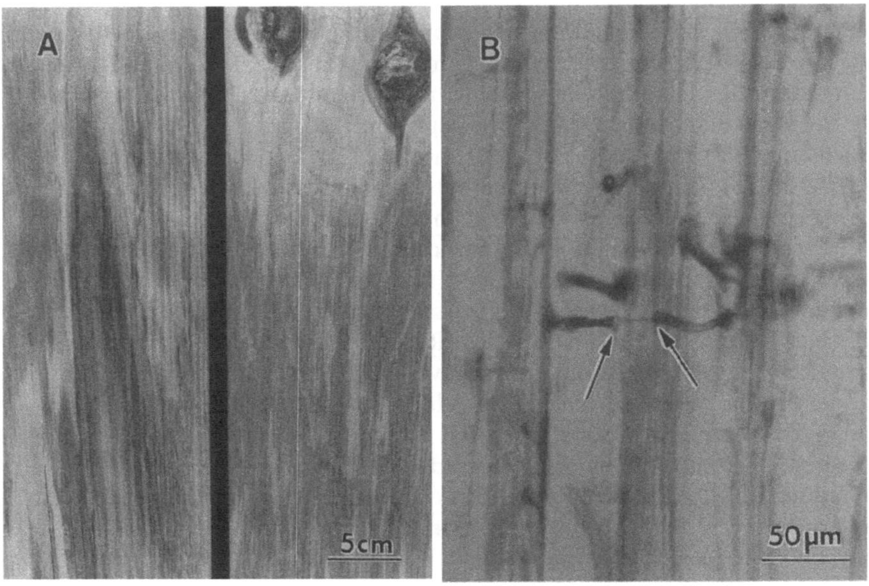

Fig. 1.17. A Boards from *Pinus strobus* with brown (*left*) and blue (*right*) stain in the sapwood caused by *Valsa* and *Ceratocystis* species, respectively. **B** Hyphae of stain fungi move from cell to cell by forming a fine hypha (*arrows*) that penetrates the wall. Once the fine hypha has reached the lumen of adjacent cell, the hyphal diameter resumes normal size. Stain fungi may grow through any cell wall but hyphae primarily colonize the ray parenchyma cells and utilize ray cell contents. **A** Radial and tangential sections of stained log. **B** Radial section; LM

lage appears to house and transport depolymerization agents for cell wall attack by decay fungi or, in the case of stain fungi, to utilize the cell contents of ray parenchyma cells. The extracellular slime layer also seems to move nutrients and degradation products efficiently from the wood back to the hyphae. Bes et al. (1987) have characterized an extracellular glucan produced by *Phanerochaete chrysosporium*. The production of a polysaccharide envelope around hyphae appears to regulate glucose levels in the medium (Eriksson et al. 1986) and forms a microenvironment where H_2O_2, needed for lignin degradation (Chap. 4), is maintained. It might also provide optimum pH levels for lignin degrading enzymes.

1.2 Wood Degradation by White-Rot Fungi

Fungi that cause white rot degrade extensive amounts of lignin from cell walls of angiosperms and gymnosperms and they also degrade cellulose and hemicellulose. The order or proportion in which these wood components are decomposed is not uniform, and great variation may be found in the type of white rot produced. The largest group of fungi that cause white rot belong to the Basidiomycotina with several thousand species causing decay in hardwoods and softwoods worldwide (Gilbertson 1980). There are also many Xylariaceous and Diatrypaceous fungi in the Ascomycotina that produce a white rot in hardwoods, but their role in the decay of conifers is not known (Rogers 1979).

Robert Hartig (1878) characterized white rots by macroscopic and microscopic differences present in the decayed wood. The bleached white color of advanced decay and the ability to degrade lignin suggested that different enzyme systems were functioning in white- and brown-rot fungi. Methods to differentiate brown- from white-rot fungi were quickly developed (Bavendamm 1928). One of the easiest methods that is still used today for identification of wood-rotting fungi in culture (Davidson et al. 1938, Nobles 1958) is an oxidase reaction indicated by the formation of a colored zone in agar containing polyphenolic compounds. Most white-rot fungi test positive while most brown-rot fungi do not form the colored zone. This test successfully separates most white-rot fungi from brown-rotters, but it has also been modified to take into account differences in phenol oxidase activity among white-rot fungi (Ander and Eriksson 1977, 1987). The intensity of reaction by different white-rot fungi varies, and several species, such as *Phanerochaete chrysosporium* and *Xylobolus frustulatum*, which agressively attack lignin, have a negative or weak oxidase reaction (Kirk and Kelman 1965, Ander and Eriksson 1977, Setliff and Eudy 1980). Since fungi such as *Phanerochaete chrysoporium* produce at least two quinone reducing enzymes (Chap. 4), routine spectrotometric assays dependent on the production of colored quinones and using syringaldazine as a substrate, may not be reliable. Another method, which is a more sensitive assay of phenoloxidase activity, measures the release of $^{14}CO_2$ from decarboxylation of vanillic acid (Ander and Eriksson 1987). This assay is not dependent on the formation of colored quinones but is contingent on the irreversible formation of CO_2 from the carboxylic group during enzymatic reaction.

Macroscopic and microscopic differences in wood decayed by white-rot fungi also portray the diversity of attack by this large group of fungi. Separate categories were used by early investigators to accommodate fungi that degraded various cell wall components at different rates and to various extents. Some fungi were found to remove all cell wall components simultaneously, in proportion to their existence in the cell wall, but others preferentially degraded lignin (Björkman et al. 1949, Jurásek 1955, Meier 1955, Schmid and Liese 1964, Von Aufsess et al. 1968). The terms corrosive rot (Falck 1926, Björkman et al. 1949) and simultaneous rot (Liese 1970) have been used to differentate the removal of all cell wall components, and the term white rot was reserved for the preferential attack on lignin. Decay by white-rot fungi has also been categorized as white-pocket, white-mottled, white-stringy, etc. based on macroscopic characteristics (Hubert 1931). Whatever the method of organization that is used, the decay caused by these fungi should be first classified as a white rot (Blanchette 1980a, Nilsson 1985, Otjen and Blanchette 1986b).

White rot fungi colonize wood quickly and may become established in all cells of the xylem (Cowling 1961, Greaves and Levy 1965, Liese 1970, Wilcox

Table 1.2. Percent weight, lignin, and sugar residues in wood polysaccharides lost in paper birch (*Betula papyrifera*) and Eastern white pine (*Pinus strobus*) after 12 weeks of decay by white-rot fungi

Fungus	Wood	Percentage loss in				
		Weight	Lignin	Glucose	Xylose	Mannose
Basidiomycotina						
Coriolus versicolor	Birch	65.3	64.6	65.4	68.8	71.7
	Pine	25.3	35.4	22.1	46.7	11.6
Dichomitus squalens	Birch	44.4	71.2	43.8	43.5	40.4
	Pine	43.1	53.0	48.3	56.9	64.9
Heterobasidion annosum	Birch	24.0	54.6	4.1	25.6	32.2
	Pine	15.1	24.0	12.6	14.7	19.6
Phanerochaete chrysosporium	Birch	39.1	72.9	15.1	55.1	0
(BKM-F-1767)[b]	Pine	19.5	30.5	3.9	44.1	0
Phanerochaete chrysosporium	Birch	46.5	51.5	48.8	58.0	44.5
(HHB-11741)[b]	Pine	5.8	4.7	10.1	12.4	20.1
Phellinus pini	Birch	17.0	53.9	5.2	12.9	11.4
	Pine	20.2	43.5	10.7	34.5	42.5
Phlebia tremellosa	Birch	34.3	75.2	4.1	39.4	28.5
	Pine	17.4	38.6	20.2	61.2	41.2
Poria medulla-panis	Birch	30.6	73.1	0.1	31.9	11.5
	Pine	19.0	38.5	28.6	62.9	35.3
Ascomytonia						
Ustulina vulgaris	Birch	22.9	6.3	25.0	14.9	25.5
	Pine	3.3	–[a]	–	–	–
Xylaria hypoxylon	Birch	18.6	13.5	18.4	12.8	22.6
	Pine	3.9	–[a]	–	–	–

[a] Not determined due to low weight loss after 12 weeks.
[b] Isolate number of strain used.

1968). The ray parenchyma cells are frequently the first to be colonized. Bore holes may be numerous in early stages of decay, and hyphal penetration from cell to cell via pit structures or directly through the wall is easily accomplished. The order in which various amounts of lignin, cellulose, and hemicellulose are degraded is different among species of white-rot fungi and may vary depending on the type of wood substrate being attacked (Campbell 1932, Kirk and Highley 1973, Blanchette 1984a, b). The amounts of lignin and sugars in wood polysaccharides lost from birch and pine wood after degradation by several white-rot fungi are shown in Table 1.2. The attack results in either a simultaneous loss of lignin, cellulose (represented in Table 1.2 as glucose lost) and hemicellulose (xylose and mannose lost), or a preferential removal of one or more components.

One frequently observed method of cell wall degradation in gymnosperms is an erosion of the cell wall that occurs from the cell lumen toward the middle lamella (Fig. 1.18). Lysis zones occur immediately surrounding the hypha, and all wall components are degraded (Fig. 1.18A, B). Erosion troughs beneath hyphae extend deeply into the secondary wall, destroying the S_1, S_2, and S_3

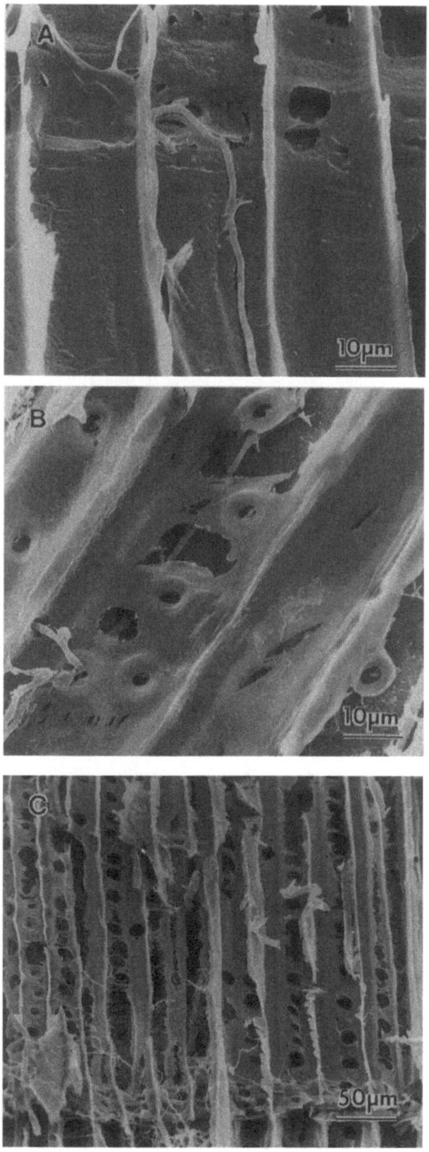

Fig. 1.18. White rot by *Coriolus versicolor* in *Pinus monticola*, showing degradation of all cell wall components. **A** Erosion troughs form beneath and around hyphae. The cell wall is eroded from the lumen toward the middle lamella. **B** Simultaneous degradation of all cell wall components results in localized erosion of the cell wall with holes forming in tracheids. **C** Advanced stage of white rot with extensive degradation of tracheids. The cell walls are degraded, and holes have enlarged to form widespread cavities. **A–C** Radial surface; SEM (**A** from Blanchette 1980b)

layers in succession. Bore holes also become progressively larger. Erosion of the wall continues into the middle lamella after the secondary wall layers have been removed. Holes develop in areas where the entire cell wall was destroyed (Fig. 1.18B). As decay progresses, the erosion of the secondary wall may broaden around the entire circumference of the lumen, causing a widespread thinning of the cell wall. The cell corner regions resist degradation the longest, and in advanced stages of decay these areas of the cell are all that remain. The

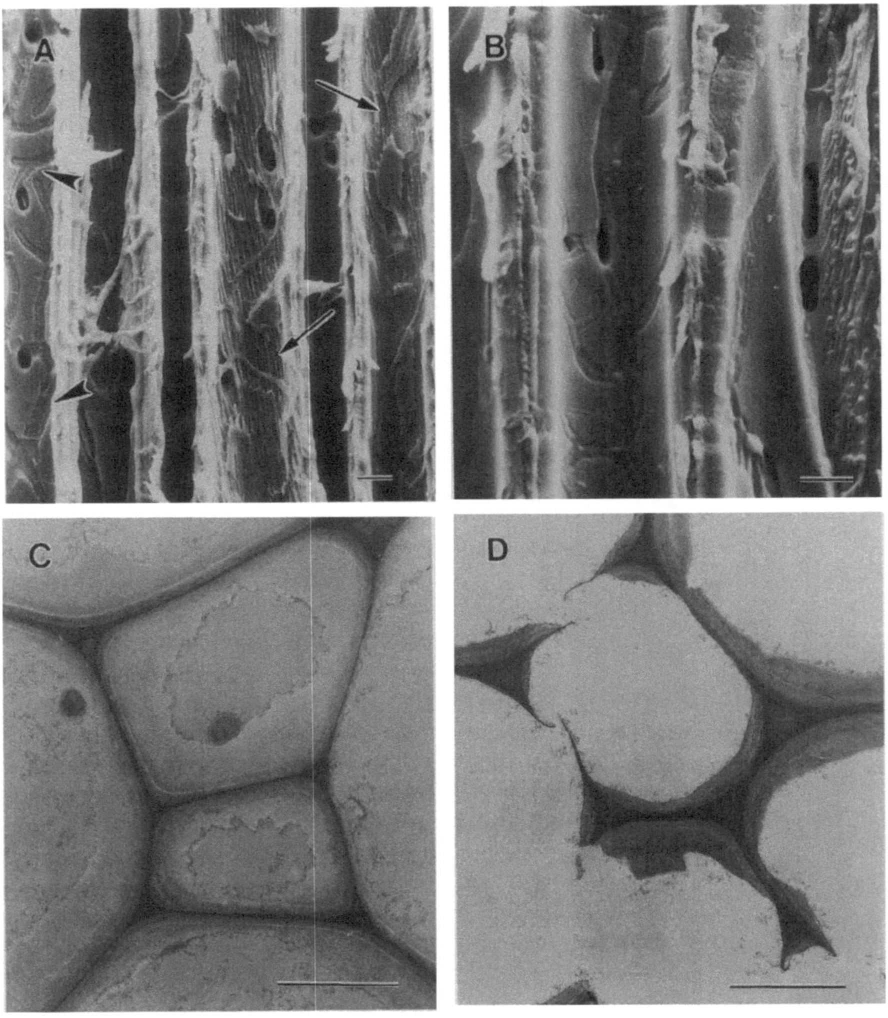

Fig. 1.19. White rot in *Pinus strobus* caused by *Coriolus versicolor* after 12 weeks of degradation in the laboratory. **A** Tracheids with incipient (*arrowheads*) and more advanced stages of degradation (*arrows*). The fungus causes a localized degradation of the cell wall, and attack is restricted to the immediate vicinity of hyphae. As degradation continues, the entire cell wall becomes affected. Some tracheids are severely attacked while others are not. **B** The attack begins with an erosion directly adjacent to fungal hyphae. **C** Attack on the tracheids erodes the cell wall from the lumen toward the middle lamella. The attack is first observed in the cell wall under the hyphae, and as decay progresses the entire circumference of the wall is affected. **D** Some parts of the cell wall are attacked to a greater extent than others, and complete degradation occurs in some areas of the wall but not in others. **A, B** Radial surface; SEM. **C, D** Transverse sections; TEM

progressive attack on the cell wall from the lumen toward the middle lamella has been observed by many investigators (Greaves and Levy 1965, Wilcox 1968, Liese 1970, Dirol 1976, Messner and Stachelberger 1984). All cell wall components are removed as they are encountered within the wall. The progressive stages of attack and loss of crystalline cellulose have been examined using polarized light microscopy (Greaves and Levy 1965, Wilcox 1968). The results suggest that crystalline cellulose of each wall layer is destroyed in succession from the S_3 to the S_1 layer. Along with cellulose degradation, all cell wall components are destroyed. Lignin and hemicellulose losses follow a similar pattern of removal as cellulose when the individual layers are being attacked (Bravery et al. 1974, Messner and Stachelberger 1984, Blanchette et al. 1987).

The same basic patterns of cell wall lysis may be found in angiosperms (Cowling 1961, Wilcox 1968, Liese 1970). Erosion troughs develop under hyphae with a sequential removal of secondary wall layers (Fig. 1.19A, B). The attack of the secondary wall layers may not be localized to only the immediate vicinity of the hypha. Although channels beneath hyphae are evident during incipient stages of decay (Fig. 1.19B), a thinning of the wall around the circumference of the lumen may occur (Cowling 1961, Wilcox 1968). The S_3 and S_2 layers may be entirely removed, leaving the S_1 layer and compound middle lamella, or all secondary wall layers and even parts of the middle lamella can be eroded (Fig. 1.19C). Degradation, however, may not be uniform throughout the wood, and one fiber may be severely eroded, while others have incipient decay or no damage (Fig. 1.19A). In advanced stages of decay, the microstructural changes become more uniform from cell to cell, but variations can still be detected on the ultrastructural level (Fig. 1.19C, D).

Another type of white rot that causes very different morphological changes in the woody cell wall has also been repeatedly observed. This occurs when lignin is preferentially degraded within all cell wall layers. The microstructural characteristics associated with this type of decay were first observed by Hartig (1878) and have received considerable study (Jurásek 1955, Meier 1955, Liese and Schmid 1966, Henningsson et al. 1972, Peek et al. 1972, Blanchette 1980a, 1984a, b, Dirol 1980, Eriksson 1981a). Lignin is preferentially removed from the entire cell wall, causing a loosening of cells. The extensive loss of middle lamella results in a separation of cells from one another (Fig. 1.20). The secondary wall remains, and, although lignin has been selectively removed from it, there are no lysis zones, erosion troughs, or thinned areas (Fig. 1.20). In conifers, the tracheids may be totally without middle lamella, but cell wall integrity is maintained (Fig. 1.20). Often, in wood with delignified tracheids the ray parenchyma cells are completely destroyed (Fig. 1.20). All cells of hardwood species may be delignified, causing a separation of fibers, vessels, and ray parenchyma cells. The entire compound middle lamella may not always be degraded. Some fungi that selectively attack lignin will remove the middle lamella between cells but do not totally degrade the cell corner regions (Fig. 1.20). Lignin removal from the secondary wall frees the cellulose, and greater birefringence may be observed when the wood is viewed with polarized light (Peek et al. 1972). The crystalline nature of cellulose is not destroyed. This is

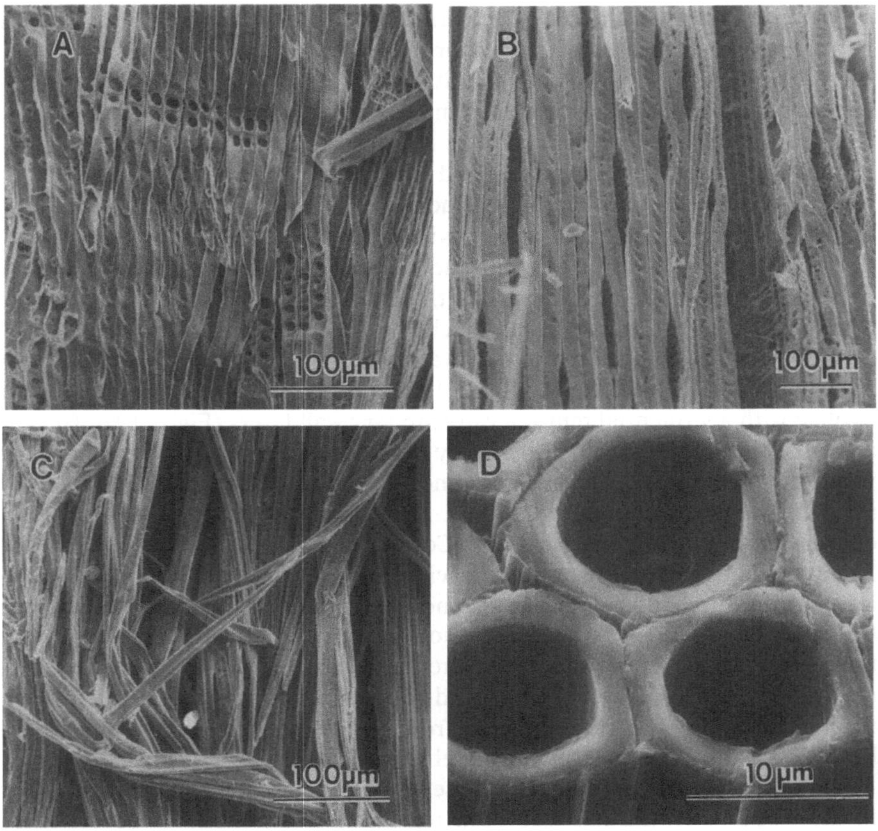

Fig. 1.20. A Wood from *Pinus resinosa*, **B** *Pinus monticola*, and **C**, **D** *Betula papyrifera* with white rot resulting in a preferential degradation of lignin. **A** Decay caused by *Heterobasidion annosum*, showing a delignification of wood and separation of tracheids due to the removal of middle lamella. **B** Decay caused by *Phellinus pini* with delignified tracheids. Ray parenchyma cells were completely removed in advanced stages of decay. Erosion throughs, holes, or cavities are not present in the tracheids. **C** Delignified fibers resulting from selective lignin degradation by *Ischnoderma resinosum*. **D** Delignification by *Hymenochaete rubiginosum*. The secondary wall is intact, but the middle lamella has been severely degraded. Without the presence of middle lamella the cells readily separate from one another. **A** Radial surface; SEM. **B** Tangential surface; SEM (Blanchette 1980a). **C** Radial surface; SEM (Otjen and Blanchette 1986b). **D** Transverse surface; SEM

in contrast to removal of the cellulose from the S_3 to S_1 layers by white-rot fungi causing simultaneous rot.

The degradation of lignin by fungi that selectively delignify wood appears concomitant with a reduction in hemicellulose content (Table 1.2). The close spatial relationship of lignin and hemicellulose in the cell wall proposed by Kerr and Goring 1975; see Fig. 1.1) helps to explain how lignin can be extensively degraded from the wall without appreciable loss of cellulose. The hemicellulose appears to serve as a carbohydrate source for the fungus as lignin

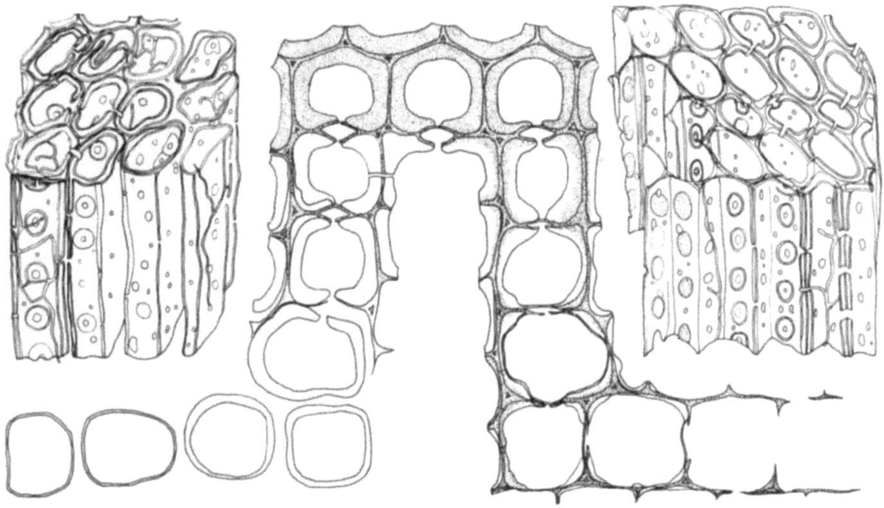

Fig. 1.21. Drawing by Robert Hartig (1878) of *Pinus sylvestris* wood decayed by *Heterobasidion annosum*. Two different forms of white rot were found. Preferential degradation of lignin resulting in the removal of lignin from the secondary wall and middle lamella is represented on the *left* side of drawing, while a simultaneous degradation of all cell wall components is shown on the *right* (*stippled areas* represent lignin). This drawing represents the first observations showing that two different forms of white rot can be produced by the same fungus. These observations were apparently overlooked by researchers for many years

is depleted. It is only after lignin has been removed that cellulase activity is observed. Erosion of delignified secondary walls is now evident, and the loss occurs from the lumen toward the S_1 layer (Fig. 1.20).

The chemical and morphological characteristics in wood with selective lignin removal are distinctly different from a simultaneous attack of all wood components. Nevertheless, these two forms of decay must be both considered a type of white-rot degradation since one fungus may produce both types of attack on the same substrate. Hartig (1878) illustrated the decay by *Trametes radicipera (Heterobasidion annosum)* and showed the same substrate may suffer either a selective removal of lignin with extensive loss of middle lamella or an attack on the entire cell wall. His precise illustrations show the details of the progressive stages of attack (Fig. 1.21). These results have been confirmed by Henningsson et al. (1972), who found two types of attack in birch chips. In this study, the outer areas of the chips contained cell wall erosions, whereas the interior portions were delignified. Recent investigations have demonstrated that two types of attack can be produced by many white-rot fungi (Blanchette 1984a, c, Blanchette et al. 1985, Adaskaveg and Gilbertson 1986, Otjen and Blanchette 1986b, Dill and Kraepelin 1986).

Investigations of decay caused by *Ganoderma tsugae* in *Tsuga canadensis* (Blanchette 1984a) and *Ganoderma applanatum* in *Betula papyrifera* (Blanchette et al. 1985) show the relationship of the two forms of white rot within

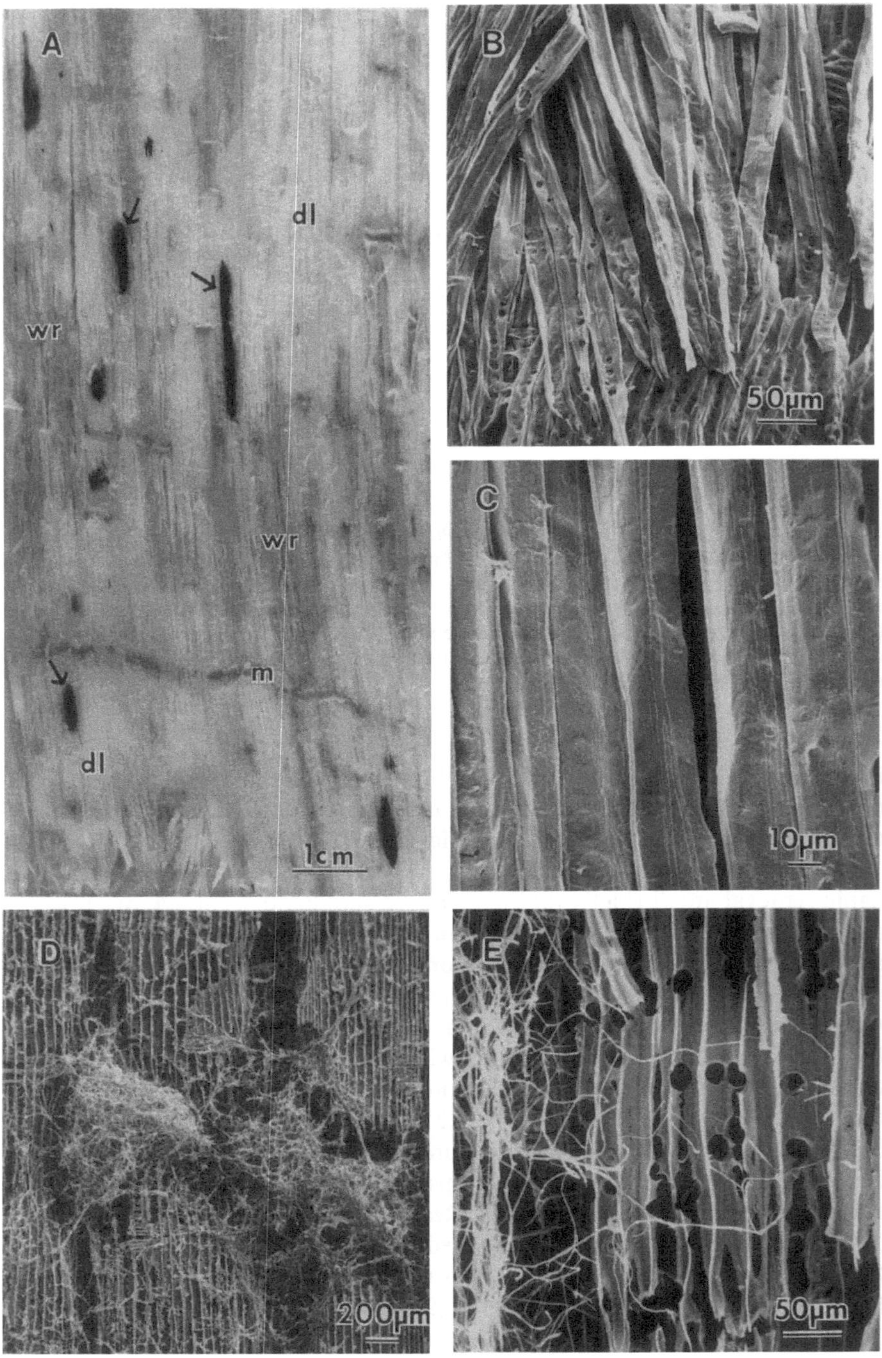
A
dl
wr
wr
m
dl
1cm
B
50µm
C
10µm
D
200µm
E
50µm

a conifer and hardwood substrate. The two forms of decay caused a mottled appearance of white and tan areas in the wood. (Fig. 1.22 A). The white zones had no reaction to phloroglucinol-HCl or other reagents specific for lignin. However, adjacent tan areas reacted positively for lignin. The white zones contained delignified cells where over 98% of the lignin had been removed. The secondary wall substances that remained were composed mostly of cellulose. In wood decayed by *Ganoderma tsugae*, tracheids from white zones had no middle lamellae, and cells separated readily from one another (Fig. 1.22 B, C). Areas that reacted positively for lignin contained degraded cells with erosion throughs and holes in the walls (Fig. 1.22 D, E). Degradation was advanced in some areas, and tracheids were severely degraded. As voids in the cells developed from merging erosion troughs, masses of mycelium filled these regions (Fig. 1.22 D).

In birch wood decayed by *Ganoderma applanatum*, the white zones were associated with delignified cells and the tan areas with a simultaneous attack (Fig. 1.23). Large masses of mycelium were found in cells with simultaneous degradation (Fig. 1.23 A), while in delignified areas mycelium was scarce. White zones that did not stain with phloroglucinol-HCl contained delignified cells (Fig. 1.23 B, C). The fiber cell walls had no middle lamella and cells were loosely arranged (Fig. 1.23 B). The secondary wall layers remained relatively unaffected (Fig. 1.23 C). Tan areas that surrounded the large zones of delignified wood contained cells that had a positive reaction with phloroglucinal-HCl. A simultaneous decay of all cell wall components was found in cells from these areas, erosion throughs and holes were evident (Fig. 1.23 D). The cell wall was attacked from the lumen toward the middle lamella, and the wall layers were eroded in a localized area (Fig. 1.23 E). As coalition of eroded areas occurred, the fungus filled these voids with mycelia.

There are many white-rot fungi that may cause a similar mottled type of white rot with both forms of decay present (Blanchette 1984 a, Adaskaveg and Gilbertson 1986, Dill and Kraepelin 1986). The amount of selectively delignified and simultaneously decayed wood varies tremendously among strains of the same fungus, different fungal species, or substrates (Blanchette 1984 c). Some fungi, such as *Ischnoderma resinosum*, often form white patches of decayed wood that macroscopically resemble white areas of delignified cells (Fig. 1.24 A, B). These white patches, however, are composed of abundant

---

Fig. 1.22. Two different types of white rot, selective lignin degradation, and simultaneous degradation of all cell wall components by *Ganoderma tsugae*, causing a mottled-rot in *Tsuga canadensis*. **A** Mottled appearance of the decayed wood with *white* and *tan areas* representing delignified (*dl*) and simultaneous-rotted (*wr*) wood, respectively. Masses of mycelia are evident within the tan areas (*m*). **B, C** Wood from white areas (*dl*), showing the morphology of delignified cells. The cells move apart from one another, and complete degradation of the middle lamella is evident. **D, E** Wood from tan areas with tracheids that have erosion troughs and holes in cell walls. A simultaneous removal of all cell wall components results in large cavities. As the voids in the wood form, mycelia fill the space. **A** Radial surface of decayed log. **B–E** Radial surface; SEM (Blanchette 1984 a)

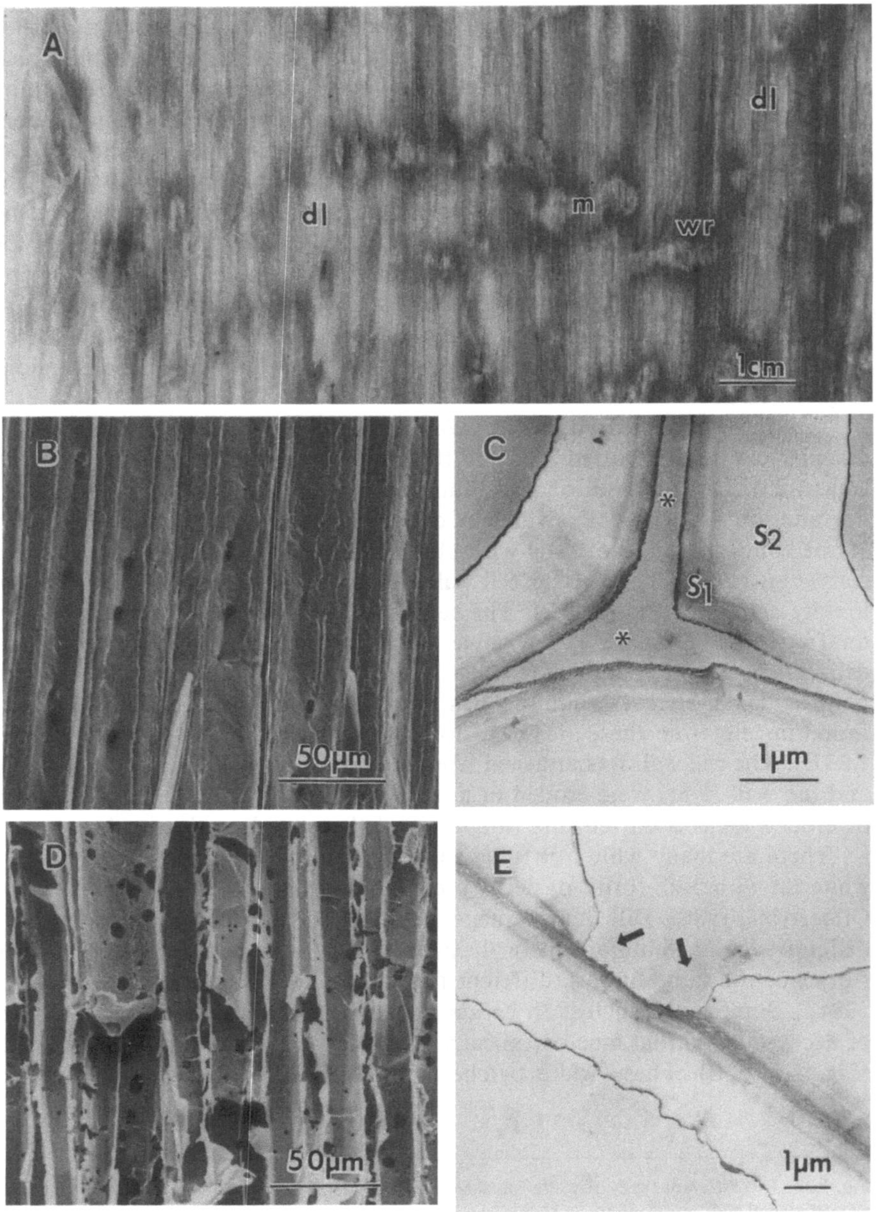

Fig. 1.23. Two types of white rot in *Betula papyrifera* caused by *Ganoderma applanatum*. **A** Mottled appearance of decayed wood with white (*dl*) or tan (*wr*) areas. Mycelia (*m*) filled cavities that formed in the tan wood. **B, C** Cells from white wood with delignified cells. The middle lamella is removed (*) but the S_1 and S_2 remain unaltered. Lignin is removed from the cells without destroying the integrity of the secondary wall. **D, E** Cells from tan areas with erosion of all cell wall layers. The attack results in degradation of all cell wall layers successively from the lumen toward the middle lamella (*arrows*). As erosion troughs continue to enlarge, holes form in the wall. **A** Longitudinal section of decayed log. **B, D** Radial surface; SEM. **C, E** Transverse section; TEM (Blanchette et al. 1985)

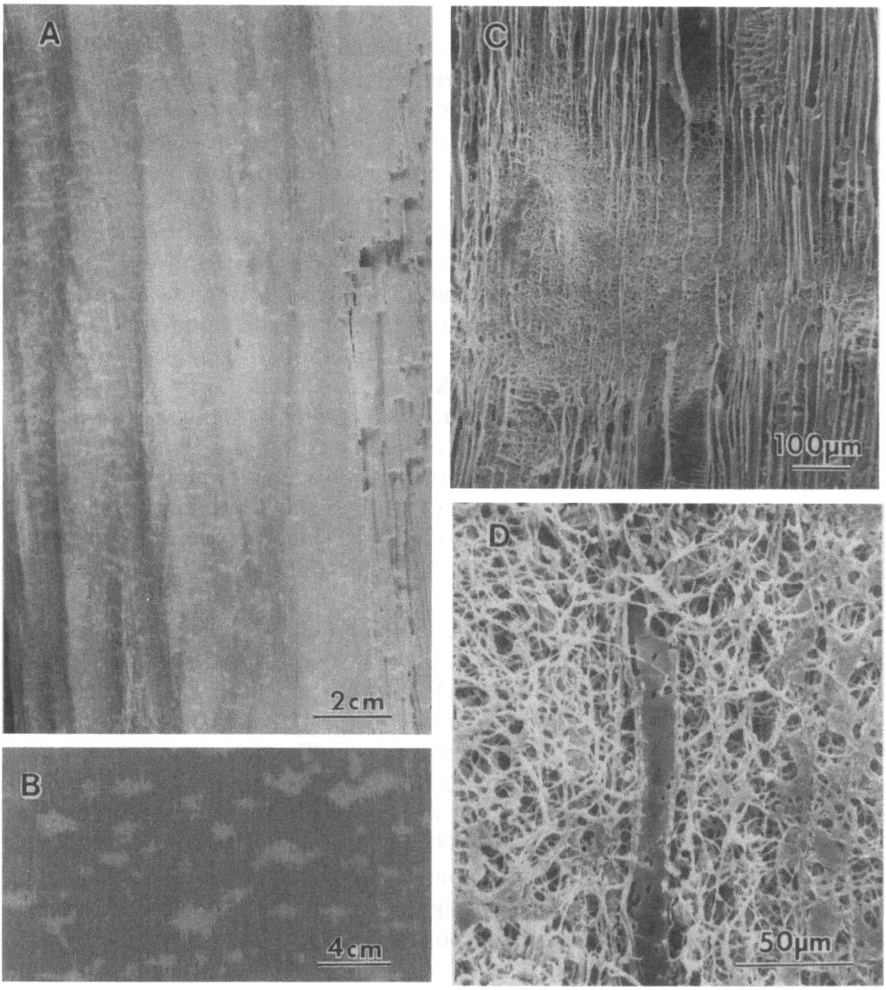

Fig. 1.24. White rot in *Populus deltoides* caused by *Ischnoderma resinosum*. **A, B** Decayed wood with irregular white patches of wood that resemble white-pocket rot (see Fig. 1.13). **C** Wood from white patches, showing cells with masses of mycelia in degraded areas. **D** Mycelium completely filling a cavity that was formed when degraded areas, resulting from a simultaneous white rot attack, coalesced. Remnants of partially degraded cells can be seen within the mass of mycelium. **A, B** Longitudinal section of decayed log. **C, D** Radial surface; SEM (Blanchette et al. 1985; Otjen and Blanchette 1986b)

mycelia that fill voids in the simultaneously decayed wood (Fig. 1.24C, D). No delignification is evident.

Factors that determine different modes of fungal degradation, delignification, or simultaneous decay, side by side in the same substrate, remain obscure. The relatively widespread occurrence of these different forms of white rot have only recently been elucidated (Kawase 1962, Blanchette 1984c, Gonzalez et al.

1986), and the partitioning of the two decays in wood is just beginning to be examined. A recent hypothesis by Dill and Kraepelin (1986) suggested that environmental factors and nitrogen concentrations in wood may be responsible for governing the type of white-rot degradation produced. In forests of southern Chile, extensive delignification was found throughout the whole interior of tree trunks on the forest floor. The low nitrogen content of the wood was considered a major factor that contributed to complete delignification of the wood. Since high nitrogen concentrations stimulate polysaccharide breakdown, low nitrogen levels could promote selective lignin removal (Fenn and Kirk 1981, Reid 1983a). Lack of desiccation and frost during the year in tropical forests may also reduce the mechanical stress on the wood and maintain conditions (moisture, oxygen, and CO_2 relations, etc.) that promote delignification. All of these factors are known to affect decay by white-rot fungi (Reid and Seifert 1982, Highley et al. 1983). Alternating areas of selective and simultaneous lignin degradation in mottled rots found in temperate forests could be due to a nonuniform distribution of nitrogen within wood (Dill and Kraepelin 1986). In addition, the formation of pseudosclerotial plates around the outer perimeter of the decaying wood (Chap. 1) may produce the appropriate environmental conditions that are favorable for selective lignin degradation.

The ability of a fungus to selectively attack lignin from the cell wall and to completely destroy the compound middle lamella without extensive deterioration of the secondary wall is a remarkable phenomenon. An ultrastructural study by Blanchette and Reid (1986) using *Phlebia* (*Merulius*) *tremellosa* demonstrated the progressive stages of selective delignification. Wood from aspen and birch that had been decayed for 12 weeks had approximately 30% weight loss and 70% lignin loss. Cellulose loss was insignificant but hemicellulose was degraded as reflected in the loss of 45% of the xylose and 19% of the mannose. Lignin removal from the wall was first evident in the S_3 and S_2 layers immediately beneath the hyphae. As lignin was lost, the cellulose in samples fixed with OsO_4-glutaraldehyde and post-stained with uranyl acetate became more electron dense (Fig. 1.25A). Apparently the stain penetrated the delignified microfibrils and reacted with the freed cellulose. The secondary wall layer gradually became more electron dense, indicating that lignin had been removed throughout the S_2 and S_1 layers. This occurred around the entire circumference of the cell (Fig. 1.25B). Inner regions of S_2 and S_1 near the cell corners remained free of stain longer than other parts of the secondary wall (Fig. 1.25B). Once lignin was removed from the secondary wall, the middle lamella between cells was degraded (Fig. 1.25C). Wood fixed with $KMnO_4$ showed the distribution of lignin in the cell wall. As lignin was removed, the cell wall became less electron dense. Lignin was removed from the secondary wall from the S_3 to the S_1 layers. The first effects on the middle lamella were located in the area between the cells (Fig. 1.25D). The middle lamella took on a granular appearance and gradually lost all electron density. The cell corners were attacked last. Even after most of the middle lamella had been removed, remnants of cell corners persisted (Fig. 1.25E, F).

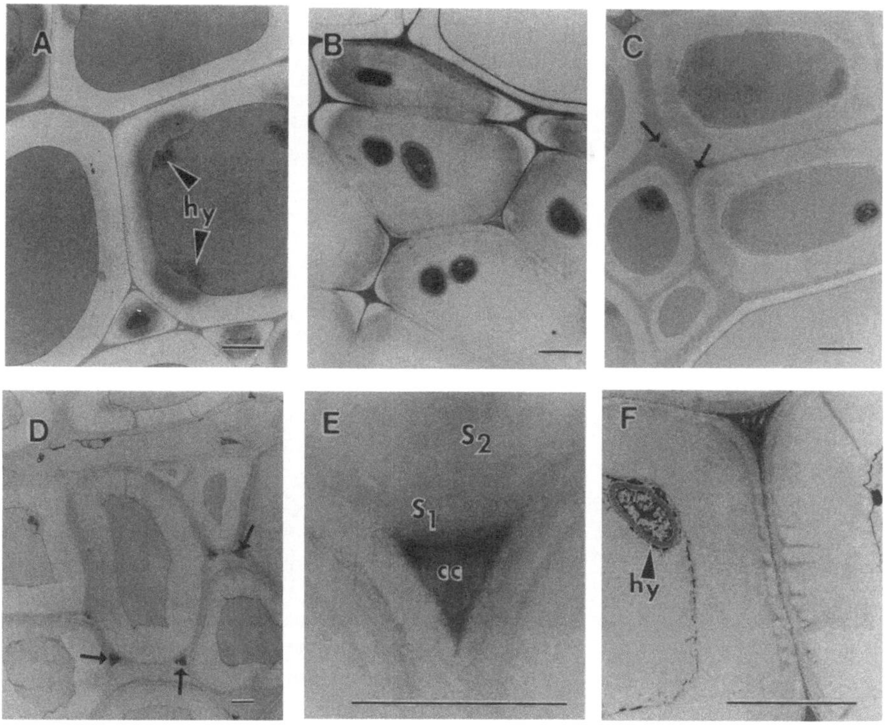

Fig. 1.25. Preferential lignin degradation in *Betula papyrifera* (**A**, **D**, **E**) and *Populus tremuloides* (**B**, **C**, **F**) by *Phlebia tremellosa*. Decayed wood fixed with OsO_4-glutaraldehyde and poststained with uranyl acetate (**A**–**C**) have cells with hyphae (*hy*) in lumina. The hyphae are surrounded by a hyphal sheath. **A** Cells from areas with incipient decay have densely stained secondary walls beneath hyphae (*arrow heads*). **B** Cells from wood with more advanced decay are seen with no middle lamella between cells. The entire secondary wall contains a diffuse electron dense stain. **C** Cell corners persist (*arrows*) in the delignified cells even after the middle lamella between cells is degraded. **D**–**F** Decayed wood fixed with $KMnO_4$, showing extensive degradation of middle lamella between the cells (**D**). Cell corners are not degraded (*arrows*). **E** Secondary wall layers have little electron density but S_1 and S_2 layers are still apparent. **F** Cells from an area of decayed wood with less advanced decay showing that the middle lamella region between cells is partially degraded. This region contains a discontinuous area of electron-dense granules. Cell corners are only partially degraded. The lack of electron density in the secondary wall indicates that lignin is removed from the S_1, S_2, and S_3 layers before it is eliminated from the middle lamella. **A**–**F** Transverse sections; TEM (Blanchette and Reid 1986)

To precisely map the distribution of lignin in sound and white-rotted wood, the methods described by Saka and Thomas (1982a, b), coupling bromine with lignin, have been used (Blanchette et al. 1987). X-ray microanalysis of brominated wood shows that lignin is distributed throughout the entire cell wall of sound birch, but the highest concentrations are found in the middle lamella (Fig. 1.26A, B). The distribution observed was consistent with other investigations using brominated wood and X-ray microanalysis (Hardell et al.

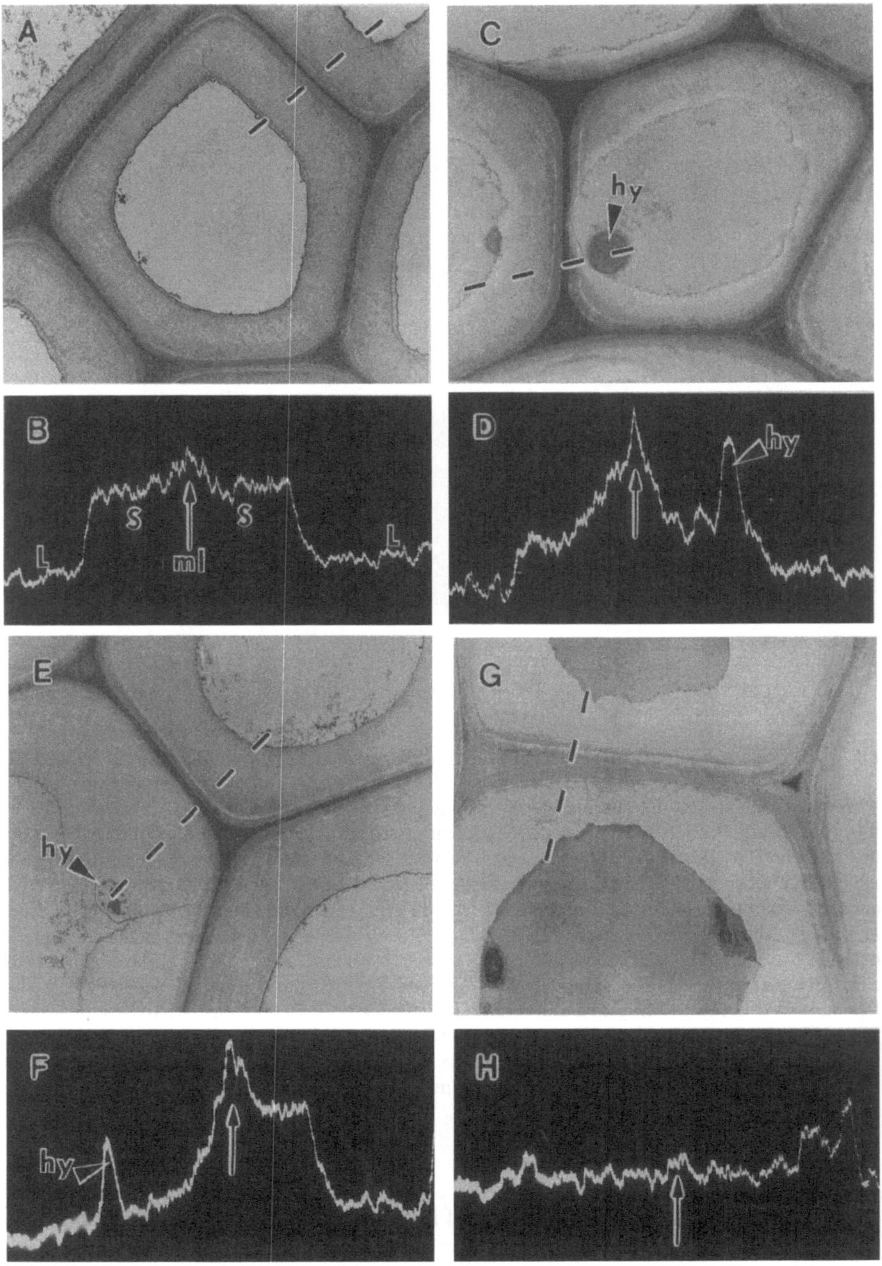
A
C
hy
B
S
S
L
L
ml
D
hy
E
hy
G
F
hy
H

1980, Saka and Thomas 1982a, Westermark 1985) as well as ultraviolet microscopy (Fergus and Goring 1970).

In advanced stages of wood decay by *Coriolus versicolor*, a simultaneous attack on all cell wall components was evident (Cowling 1961, Wilcox 1968, Liese 1970, Dirol 1976). Lignin, however, was removed from the secondary wall beneath hyphae and around the outer circumference of the secondary wall near the lumen before any cellulose loss was detected (Fig. 1.26C, D). Lignin lost from the cell wall was restricted to the secondary wall adjacent to hyphae or a shallow zone around the perimeter of the S_3 layer. As decay by *Coriolus versicolor* progressed, lignin in the cell wall was continually removed in advance of erosion trough formation. Once Lignin was removed, rapid degradation of cellulose occurred. When cell wall erosion reached the middle lamella, a localized attack on only the exposed portions of the middle lamella took place. This attack did not progress laterally into the wall, and surrounding cell wall layers remained lignified.

Wood delignified by *Phlebia tremellosa* and *Phellinus pini* was subjected to a different type of attack in which lignin was removed sequentially from the S_1 and S_2 wall layers near hyphae (Fig. 1.26E, F). As decay progressed, a diffuse attack on lignin throughout the secondary wall occurred. The middle lamella between cells was degraded before cell corner regions. No lignin was detected in cells from areas with advanced decay except for a few remaining cell corners that had not been completely degraded (Fig. 1.26G, H).

Lignin serves as a physical and chemical barrier to enzymatic degradation of wood polysaccharides (Kirk 1971, 1973). It has been suggested that enzymes move into the secondary wall and degrade the lignin that encrusts the cellulose fibrils (Cowling 1961, Peek et al. 1972, Messner and Stachelberger 1984a, b, Eriksson 1985). The rate and extent of lignin degradation is different among white-rot fungi. Selective delignifying fungi have the capacity to remove large

Fig. 1.26. X-ray microanalysis of sound (**A, B**) and white-rotted *Betula papyrifera* treated with bromine to determined the distribution of lignin in the cell walls. The *dotted line* represents the scan line for bromine analysis. **A, C, E,** and **G** are sections of $KMnO_4$-fixed wood, showing a representative area where the scan line was made on thick sections of brominated wood. **A, B** Sound wood, showing bromine distribution profile and quantity of lignin in the secondary walls (*S*) and middle lamella (*ml*). Background levels of bromine are indicated by the scan line analysis as it passed through the lumen (*L*). Lignin-bromine is found throughout the cell wall with highest concentrations in the middle lamella (*arrow*). **C, D** *Coriolus versicolor* removed lignin from the secondary wall before an erosion of the cell wall was evident. Lignin in the middle lamella remained at high concentrations (*arrow*). As the scan line passed through a hypha (*hy*), a high peak of bromine can be seen. **E, F** Incipient stage of decay by *Phlebia tremellosa* showing selective lignin degradation. The scan line passes through hypha (*hy*) and two fiber cells. Bromine distribution profile shows low concentrations in the secondary wall of a cell with hypha. The bromine concentrations in the middle lamella and secondary wall of an adjacent cell are not different from those found in sound wood. **G, H** Cells decayed by *Phellinus pini* in advanced stages of selective lignin removal with the middle lamella completely destroyed. Secondary cell wall layers contain little electron-dense stain. The bromine distribution profile shows low concentrations of bromine throughout the cell walls. **A, C, E,** and **G** Tranverse sections; TEM. **B, C, D,** and **F** Bromine distribution profile analyzed by EDXA-STEM. *Bar* = 5 µm (Blanchette et al. 1987) quantities of lignin from throughout the cell wall without destruction of cellulose (Blanchette 1984a, Blanchette et al. 1985, 1987). In contrast, the removal of lignin by *Coriolus versicolor* is not a diffuse attack; cellulase activity is observed soon after the cellulose microfibrils have been uncovered by removal of their lignin shield. There does appear to be movement of lignin degrading enzymes of *Coriolus versicolor* (Blanchette et al. 1987) and other white rots (Messner and Stachelberger 1984a, b, Murmanis et al. 1984, Blanchette et al. 1985, Highley and Murmanis 1987) around the cell lumen and into the S_3 layer. However, this attack does not penetrate very deeply into the secondary wall. As lignin is removed, cellulase activity causes a slow erosion of the entire secondary wall. The degraded cell wall surface adjacent to the lumen may become irregularly eroded and appear serrate (Highley and Murmanis 1987), or an overal thinning of the secondary wall results. When the cell wall is uniformly thinned all around, the surface adjacent to the lumen may appear smooth. (Wilcox 1968, Liese 1970, Messner and Stachelberger 1984a, b, Ruel et al. 1986).

Some different micromorphological and ultrastructural aspects of cell wall degradation have been observed when *Phanerochaete chrysosporium* was used to degrade spruce (*Picea abies*) (Ruel et al. 1981). An inward attack of the cell originated at the pit membrane, and degradation was evident across the transition region between the S_1 and S_2 layers. A lateral progression of the degradation occurred within the wall, and disintegration became evident. Decay was most obvious at transition areas between the layers (S_1-S_2 and S_2-S_3). Attack was also observed from the lumen, and degradation progressed toward the middle lamella. The S_3 layer of the tracheids, however, remained relatively unaltered throughout the decay process. Most of the degradation was in the inner secondary wall layers, and broad areas of disintegration formed in the S_2 layer. The middle lamella was degraded after the secondary wall was severely eroded. Hyphae were not located in the S_2 layer, but enzymes apparently moved into the cell wall from the lumen or pit chambers.

A more selective attack of lignin was observed in wood after decay by cellulase-less mutants of *Phanerochaete chrysosporium* (Eriksson et al. 1980) or after wood had been impregnated with glucose and decayed (Ruel et al. 1984). Extracellular enzymes moved into the wall and caused extensive degradation of the middle lamelle and a loosening of the microfibrils. No cell wall thinning was observed. Cellulase-less mutants also lacked the ability to bore holes into the cell wall. The mutant did, however, spread uniformly through the wood and was able to penetrate pit membranes, moving from cell to cell. The micromorpholigical aspects of decay by the cellulase-less mutants appear very similar to the selective delignification process previously described.

In a recent paper by Ruel et al. (1986), isolates derived from intercrossing mutant strains of *Phanerochaete chrysoporium* were used to decay birch and spruce woods. A nonselective attack was observed in cells with advanced stages of decay, and both lignin and polysaccharides were degraded. Micrographs of the decayed cells showed that fragments of middle lamella and cell corner regions were the only parts that remained. This selective white-rot decay was different from the selective lignin removal previously reported for other cellulaseless mutants (Eriksson et al. 1980; Ruel et al. 1981, 1984). Failure of intercrossed strains to remain cellulase-deficient may be due to the ability of the fungus to revert back to the physiological characteristics of the wild-type fungus. The results clearly demonstrated that the particular strains used by Ruel and others (1986) were not entirely cellulase-deficient after prolong periods of degradation of wood.

The patterns of attack by *Schizophyllum commune*, a weak wood-destroying fungus causing low wood substance loss (Schmidt and Liese 1980), has peculiar decay features (Nilsson and Daniel 1983b). The fungus caused a loosening of the S_3 layer from the rest of the cell wall and a weakening of the S_2-S_1 interface. The concentric layers often associated with the S_2 were also more distinct after fungal activity. The hyphae of *Schizophyllum* were found in cell lumina, and degrading enzymes seemingly penetrated the cell wall to affect inner regions. The S_3 layer persisted until very late stages of cell wall breakdown, but it was partially degraded and had an increased porosity. Partial dissolution of the thin concentric layers of the S_2 was also observed. The attack at the interface of cell wall layers or at the concentric bands of the S_2 suggested that a nonhomogeneous distribution of wall components exists across the secondary wall. The degrading agents from *Schizophyllum* diffuse into the wall and preferentially attack these regions. Laboratory decay studies (Blanchette, unpublished) using *Schizophyllum commune* have resulted in insignificant amounts of weight loss in birch and pine woods. The fungus colonized all cell types, but no loss of wood substance was found after 12 weeks of incubation. Similar results have been reported by other investigators (Schmidt and Liese 1980). The differences between these studies and those of Nilsson and Daniel may be due to variation among strains of *Schizophyllum commune.* Differences among *Schizophyllum commune* strains have been reported from tropical areas, where this fungus is often associated with a high wood-destroying capacity (Abdurachim 1965).

A number of species of Ascomycetes in the family Xylariaceae and Diatrypaceae have been reported to cause a white rot in wood (Campbell and Wiertelak 1935, Hinds 1981, Kirk 1983). The degradation is usually restricted to angiosperms, and loss of wood substance is relatively low when compared to the decay of some aggressive Basidiomycetes such as *Coriolus versicolor* (Table 1.2). The percent weight loss in birch after 12 weeks of decay by *Ustulina vulgaris* and *Xylaria hypoxylon* was 22.9% and 18.6%, respectively, but in pine the weight loss was insignificant (Table 1.2). Merrill et al. (1964) found that several different Xylariaceous species caused approximately the same extent of decay in different hardwood species (12% to 25% weight loss). Although the extent of decay was lower than that caused by *Coriolus versicolor*, it is comparable to the rate of decay produced by many other species of white-rot Basidiomycetes (Table 1.2). The chemistry of wood decayed by *Ustulina vulgaris* was found to be similar to that of wood with a typical white-rot attack caused by species of Basidiomycetes (Campbell and Wiertelak 1935). The morphological aspects of cell degradation confirm these previous results. Cell wall

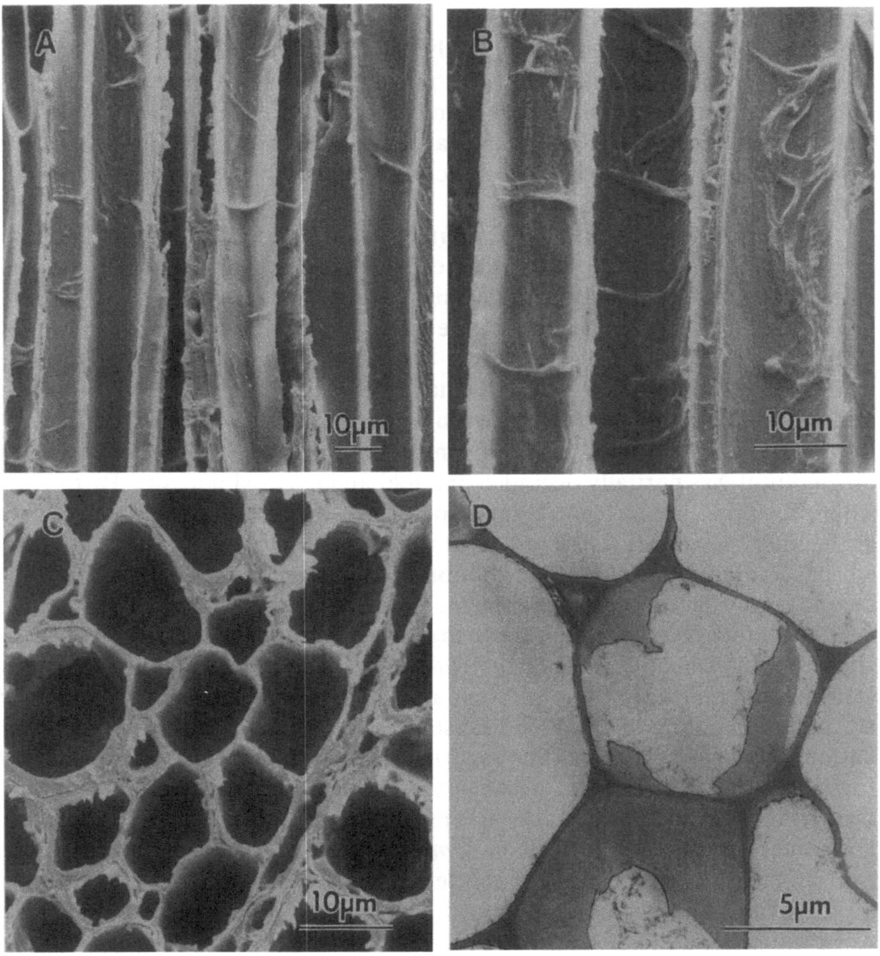

Fig. 1.27. White-rot degradation of *Betula papyrifera* by *Ustulina vulgaris*. **A, B** Cell wall erosion resulting from a thinning of the secondary wall around the entire circumference of the lumen. **C, D** The secondary wall is severely degraded in some cells but the middle lamella is not destroyed. Secondary wall layers are eroded to various extents. In some fiber cells the entire secondary wall is removed, while others have parts of secondary walls still remaining. The middle lamella is not degraded in any of the cells. **A, B** Radial; SEM. **C, D** Transverse section; TEM

erosion of birch wood occurred from the lumen toward the middle lamella. The entire secondary wall was attacked, but the middle lamella was not destroyed (Fig. 1.27A–D). In some cells, the entire secondary wall was destroyed, while other cells displayed severely eroded S_3 and S_2 layers (Fig. 1.27C, D). These observations suggest that the fungus cannot successfully degrade the compound middle lamella. Histological examination of many Xylariaceous fungi by Merrill et al. (1964) also showed reduced thickness of the secondary wall. After 32 weeks of decay, and as much as 40% weight loss, the secondary wall was consumed but the middle lamella remained intact.

The inability of Xylariaceous fungi to degrade middle lamella regions of cells from hardwoods and their low capacity for decay in softwoods demonstrate that the degradation produced is different than that caused by other white rots. The appropriate classification for this type of decay is difficult to ascertain on the basis of current knowledge of these fungi. Pseudo-soft rot (Kirk 1984) or a subgroup of the true soft rots (Nilsson 1985) has been proposed to include these fungi. The definition proposed for a true soft-rot fungus (Sect. 1.4) requires the production of discrete cavities within the secondary wall. Fungi, such as *Hypoxylon mammatum* and *Xylaria digitata*, apparently produce soft-rot cavities in cell walls of various hardwoods (Nilsson, personal communication), and may appropriately be placed within the soft-rot group. However, the macroscopic appearance of the decayed wood and extensive erosion of the entire secondary wall that occurs by Xylariaceous fungi does not resemble a typical soft-rot attack. There are also some white-rot Basidiomycetes that produce a decay pattern similar to that produced by Xylariaceous fungi. Birch wood decayed by *Fomes fomentarius* suffered extensive loss of all secondary wall layers, but the middle lamella was not severely degraded (Fig. 1.28A, B). Apparently, species of Ascomycetes as well as some Basidiomycetes that cause a simultaneous degradation of all wood components have varying capacities to attack the highly lignified middle lamella. Until additional research has been completed, the most appropriate classification appears to be the retention of Xylariaceous fungi and Basidiomycetes that cause this type of decay (Figs. 1.27 and 1.28) within the white-rot group.

In a recent study by Francis and Leightley (1984), an unusual decay pattern was caused by *Hypoxylon mediterraneum* on *Pinus* and *Eucalyptus* species.

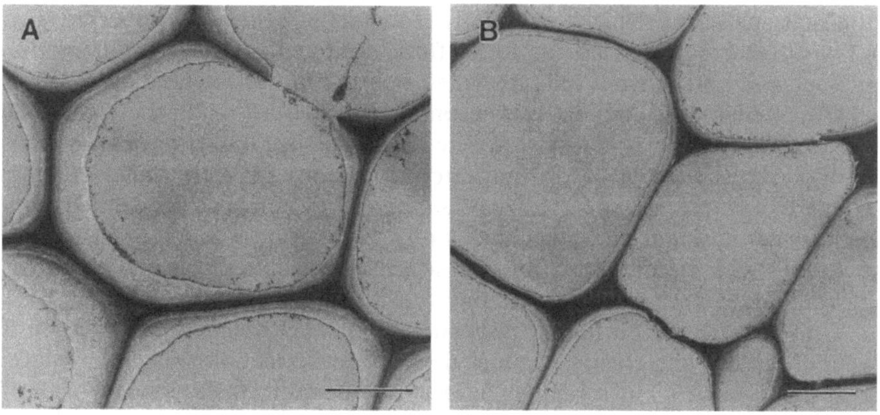

Fig. 1.28. White-rot degradation of *Betula papyrifera* by *Fomes fomentarius* **A, B** Secondary wall layers are seen with an eroded appearance due to the destruction of the secondary wall. The entire cell wall is affected and an overall thinning of the secondary wall can be observed. The middle lamella remains in all cells. Transverse section; TEM. *Bar* = 5 µm

Cell wall erosion and prolific cavity formation were observed in cells of *Eucalyptus*. Chains of rhomboid cavities formed from the lumen into the secondary wall. These cavities differed from those produced by soft-rot fungi since they were initiated directly from the lumen. The symmetrical cavities produced along the lumen, which extended into the S_2 layer, may be regulated by the same factors responsible for soft-rot cavity formation. Decay by *Hypoxylon mediterraneum* in pine was considered by Francis and Leightley (1984) to be very similar to a brown rot with loss of cellulose from within the cell wall. The S_2 layer was disintegrated, leaving a spongy matrix, but the S_3 remained free of attack. These unusual decay patterns in *Eucalyptus* and *Pinus* apparently result from the type of wood substrate being degraded. Although additional information is needed to appropriately elucidate the characteristics of the decay, the "brown rot" symptoms are similar in many respects to the localized loss of cellulose within the secondary wall by *Phanerochaete* and *Schizophyllum*. Since extensive depolymerization of cellulose and micromorphological characteristics of brown rot (presented in Sect. 1.3 on brown rot degradation of wood) were not observed, the decay by *Hypoxylon mediterraneum* is not a typical brown rot.

Residual wood found in advanced stages of a white rot may consist of large amounts of cellulose left after selective delignification (Blanchette 1984a, b, Blanchette et al. 1985, Dill and Kraepelin 1986), remnants of decayed cells with only a skeleton of middle lamella remaining, or fragments of cell corners left after simultaneous lignin-cellulose removal. Another type of advanced decay has a white, stringy appearance. This decay of dicotyledonous angiosperms causes a separation along the radial and tangential planes into string-like fragments of wood (Fig. 1.12D). A simultaneous attack of all wood components was evident in the wood, but only certain cells were degraded (Blanchette et al. 1988). Fibers, fiber tracheids, and parenchyma cells were completely removed, but vessels were not destroyed. The resulting decayed wood consisted of a mass of vessel elements with only remnants of other cells adhering (Fig. 1.29). Often entire logs on the forest floor exhibited complete disintegration of fiber and parenchyma cells, but the vessels were not attacked. As various cells were degraded, the vessels that remained coalesced to fill the voids. The loose aggregates of vessel elements were responsible for the stringy characteristic of the rotted wood. To demonstrate the capacity of some white-rot fungi to cause a white stringy rot, *Acer saccharum* wood was inoculated with *Dichomitus squalens* and incubated for 12 weeks. The wood showed various stages of a stringy rot, consisting of a selective attack on fibers and parenchyma cells but not vessel elements (Fig. 1.30A, B).

One possible reason for the resistance of vessel elements to degradation is the morphology and chemical composition of the vessel wall. Vessels have larger S_1 and S_3 layers but a smaller S_2 layer compared to the secondary wall of fibers. Differences in the microfibrillar orientation of the secondary wall layers and increased concentrations of lignin have also been reported (Harada 1965, Fergus and Goring 1970). The type of lignin that occurs in the vessels also appears to influence the ability of some white-rot fungi to degrade them. Syringyl

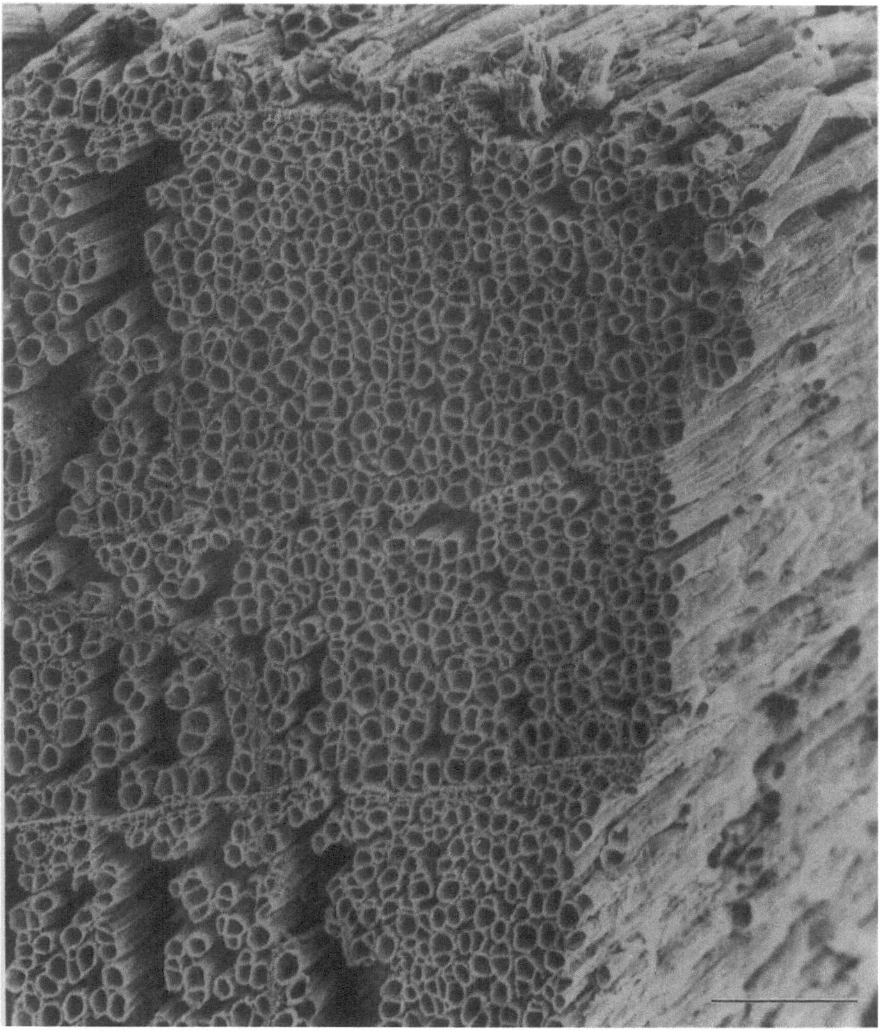

Fig. 1.29. Wood of *Acer saccharum* after degradation by fungi that cause a "white-stringy" type of white rot. Remarkably, only vessel elements are left in the decayed wood. All fibers and parenchyma cells are completely degraded. Vessel elements resist this type of white-rot attack. Entire logs that have been decayed by white stringy-rotters consist solely of vessel elements. Transverse surface; SEM. *Bar* = 100 μm

units of lignin are degraded more rapidly than guaiacyl units (Kirk et al. 1975, Highley 1982a, Ander et al. 1984, Faix et al. 1985), and vessel elements have greater concentrations of guaiacyl versus syringyl lignin than adjacent fibers (Fergus and Goring 1970, Saka and Goring 1985). The resistance of hardwood vessels to degradation by fungi that cause a stringy rot is probably due to the combined anatomical and chemical composition of the vessels (Blanchette et

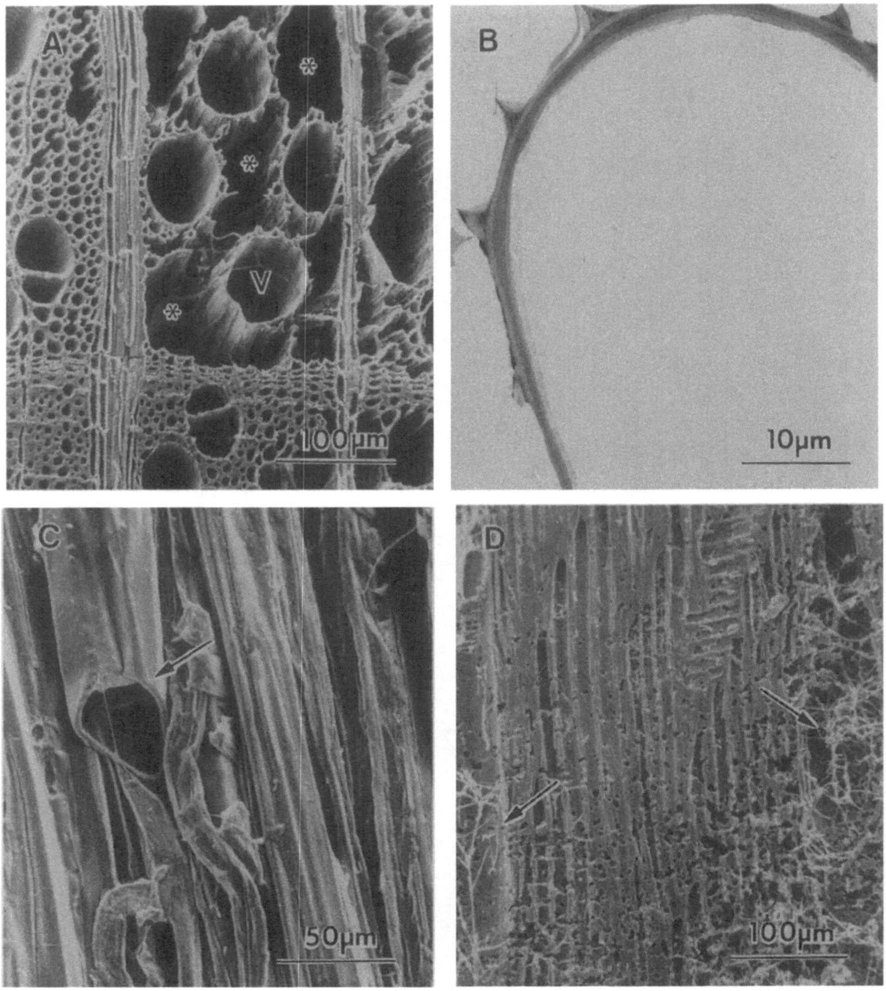

Fig. 1.30. **A** Wood of *Acer saccharum* decayed in the laboratory by *Dichomitus squalens*. A selective attack on fibers and parenchyma cells is evident. Voids (*) among the vessel elements result from the loss of cells. **B** Remaining vessels have relatively intact cell wall layers. Cell corner regions of the middle lamella that were located between the vessel and the adjacent fiber are seen adhering to the vessel. **C, D** Other species of white-rot fungi have the ability to attack vessels. *Phlebia tremellosa* in *Populus tremelloides* caused a selective delignification of vessels **(C)**, and the vessel elements separated at the perforation plate (*arrow*). *Inonotus obliquus* causes an erosion of vessel walls, and all cell wall layers are attacked. Typically, holes develop in the vessels (*arrows*) and gradually enlarge to form extensive cavities. **A** Transverse surface; SEM. **B** Transverse section; TEM. **C** Tangential surface; SEM (Blanchette and Reid 1986). **D** Radial surface; SEM (Blanchette 1982b)

al. 1988). Not all white-rot fungi cause a stringy rot, and many species of white-rotters may degrade vessels (Fig. 1.30C, D). *Phlebia tremellosa* has been shown to remove lignin selectively from the cell walls of vessels (Blanchette and Reid 1986), and *Inonotus obliquus*, *Ischnoderma resinosum*, and many others may severely erode vessel walls (Blanchette 1982b, Blanchette et al. 1985). These different degradation patterns reflect the great diversity found among white-rot fungi. Since relatively few species of white-rotters have been investigated, and hundreds of different species exist, there is undoubtedly more variation to be revealed.

1.3 Wood Degradation by Brown-Rot Fungi

Brown-rotted wood is characterized by extensive degradation of cellulose and hemicellulose but limited degradation of lignin. The most suitable definition, reflecting the general characteristics of all brown rots, has been suggested by Nilsson (1985). His definition states that brown rot is a form of fungal decay caused by Basidiomycetes that mainly degrades the polysaccharides through extensive depolymerization. Although some degradation of lignin is apparent, the residual substrate after advanced brown-rot decomposition consists of an amorphous, crumbly residue that is composed primarily of lignin.

The growth of a brown-rot fungus within wood and subsequent cell wall attack were first observed by Hermann Schacht in 1863. He demonstrated that the decay of wood in living trees and in rotted wood from ships was directly related to the degradation produced by mycelia of various fungi. A comprehensive study by Robert Hartig (1878) illustrated the progressive stages of decomposition by several brown-rot fungi in coniferous and deciduous woods. The term he used for this type of decay was red rot. He showed that degradation started at the cell wall near hyphae located in the lumen, and continued inward toward the middle lamella. A loss of integrity was exemplified by shrinkage cracks and checks throughout the secondary wall (Fig. 1.31). In advanced stages of decay, the cells were altered microscopically, but their structure was still distinguishable.

Falck (1926) also noted that the shape of the cell and thickness of the cell wall appeared unchanged after wood had been brown-rotted. A rapid and uniform loss of cellulose was found to occur during the decay process. Shrinkage of the extensively attacked wood resulted in the formation of cracks and an overall cubical appearance of the rotted wood. He preferred the term destruction rot instead of red rot or brown rot to characterize the decay.

Differences in the rate and extent of decomposition within various cell wall layers was reported by Jurásek (1958, 1964). The S_3 layer of the secondary wall resisted attack but the S_2 was degraded rapidly. The compound middle lamella was the last layer attacked. The observed changes in the microstructure of beech and fir wood were related directly to the destruction of those layers that were rich in polysaccharides.

Fig. 1.31. Illustrations by Robert Hartig (1878) of *Pinus* wood decayed by the brown-rot fungus *Tyromyces mollis* (*Polyporus mollis*). The brown-rotted wood cracks into cubical segments with mats of mycelium forming within the spaces (*top* to figure). Hyphae penetrate the latewood and earlywood cell walls by boring through the wall in any direction. Decayed tracheids show countless splits that begin at bore holes or bordered pits. The splits in the wall follow one direction. Hartig suggested that the splits are due to contraction when the brown-rotted wood was drying. *Bottom right*, Incipient stage of decay; *bottom left*, advanced stage of brown rot

The effect of brown-rot fungi on the polysaccharide constituents of the cell wall, specifically the crystalline cellulose component, has been studied with polarized light microscopy (Schulze and Theden 1938). Crystalline cellulose appeared as a bright image between crossed polarized filters. A rapid loss of birefringence in wood was observed after brown-rot degradation. Not all areas of the wood were equally affected, however, and differences in birefringence occurred between regions of earlywood and latewood. There were also differences observed between ray parenchyma cells and tracheids as well as between fibers and vessels. Observations made by Wilcox (1968) from polarized-light photomicrographs of brown-rotted sweetgum and pine clearly demonstrated a selective attack on the S_2 layer, whereas the birefringence was retained in S_3 and S_1 layers during early stages of decomposition (Fig. 1.32A). As decay progressed, all layers of the secondary wall lost birefringence, indicating that crystalline cellulose was removed extensively from these cells (Fig. 1.32C). The cellulolytic enzymes or other agents produced by brown rot fungi apparently were able to diffuse throughout the secondary wall in both sweetgum and pine. The order in which deterioration among layers of the cell wall progressed was thought to be related directly to their lignin content. In advanced stages of decay only vessels and parenchyma cells retained their birefringence in wood of sweetgum (Fig. 1.32C), and an uneven degradation of tracheids was found in pine (Fig. 1.32D).

Fluorescence microscopy has also been used to monitor the rapid changes in the cell wall during early stages of decomposition by brown-rot fungi (Jurásek 1964, von Aufsess et al. 1968, Wilcox 1968, Fukazawa et al. 1975, Krahmer et al. 1982). When acridine orange was used as a stain, sound cell walls appeared yellow-green. After colonization of lumina by brown-rot fungi, the polysaccharide component of the cell wall was attacked rapidly, and the cells fluoresced orange-red. This change in fluorescence occurred after only a 1%–3% weight loss (Krahmer et al. 1982). The extensive zone of fluorescence demonstrated the diffuse nature of the attack. This method provided a rapid technique to detect early stages of decay by brown-rot fungi and was more sensitive than other commonly used histological reagents.

Brown-rot fungi caused a depolymerization of holocellulose early in the decay process (Cowling 1961). Strength properties were reduced significantly after only a slight amount of weight loss. In these early stages of decay, the cells did not appear altered, and polarized light microscopy or fluorescence microscopy had to be used to detect changes. In a recent study of advanced brown-rotted wood (Cease and Blanchette, unpublished data), tracheids appeared intact, and massive erosion of the cell wall was not evident. There was, however, visual evidence that some loss of cell wall integrity had taken place (Fig. 1.33A, B). Tracheid walls were no longer rigid, and the cells assumed distorted shapes or collapsed during specimen preparation. Even after extensive weight loss and polysaccharide degradation (Table 1.3), the residual cell wall did not appear altered, and many bordered pit membranes were unimpaired (Fig. 1.33C). Hyphae in lumina were sparse during early stages of decay, but mycelia were abundant in tracheids with advanced decay (Fig. 1.33D).

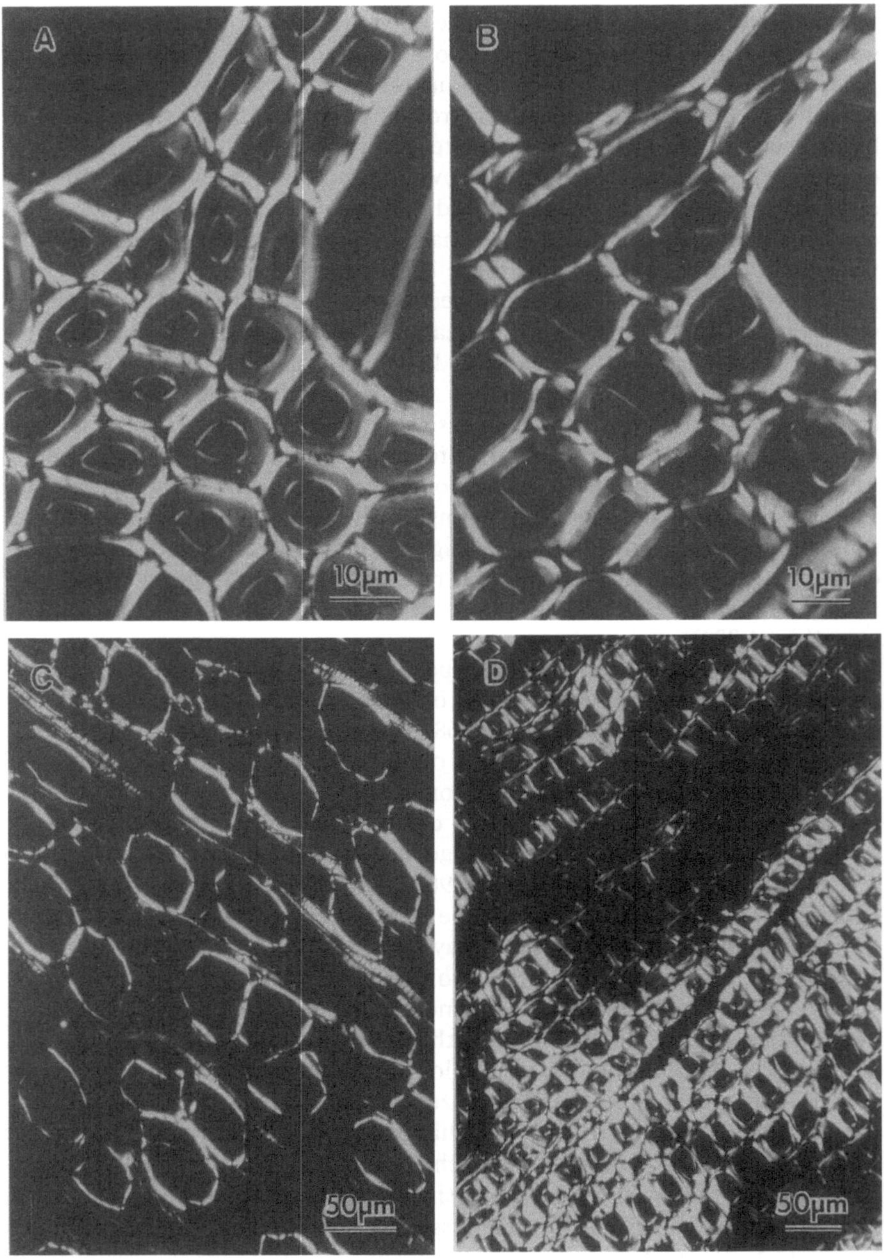
A
10µm
B
10µm
C
50µm
D
50µm

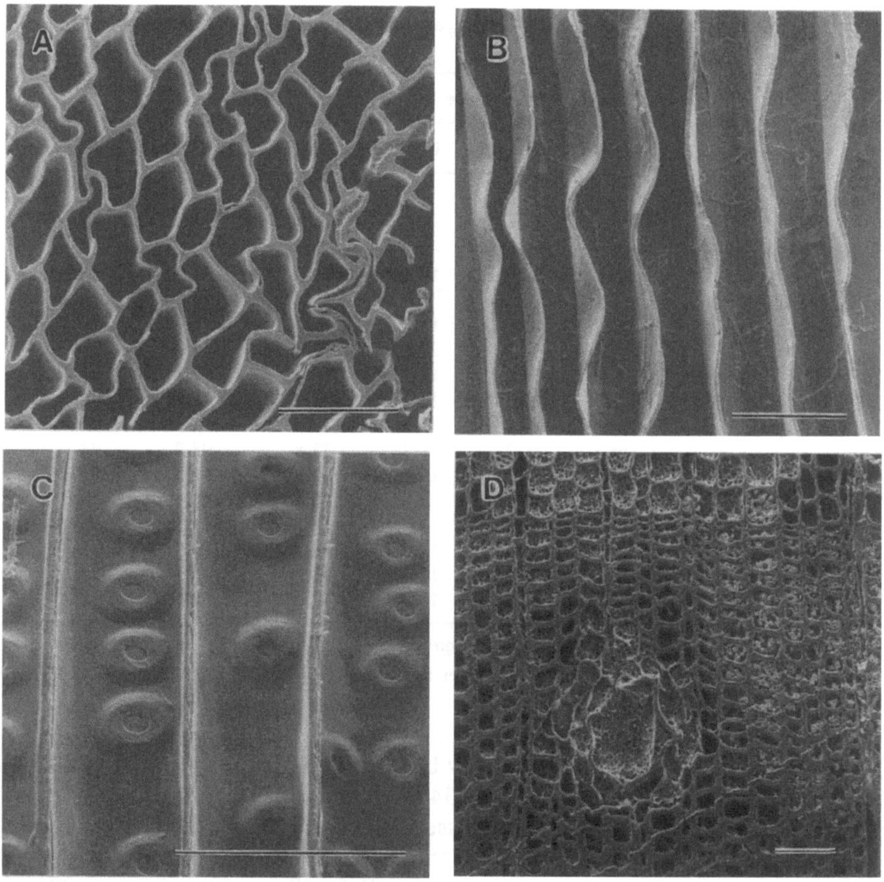

Fig. 1.33 A–C. *Pinus strobus* wood after 12 weeks of degradation and **D** wood decayed in the field by the brown-rot fungus *Fomitopsis pinicola*. **A, B** Degraded tracheids show a lack of cell wall integrity. Cells have collapsed and have a "wavy" appearance. Cell walls do not have erosion troughs or cavities. **C** Even after severe degradation, over 50% weight loss, the cell walls are intact, and bordered pit membranes are evident. Hyphae are not prevalent. **D** Advanced stages of brown rot, showing mycelia filling the latewood and earlywood tracheids and resin duct. Masses of mycelia are often observed during the late stages of degradation. The cell wall structure remains intact. **A, D** Transverse surface; SEM. **B, C** Radial surface; SEM. *Bar* = 50 μm

Fig. 1.32. Changes in cell birefringence after brown-rot degradation by *Poria placenta*. **A** Sound wood of *Liquidambar styraciflua*, showing bright cell walls due to the presence of crystalline cellulose. The S_1 and S_3 layers appear brighter than the S_2 due to the orientation of the cellulose microfibrils. **B** *Liquidambar styraciflua* wood after 19% weight loss. The S_2 layer is extensively degraded and the S_1 and S_2 layers continue to show birefringence. A diffuse attack on the S_2 layer in all fibers is evident. **C** After 68% weight loss no birefringence is seen in any of the fiber walls, but vessels continue to appear bright with intact cell walls. **D** Wood of *Pinus* sp. with 45% weight loss. Large areas of brown-rotted tracheids lack birefringence. The bright cells represent areas in less severe stages of brown-rot degradation. **A–D** Transverse sections, polarized LM. **A, C** from Wilcox (1968) coutesy of US Forest Products Laboratory. **D** Courtesy of W. W. Wilcox

Table 1.3. Percent weight, lignin, and wood sugars lost in birch and pine woods after 12 weeks of decay by brown-rot fungi

Fungus	Wood	Percentage loss in				
		Weight	Lignin [a]	Glucose [b]	Xylose [b]	Mannose [b]
Coniophora puteana	Birch	41.4	4.9	51.6	63.7	95
	Pine	44.4	4.5	63.7	77.6	88.7
Fomitopsis pinicola	Birch	52.8	9.6	72.7	72.7	95
	Pine	52.2	17.4	74.0	83.7	86.5
Gloeophyllum trabeum	Birch	43.7	13.1	48.6	71.1	95
	Pine	57.2	27.3	73.9	90.7	91.8
Laetiporus sulphureus	Birch	33.5	12.3	40.6	58.3	95
	Pine	37.1	14.6	41.5	78.1	94.7
Lentinus lepidius	Birch	59.2	7.9	78.6	79.8	95
	Pine	59.7	11.4	83.8	98.2	94.1
Piptoporus betulinus	Birch	37.7	16.5	42.0	56.5	81.8
	Pine	5.7	3.9	19.8	5.6	28.8
Poria placenta	Birch	57.9	4.7	81.2	81.8	95
	Pine	55.1	12.5	78.7	81.1	92.1
Tyromyces balsameus	Birch	41.3	12.7	50.3	65.1	95
	Pine	49.1	19.8	61.6	82.0	81.2

[a] Determined using the methods of Effland (1977).
[b] Determined using the methods of Pettersen et al. (1985).

Degradation of deciduous wood by brown-rot fungi also resulted in a relatively unchanged cell structure (Fig. 1.34A). After 40% to 50% of the original weight of the wood had been lost, vessels, fibers, and parenchyma cells were still discernable. Transverse sections of the cell wall, however, revealed a porous appearance of the secondary wall (Fig. 1.34B,C). Other investigators (Meier 1955, Cowling 1961, Jurásek 1964) have also reported a loose, porous framework remaining within the secondary wall layers of brown-rotted deciduous wood. The extensive degradation of the cell wall, and apparent swelling and drying during moisture fluctuations, caused cracks and cavities to form (Liese

Fig. 1.34. Brown rot of *Betula papyrifera* after 12 weeks of degradation by *Fomitopsis pinicola*. **A** Vessels, fibers, and ray parenchyma cells can be seen after approximately 50% weight loss. Cell organization is relatively unchanged. **B** The S_2 layer of the secondary wall has a porous appearance after extensive degradation of the cells. In some cells the degraded S_2 has collapsed forming small holes in the wall. **C** Crack and checks that develop in the brown-rotted wood are filled with fungal mycelia. **A, B** Transverse surface; SEM. **C** Radial surface; SEM (Courtesy of K. Cease)

Fig. 1.35. Brown-rot degradation in *Pinus monticola*. **A** Bore holes formed when hyphae have penetrated the cell wall (*arrows*). At this stage of decay the hypha have lysed and are no longer evident. Cracks in the cell wall are evident; bordered pit pairs have intact membranes. **B** Clamp connection on hyphae in cell wall in an advanced stage of brown rot. The cracks have elongated, and the cell wall appears porous. **C** A very advanced stage of brown rot showing an extremely fragile and porous cell wall with cracks. **A–C** Radial surface; SEM. *Bar* = 10 μm

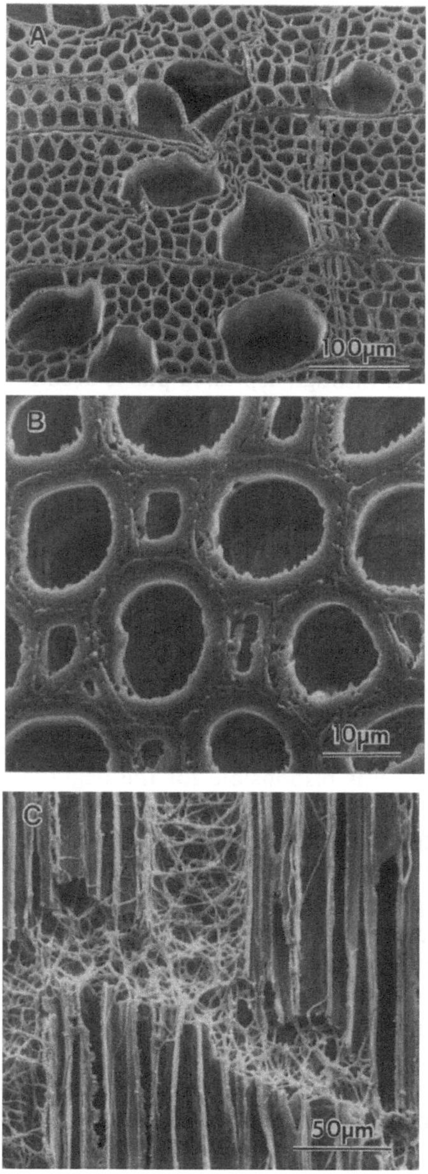

Fig. 1.34

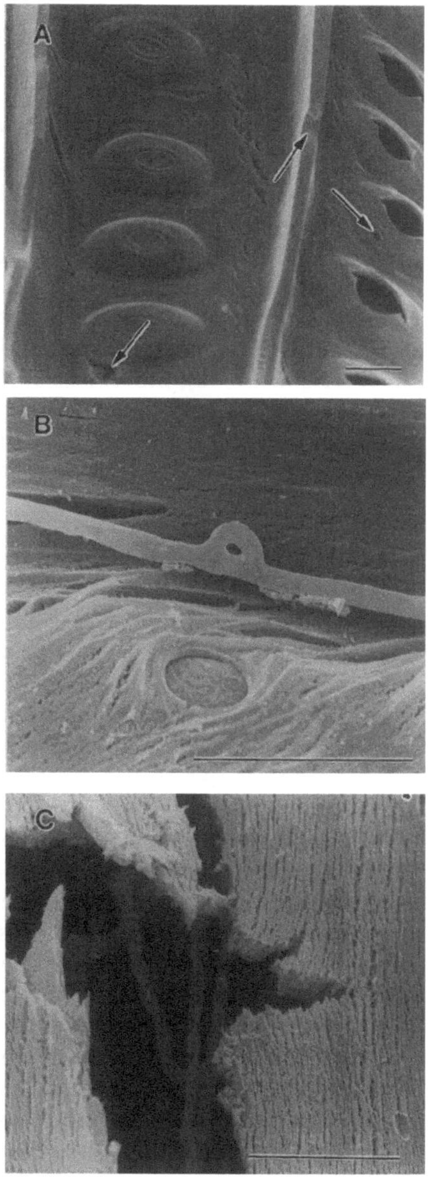

Fig. 1.35

1970). These cavities are usually only distinguishable when the cell wall is observed at high magnifications (Fig. 1.34B). Hyphae, usually found to be sparce within the lumina, can become much more extensive as decay progresses and may fill voids produced in the wood as cracks and checks develop (Fig. 1.34C).

Movement of the fungus from cell to cell is accomplished by penetration through pit membranes and aperatures or by bore holes that allow direct penetration through the wall (Proctor 1941). Bore holes are not frequently observed in early stages of brown rot (Greaves and Levy 1965, Wilcox 1968, Liese 1970, Messner and Stachelberger 1984a), and there does not appear to be a preference for brown-rot fungi to move via pits. Direct penetration through the wall occurs randomly at infrequent intervals (Fig. 1.35A). Aspiration of bordered pits does not have an effect on colonization or subsequent decay by brown-rot fungi (Highley 1978); movement of the fungi through aspirated pits is accomplished easily. Intact pit membranes may be found in cells, even after decay resulting in weight loss of over 40% and a concomitant loss of over 50% of the wood sugars (Fig. 1.35A). Even in late stages of decay some cells have intact tori.

An unusual feature of brown rot in softwoods or hardwoods is the extensive loss of polysaccharides from wood when only a few hyphae are present in cell lumina. Several different brown-rot fungi caused substantial losses of glucose, xylose, and mannose residues during the decay process (Table 1.3). This degradation occurred when only a few hyphae were observed in the wood. In some cells only a single hypha colonized the cell lumen, but degradation was evident throughout the entire circumference of the cell wall (Fig. 1.35B). The degradation also appeared to affect adjacent cells that had no hyphae. This suggests that an extremely diffusible agent was responsible for the degradation.

The porosity of the cell wall increased as decay continued. The fragile, highly lignified framework that remained in cells of brown-rotted wood was observed in advanced stages of decomposition (Fig. 1.35C). Without the bonding strength of cellulose fibrils, the brown-rotted cell walls cracked or ruptured easily (Fig. 1.35B, C). The profound loss of cell wall integrity occurred when a small amount of pressure was applied to brown-rotted wood (Fig. 1.36). The cells fractured and broke apart into minute fragments.

The cell wall ultrastructure of wood degraded by brown-rot fungi provides additional evidence for a diffuse attack originating from the surface of the lumen. After a 28% weight loss caused by *Fomitopsis pinicola*, tracheids of white pine had cell wall layers that were still distinguishable (Fig. 1.37A, B). The attack on the polysaccharide-rich microfibrils had not advanced to a stage where ultrastructural changes were evident. Dramatic alterations, however, were seen in cells from wood with approximately 50% loss of original weight. Secondary wall layers were no longer distinct, and a fine network of residual lignin was distributed throughout the cell wall. The middle lamella remained clearly defined. Electron-dense areas, observed on micrographs of sections stained with $KMnO_4$, indicate the location of lignin within the cell wall (Bland et al. 1971). Extensive loss of polysaccharides from the secondary wall

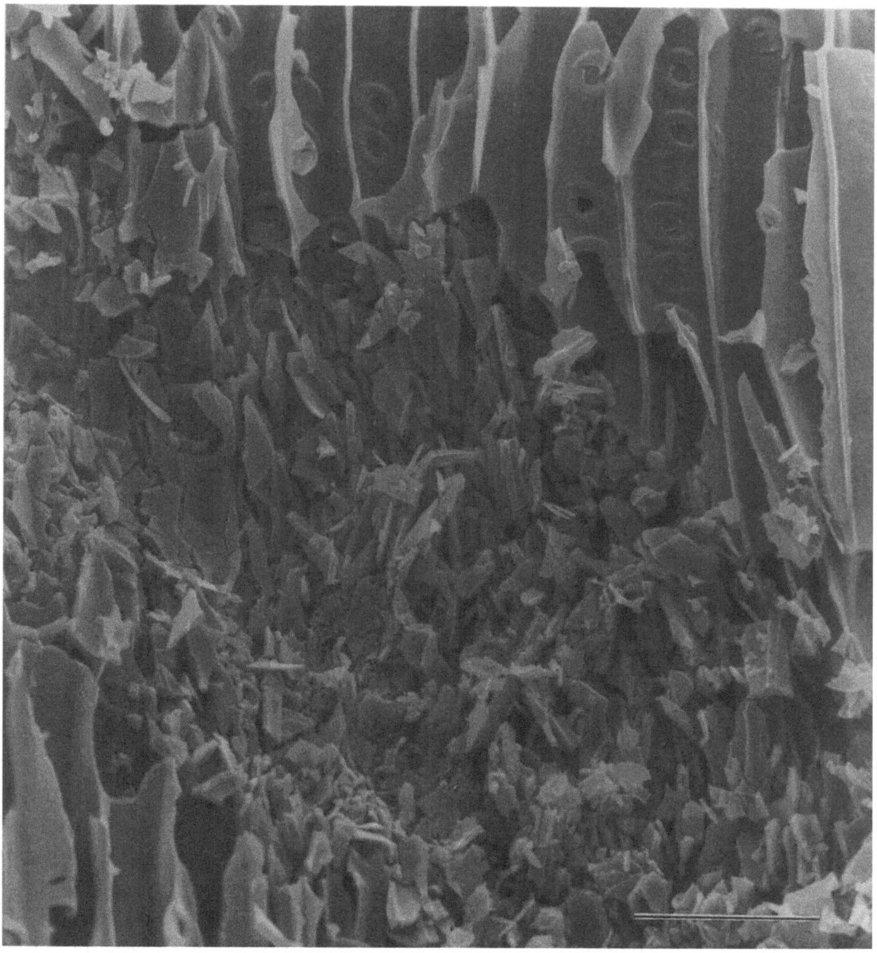

Fig. 1.36. Advanced brown-rot degradation in *Pinus* sp. The loss of wood strength by brown-rot attack is evident when slight pressure is applied to the wood and the remaining lignin skeleton crumbles into minute fragments. Radial surface; SEM. *Bar* = 100 μm

resulted in a loose framework of lignin. The morphology of the degraded cell wall was similar to the lignin skeletons of tracheids observed by Côté et al. (1966, 1968) and by Parham and Côté (1971) when wood was treated with hydrofluoric acid. The dense deposits of unhydrolyzed lignin of acid-treated wood in previous studies and in brown-rotted wood presented here (Fig. 1.37 A, B) were found throughout the cell wall, with the greatest amount in the middle lamella. In the secondary wall, a greater intensity of stain and a more compact layer of residual lignin more often occurred in the S_3 layer than in the S_2 and S_1 layers.

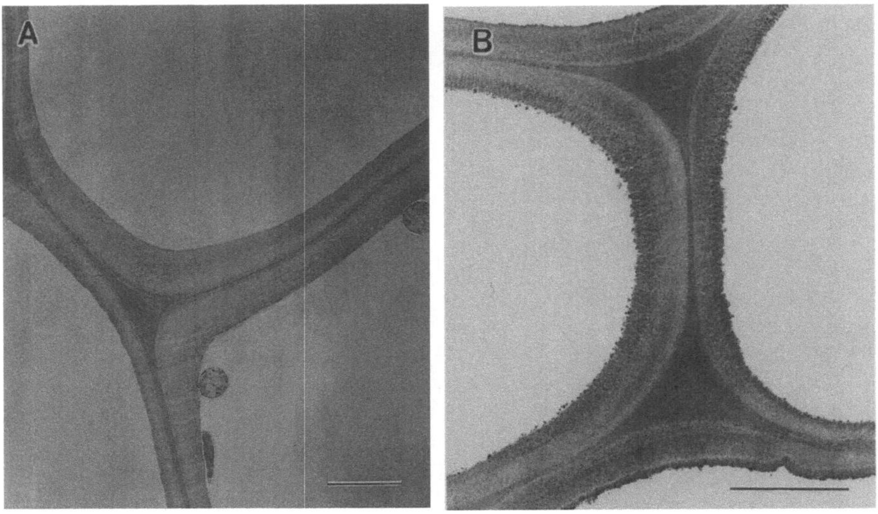

Fig. 1.37. Cell wall ultrastructure of *Pinus strobus* wood decayed by *Fomitopsis pinicola*. **A** Wood fixed with OsO_4. After 28% weight loss tracheids have cell wall layers that are intact, but extensive depolymerization of cellulose has taken place. **B** Wood fixed with $KMnO_4$. After approximately a 50% weight loss the secondary wall is severely altered. The middle lamella remains clearly defined. **A, B** Transverse section; TEM. *Bar* = 5 μm (Courtesy of K. Cease)

In contrast to brown-rot degradation in pine, birch wood decayed by *Fomitopsis pinicola* to only a 30% weight loss had cells with degraded secondary walls (Fig. 1.38A, B). Demarcation of the S_3 and S_2 layers was not observed, but the S_1 and middle lamella were distinct. The loss of polysaccharides from the cell wall occurred first in the S_2 layer (Fig. 1.38B). This area showed evidence of degradation before other secondary wall layers displayed evidence of attack. Degradation of the S_2 layer by brown-rot fungi without causing noticeable loss of the S_3 has been reported previously (Dirol 1978, Highley et al. 1985). Although hyphae were in direct contact with the S_3 layer, ultrastructural changes were most obvious deep within the secondary wall. Localized areas of erosion in the secondary wall have been observed by Liese (1970). When this occurs, the S_3 and S_2 layers are affected, and an eroded appearance of the cell wall results. Cracks and crevises found in the secondary wall may vary in size and form depending on the fungal species involved and the extent of decomposition. They may also be influenced by the preparation procedures used for electron microscopy (Liese 1970).

From the first investigations using transmission electron microscopy, evidence has mounted to suggest that a change in lignin occurs and some lignified portions of the cell wall are destroyed during brown-rot decay. Meier (1955) found that after prolonged periods of decay some lignin was lost from the loose residual structure of remaining cells. Jurásek (1964) and Liese (1970) also noted that lignified areas of the cell wall were altered, and Wilcox et al. (1974) showed that occasionally there were areas of middle lamella, including cell cor-

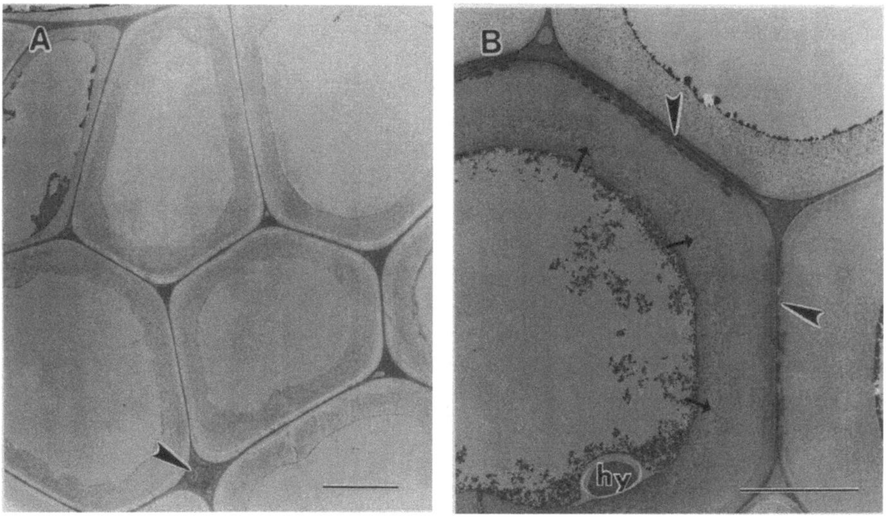

Fig. 1.38. Cell wall ultrastructure of *Betula papyrifera* wood after a 30% weight loss caused by *Fomitopsis pinicola*. **A** Wood fixed with OsO_4, showing cells with degraded secondary walls. **B** Wood fixed with $KMnO_4$. Degradation of the S_2 layer (*arrows*) can be seen in the cell containing hyphae (*hy*). There is also evidence of middle lamella alteration (*arrowheads*). Transverse section; TEM. *Bar* = 5 μm

ner regions, that were completely destroyed. Recent investigations (Messner and Stachelberger 1984a, Highley et al. 1985) strongly support the view that the middle lamella and cell corner regions of some cells become less electron dense. Two physiological types of hyphae have been proposed to account for the limited attack on lignin by brown-rot fungi (Highley et al. 1985). Some hyphae have the capacity to degrade and metabolize all cell wall components, including lignin, while other hyphae merely modify the structure of the lignin.

Evidence for a modification of lignin by demethylation of both phenolic and nonphenolic units is strong (Kirk et al. 1971, Kirk 1975). Demethylation of lignin in wood has been monitored by electron microscopy using osmium tetroxide fixation (Messner and Stachelberger 1984, Messner et al. 1986). When osmiophilic particles were found throughout the cell wall of brown-rotted wood. These particles are reaction products of dihydroxyphenolic substances (formed by demethylation of phenolic units) and OsO_4. Particles were first found to be distributed in the S_3 layer, but early in the decay process they extended throughout the cell wall. Demethylation of the cell wall lignin appears to progress as rapidly as the attack on polysaccharrides.

Ultrastructural evidence that lignin is substantially metabolized by brown-rot fungi occurs when middle lamella and cell corner regions of decayed wood appear less electron dense when fixed with OsO_4 or $KMnO_4$. Figures 1.39 A and B show examples of the changes that may occur in some parts of the cell wall after degradation of birch wood by *Fomitopsis pinicola*. A loss of continuous contrast in the middle lamella and attack on cell corners can be

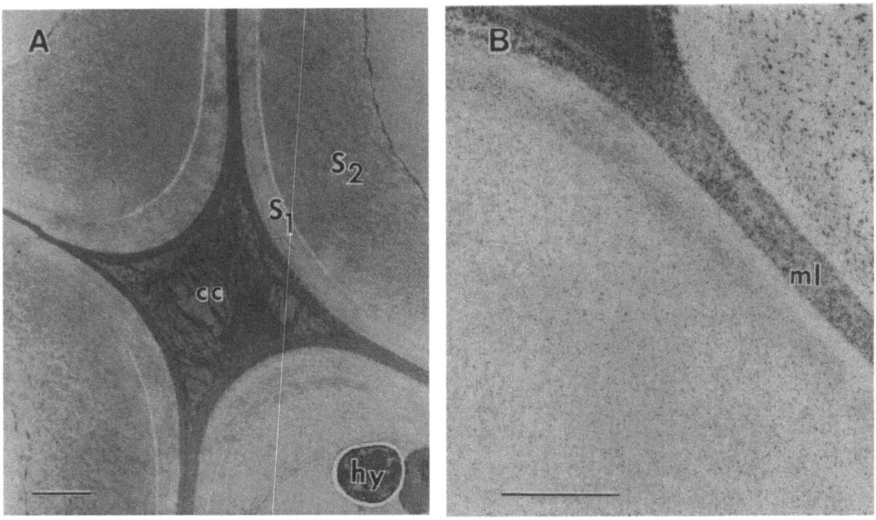

Fig. 1.39. Evidence for alteration of the lignified middle lamella in *Belula papyrifera* after degradation by *Fomitopsis pinicola*. **A** Wood fixed with OsO_4. Cell corner region (*cc*) of some cells have a degraded appearance. The S_1 and S_2 layers are still discernible; *hy* hypha. **B** Wood fixed with $KMnO_4$. Alteration of the middle lamella between cells is also evident in some areas. Instead of a continuous zone of electron-dense middle lamella, a disrupted middle lamella with little stain is apparent. Transverse sections; TEM. *Bar* = 1 µm

observed only infrequently in the decayed wood. Interpretation of these micrographs and those of other investigators (Messner and Stachelberger 1984a, Wilcox et al. 1984, Highley et al 1985), suggests that substantial degradation of lignin takes place in localized areas of the wood. Since a slightly modified lignin is more reactive to both OsO_4 and $KMnO_4$ fixation (Bland et al. 1971, Messner and Stachelberger 1984a), a partial degradation would result in greater electron density. This, however, was not the case. Instead, there was a loss of electron density in these areas. Chemical analyses also support the view that some lignin is lost from the wood (Table 1.3). Decreases in lignin content by brown-rot fungi have been previously reported (Kirk and Highley 1973), but the accuracy of the assay technique has been questioned due to the acid solubility of modified lignin. However, Crawford (1981a) provides conclusive evidence that some lignin is degraded by brown-rot fungi. The large amount of lignin that remains in advanced stages of brown-rotted wood (Table 1.3) demonstrates that if some lignin is degraded it is not removed extensively, and degradation does not occur uniformly throughout all cells.

Loss of some lignin from the cell wall is most readily observed in the highly lignified middle lamella at the cell corners (Fig. 1.39A, B), but it may occur also to a limited extent throughout the cell wall. The difficulty with interpretation of many previous ultrastructural investigations using brown-rotted wood is the change that occurs in the cell wall due to excessive loss of polysaccharides, resulting in artifacts during specimen preparation. Cracks and cavi-

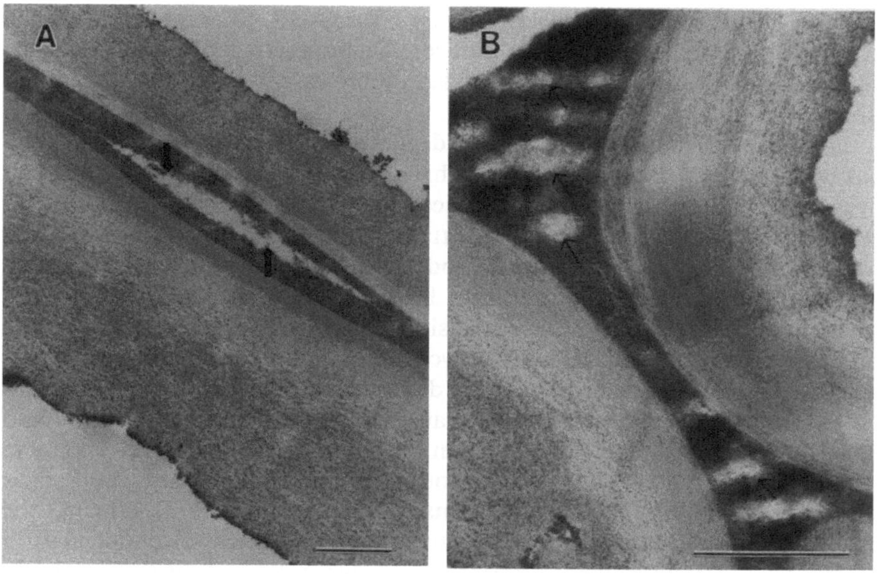

Fig. 1.40. Brown-rotted wood of ***Betula papyrifera*** fixed with $KMnO_4$, showing artifacts produced during specimen preparation and sectioning. **A** Splits in the middle lamella (*arrows*) due to drying. **B** Chatter produced in the direction of cutting (*arrows*), resulting in clear patches in the middle lamella. These forms of preparation artifacts have been interpreted by many researchers as evidence of lignin degradation by brown-rot fungi. Transverse sections; TEM. *Bar* = 1 μm

ties observed in the cell wall (Fig. 1.40A) and chatter marks of middle lamella produced during sectioning (Fig. 1.40B) are examples of artifacts that some investigators have used to show that lignin was degraded. The possibility that some morphological features observed in electron microscopy were due to swelling and drying of cells during decay or specimen preparation had been noted by Liese (1970). Despite the fact that artifacts may still be observed in micrographs of recently published studies, even when improved specimen preparation techniques are used, there does appear to be good evidence that a progressive attack on lignin occurs in some cells of brown-rotted wood (Fig. 1.40A, B) (Meier 1955, Jurásek 1964, Liese 1970, Wilcox et al. 1974, Messner and Stachelberger 1984a, Highley et al. 1985).

Most brown-rot fungi do not have the ability to degrade pure isolated cellulose (Nilsson 1974a, Highley 1977), but a few species, several of which are important forest product degraders, are unusual in that they have this capacity (Nilsson and Ginns 1979). Those brown-rot fungi may begin to degrade cellulose when the fibers are in contact with wood (Nilsson 1974a). This indicates that wood components other than cellulose may be necessary for the initiation of cellulolytic action. It may be dependent upon prior or concomitant removal of the hemicellulose component (Highley 1973, 1987a, Kirk and Highley 1973). Extensive loss of hemicellulose is evident in deciduous and coniferous wood that has been decayed by brown-rot fungi (Table 1.3). Lignin may also have an influence on cellulolytic activity. Nilsson (1974a) suggested that the presence of lignin in some way induces the production of cellulases since lignified jute fibers were easily degraded by brown-rot fungi, but delignified fibers were not.

Recently, an unusual type of wood degraded by both white and brown rots has been reported (Blanchette 1983). This naturally decayed wood provided an opportunity to test the proposed concepts, formulated from laboratory investigations, of how brown-rot fungi attack cellulose. A living spruce tree that had a white-pocket rot in the heartwood, caused by *Phellinus pini*, was blown over and subsequently colonized by a brown-rot fungus *Fomitopsis pinicola* (Blanchette 1983). This type of succession where a brown rot follows a white rot has been reported occasionally (Boyce 1961). The decayed spruce wood had macroscopic features of brown cubical decay but also had white-pockets, composed primarily of delignified cells that retained their white appearance and integrity (Fig. 1.41 A, B). The cells from the white-pocket areas had 3% lignin and 78% total sugars when assayed for sulfuric acid lignin and wood sugar content, respectively (Moore and Johnson 1967, Effland 1977), whereas the brown-rotted cells immediately adjacent to the white-pockets contained 70% lignin and 17% total sugars. Although the brown-rotted tracheids completely surrounded the delignified cells, no degradation of the cellulose was observed within the white-pockets (Fig. 1.41 A, B). The lignin and hemicellulose content of the wood between the white-pockets had evidently not induced the degradation of the cellulose in the pocket, even though delignified tracheids were less than 1 μ away from the brown-rotted cells (Fig. 1.41 C, D). Fungi that cause a white-pocket rot remove lignin and hemicellulose from the wood (Chap. 1.2). The depletion of lignin and hemicellulose appears responsible for restricting cellulose degradation and supports the laboratory studies of Highley (1973, 1987a) and Nilsson (1974a). The loss of lignin from the white-pockets caused a separation of cells at the middle lamella (Fig. 1.41 C, D). This physical alteration of the cell wall may also be responsible for inhibiting the movement and diffusion of cellulolytic enzymes. Whatever the mechanism, the end result is a very unusual type of decay consisting of brown-rotted wood with islands of pure cellulose distributed throughout (Fig. 1.41 A, B).

1.4 Wood Degradation by Soft-Rot Fungi

Wood in contact with excessive amounts of moisture may be attacked by species of fungi in the Ascomycotina and Deuteromycotina that grow within the cell wall, causing cavities in the S_2 layer. The term soft rot was first used by Savory (1954) to characterize the decay due to the soft appearance of the wood surface. When dry, the decayed wood is brown with cracks and checks that give it a macroscopic appearance that is similar to brown-rotted wood (Fig. 1.11). A unique feature of soft-rot attack is the chains of biconical and cylindrical cavities produced within the secondary wall. This unusual microscopic pattern

Fig. 1.41. *Picea engelmannii* wood decayed first by a white-pocket rot fungus (*Phellinus pini*) and then by a brown-rot fungus (*Fomitopsis pinicola*). **A, B** Wood with brown-cubicle rot and a white-pocket rot. The brown wood was composed of 70% lignin and the white-pockets 3% lignin. **C, D** Abrupt borders can be seen between brown-rotted (*br*) and delignified (*dl*) cells (*arrows*). Tracheids within white-pockets lacked middle lamella, and secondary walls were delignified. Brown-rotted tracheids appear intact after cellulose was removed, leaving only a framework of lignin. Although brown-rot fungi can extensively degrade cellulose within normal cell walls, the fungus did not attack the relatively pure cellulose found in cells of white-pockets. **A, B** Longitudinal sections of decayed wood. **C** Tangential surface; SEM. **D** Radial surface; SEM (Blanchette 1983)

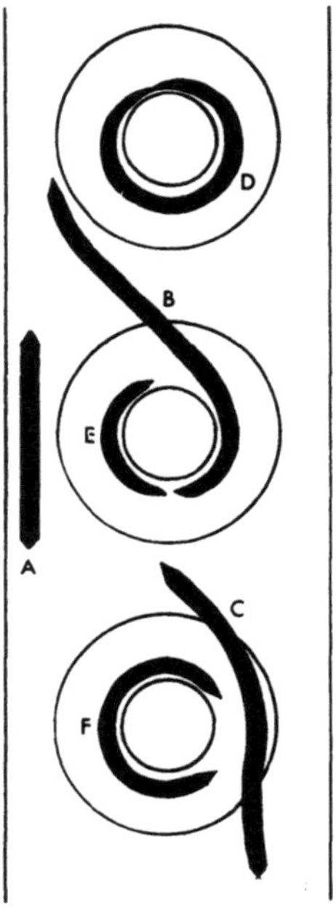

Fig. 1.42. Drawing from Nilsson (1974b) illustrating the orientation of soft-rot cavities, (**A–F**) in different areas of the cell wall. The cavities always follow the microfibrillar structure of the S_2 layer. At sites of bordered pit pairs the cavities crossed over the pit border or completely encircled the pit aperature. At least two different wall layers, with different orientation of the microfibrils, are affected, indicating that the S_1 layer of pit borders may also permit cavity formation. Radial surface

of decay in the S_2 layer is characteristic of the soft-rot type of deterioration (Nilsson 1985). Cavities within the cell walls of decayed wood were observed by many investigators before the term soft rot had been used (Schacht 1863, Bailey and Vestal 1937, Braghoorn and Linder 1944, Findlay and Savory 1954), and a large number of different species have been shown to produce cavities when grown on wood in the laboratory (Savory 1954a, b, Duncan 1960, Lundström 1972, Nilsson 1973, 1974b).

Soft-rot cavity morphology may vary depending on the fungus and wood substrate. As many as 14 different groups of soft-rot cavities have been proposed based on cavity origin and morphology (Courtois 1963). The orientation of all cavities, however, depends upon the microfibrilar structure of the various cell wall layers (Fig. 1.42). In addition to cavities, most soft-rot fungi produce an erosion of the cell wall from the lumen inward toward the middle lamella. Corbett (1965) separated the types of attack into two groups. Type 1 attack resulted in the formation of cavities within the cell wall (Fig. 1.43) and type 2 caused an erosion of all secondary wall layers directly beneath the hyphae. Soft-rot fungi may produce type 1 attack or a combination of types 1 and 2. The proposed definition of soft rot (Nilsson 1985) does not include fungi that produce only type 2 attack. Fungi that cause only an erosion of the secondary wall result in decayed wood that has morphological characteristics similar to white-rot degradation (Sect. 1.2). Detailed investigations are needed, however, to ascertain whether cavity formation is produced in addition to the erosion (type 2 attack). Usually, microfungi that produce cavities within the secondary wall have the capacity to form cavities in a variety of different sub-

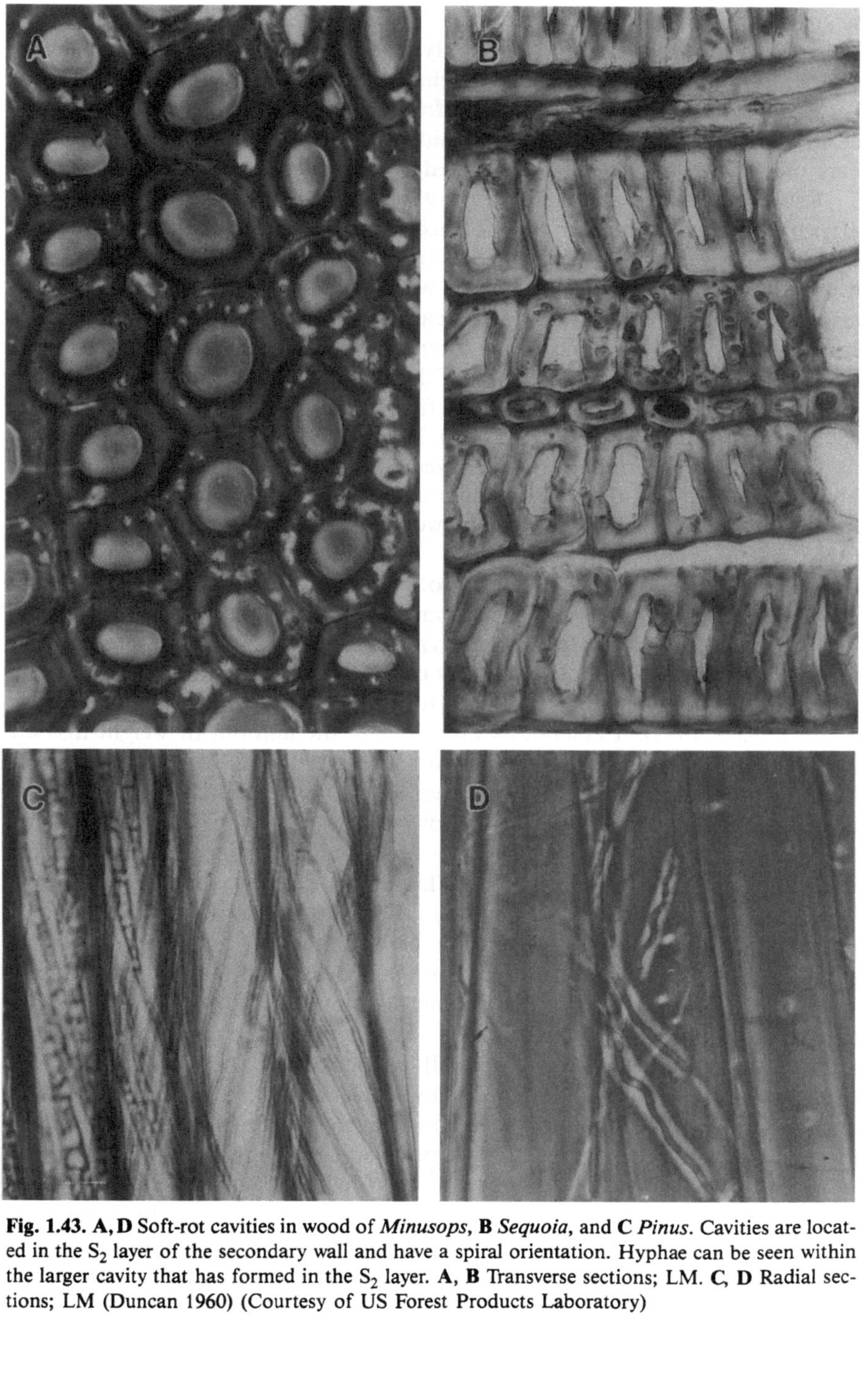

Fig. 1.43. A, D Soft-rot cavities in wood of *Minusops*, **B** *Sequoia*, and **C** *Pinus*. Cavities are located in the S_2 layer of the secondary wall and have a spiral orientation. Hyphae can be seen within the larger cavity that has formed in the S_2 layer. **A, B** Transverse sections; LM. **C, D** Radial sections; LM (Duncan 1960) (Courtesy of US Forest Products Laboratory)

strates (Nilsson 1973). Most fungi that produce both types 1 and 2 attack in cells of various hardwoods produce only type 1 attack in softwoods. Since cavity formation is a distinct, relatively reliable character of soft-rot fungi, it can be used readily to identify this decay from other types of degradation. Zabel et al. (1985) have used the method to identify soft-rot fungi from large numbers of cultures obtained from decayed utility poles.

Recently, Francis and Leightley (1984) found a type 2 attack caused by *Hypoxylon mediterraneum* that formed rhomboid cavities in the wall. The cavities, however, originated by an erosion of the wall from the lumen and were not typical soft-rot cavities that occur within the secondary wall. This fungus also produced other unusual decay features (i.e., degraded cell walls with a spongy matrix). For this particular fungus, additional investigations are needed to appropriately classify the type of decay it produces in various substrates. Although some brown- and white-rot fungi have also been found to produce cavities within the S_2 layer (Liese 1970, Dirol 1978), these observations are infrequent and may result from a localized collapse of the degraded S_2 layer by brown-rotters or an odd type of simultaneous erosion by white-rotters. Cavity formation is not typically found in wood attacked by brown- or white-rot fungi.

The rate and extent of decay by soft-rot fungi depend on the wood substrate (Tables 1.4 and 1.5). In general, hardwoods are attacked by soft-rot fungi to a greater extent than softwoods (Eslyn and Highley 1976). The rate of decay in softwoods may be low and the extent of the degradation minimal when compared to the decay by white- or brown-rot fungi incubated for the same period of time (Kirk 1971). Tremendous variations in the amount of weight loss are also evident among different species of soft-rot fungi (Table 1.4). The loss of major structural components from test blocks indicates that both carbohydrates and lignins are depleted. In hardwoods, the carbohydrate fraction is removed faster than lignin, whereas in softwoods lignin appears to be removed faster than cellulose or hemicellulose (Table 1.5). The amount of lignin actually degraded by soft-rot fungi is uncertain. The Klason lignin assay used in these investigations is a crude determination method that may not be accurate. This situation is similar to that observed for lignin determinations of brown-rotted wood (Sect. 1.3). Microspectrophotometry using UV light has also been applied to measure lignin breakdown by soft-rot fungi; the results indicate that lignin is degraded in the secondary wall, particularly in the vicinity of cavities (Bauch et al. 1976). The investigations by Haider and Trojanowski (1975, 1980) using ^{14}C-labeled compounds have conclusively demonstrated that aromatic moieties of lignin are degraded by various soft-rot fungi.

Colonization of hardwoods by soft-rot fungi is observed first in the rays and vessels from where hyphae grow into the fiber lumina. In softwoods, hyphae colonize ray parenchyma cells and move from the rays into the lumina of tracheids (Courtois 1963, Corbett 1965, Liese 1970). The first evidence of cavity formation is observed in the cells adjacent to the ray parenchyma (Hale and Eaton 1986a). Hyphae from the fiber or tracheid lumina bore into the secondary wall. These microhyphae or fine penetrating hyphae of less than

Table 1.4. Weight losses (%) caused by soft-rot fungi in sapwood of 15 species in a serial-block test after 12 weeks of decay[a] (After Eslyn and Highley 1976)

Wood species	Soft-rot fungi						
	Papulaspora sp.	*Monodictys* sp.	*Graphium* sp.	*Dactylomyces crustaceous*	*Allescheria* sp.	Penicillium sp.	*Thielavia terrestris*
Pseudotsuga menziesii	0.6	1.2	2.1	0.0	2.5	0.0	0.8
Picea engelmannii	5.9	6.4	5.6	2.8	4.0	4.6	6.9
P. glauca	7.2	3.3	4.0	3.0	6.2	5.0	2.7
P. sitchensis	6.5	4.3	2.4	4.0	8.1	2.7	2.2
Pinus ponderosa	12.6	4.0	3.2	0.7	6.0	1.6	1.8
P. monticola	12.8	4.4	4.0	5.9	8.4	8.0	6.8
P. contorta	9.0	3.6	3.5	3.6	8.4	5.6	5.3
P. sp. (Southern pine)	4.2	5.9	4.1	4.5	4.3	4.4	3.9
P. resinosa	4.0	5.8	2.1	3.0	3.6	3.7	3.5
Tsuga heterophylla	3.9	2.6	2.6	3.1	2.4	2.3	2.0
Thuja plicata	0.4	2.6	1.1	1.5	0.3	0.7	0.8
Sequoia sempervirens	4.1	5.2	4.6	3.7	4.1	4.0	4.2
Populus balsamifera	16.7	20.9	23.7	5.4	37.8	37.0	17.4
Eucalyptus sp.	12.2	23.0	22.9	11.2	24.3	18.1	15.6
Alnus rubra	10.9	27.4	32.4	2.1	34.8	34.0	21.6

[a] Data are maximum weight losses in four replicates.

Table 1.5. Loss of major structural components from *Populus balsamifera* and *Pinus monticola* wood test blocks decayed to various extents by six soft-rot fungi (Eslyn et al. 1975)

Wood	Fungus	Extent of decay (%)[a]	Structural component loss (%)			
			Lignin	Glucan	Xylan	Mannan
Populus balsamifera	*Graphium* sp.	16	10	20	25	12
		22	15	27	34	16
	Monodictys sp.	10	0	8	12	19
		22	13	27	26	38
	Papulaspora sp.	10	0	14	17	14
		21	4	27	29	25
	Paecilomyces sp.	14	10	15	35	28
		28	11	41	37	30
	Thielavia terrestris	10	6	11	12	29
		23	13	25	27	43
	Allescheria sp.	11	6	15	17	22
		28	9	32	29	30
Pinus monticola	*Papulaspora* sp.	15	12	18	18	13
	Paecilomyces sp.	10	14	7	8	6
	Thielavia terrestris	7	14	3	9	−3

[a] Percent of total original weight lost by test block.

0.5 μm in diameter penetrate the S_3 layer and move into the S_2 (Fig. 1.44). They align their direction of growth parallel to the direction of the cellulose microfibrils and form either an "L-bending" hypha in one direction or a "T"-shaped branch where hyphae grow in opposite directions (Corbett 1965, Levi 1965; Fig.1.44). The fine hyphae may also penetrate directly through the secondary wall and middle lamella. Growth continues into the adjacent cell wall and either follows the orientation of the microfibrils in the S_2 layer, forming an L-bending or a T-branching hypha, or passes into the next cell lumen. Once in the lumen of the adjacent cell, the diameter of the hypha resumes its normal size (Fig. 1.44).

Cavity formation begins after the L- or T-branching hypha extends within the S_2 layer and temporarily stops growth (Hale and Eaton 1985b). Enzymes are released along the length of the fine hyphae, and cell wall dissolution begins. After a period of cavity enlargement, a new fine hypha or proboscis hypha (Corbett 1965, Hale and Eaton 1985b) emerges from the end of the cavity. Apical extension at a linear growth rate occurs, and the proboscis hypha may reach a maximum of 40 μm before growth ceases and a new cavity starts to form (Hale and Eaton 1986a). A series of cavities results by repeated emergence of proboscis hyphae, cessation of hyphal growth, and subsequent cavity formation.

In advanced stages of decay, the S_2 layer is severely degraded with many cavities that coalesce (Fig. 1.45A, B). The degraded areas of the S_2 appear as cylindrical cavities with conical ends or as biconical (diamond-shaped) cavities (Fig. 1.46A). These cavities appear round or oval when the cross-sectional area

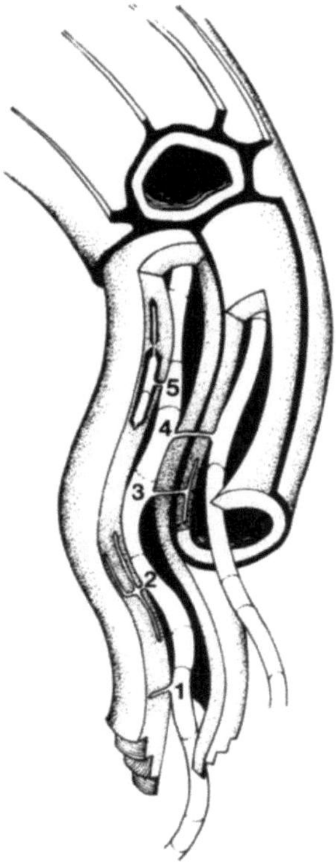

Fig. 1.44. A soft-rot fungus attacking woody cells, causing cavities in the S_2 layer of the secondary wall. Large diameter haphae in the lumen form a fine hypha that penetrates the S_2 layer and moves into the S_2 (*1*). The fine hyphae align their growth to the orientation of the microfibrils and form an "L"-bending or "T"-branching hypha. After a period of extension in the S_2 layer growth stops (*2*). A fine penetrating hypha may move through the entire secondary wall and middle lamella into the adjacent cell. The hypha may change directions in the S_2 layer forming a "L"-bending or "T"-branching hypha (*3*) or continue to penetrate through the cell wall and into the lumen of an adjacent cell (*4*). Once in the lumen, the hypha resumes normal diameter size. Cavities in the S_2 layer arise from enzymatic activity along the length of the fine hyphae that form the "L"- or "T"-branch (*5*). As the cavity becomes larger, the hyphal diameter increases and a new fine hypha or proboscis hypha emerges from the end of the cavity. Proboscis hyphal growth ceases, and a new cavity forms (*5*)

of the cells is observed (Fig. 1.46B). The preference of soft-rot fungi for the S_2 layer was demonstrated by Nilsson (1974b) in a laboratory experiment. Sections of wood inoculated with *Humicola alopallonella* showed that the fungus preferentially penetrated the S_2 layer to form cavities (Fig. 1.46C). As soft-rot degradation continued in the section, lyzed zones in the S_2 layer enlarged, and new cavities developed (Fig. 1.46D). Although all of the cell wall layers were equally exposed to the fungus, only the S_2 layer was attacked.

The chains of cavities that form in the S_2 layer always follow the helical orientation of the microfibrils. Both cylindrical cavities with conical ends and biconical cavities may form in the same substrate (Fig. 1.47). Great variation, however, may be found in the shape of individual cavities. The species of fungus, growth conditions, and substrate all influence cavity formation (Corbett 1965, Eslyn et al. 1975, Hale and Eaton 1986a). Soft-rot cavities in early stages of development are circular in transverse section (Fig. 1.47). Cavity widening occurs solely on the surface of the cyclindrical portion (Hale and Eaton 1986a). The cavity becomes wider, but the length does not change. There is a lack of enzyme activity at the exposed microfibril ends of the cone surface. The widening process follows the angle of the cone, which ultimately forms a diamond-shaped cavity (Fig. 1.47). Often, cylindrical cavities that are widening fuse with another cavity before the diamond shape can be completed (Fig.

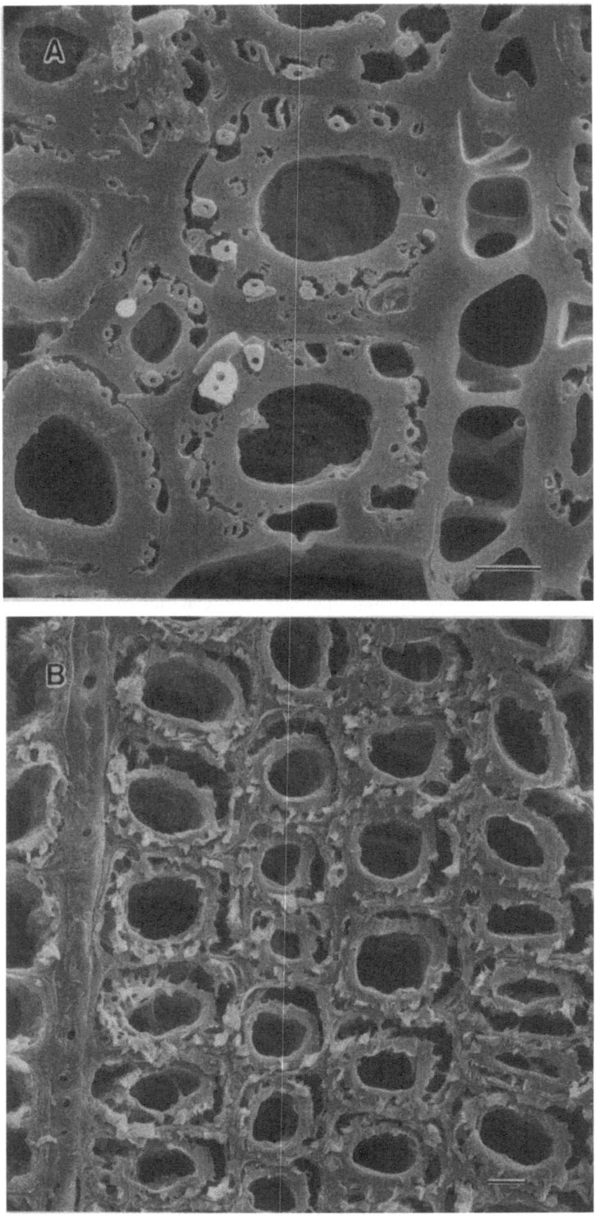

Fig. 1.45. Extensive cell wall damage associated with soft-rot attack in creosote-treated southern pine poles. **A, B** Cavities and hyphae in the S_2 layer of the secondary wall. Transverse surface; SEM. *Bar* = 10 μm (Zabel et al. 1985)

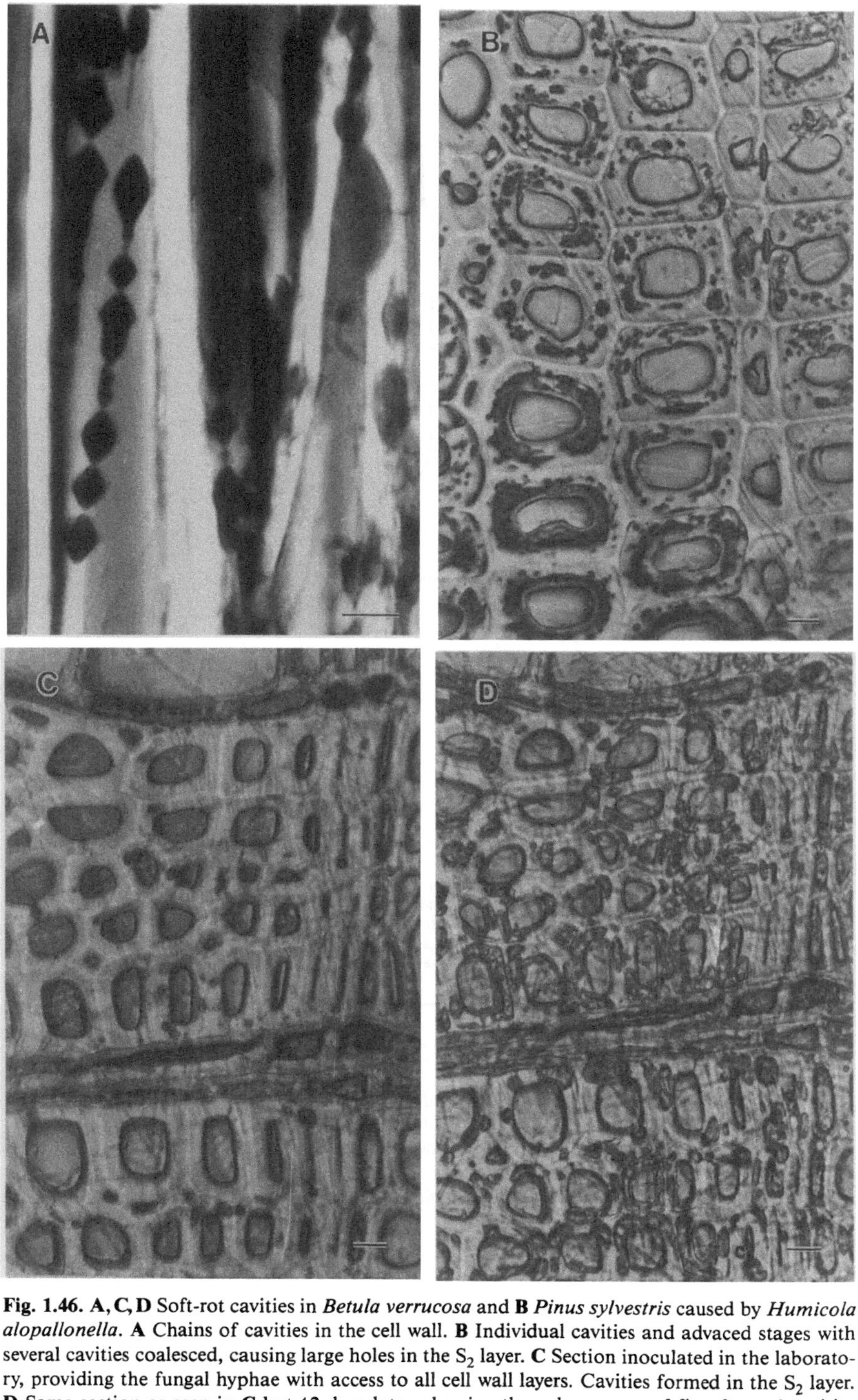

Fig. 1.46. **A, C, D** Soft-rot cavities in ***Betula verrucosa*** and **B** *Pinus sylvestris* caused by *Humicola alopallonella*. **A** Chains of cavities in the cell wall. **B** Individual cavities and advaced stages with several cavities coalesced, causing large holes in the S_2 layer. **C** Section inoculated in the laboratory, providing the fungal hyphae with access to all cell wall layers. Cavities formed in the S_2 layer. **D** Same section as seen in **C** but 12 days later, showing the enlargement of first-formed cavities and new cavities in the S_2 layer. **A** Radial section, polarized light microscopy. **B–D** Transverse section; LM. *Bar* = 10 µm (Nilson 1974b)

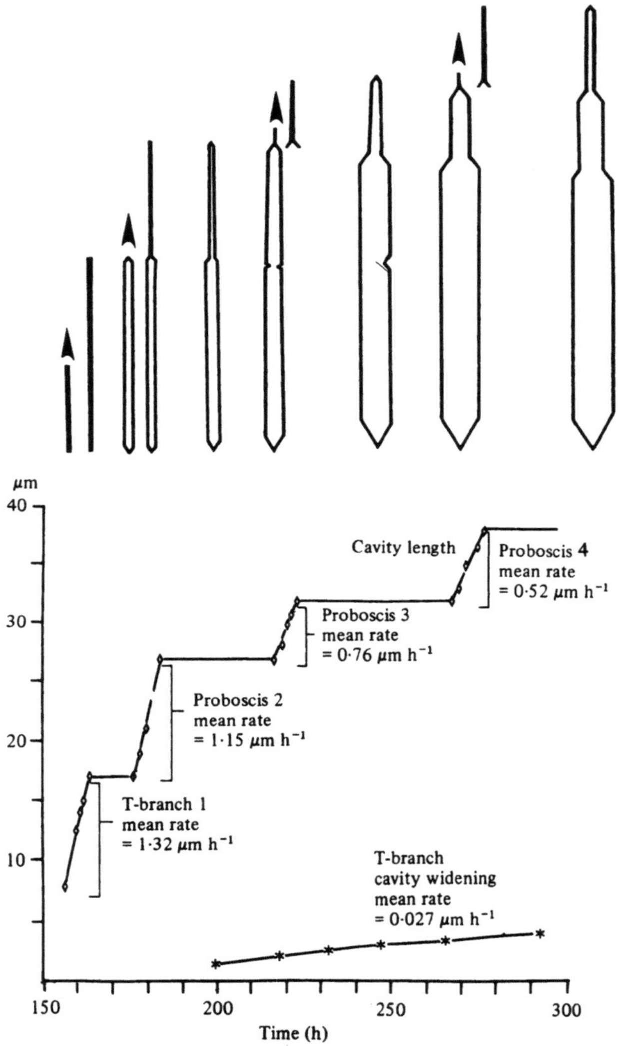

Fig. 1.47. Oscillatory growth of *Phialophora hoffmannii* hyphae in cell walls of *Pinus sylvestris*. Time-lapse microscopy, exposing one frame of film every 4 min, shows growth of the soft-rot fungus (*lower diagram*) and cavity formation during four phases of apical hyphal growth, initial T-branch growth (*1*) and three phases of proboscis development (*2, 3, 4*); fine hyphae followed by enzyme release along the hyphal surface. Filming was carried out for 145 h (Hale and Eaton 1985a)

1.47). Cone angles of soft-rot cavities differ among different types of wood. Hale and Eaton (1986a) suggest that the lack of attack at the cavity ends is due to an inability of endocellulases produced by the soft-rot fungus to diffuse longitudinally into the microfibrils. The exposed cellulose microfibrils at the end of the cavity would not be readily attacked by the endocellulase. The structure of the cellulose and size of the cellulose enzymes appear to be important aspects that determine the angle of the cone. Hale and Eaton (1986a) conclude that the different microfibrillar spacing among species of wood and "angled attack" on the ends of the microfibrils determines the overall cone angle.

Recently, a mathematical model has been presented to describe the cavity widening process (Hale and Eaton 1986a). The initial stages of cavity widening show linear kinetics, which suggests that enzyme concentration is not limiting. If enzyme concentration was a limiting factor, a decline in rate would occur in the early stages of cavity widening when the available substrate increases. The limiting factor, therefore, appears related to substrate susceptibility. When cavity widening in birch and pine was compared (Hale and Eaton 1985a) the values obtained were $0.061-0.206\ m^{-1}$ and $0.009-0.038\ mh^{-1}$, respectively. These differences in rates of cavity widening demonstrate the effect of cell wall composition on the rate of enzyme activity.

A rhythmic type of behavior by the fungus or oscillatory growth of hyphae in the S_2 layer has been suggested by Bailey et al. (1968) and Leightley and Eaton (1977) and confirmed by Hale and Eaton (1985a) using time-lapse cinemicrography. The development of cavities can be seen in a diagram drawn from a film series taken over 145 h (Fig. 1.48). The "stop-start" oscillatory growth of fine hyphae and subsequent degradation form the chains of discrete cavities (Hale and Eaton 1985a). The oscillatory growth activity observed in soft-rot fungi has been explained in terms of nutritional support from parent hyphae in the lumen during "T-branch" formation and from parent cavity hyphae during apical growth of fine proboscis hyphae (Hale and Eaton 1985a). When nutrients are available following cavity enlargement, enzyme activity is greater at the growing tip, and proboscis extension occurs. The stored nutrients are soon depleted and growth ceases, causing enzymes to be released along the surface of the hyphae. The dissolution of the S_2 layer results, and a new cavity forms. In advanced stages of soft-rot decay, the individual cavities enlarge and coalesce until the entire S_2 layer is depleted.

Investigations concerning the ultrastructure of soft-rot attack in wood cell walls have monitored the developmental stages of cavity formation (Hale and Eaton 185b, c). Fine hyphae that penetrate the cell wall have numerous ribosomes in the apical region. In the cell wall, an electron transparent zone is formed immediately surrounding the hypha (Fig. 1.49A, B). This area represents a zone of cell wall dissolution. Extracellular enzymes diffuse into the wall and degrade it, resulting in a clear zone. Fine strands of amorphous material extend from the hypha to the wood cell wall, representing a layer of muscilage and/or degraded cell wall substances (Fig. 1.49A, B). Protrusions of the hyphal plasma membrane are also found extending across the electron transparent area. These structures appear to provide contact between the hypha and wood cell wall and may aid in the movement of enzymes. A dark halo of electron-dense stain appears around the clear zone. This dark area contains partially degraded cell wall components that react with greater intensity to treatment with glutaraldehyde and OsO_4 fixatives and uranyl acetate stain. T-branch hyphae and young proboscis hyphae also consistently have regions that stain

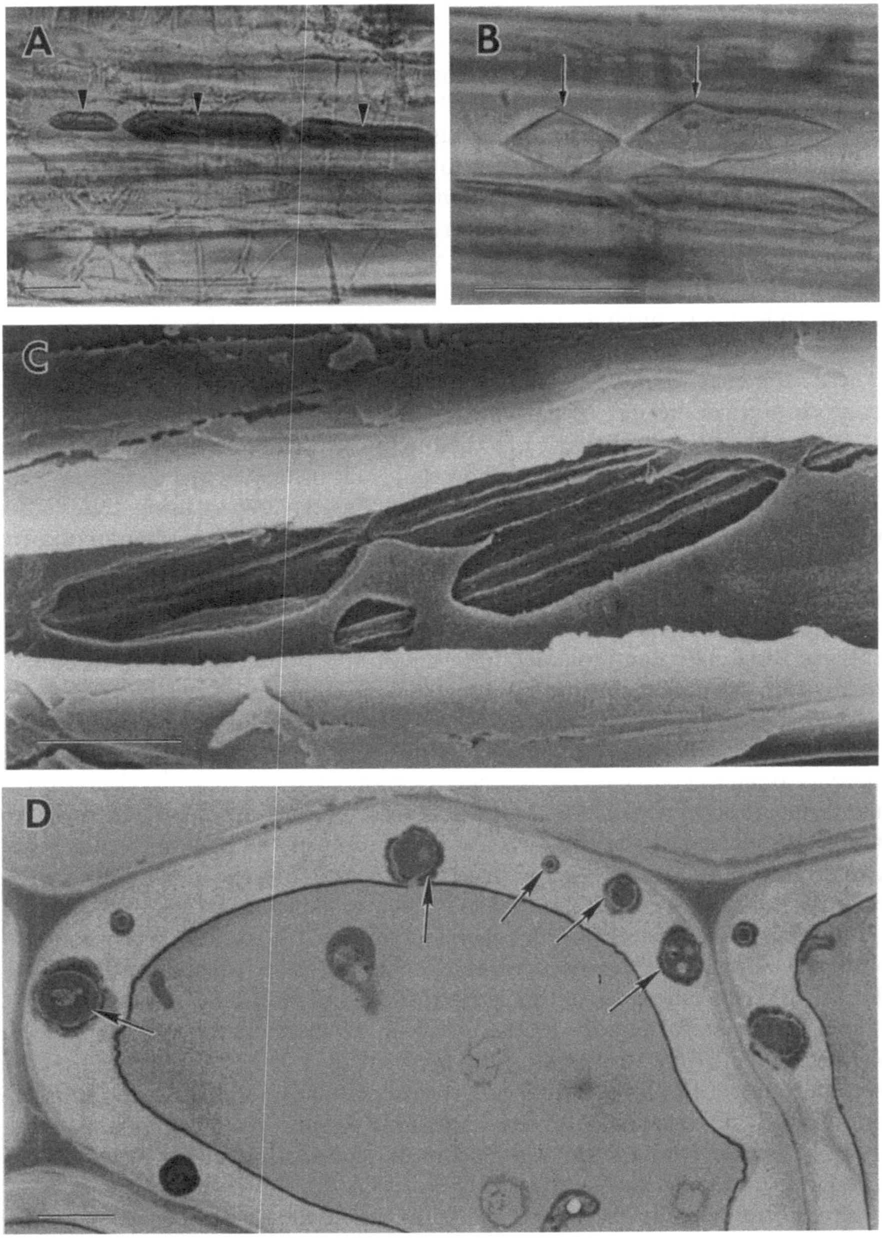

Fig. 1.48. Biconical and cylindrical cavities produced by soft-rot fungi. **A** Chain of cylindrical cavities with conical ends in *Betula* inoculated with *Phialophora fastigiata* (*arrowheads*). **B** Biconical soft-rot cavities in *Betula* inoculated with *Humicola alopallonella* (*arrows*). **C** Biconical and cylindrical cavities in fiber of *Albizia falcataria*. **D** Circular soft-rot cavities in the S_2 layer of wood from *Eucalyptus deglupta* inoculated with *Lecythophora mutabilis*. **A** Radial section; polarized LM. **B** Radial section; LM. **C** Radial surface; SEM. **D** Transverse section; TEM. *Bar* = 1 µm (Hale and Eaton 1986a)

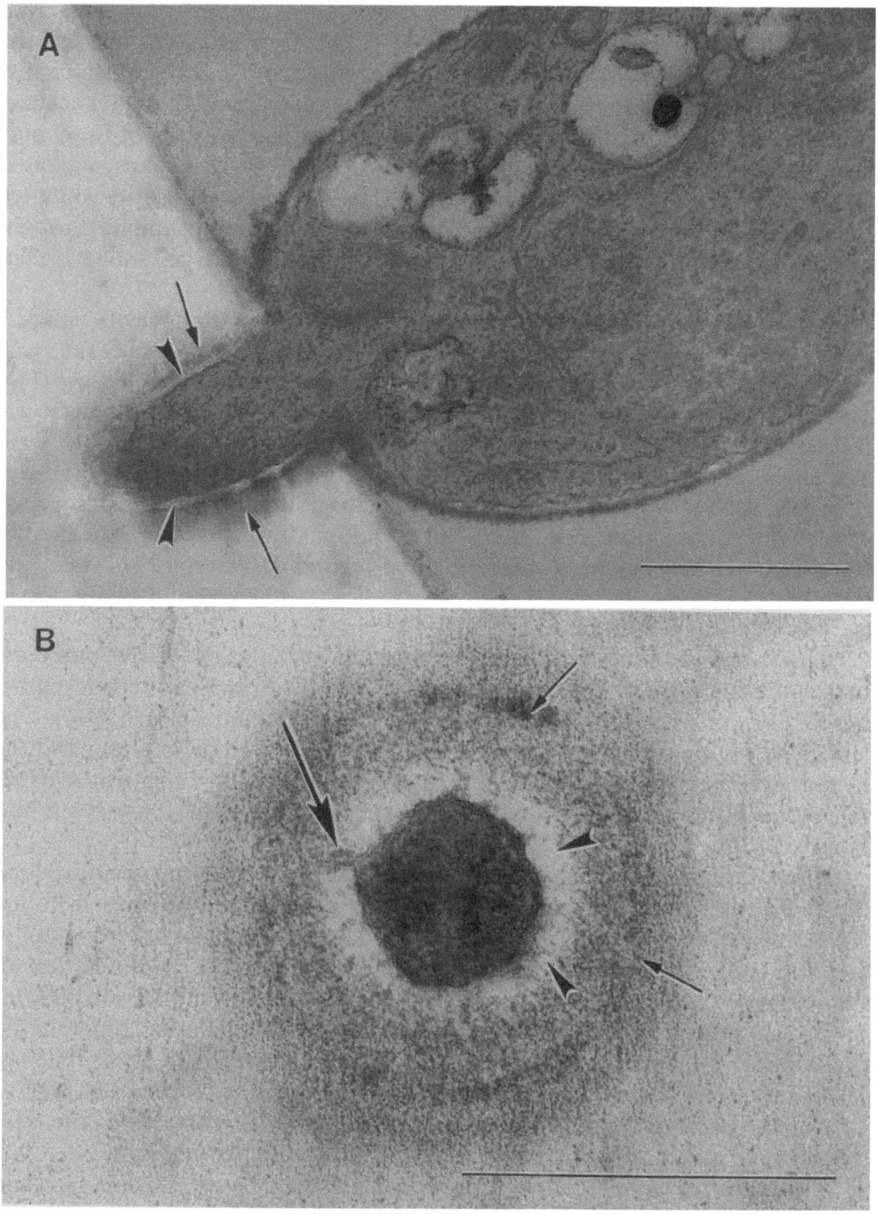

Fig. 1.49. A Fine penetrating hypha of *Humicola alopallonella* and **B** *Monodictys putredinis* in *Betula pendula*. **A** Fine hypha, originating from a hypha in the lumen, is penetrating the cell wall. An electron transparent area is seen immediately surrounding the fine hypha (*arrowheads*); a halo of a dark electron-dense region occurs around it (*arrows*). **B** Fine penetrating hypha with circular zones of electron transparent (*arrowheads*) and stained halo regions (*arrows*). The membrane of a fine hypha is evident, and a protusion of the plasma membrane is visible extending across the electron transparent zone to the cell wall (*large arrow*). **A** Radial section; TEM. **B** Transverse section; TEM. *Bar* = 1 μm (Hale and Eaton 1985b)

darkly in the surrounding cell wall. The dark stain is similar to that observed in the cell wall under white-rot fungi (Fig. 1.26). Degrading enzymes of the white-rot fungus, ***Phlebia tremellosa***, diffused into the cell wall, causing a disorganization of the structure that allowed greater reaction with uranyl acetate. The close association of the dark halo with fine hyphae of soft-rot fungi may also indicate that degrading agents, or a "precellulolytic system" as suggested by Hale and Eaton (1985c), are functioning. In contrast to attack by white-rot fungi, the dark areas around hyphae of soft-rot fungi are not found throughout the cell wall. Instead, the degradation is localized to the cell wall adjacent to the hyphae.

After penetration into the S_2 layer and T-branch formation, cavity enlargement begins. The width of the cavity increases but the length remains the same (Hale and Eaton 1985c). Hyphal diameter increases as the cavity enlarges (Fig. 1.50A). Cavity enlargement and subsequent increase of hyphal diameter are followed by the emergence of proboscis hypha, which extends out into the wall. The angles at the end of cavities coincide with constrictions in hyphal diameter (Fig. 1.50B). The position of septa in hyphae has often been found associated with cavity constriction (Levi 1965), but additional investigations indicate that septa may not always be located in these regions (Nilsson 1974, Hale and Eaton 1985c); (Fig. 1.50A–C). The constrictions at the end of a cavity persist even in late stages of attack (Fig. 1.50C).

The increased amount of enzymes released near the tip of the proboscis hypha and the reduced amounts at the proximal portion appear to be responsible for the constricted regions between cavities (Hale and Eaton 1985c). The less active region of enzyme release will have less cell wall dissolution, and the hyphal diameter remains small. This suggests that soft-rot fungi can regulate the release of enzymes at the tip and along the length of fine hyphae (Hale and Eaton 1985a, c).

Many other theories have been offered to explain the constrictions along the chain of cavities. The accumulation of toxic phenolic substances from the breakdown of lignin has been suggested by Liese (1970). Another possibility is a catabolic repression of enzyme production due to the accumulation of cellobiose or glucose released during the lysis of the cell wall (Nilsson 1974b). It has also been suggested that the hypha must be oriented exactly parallel to the cellulose microfibrils within the S_2 layer for degradation to occur (Nilsson 1974b). It is very likely that a combination of factors, nutritional as well as catabolic repression, may be responsible for the constricted areas of the cavities and the small hyphal diameter that persists in these locations.

The rate of decay by soft-rot fungi appears to be influenced by the type and quantity of lignin in the wood (Jutte and Wardrop 1970, Kirk 1971, Eslyn et al. 1975, Zainal 1976, Butcher and Nilsson 1982, Hale and Eaton 1986b). Timbers with low lignin content are more susceptible than those with high concentrations of lignin (Butcher and Nilsson 1982). Guaiacyl components of softwood lignin present a greater obstacle to degradation by soft-rotters than syringyl-guaiacyl components of hardwood lignin. The microdistribution of lignin within a cell wall can also affect soft-rot attack. The S_2 region of a

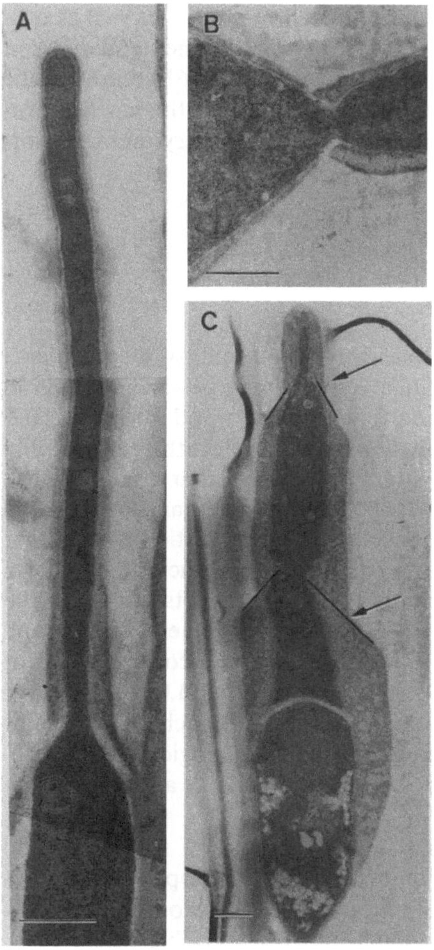

Fig. 1.50. Ultrastructural aspects of cavity formating by **A** *Phialophora hoffmannii* and **B, C** *Humicola alopallonella* in *Betula*. **A** *Proboscis* hypha emerging from a larger diameter hypha in a previously formed cavity. Hyphal diameter increases as cavities enlarge. **B** Constriction between cavities at a late stage of attack. **C** A continuous chain of three cavities. The angles at cavity ends (*arrows*) extrapolated to the hypha coincide with constrictions in the hypha. **A–C** TEM; *bar* = 1 μm (Hale and Eaton 1985c)

large number of timbers exhibit thin, concentric zones (Hartig 1878, Ruel et al. 1978, Nilsson and Daniel 1983a, Daniel and Nilsson 1984). The concentric zones appear to have an increased lignin content and these lignin layers act as barriers to cavity formation (Nilsson and Daniel 1983a). Enzymatic action is inhibited at the layer, and degradation progresses in the opposite direction. This results in a cavity with the shape of a "half moon" when observed in transverse section.

To demonstrate the effect of lignin on soft-rot fungi Takahashi and Nishimoto (1976) and Zainal (1976) used chemically delignified pine wood as a substrate for soft-rot attack. Large weight losses occurred in the delignified wood as compared to that in sound, normal wood. The removal of lignin, however, greatly influenced the mode of action by the fungus, and cavities typical of soft-rot attack were not observed. The removal of lignin apparently allowed enzymes to move freely within the secondary wall. These studies as well as others (Cassagrande and Ouellette 1971, Nilsson 1974b, Butcher and Nilsson 1982, Frances and Leightley 1984, Hale and Eaton 1985a, 1986b) reinforce the conclusions made by Liese (1964) that the size and form of the cavities produced by soft-rot fungi are greatly influenced by the chemistry and the morphological structure of the cell wall.

1.5 Wood Degradation by Bacteria

Bacteria cause many different types of attack on the cell walls of woody plants and may adversely or synergistically influence the growth of various fungi in wood. Greaves (1971a) placed the bacteria that colonize wood into four categories; bacteria that utilize the cell contents of rays and affect the permeability of wood to liquids but do not alter the strength, bacteria that attack the cell walls directly, those that are associated with other microorganisms in wood and contribute to the decomposition process, and antagonistic bacteria that are inhibitory to other microorganisms that colonize wood. Bacteria have been observed as primary colonizers of exposed wood products in terrestrial (Greaves 1968, Liese and Greaves 1975) and aquatic environments (Eaton and Dickinson 1976, Leightley and Eaton 1977) as well as in discolored wood of living trees (Shigo 1965, Hallaksela 1984). They are also found throughout the various sequential stages of decomposition (Shigo 1967, Blanchette and Shaw 1978). The great diversity among bacteria and different ecological and physiological requirements of these ubiquitous organisms result in an array of decomposition patterns in wood.

Bacteria appear to colonize the ray parenchyma cells of softwoods and hardwoods before attacking other cell types (Liese and Karnop 1968, Greaves 1970a). They utilize the cell contents and may attack the walls of the parenchyma cells. The parenchyma end walls are particularly susceptible (Liese 1970, Liese and Greaves 1975). In conifers, bacteria are often observed in resin ducts and epithelial cells as well as in other parenchymatous tissue (Boutelje and Kiessling 1964, Knuth 1964, Courtois 1966, Liese and Karnop 1968). The sapwood is the site primarily affected (Liese 1970) but colonization of ray parenchyma and degradation of cell contents and cell walls in the heartwood have also been reported (Greaves 1970a, Liese 1970). Bacteria move into adjacent cells quickly via the cross-field pits between ray parenchyma cells and tracheids, with rapid destruction of pit structures (Levy 1975).

The lytic action of bacteria on woody cell walls was first observed by Liese (1950). Bacteria in pit chambers at the annulus region attacked the cell wall, causing perforations in the outer margins of the bordered pits. Liese and Karnop (1968) showed a progressive bacterial attack that first appeared as a reduction in uptake of stain in the pit membrane, followed by a granulated appearance and complete distintegration. The morphology of bacterial attack on pit structures in wood also may take additional forms. Greaves (1969) identified

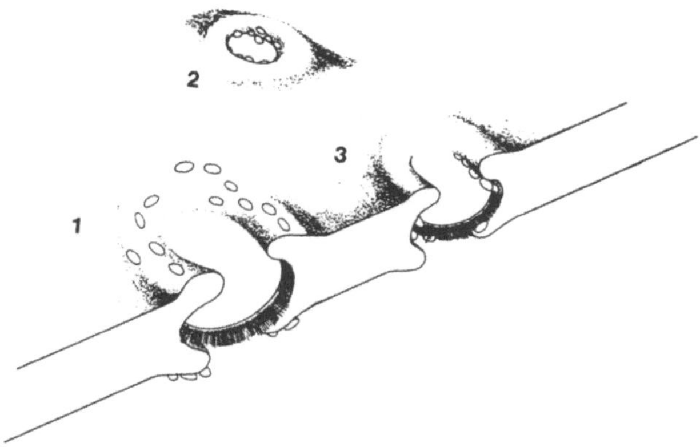

Fig. 1.51. Illustration, redrawn from Greaves (1969), showing three types of attack on pit borders caused by bacteria. Type *1* represents attack on the pit border on the lumen surface of the pit or from the inside surface of the border. A perforated appearance of the pit border results. Type *2* attack takes the form of erosion at the edge of the pit borders that are in an aspirated condition. The attack is limited to the periphery of the aperture. Type *3* results from bacteria that have entered the pit chamber and attacked the border. The eroded border attack has a frilled appearance and has been referred to as "lacy border". The border may be completely freed from the pit structure after this type of degradation

three different types of pit border attack in *Pinus*. Bacteria randomly eroded the border, causing a perforated appearance (Fig. 1.51 – type 1 attack). Attack may be from the pit chamber or from the lumen surface of the border. As this type of attack progresses, the border may become so severely degraded that it is removed. Type 2 attack is found when bacteria are concentrated at the edge of the border, where an aspirated pit has closed off the pit aperture. Bacterial action causes an erosion at the edge of the pit border near the torus. The margo and torus may also be attacked, resulting in a deaspiration of the pit (Fig. 1.51). The third form of attack occurs when bacteria enter the pit chamber and move to the region of the annulus. A selective attack in this region causes a frilled appearance or "lacy border", as described by Greaves (1971 a). This concentrated attack around the perifery of the border may result in loss of the border and removal of the margo and torus (Fig. 1.51). This "lacy border" type of attack corresponds to the degradation observed by Liese in 1950.

The attack of bordered pit membranes can be observed in micrographs of *Picea exelsa* wood inoculated with *Bacillus macerans* (Blanchette, unpublished data using isolate number DSM–24 obtained from Dr. O. Schmidt, Hamburg). Twelve and 20 days after inoculation, samples were sectioned, causing the outer pit border to be removed. Bacteria were seen accumulating in the pit chambers (Fig. 1.52 A). The bacterial cells attached to the microfibrils of the margo caused a dissolution of the membrane (Fig. 1.52 B, C). In some cells, degradation of the torus was observed also 20 days after inoculation (Fig. 1.52 D). In samples of *Pinus monticola* from the field, bacteria had destroyed the margo completely, leaving remnants of the torus in the pit chamber (Fig. 1.53 A). Removal of the margo and torus by this mixed population of bacteria

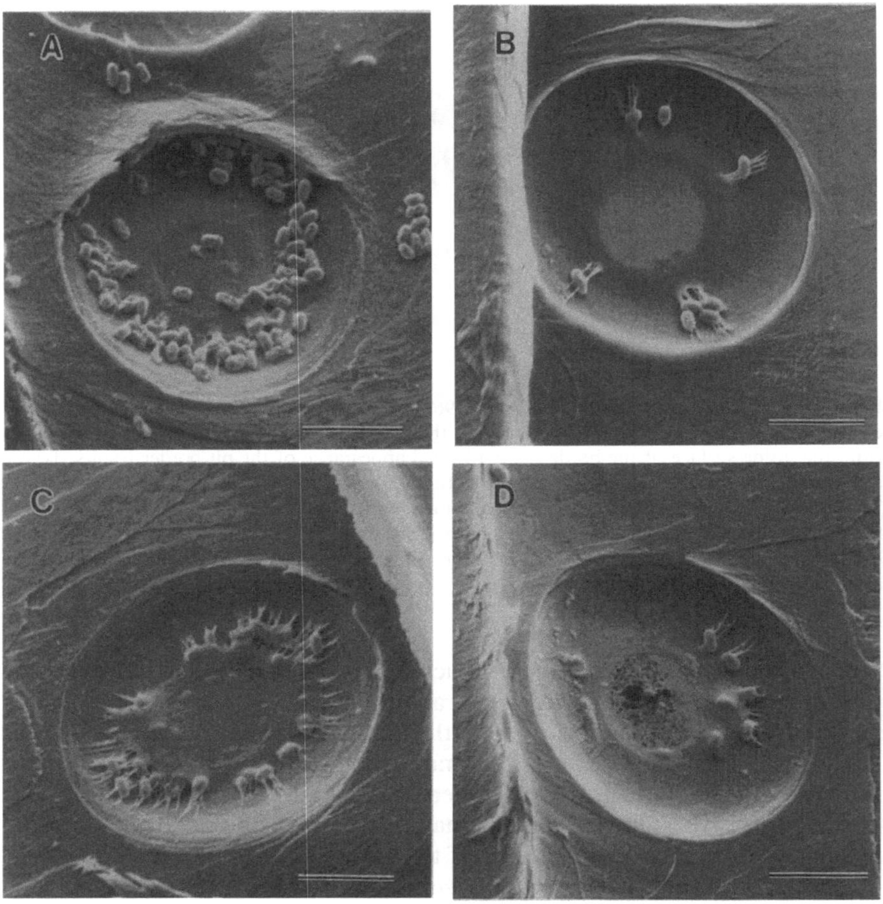

Fig. 1.52. Bacterial degradation of torus and margo in bordered pit pairs. Twelve (**A, B**) and 20 (**C, D**) days after inoculation of *Picea abies* wood with *Bacillus macerans*. The pit border has been removed during sectioning to expose the pit membrane. **A** Bacteria in pit chamber. **B, C** Bacteria adhering to the margo, causing a dissolution of the membrane. Microfibrils of the margo are distinct around the bacterial cells (**B**), which are subsequently degraded (**C**). **D** Erosion of the torus is seen causing an opening in the membrane. **A–D** Radial surface; SEM. *Bar* = 5 μm

completely opened the pit aperture, allowing free movement of bacteria between cells (Fig. 1.53 B).

Large increases in permeability to liquids have been reported for ponded wood (Elwood and Ecklund 1959, Knuth and McCoy 1962, Liese and Karnop 1968, Highley and Lutz 1970). Highley and Lutz (1970) indicated that the cause of the increased permeability in southern pine and yellow poplar after ponding was due to a pronounced bacterial attack on pit membranes. The presence of aspirated pit membranes, especially in conifers, restricts uptake of various preservatives. Increased rates of permeability in wood allow better uptake and

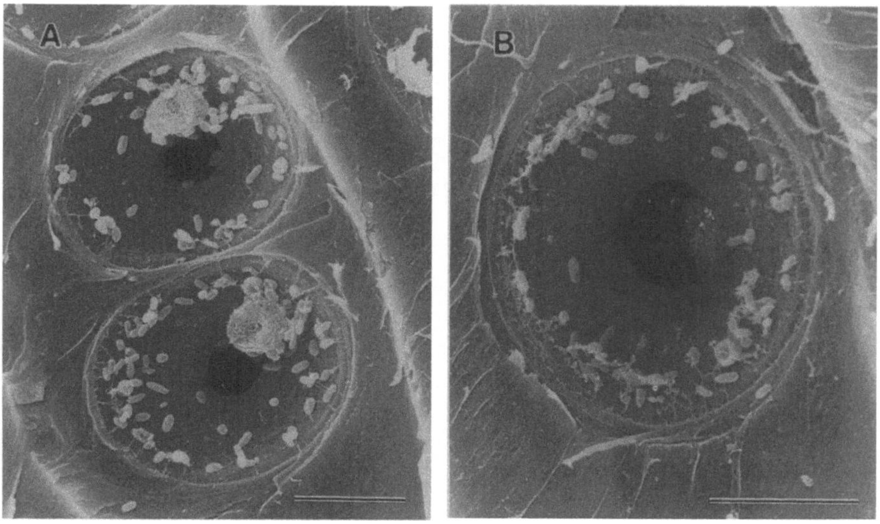

Fig. 1.53. Bacterial degradation of pit membrane in bordered pit pair of *Pinus monticola*. Pit border was removed during sectioning revealing the pit chamber. **A** Bacterial cells in pit chamber. The margo is destroyed, and remnants of the degraded torus can be seen. **B** The margo and torus are completely degraded and free access of bacteria to the adjacent cell is possible. **A, B** Radial surface; SEM. *Bar* = 5 μm

penetration of preservative. Submersion of logs in water for only 5 weeks resulted in a 155% increase in preservative uptake (Unligil 1972). After several years of storage in water, the permeability of spruce wood was seven- to tenfold greater than that of nonponded bolts (Adolph et al. 1972). This type of attack is by no means restricted to wood that is completely saturated, and the same results may occur in many diverse environments. In terrestial environments, bacteria have also been found to attack pit membranes and increase permeability in a short time (Greaves 1970, Levy 1975). Although few investigations have been completed using hardwood substrates, fibers, vessels, and parenchyma cells are colonized and pit membranes are degraded (Greaves 1970a, b, Highley and Lutz 1970).

Fibers, vessels, and tracheids can be also attacked directly by bacteria. However, not all species or strains can degrade cell wall components. Of the 66 different strains of bacteria tested previously (Schmidt 1980, Schmidt and Liese 1982) none of them could attack the lignified cell walls of beech or pine sapwood. In another study of 57 bacterial strains, no cell wall degradation was observed in the laboratory (Schmidt et al. 1987). Only after treatment of wood with chlorite to cause a mild delignification could the bacteria attack the cell wall and cause substantial weight losses. Holt and Jones (1978) also found that timber was resistant to attack by bacteria, and that the cell walls were degraded in the laboratory only after chemical pretreatment to remove or alter the structured of the lignin. Preconditioning the wood with cellulase-less mutants of

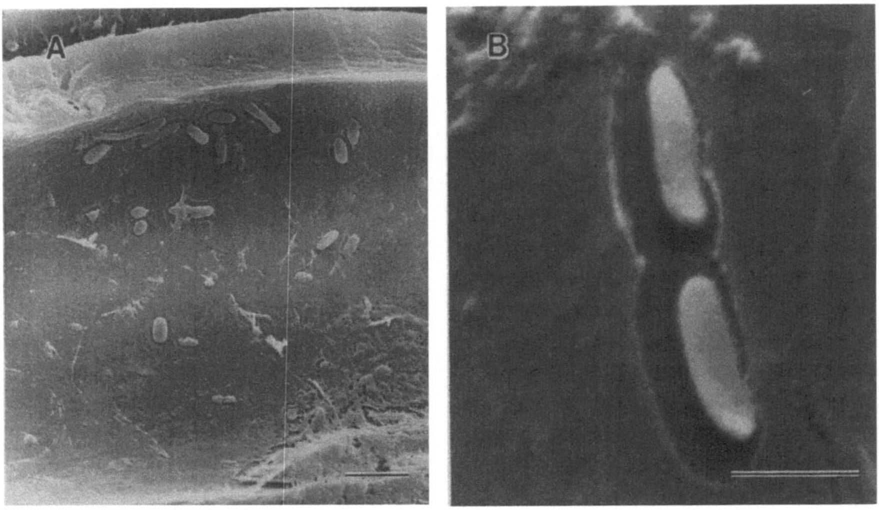

Fig. 1.54. Erosion type of degradation by bacteria in a tracheid of *Pinus monticola*. **A, B** Shallow erosion troughs are evident in the tracheid cell wall around bacterial cells; SEM. *Bar* = 2

Phanerochaete chrysosporium and *Phlebia gigantea* to remove lignin also had a significant effect on the ability of bacteria to attack the remaining wood components (Holt 1982). Pretreatment with brown, soft, and white rot fungi, however, had only slight effects on the susceptibility to degradation by the cellulolytic prokaryotes.

There are numerous reports of bacteria causing an erosion of the tracheid or fiber wall, starting at the cell lumen and progressing into the secondary wall (Harmsen and Nissen 1965, Courtois 1966, Boutelje and Bravery 1968, Greaves 1968, Liese and Karnop 1968, Kohlmeyer 1980, Singh and Butcher 1985). Morphological studies have demonstrated the capacity of some bacterial species or mixed populations to alter greatly the lignified cell wall in a number of different ways. Greaves (1971) found erosion troughs, wall pitting, and a diffuse attack on the crystalline structure of cellulose within tracheids. The lysis troughs were either shallow erosions confined to the S_3 layer of the wood, and usually many times wider than the bacteria, or deep troughs which progressed from the lumen through the entire secondary wall. Erosion troughs under bacterial cells can be seen in micrographs of coniferous wood samples collected from the forest floor (Fig. 1.54 A, B). A stippling of the lumen surface with depressions that do not progress very deeply into the wall have also been observed around colonies of bacteria in tracheids (Greaves 1971 a). Apparently each bacterial cell in the colony produces a small eroded pit. The pits remain small and do not develop into troughs. In addition to these examples of direct cell wall attack, a change in birefringence may occur in the secondary wall. The localized loss of birefringence is similar to the type of deterioration produced by brown-rot fungi, resulting in a diffuse attack on the crystalline arrangement of cellulose microfibrils in the secondary wall. In wood attacked by bacteria for long periods of time, cell wall erosion accompanied by the destruction of crystalline cellulose can be extensive. As deterioration continues, erosion troughs coalesce causing a complete loss of the secondary wall. All that may be left is a skeleton consisting of middle lamellae and primary walls (Greaves 1969, Willoughby and Leightley 1984).

Ultrastructural studies of mixed populations of bacteria attacking birch and pine wood have been presented by Daniel and Nilsson (1986). In birch, the erosion bacteria form narrow troughs through the S_3 layer and align along microfibrils of the S_2 (Fig. 1.55). All secondary wall layers are degraded and attack on parts of the middle lamella and primary wall begins once the S_1 has been removed. Eroded areas of the middle lamella were found together with areas that displayed a loss of electron density. The loss of electron density and degradation of middle lamella suggest that the bacteria possess strong ligninolytic activity. Lignin was degraded to such an extent that no degradation products in the form of residual particulate matter were detected. Attack of pine wood by the same bacteria resulted in conical-shaped cavities in the S_2 layer, aligned along the microfibrils. The entire S_2 and S_1 layers were gradually degraded, but the S_3 layer persisted. A limited degradation of middle lamella was also observed, and a loss of electron density was evident in the middle lamella. The loss of electron density between cells and at cell corners suggests that lignin degrading enzymes may diffuse into the wall. Several investigations have shown that these populations of bacteria can degrade lignin model compounds (Nilsson and Daniel 1986, Daniel et al. 1987).

The differences observed in bacterial attack of birch and pine suggest that the type of attack may be significantly influenced by the composition and structure of the wood. This is further demonstrated when wood that has been treated with copper-chrome-arsenic preservative is attacked by bacteria, and a different type of degradation of the S_2 layer was found (Nilsson and Singh 1984). Extensive cavities that coalesced together were usually located within the S_2 layer. The S_3 layer was not affected but it collapsed after severe degradation of the underlaying S_2 layer. The altered form of cell wall degradation most likely resulted from the presence of toxic preservatives in the wood: Once they entered into the S_2 layer of the wall, the bacteria avoided the large amounts of preservatives in the cell lumina and confined their attack to the inner wall layers. Large, eroded cavities are the end result.

Other patterns of cell wall decay, cavitation and tunneling, also have been identified in wood colonized by mixed cultures of bacteria in natural environments and in the laboratory (Nilsson and Daniel 1983c, Nilsson and Holt 1983, Daniel et al. 1987, Singh et al. 1987). In contrast to erosion bacteria, which attack the cell wall from the lumen toward the middle lamella, cavitation and tunneling bacteria grow within the wall (Singh et al. 1987). Bacteria that tunnel through the cell walls have been found in hardwoods and softwoods, as well as in wood usually considered decay resistant. This includes high-lignin tropical hardwoods, such as *Alstonia scholaris* and *Homalium foetidum*, and preservative-treated wood (Nilsson and Singh 1984, Singh et al. 1987).

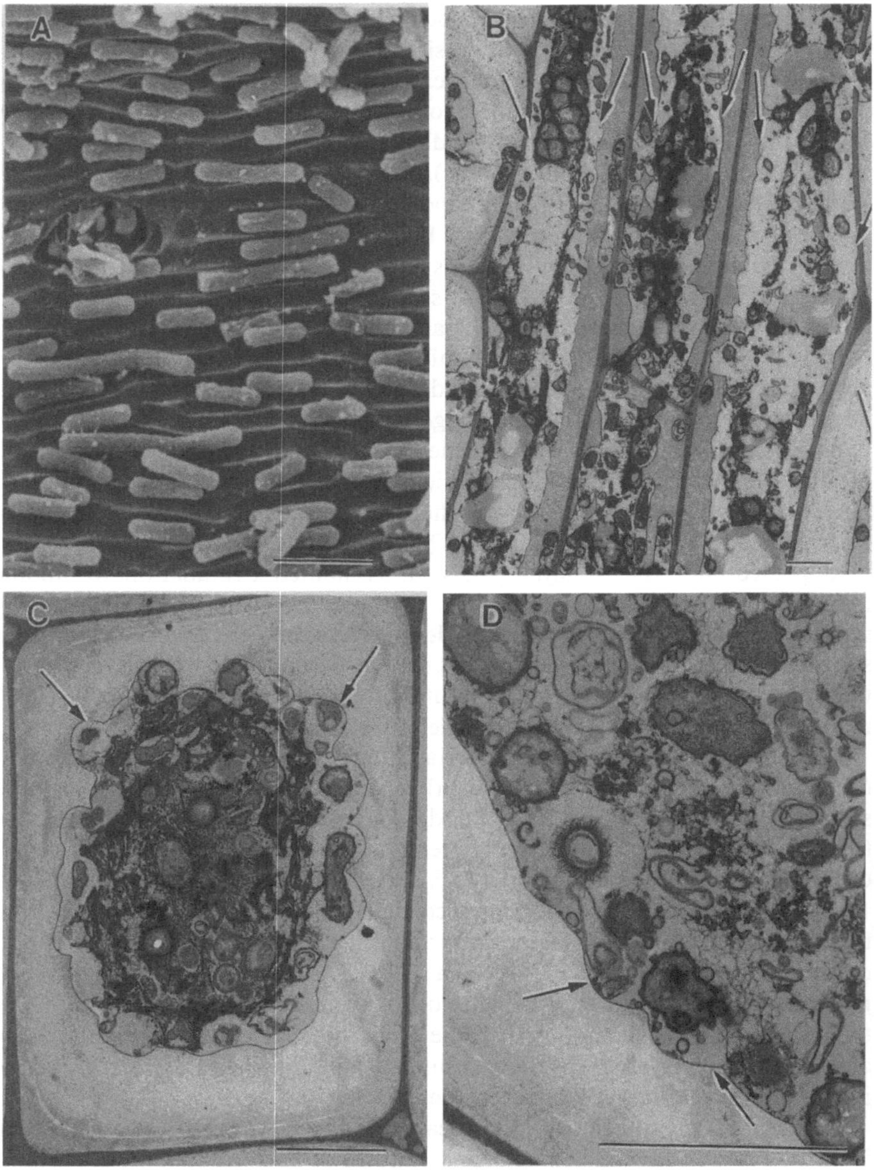

Fig. 1.55. Degradation of *Betula verrucosa* fiber and parenchyma cells by erosion bacteria. **A** Erosion of wood fiber wall by rod-shaped bacteria and longitudinal alignment of erosion troughs in the S_2 layer. **B** Advanced decay of fiber and parenchyma cells with large numbers of bacteria eroding cell walls (*arrows*). **C** Wood fiber with lumen filled with bacteria and well-formed erosion troughs (*arrows*). **D** Bacterial erosion of fiber (*arrows*) showing characteristic features of fibrillar material (slime layers) produced by the bacteria. **A** Radial surface; SEM. **B, C, D** Transverse sections; TEM. *Bar* = 5 µm (Daniel and Nilsson 1986)

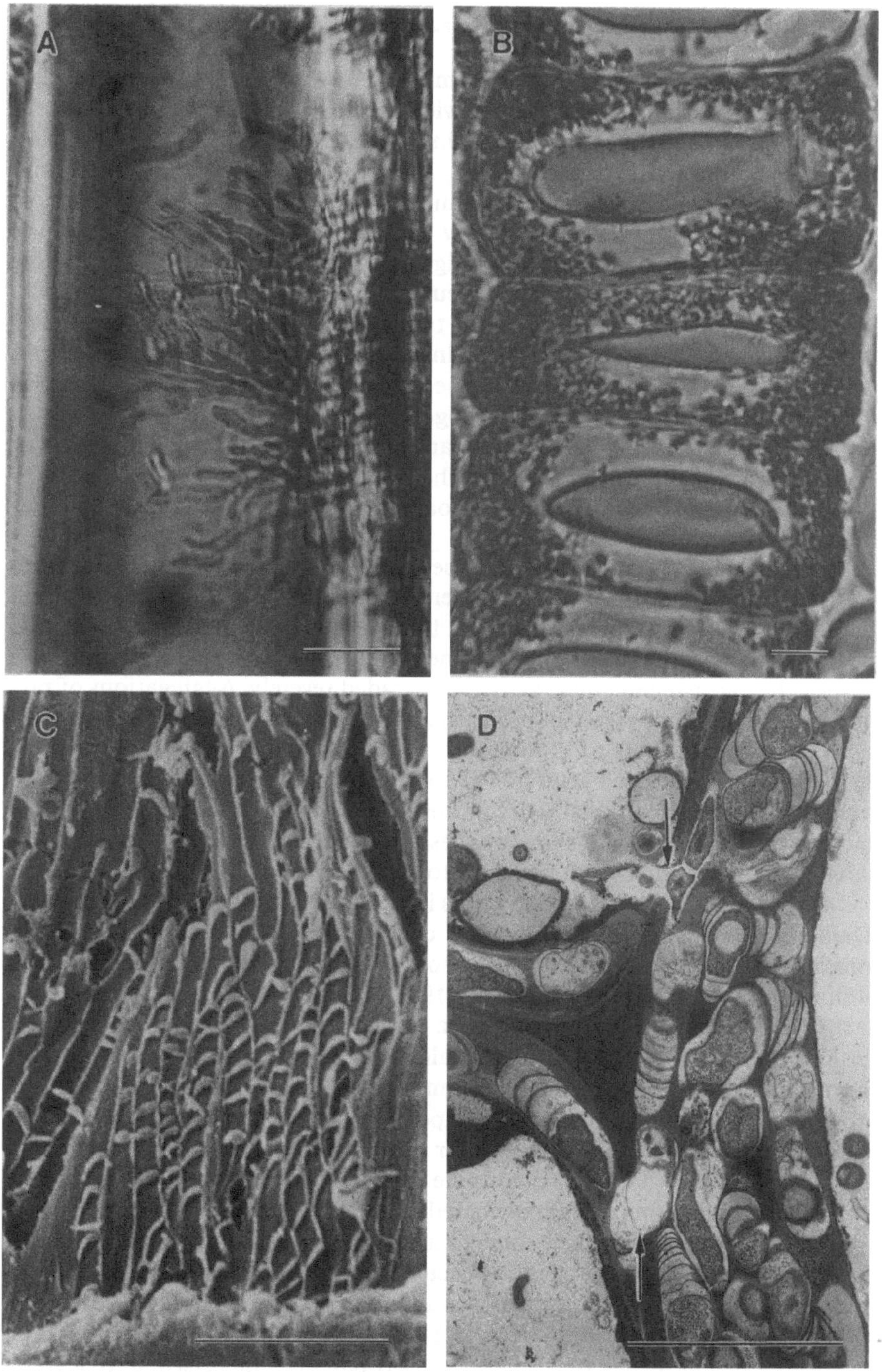

Fig. 1.56. Degradation of wood by tunneling bacteria. **A, B** Tunneling bacteria in tracheids of *Pinus ponderosa*. **C** Tunnels in S_1 layer of a *Picea abies* tracheid, showing polysaccharide slime deposits within the tunnels. **D** Tunneling bacteria within cell walls of *Homalium foetidum* wood. Bacteria can be seen tunneling through the secondary wall and middle lamella (*arrows*). **A** Radial section; LM. **B** Transverse section; LM. **C** Radial surface; SEM. **D** Transverse section; TEM. *Bar* = 5 m (Nilsson and Daniel 1983)

The tunneling bacteria attach themselves to the S_3 layer and produce a hole in the S_2, forming small cavities with pointed ends. As the bacterial cells divide, they form tunnels that radiate into the S_2 layer (Fig. 1.56A). The tunneling action occurs throughout the S_2 layer and may extend into the S_1 and S_3 (Fig. 1.56B). The direction of the tunnel is influenced by the orientation of the microfibrils within each secondary wall layer (Fig. 1.56C). However, the tunnels often go in any direction (Fig. 1.56D). Penetration of the middle lamella also is common, and may occur at any angle (Fig. 1.56D). Polysaccharide secretions are deposited in the tunnels as bacteria move to new areas, and characteristic bands are produced in the tunnels (Fig. 1.56C). These extracellular secretions produced within the tunnels are always found associated with tunneling bacteria. As decay progresses, the S_3 layer nearest the lumen remains intact except for bacterial entrance and exit holes. The degradation of the inner S_3 layer region occurs from the S_2 layer of the wall (Fig. 1.56D). In advanced stages of decay, the tunnels coalesce and the middle lamella and outermost S_3 layer are all that remain.

A remarkable characteristic of these bacteria is their ability to attack impregnated wood that is usually considered resistant to microbial degradation. Daniel and Nilsson (1985) have shown that wood of *Pinus radiata*, treated to more than double the retention recommended for the in-service condition with copper-chrome arsenate preservative, is degraded severely. Aggregations of precipitates, analyzed by X-ray microanalysis (Daniel and Nilsson 1985) containing high concentrations of chromium, arsenic, and copper were found along the tunnels made by the bacteria. The toxic compounds apparently do not adversely affect the bacteria, and decomposition continues uninhibited. The production of polysaccharide slime by these bacteria encapsulates them and apparently prevents direct contact with toxic substances. The preservative that was encountered in the wood cell wall is left behind in the residual slime matrix (Daniel and Nilsson 1985). Mucilage coating the hyphae of soft-rot fungi found in preservative-treated wood also has been suggested to function in a similar fashion (Francis and Leightley 1983). The mucilage produced may aid in the initial attachment of the bacteria to the cell walls and provide a matrix for enzymes to be directed into the wall. Under wet conditions, the polysaccharide slime would be an essential component for enzymes to move into the substrate, and bind degraded wall components so that they would be directed back to the bacterium. The extracellular slime also provides a medium for bacterial mobility. The bacteria responsible for the attack are not flagellated, but most likely glide along the slime coated channels (Daniel and Nilsson 1985).

Relatively few studies have focused attention on the role of *Actinomycetes* in wood deterioration. Filamentous bacteria in the order Actinomycetales may colonize wood and cause a gradual degradation of wood components. In wood of *Acer saccharinum*, three different *Streptomyces* species (Figs. 1.57A–C) were found to colonize vessels, fibers, and ray parenchyma cells and to degrade cell contents as well as vessels occlusions (Blanchette et al. 1981). *Streptomyces* species have extremely small diameter mycelium that may colonize all types of cells (Fig. 1.58A, B). Sporulation can occur within any cell of the wood but

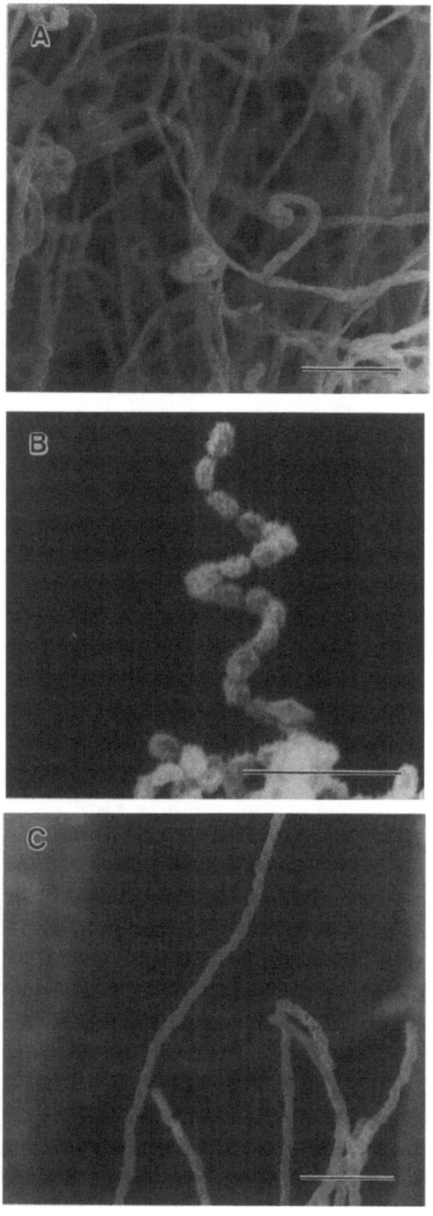

Fig. 1.57. Sporulation of three Actinomycete species isolated from discolored wood of *Acer saccharinum*. **A** *Streptomyces parvulus* with spiral chains of smooth spores. **B** *Streptomyces sparsogenes* with spiral chains of spiny spores. **C** Unidentified *Streptomyces* species, strain 870, with long, flexuous chains of smooth spores; SEM (Blanchette et al. 1981)

is most profuse on the surface of the substrate (Fig. 1.58C). Vessel occlusions, rich in phenolic substances (Shigo 1967, Shortle 1979), are readily colonized by the *Streptomyces* mycelia (Fig. 1.58D). In liquid media, *Streptomyces* tolerated several phenols tested and utilized p-hydroxybenzoic acid as a carbon source (Blanchette et al. 1981). No evidence, however, was found for cell wall

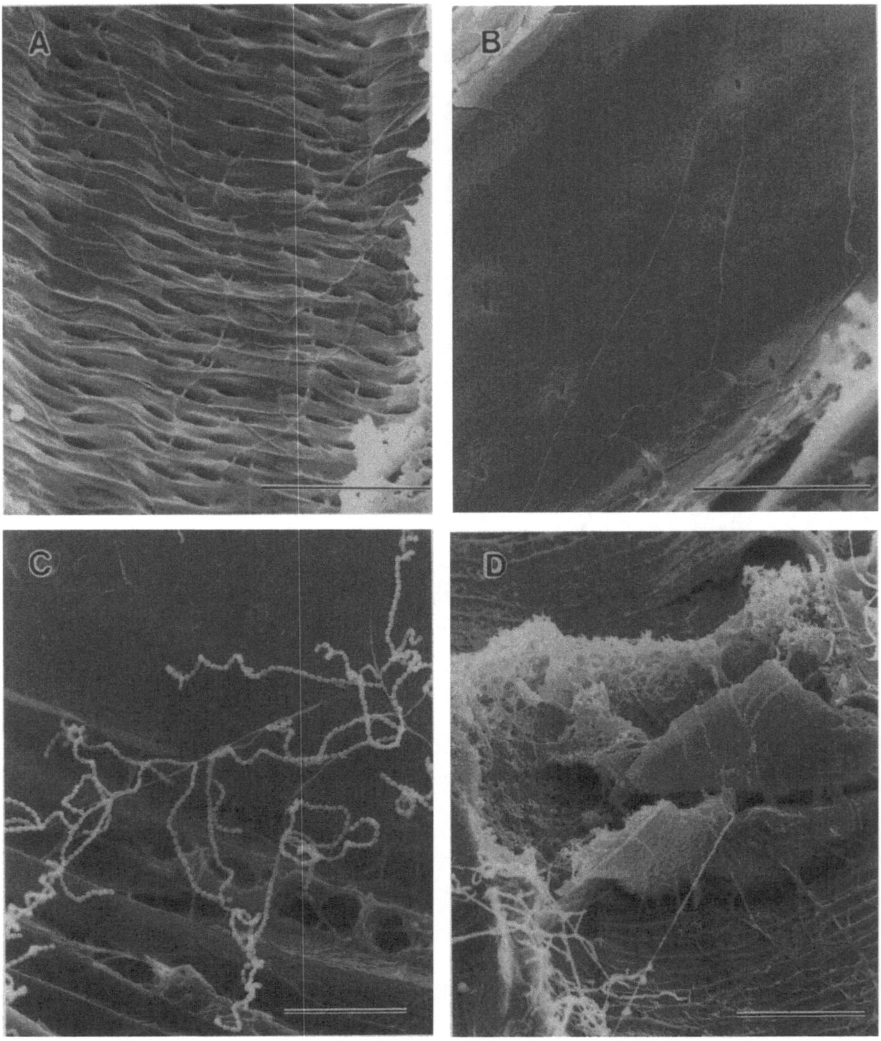

Fig. 1.58. *Streptomyces* in sapwood and discolored wood of *Acer saccharinum*. **A, B** Hypha of *Streptomyces sparsogenes* colonizing vessel (**A**) and fiber (**B**). The actinomycete readily colonize the cell lumina, but no cell wall degradation is apparent. **C** Sporulation of *Streptomyces sparsogenes* within wood. **D** Colonization of occlusions in vessel by *Streptomyces* strain 870. The hypha colonized and degraded the vessel plugs. **A, B, D** Radial surface; SEM. **C** Tangential surface; SEM. **A, B** from Blanchette et al. (1981)

degradation by the *Streptomyces* species in sapwood or discolored wood of inoculated wood blocks (Fig. 1.58 A, B).

The discolored wood of living trees contains morphological and chemical barriers to colonization by microorganisms. Vessel occlusions and tyloses, combined with high concentrations of phenols, restrict activity of most microorganisms in discolored wood. The *Actinomycetes* isolated by Blanchette et al. (1981) are pioneer organisms that colonize discolored wood and remove or alter phenols and vessel plugs. Once these substances have been removed from the wood, wood destroying fungi rapidly colonize and degrade the cells (Shigo 1967).

King and Eggins (1977) described penetration and colonization of *Picea sitchensis* wood with a *Streptomyces* species. Rapid colonization of ray parenchyma cells and tracheids was observed. The streptomycete mycelium penetrated bordered pit pairs and degraded pit membranes. No direct attack was found on the cell walls of xylary tissues. The *Streptomyces* species were found to colonize wood rapidly and penetrate to the same extent and depth as other microorganisms that aggressively degrade wood. In subsequent studies, other *Streptomyces* species have been shown to degrade various woody substrates. Breakdown of Douglas-fir (*Pseudotsuga menziesii*) phloem was observed after inoculation with *Streptomyces flavovirens* (Sutherland et al. 1979). The actinomycete rapidly colonized the phloem and degraded parenchyma cells as well as thick-walled, heavily lignified sclereids (Fig. 1.59A–D). As much as 45% of the carbohydrate content was removed from the substrate while only small quantities of lignin were lost. Parenchyma cell walls were extensively disrupted, and a general erosion of the cell wall was observed (Fig. 1.59B). In advanced stages of degradation, parenchyma cells were completely destroyed, and sclereids showed evidence of eroded cell walls (Fig. 1.59C). The thick-walled, lignified sclereids resisted attack by *Streptomyces flavovirens* longer than adjacent parenchyma cells. However, after 9 weeks of degradation, substantial attack of the sclereid wall also had occured.

Many different forms of deterioration have been observed and characterized by King et al. (1978) and by Baecker and King (1980a). When an unidentified *Streptomyces* species was used to decay *Tilia vulgaris* and *Pinus sylvestris*, bore holes and an etching of the cell wall were observed in both types of wood. In addition, cavities formed in the S_2 layer of cells from *Tilia*. These patterns of deterioration were similar to some of those caused by species of Eubacteria. The large number of investigations where actinomycetes have been isolated from wood suggest that the possible deterioration of wood by these organisms may be more widespread than is currently realized (Eslyn 1967, Butcher and Howard 1968, Eaton and Dickinson 1976, Leightly and Eaton 1977). Some species also have the capacity to degrade lignin. Crawford and Sutherland have shown from a number of investigations that actinomycetes play a significant role in the decomposition of lignocellulosic materials (Crawford and Sutherland 1979, 1980).

Actinomycetes and species of Eubacteria may exert strong antagonism toward other wood-inhabiting microorganisms. Many species produce antibiotics (Waksman 1967), and some of these species have been shown to inhibit wood-rotting fungi in laboratory wood-block tests (Greaves 1970b, Blanchette et al. 1981). In addition to antibiotic production, nutrients may be depleted in ray parenchyma cells and subsequent colonization by other organisms restricted. Whatever the reason for the antagonistic properties, a large number of dif-

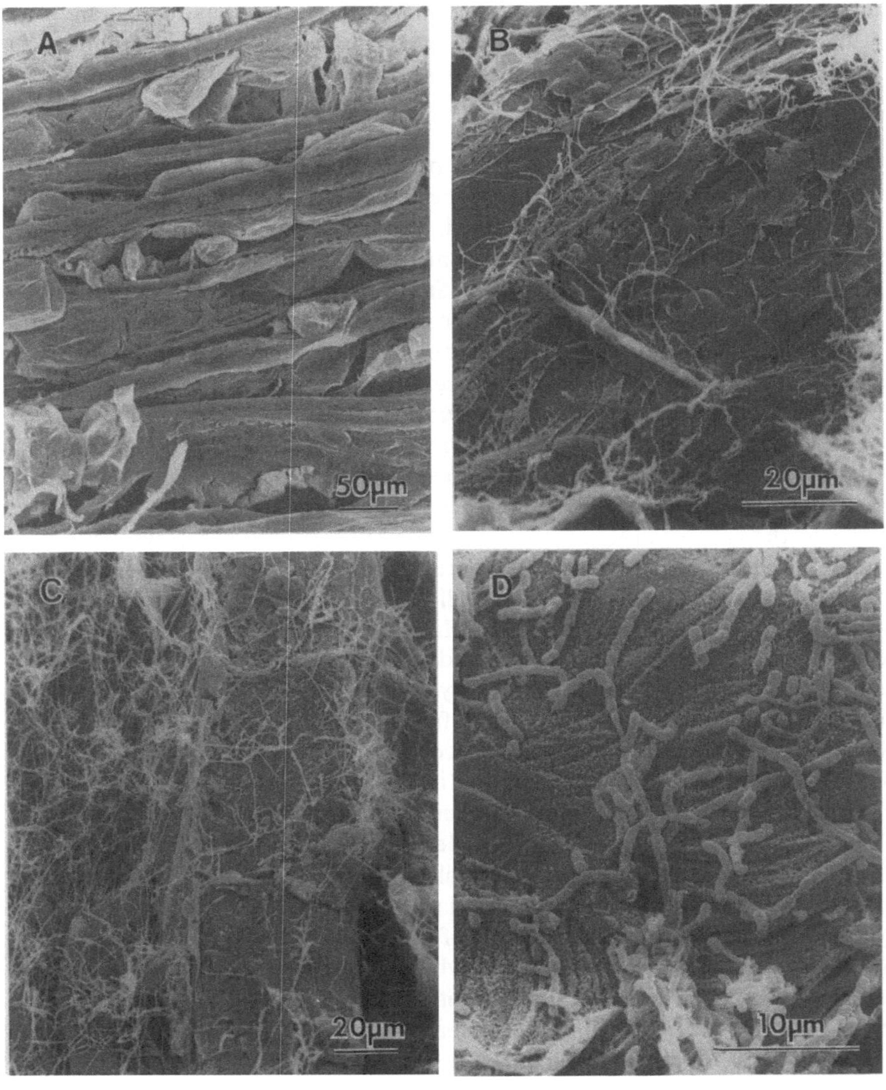

Fig. 1.59. Sound phloem cells of *Pseudotsuga menziesii* (**A**) and phloem cells after 2 (**B**) and 5 (**C**, **D**) weeks of degradation by *Streptomyces flavovirens*. **A** Intact cells from sound phloem. **B** Parenchyma cells with erosion of the cell wall after 2 weeks of degradation. **C** Many of the parenchyma cells are completely destroyed, and erosion of the sclereid cell wall has begun. **D** Sporulation of *Streptomyces flavovirens* on a degraded cell wall. **A–D** Radial surface; SEM. **A–C** from Sutherland et al. (1979)

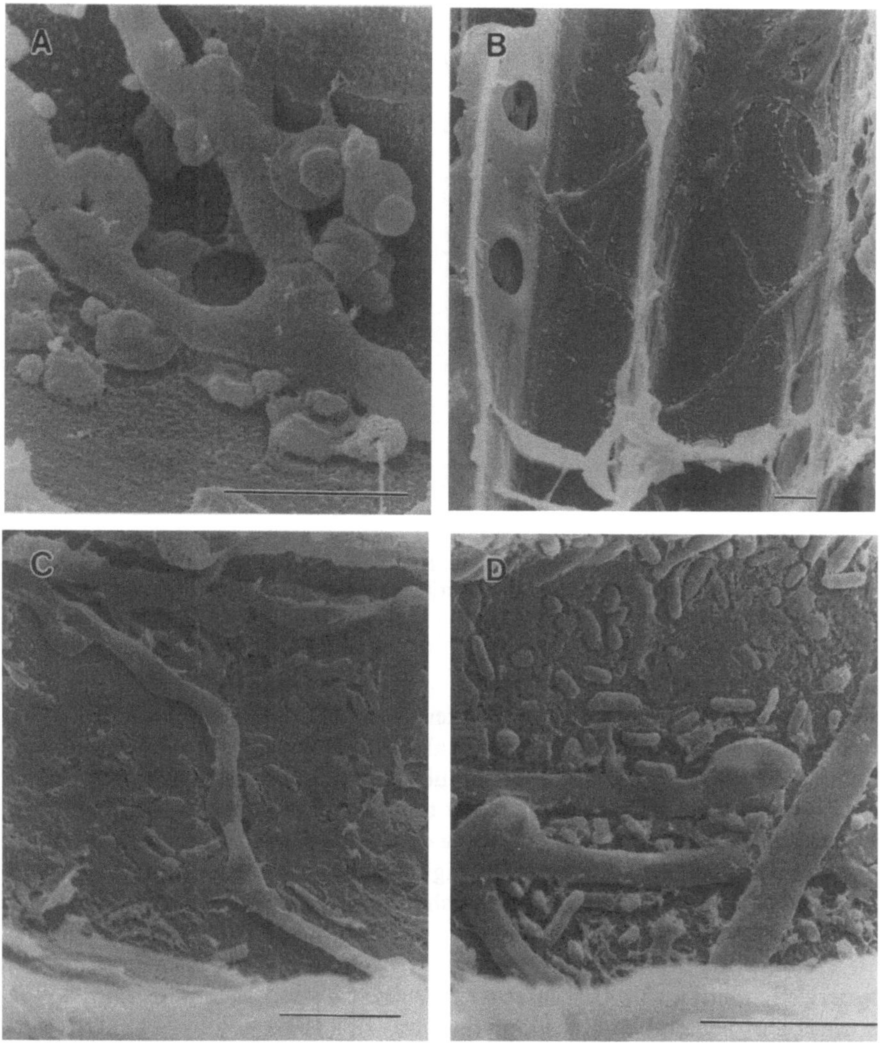

Fig. 1.60. Symbiotic association of mixed microbial population during wood decomposition. **A** Hypha of *Coriolus versicolor* in tracheid with budding yeast cells clustered around the hypha. **B** Bacteria around hyphae of *Cryptoporus volvatus* during incipient stages of a white-rot decay in sapwood tracheids. **C** Bacteria (*Enterobacter* sp.) associated with lysed zones of cell wall around hypha of *Coriolus versicolor.* **D** Bacteria associated with hyphae of *Hirschioporus abietinus*. Lysis zones are seen around bacteria in the tracheid wall modified by the white-rot fungus. Excess cell wall degradation products resulting from action of extracellular enzymes of the white-rot fungi are utilized by bacteria and yeasts. The bacteria and yeasts provide growth-promoting substances that enhance fungal growth and wood decomposition. **A–D** Radial surfaces; SEM. *Bar* = 5 μm (Blanchette and Shaw 1978)

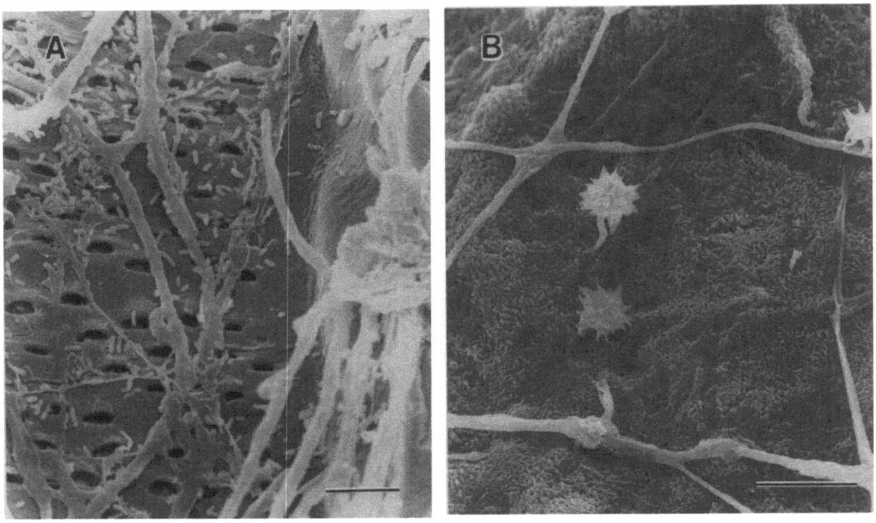

Fig. 1.61. Bacteria associated with advanced stages of white-rot decay. **A** Hypha of unidentified white-rot fungus and bacteria in vessel of *Malus*. **B** Hyphae of *Hirschioporus abietinus* with calcium oxalate crystals and large numbers of bacterial cells in decaying wood of *Pseudotsuga menziesii*. Radial surface; SEM. *Bar* = 10 μm

ferent strains have been shown to exert various degrees of inhibition (Hejmanek and Rypacek 1954 in Rossel et al. 1973, Henningsson 1967, Cavalcante and Eaton 1981). As isolation techniques are improved, such as those used by Baecker and King (1980b), and greater numbers of organisms are tested for inhibitory effects, a better understanding of how bacteria interact with other microorganisms will evolve. The use of highly antagonistic organisms for biological control of wood destroying fungi should be an active area of study in the future.

Interactions between some species of bacteria and fungi may result in a synergistic relationship that increases the rate of wood decomposition. An association of bacteria and basidomycetes was first suggested by Schmitz (1919). He showed that the combined actions of mixed microbial populations caused greater deterioration in wood than decay by the basidiomycete alone. The stimulatory influence of bacteria on aggressive wood destroying organisms has been repeatedly demonstrated (Shigo 1965, Henningsson 1967, Blanchette and Shaw 1978). As extracellular enzymes of wood destroying fungi attack the cell wall, degraded cell wall components are released and carbon sources that would otherwise not be available to the bacteria are abundant. Growth of the bacteria depletes concentrations of cellobiose, glucose, etc. around the hypha and cellulases are not repressed (Henningsson 1967, Smith 1975, Eriksson and Hamp 1978). Bacteria may also supply essential vitamins or growth-promoting substances to wood decay fungi (Blanchette and Shaw 1978, Blanchette 1979), and some wood inhabiting bacteria can fix atmospheric nitrogen

(Seidler et al. 1972, Aho et al. 1974, Larsen et al. 1982). All of these factors may promote the growth and decay capacity of the associated fungus.

There is morphological evidence for a synergistic association among bacteria and fungi as well as yeasts (Blanchette and Shaw 1978, Blanchette 1979). Increases in fungal biomass and decay capacity were observed when wood was inoculated with mixed populations in the laboratory. Bacteria and yeasts were located immediately adjacent to the hyphae in regions of degraded cell walls (Fig. 1.60A). In areas of incipient decay, bacteria were clustered near the hypha (Fig. 1.60B). As decay progressed, lysis zones around bacterial cells were observed, indicating that the bacteria utilized the modified wall components (Fig. 1.60C, D). In wood samples of advanced stages of decay from the forest floor, large numbers of bacteria may also be found associated with decay fungi (Fig. 1.61A, B) (Shigo 1967; Blanchette and Shaw 1978). The presence of bacteria in wood throughout the successional stages of decomposition suggests that interactions among mixed microbes may have significant effects on the patterns and rates of decay within the forest ecosystem (Kaarik 1975). The importance of actinomycetes and Eubacteria in wood decay is just beginning to be understood, and our current knowledge of these diverse groups of microorganisms is incomplete.

Chapter 2

Biodegradation of Cellulose

Almost half of the biomass synthesized by photosynthetic fixation of carbon dioxide is made up of cellulose. Cotton and wood fibers are the most common sources. It makes up for about 90% of cotton fibers but for only about 45% of the average wood fiber.

All biomass is eventually degraded. Hoppe-Seyler (1883) and de Bary (1886) first reported that cellulose is decomposed in nature. The main deterioration of cellulose is caused by microorganisms and, due to their abundance, microbial degradation of lignocellulosic materials is one of the most important processes in nature. Under aerobic degradation conditions cellulose is oxidized to carbon dioxide and returned to the atmosphere. Under anaerobic conditions biomass is converted to methane and CO_2. Approximately 5–10% of the cellulose is converted to methane and enters the global methane cycle. (Ehhalt 1976, Vogels 1979). Microbial fermentation of cellulose was first studied by Omelianski (1902). Lists of cellulolytic microorganisms have been presented by Siu (1951), Gascoigne and Gascoigne (1960), Mandels and Weber (1969) and Ljungdahl and Eriksson (1985). Coughlan (1985) presented a list of cellulase producing organisms of established or potential commercial use.

During World War II, the US Army was alarmed at the rate at which materials were rotting in tropical regions. The fungus *Trichoderma viride* was identified as a severe degrader of cellulosic materials. The studies of *T. viride* and its cellulose degrading enzymes were at that time undertaken mainly from the point of view to understand and to prevent cellulose degradation.

Cellulolytic enzymes, cellulases, have been defined as the enzymes hydrolyzing cellulose thereby yielding water-soluble sugars (Eriksson 1967). It seems reasonable to modify this definition since in addition to the hydrolytic enzymes, oxidative enzymes (Ayers et al. 1978, Vaheri 1982b) and an oxidoreductive enzyme (Westermark and Eriksson 1974a, b) are now known to participate in the conversion of cellulose.

During the past 20 years, extensive research has been carried out on cellulolytic enzymes of both fungal and bacterial origin. The reason for this research is, however, different from the early studies. Now the vast potential of cellulose as a source of food for ruminants and as a chemical feedstock is appreciated. Accordingly, investigators in many laboratories quickly realized the possibility to explore cellulases as a basis for saccharification of cellulose. Many review articles of these studies have been written, the two most recent ones being those of Eriksson and Wood (1985) and Ljungdahl and Eriksson (1985).

2.1 The Structure and Biosynthesis of Cellulose

2.1.1 The Chemical Structure of Cellulose

Cellulose was named by Payen (1838). He was the first to recognize that cellulose and starch were isomeric products. Cellulose viscosity measurements showed that cellulose was a macromolecule. Freudenberg et al. (1932) demonstrated that cellulose was a polymer of β-D-glucopyranose units linked together by the 1–4 carbons. Cellulose is the major component of wood and amounts to approximately 40%–45% of most wood species. The amount of the cellulose found in many different plants has been presented by Stephens and Heichel (1975).

Cellulose molecules are linear and easily form intra- and intermolecular hydrogen bonds. Glucan chains have a twofold screw axis of symmetry, stabilized and stiffened by intra- and intermolecular bonds (Gardener and Blackwell 1974; Fig. 2.1). The size of the cellulose molecule is normally given in terms of the degree of polymerization (DP), i.e., the number of glucose moieties in one cellulose molecule. A cellulose molecule in higher plants can have DP-values of up to 14000 (Goring and Timell 1962, Marx-Figini and Schulz 1966). The DP of cellulose from bacteria is usually somewhat lower with a value of 3500 obtained for cellulose produced by *Acetobacter xylinum* (Marx-Figini 1982). In the alga *Valonia*, a DP-value as high as 25000 has been reported (Palma et al. 1976).

Cellobiose, rather than glucose, is the basic structural unit of cellulose (Tönnesen and Ellefsen 1971, Blackwell 1982, Atalla 1983). It should be men-

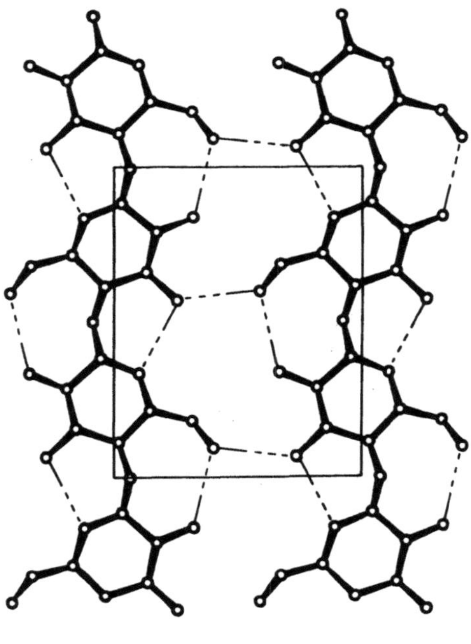

Fig. 2.1. Projection of the (020) plane in cellulose I, showing the hydrogen bonding network and the numbering of the atoms. Each glucose residue forms two intramolecular hydrogen bonds (03·H···05′ and 6···H·02′) and one intermolecular bond 06·H···03) (Slightly modified from Gardner and Blackwell 1974)

tioned in this context that cellobiose is the major product from hydrolysis of cellulose and that cellulolytic enzymes of bacterial and fungal origin appear to recognize the cellobiose unit of cellulose.

Cellulose exists in several crystalline forms (cellulose I – IV) with different X-ray diffraction patterns and spectra (Blackwell 1982, Atalla and van der Hart 1984). Cellulose I is the native form of cellulose as it occurs in plant-cell walls. Cellulose II is a regenerated form of cellulose obtained by mercerization in solid state or by dissolution. The transition from cellulose I to cellulose II is irreversible.

The crystalline forms of cellulose are still incompletely understood. Crystalline cellulose is highly resistant to microbial and enzymatic degradation while amorphous cellulose is hydrolyzed much faster. This has long been known and utilized in processes such as rotting of flax and jute. Numerous laboratory studies have also recognized these facts (Ljungdahl and Eriksson 1985).

2.1.2 Cellulose Biosynthesis: Biochemical Studies

Many different organisms have been used for the study of cellulose biosynthesis. Among the bacteria, *Acetobacter xylinum* has been a model organism used by many investigators (Hestrin 1962, Brown 1979). In *A. xylinum*, cellulose is produced as a virtually pure extracellular polysaccharide.

In algae, cellulose is often a cell wall component (Preston 1974). *Valonia* and *Oocystis* have been the most widely studied algae for cellulose biosynthesis, although few biochemical studies have been undertaken with these organisms (Delmer 1983). Several slime molds and the amoeba *Acamthamoeba castellanii* also have been used for biosynthesis studies.

The developing cotton fiber has been a useful experimental system used mainly by Delmer and co-workers (Delmer 1983). However, woody tissues have not been utilized in studies of cellulose synthesis.

It has long been known that glycosylesters of nucleoside diphosphates are the most common donors of activated sugars for polysaccharide biosynthesis (Hassid 1969). Very early the most likely donor compounds appeared to be uridine-diphosphate-glucose (UDP-glucose) and guanosine-diphosphate-glucose (GDP-glucose), (Delmer 1983). The formation of UDP-glucose is presented in Fig. 2.2. It is now generally agreed that UDP-glucose is the precursor in cellulose biosynthesis (Delmer 1987). However, it took a long time until this point could be unequivocally resolved. The story how these facts developed is exciting and worthwhile to tell.

Most of the biochemical studies have been carried out with the cotton fiber system or with etiolated seedlings of legumes such as peas or beans (Delmer 1983, 1987). The usual approach has been to follow the incorporation of radioactive glucose or UDP-GDP-^{14}C-glucose into cellulose (e.g., Carpita and Delmer 1981). A possible role for GDP-glucose emerged from studies of membrane preparations from etiolated hypocotyls of mung beans by Hassid and

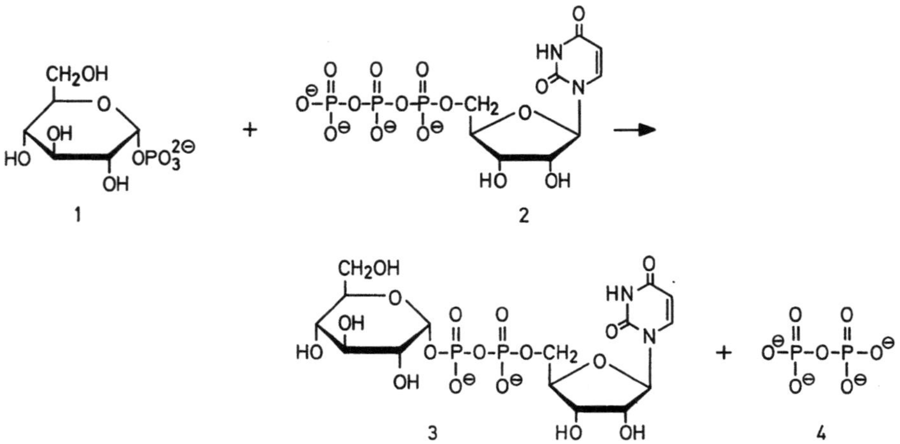

Fig. 2.2. Formation of uridine diphosphate glucose (UDP-D-glucose) (*3*) from α-D-glucopyranosyl-1-phosphate (*1*) and uridine triphosphate (*2*) under simultaneous release of pyrophosphate (*4*)

co-workers (Barber et al. 1964, Elbein et al. 1964). However, the level of GDP-glucose seems to be very low in most plants (Carpita and Delmer 1981), while UDP-glucose is present in cotton fibers and also in other plant tissues in substantial amounts (Barber and Hassid 1964, Carpita and Delmer 1981). When UDP-^{14}C-glucose was used as a precursor in homogenates of the cotton fiber system, a surprising result was obtained. Synthesis of β-1,3-glucan increased linearly almost from time zero. This result indicated that UDP-glucose might be the precursor for β-1,3-glucan but not for cellulose (Delmer 1983). The next step was obvious, namely, to investigate whether β-1,3-glucan was a precursor for cellulose, as suggested by these studies and by other work from Meier's group in Switzerland (Meier et al. 1981). However, no conversion was found from the β-1,3- to a β-1,4-glucan in short-term studies by Maltby et al. (1979). In longer-term studies in vivo, Meier et al. (1981) found a slow conversion of β-1,3-glucan, but the evidence that this polymer is a precursor of cellulose was not compelling.

It was found (Carpita and Delmer 1980) that detachment of the cotton fiber cell from the ovule results in loss of more than 90% of the in vivo capacity for synthesis of ^{14}C-cellulose from ^{14}C-glucose. If, however, polyethylene glycol (PEG) 4000 is present during this detachment and during incubation, more than 50% of the capacity for cellulose synthesis is retained. This was, of course, not a true example of in vitro cellulose biosynthesis, but it suggested that the integrity of the plasma membrane is necessary for β-1,4-glucan synthesis. In conclusion, the cellulose synthesizing complex is very sensitive to disturbances which give rise to loss of β-1,4 synthesis and a dramatic increase of the β-1,3 synthesis. Carpita and Delmer (1980) concluded from their experiments that protection of the β-1,4-glucan synthesis by PEG is an all-or-none phenomenon, requiring a certain degree of "intactness" of the synthesizing system. Several groups have suggested that the β-1,3-glucan synthase and the cellulose synthase may in fact be the same enzyme and that, upon cell breakage, unknown factors are varied which convert the enzyme to the form which synthesizes β-1,3-glucan (Delmer 1987).

Although chitin was synthesized in vitro by Bracker et al. (1976), it was only recently that similar progress could be made with cellulose in a bacterial system. Aloni et al. (1982) demonstrated an activation of β-1,4-glucan synthesis by a cooperative interaction of the *Acetobacter* enzyme system with GTP, polyethylene glycol, and a protein factor. In vitro synthesis of cellulose microfibrils has now been visualized by Lin et al. (1985) using a digitonin-solubilized cellulose synthase from *A. xylinum*. The crystallinity of these microfibrils, however, was low.

Further studies have demonstrated that the bacterial cellulose synthase has a very complex regulation. It seems to be mediated by GTP in which the synthase enzyme is solubilized in a form that still displays regulatory properties (Aloni et al. 1983). The role of GTP has recently been clarified (Ross et al. 1985). GTP appears to be converted to an unusual guanyl nucleotide by an enzyme different from cellulose synthase. The guanyl nucleotide, recently characterizied as bis (3'→5')-cyclic diguanylic acid (Ross et al. 1987), seems to be the real activator of the synthesizing enzyme.

These recent findings have clarified the situation, and at least in bacteria UDP-glucose is the true precursor in cellulose synthesis. However, the guanine base is also involved, not as GDP, but as cyclic dinucleotide, which activates cellulose synthase. Unfortunately, there is no evidence as yet for the existence of this compound in higher plants (Delmer 1987). Thus, the question of why it is so difficult to achieve in vitro synthesis of cellulose in higher plants remains unanswered.

2.1.3 The Cellulose Microfibril

As soon as long glucan chains are formed they associate in a parallel arrangement. Adjacent chains form the water-insoluble microfibril (Blackwell 1982) which can be visualized at the ultrastructural level. Different types of microfibrils can be distinguished. The subelementary microfibril which consists of only a few glucan chains (10–15) exhibits very low crystallinity. The subelementary microfibrils can associate and form the elementary microfibril, considered as the fundamental microfibril unit. According to Mühlethaler (1967), the elementary microfibril contains approximately 40 glucan chains and has a diameter of 35 Å. Elementary microfibrils can aggregate into microfibril bundles or into large microfibrils as in *Valonia* and *Oocystis* (Fig. 2.3; Brown 1981). It has been demonstrated by Giddings et al. (1980) that bundles of large microfibrils can associate into macrofibrils or ribbons. The dimensions of these can be visualized in dark-field light microscopy (Brown 1979).

Fig. 2.3. Freeze-fracture through secondary wall of *Oocystis* demonstrating ordered arrangement of microfibrils; ×77000 (Courtesy of R.M. Brown Jr. and D.L. Montezinos)

An equally exciting story to the biochemistry of cellulose synthesis is how knowledge of cellulose synthesis was revealed by various microscope investigations. Cellulose was the first biological material to be investigated with the aid of electron microscopy (Frey-Wyssling 1976). Thus, the microfibrillar and crystalline nature of cellulose was known before the use of fixation and thin sectioning techniques. New methods of this kind were developed during the mid-1960's but even with these new techniques cellulose microfibrils were often

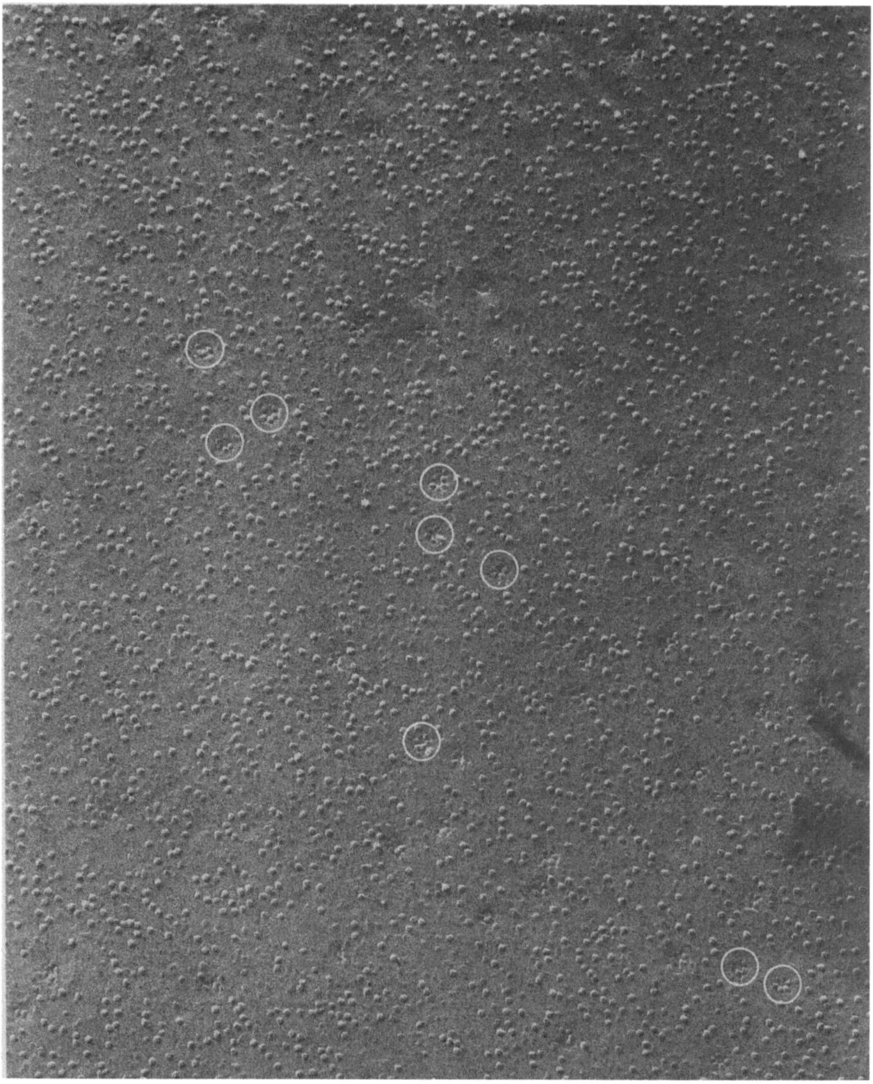

Fig. 2.4. Electron micrograph showing rosettes (*circled*) on the P-fracture face of the plasma membrane of corn root (Courtesy of R.M. Brown and S. Mueller, see Fig. 2.3)

poorly visualized. During the 1960's the procedure known as freeze etching was developed (Willison and Rowe 1980). These new techniques had their shortcomings. One of the most serious problems was intracellular damage caused by ice crystal formation. This was overcome to some extent by using cryoprotectants, but these, in turn, cause osmotic problems within the cell.

In 1976, Brown and co-workers developed techniques for freeze etching of living cells without the use of cryoprotectants or fixatives (Brown and

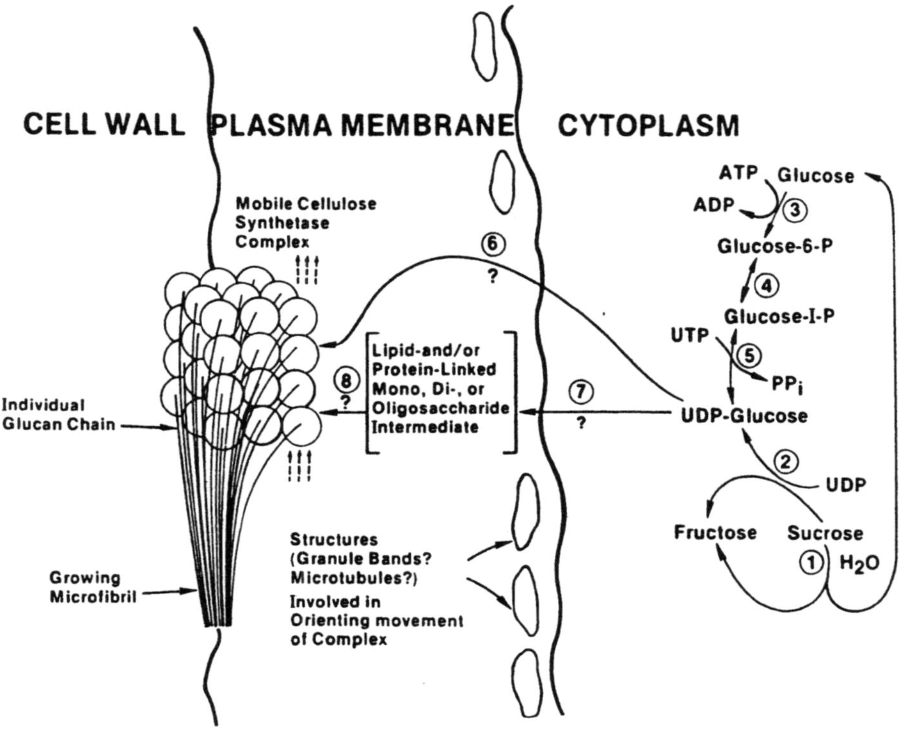

Fig. 2.5. Hypothetical model for the biosynthesis of cellulose. *Numbers* refer to reactions catalyzed by the following enzymes: *1* invertase; *2* sucrose synthetase; *3* hexokinase; *4* phosphoglucomutase; *5* UDP-glucose pyrophosphorylase; *6, 7, 8,* hypothetical reactions on the pathway to cellulose (Delmer 1982a)

Montezinos 1976). The unicellular green alga *Oocystis apiculata* was used in these studies, which revealed the impressions of microfibrils in the plasma membrane and also a distinct, linear, multiple-particulate complex which was associated with the ends of microfibril impressions. It should be pointed out that earlier studies using the old techniques had not revealed any microfibril impressions or terminal complexes. Thus, the use of rapid freezing without cryoprotectants led to new discoveries.

The new technique had revealed not only terminal complexes in *Oocystis* but that a directionality was involved in microfibril synthesis. Only one microfibril layer is synthesized at a time. When one layer is synthesized, the direction of the synthesis is changed in order to form the next layer. It seems likely that some mechanism of control is involved to obtain the exact geometrical pattern of the microfibril. Brown and co-workers (Brown 1985) found differences in the structure of the terminal synthesizing complexes. All these observations strengthen the "mobile site hypothesis", i.e., the terminal complex is moving in the plasma membrane rather than remaining stationary (Brown

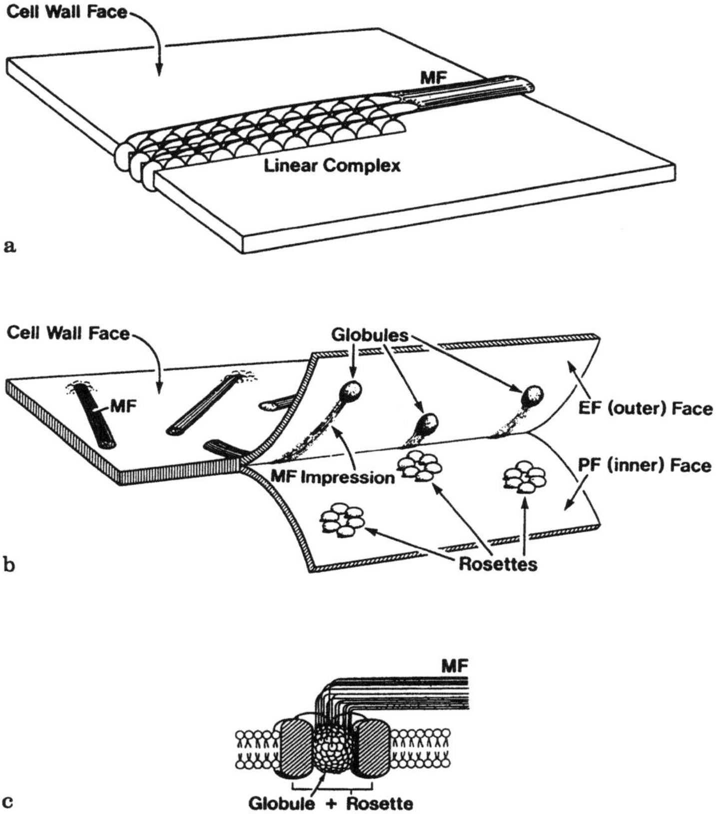

Fig. 2.6. Stylized drawings of terminal complexes observed by freeze-fracture of plasma membranes of cellulosic algae and higher plants. **a** Linear terminal complexes found in some cellulosic algae. For simplicity the number of subunits drawn is less than actually observed. **b** Rosettes and globules found in some cellulose algae and in lower and higher plants. Linear complexes as in (**a**) usually fracture with the EF-face but sometimes fracture with the PF-face; rosettes fracture with the PF-face and globules with the EF-face. In double-replica fracturing, globules are sometimes found in a complementary location to rosettes, leading to the speculation that rosettes and globules may be part of the same complex in vivo as shown in **c**. *MF* microfibril (Delmer 1987)

1979). It was envisioned that crystallization is a separate step from polymerization, and follows immediately after polymerization and that the linear terminal complex is deeply embedded in the plasma membrane (Haigler et al. 1980).

Brown and co-workers (Mueller and Brown 1980) discovered a rosette-like sturcture at the end of the microfibril. The rosette unit is located on the inner side of the plasma membrane while a globular unit is found in association with the microfibril. Complexes of similar morphology have later been observed both in monocots and in dicots (Willison and Brown 1977, Mueller and Brown 1982 a, b). In Fig. 2.4 dense rosettes on the protoplasmic fracture face of the plasma membrane of *Zinnia elegans* is presented.

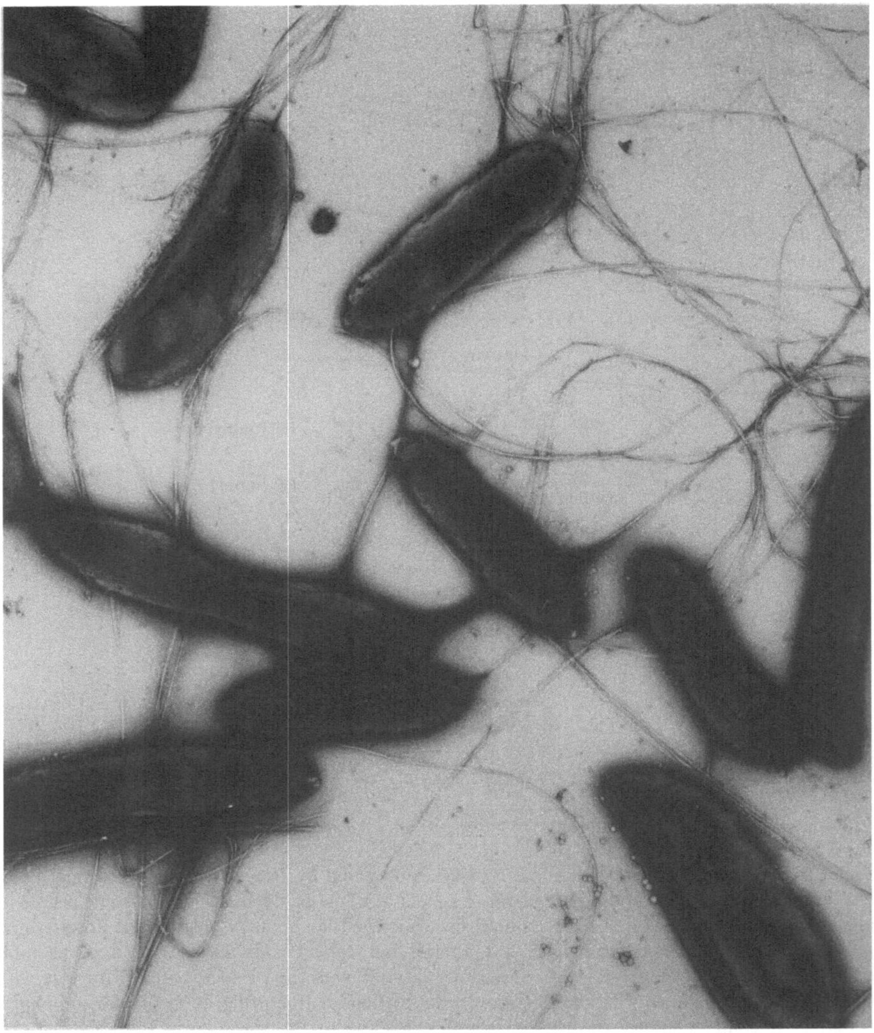

Fig. 2.7. Overview of more than 30 min of cellulose synthesis by *Acetobacter* as shown with negative staining. Note the massive twisting of interconnecting ribbons of cellulose. ×8200 (Brown 1979)

During normal conditions of growth, microfibrils are generated in clusters (Mueller and Brown 1982a, b). Microfibrils have a certain orientation which suggest that some elements in the plasma membrane are flowing or streaming, something which give the fibrils their direction. The cause of this flow is not known. The appreciation of the biochemistry and morphology of cellulose synthesis inspired Delmer (1982a) to propose a hypothetical model of the mechanisms of cellulose synthesis in plants as shown in Fig. 2.5. Delmer (1987) has also published a shematic drawing of the terminal complexes observed by freeze fracture of plasma membranes of cellulosic algae and higher plants (Fig. 2.6).

The mobile site hypothesis describes the microfibril assembly patterns for most eukaryotic cells. However, in prokaryotic cells the fixed site hypothesis for cellulose synthesis seems to dominate. The reason may be that in prokaryotes the site for the cellulose synthesis is on the outside of the cell surface (Brown et al. 1976). In *Acetobacter* cellulose synthesis has been photographed. Brown et al. (1976), with the assistance of Eastman Kodak Company, used a special "missile tracking film" and succeeded in time lapse cinematography of ribbon extrusion from *A. xylinum* (Fig. 2.7). The fibril is spun out into the medium away from the cell. These microfibrils merge into bundles which subsequently grow into ribbons when the bacteria are rotating around their longitudinal axes. There are approximately 80 microfibrils per ribbon. With the aid of the cinematography it has been possible to calculate the rate of cellulose synthesis. Approximately 1×10^8 glucose units per hour are incorporated into cellulose by one single bacterial cell. Shortly after Brown et al. (1976) had reported these results, Zaar (1979) showed that pores are present in rows parallel to the longitudinal cell axes in *Acetobacter.*

To obtain a better opportunity to film cellulose synthesis "Calcofluor White", a commercial cellulose brightener, was added. Brown and co-workers (Haigler et al. 1980) found that ribbons were not synthesized in the presence of Calcofluor. The reason seems to be that Calcofluor White forms hydrogen bonds with cellulose and prevents the normal association of glucan chains to assemble microfibrils. Calcofluor White thus influences the ability of cellulose to crystallize. In the presence of Calcofluor, the rate of polymerization increased 200–400% while glucose oxidation increased only 16%. Thus, it could be concluded that Calcofluor has no influence on the general metabolism of *Acetobacter.* However, the rate of polymerization dramatically increases when crystallization is inhibited. It thus seems reasonable to assume that crystallization is the rate-limiting step in cellulose biosynthesis.

2.2 Cellulolytic Fungi

A vast number of microorganisms degrade cellulose in nature. These organisms have in common the production of extracellular, hydrolytic enzymes, cellulases, which attack the cellulose polymer. However, relatively few microorganisms are able to produce the necessary cluster of enzymes for degradation of crystalline cellulose (Mandels and Weber 1969).

We will begin the presentation of cellulolytic microorganisms with the fungi, the most studied organisms in this respect. An extensive, if not complete, list of fungi considered in cellulase studies is presented in Table 2.1.

Fungi producing the necessary enzymes for a cell-free degradation of crystalline cellulose normally belong to the Ascomycete and Deuteromycete groups or to the white-rot Basidiomycetes (Eriksson and Johnsrud 1982). Brown-rot

Table 2.1 Examples of fungi used in cellulase studies

Agaricus bisporus	Manning and Wood (1983)
Ascobolus furfuraceus	Wicklow et al. (1980a)
Aspergillus aculeatus	Kanamoto et al. (1979)
A. fumigatus	Rogers et al. (1972), Jain et al. (1979)
A. niger	Ikeda et al. (1973), Sternberg et al. (1977); Hurst et al. (1978)
A. phoenicis	Sternberg et al. (1977)
A. terreus	Sinha et al. (1981), Garg and Neelakantan (1982)
A. wentii	Keilich et al. (1970)
Botryodiplodia theobromae	Umezurike (1979, 1981)
Chaetomium cellulolyticum	Chahal and Hawksworth (1976), Moo-Young et al. (1978), Fähnrich and Irrgang (1982)
C. globosum	Keilich et al. (1970)
C. thermopile	Fergus (1969), Romanelli et al. (1975), Eriksen and Goksöyr (1976)
Chrysosporium lignorum[a]	Eriksson and Rzedowski (1969a, b)
Cladosporium cladosporioides	Lynch et al. (1981)
Coriolus versicolor	Highley (1975)
Dichomitus squalens	Rouau and Foglietti (1985)
Eupenicillium javanicum	Tanaka et al. (1981)
Fomes fomentarius	Kozlik and Schánĕl (1974a, b)
Fusarium moniliforme	Matsumoto et al. (1974)
F. solani	T.M. Wood (1971), T.M. Wood and McCrae (1977a, b)
Fusarium sp.	Trivedi and Rao (1980, 1981)
Humicola grisea	Fergus (1969), Araujo et al. (1983)
H. insolens	Fergus (1969), Jain et al. (1979), Hayashida and Yosihoka (1980)
Hypocopra merdaria	Wicklow et al. (1980a)
Irpex lacteus	Wakabayashi and Nisizawa (1964), Kanda et al. (1983)
Lenzites trabea	Herr et al. (1978a, b)
Myceliophthora thermophila	Sen et al. (1981)
Myriococcum albomyces	Fergus (1969)
Myrothecium verrucaria	Whitaker and Thomas (1963), Selby and Maitland (1965)
Neocallimastix frontalis	Mountfort and Asher (1983), Wood et al. (1986)
Neurospora crassa	Eberhart et al. (1977), Rao et al. (1983a)
Paecilomyces fusisporus	Mishra et al. (1981)
P. variotiy	Mishra et al. (1981)
Papulaspora thermophila	Chapman et al. (1975)
Pellicularia filamentosa	Tanaka et al. (1977)
Penicillium chrysogenum	Mishra et al. (1981)
P. citrioviride	Shawky et al. (1984)
P. funiculosum	T.M. Wood et al. (1980), T.M. Wood and McCrae (1982), Deshpande et al. (1983), Joglekar et al. (1983b)
P. notatum	G. Pettersson (1968), K.-E. Eriksson and G. Pettersson (1968)
P. pinophilum	Brown et al. (1986), Wood and McCrae (1986a, c)
P. variabile	Amemura et al. (1967)
P. verruculosum	Szakacs et al. (1981)
Pestalotiopsis versicolor	Thakur and Sastry (1981), Rao et al. (1983b)
Phanerochaete chrysosporium[a]	Reid (1983a, b)
Phialophora malorum	Berg (1978)
Phoma hibernica	Urbanek et al. (1978)
Physarum polycephalum	Koevenig and Liu (1981)
Pleurotus ostreatus	Kozlik and Schánĕl (1974a, b)

Table 2.1 (continued)

P. sajor-caju	Madan and Bisaria (1983)
Podospora decipiens	Wicklow et al. (1980a)
Polyporus schweinitzii	Keilich et al. (1970)
P. versicolor	G. Petterson et al. (1963), Highley (1973)
Poria placenta	Highley (1980), Wolter et al. (1980), Highley et al. (1981), Highley and Wolter (1982)
Poronia punctata	Wicklow et al. (1980a)
Pyricularia oryzae	Sudo et al. (1977)
Saccobolus truncatus	Wicklow et al. (1980a)
Schizophyllum commune	Desrochers et al. (1981), Rho et al. (1982), Yaguchi et al. (1983), Paice et al. (1984)
Sclerotinia libertiana	Tanaka et al. (1982)
Sclerotium rolfsii	Sadana et al. (1979), Mishra et al. (1981), Sadana et al. (1983), Sadana and Patil (1985)
Scytalidium lignicola	Desai et al. (1982)
Sordaria fimicola	Wicklow et al. (1980a)
Sporotrichum pulverulentum[a]	K.-E. Eriksson and B. Pettersson (1975a, b), Almin et al. (1975), Streamer et al. (1975), Deshpande et al. (1978), K.-E. Eriksson and Hamp (1978)
S. thermopile	Fergus (1969), Romanelli et al. (1975), Coutts and R.E. Smith (1976), Canevascini and Meyer (1979), Margaritis and Merchant (1983), Grajek (1987)
Stereum sanguinolentum	Bucht and K.-E. Eriksson (1968), K.-E. Eriksson and B. Pettersson (1968), Björndal and K.-E. Eriksson (1968)
Talaromyces emersonii	Folan and Coughlan (1981), McHale and Coughlan (1981a), Moloney et al. (1983)
Thermoascus aurantiacus	Romanelli et al. (1975), Tong et al. (1980), Shepherd et al. (1981), Grajek (1987)
Thraustotheca clavata	Berner and Chapman (1977)
Torula thermopile	Fergus (1969), Jain et al. (1979)
Trichoderma koningii	Iwasaki et al. (1964), T.M. Wood and McCrae (1978), Halliwell and Vincent (1981)
T. pseudokoningii	Zhu et al. (1982), Harrer et al. (1983)
T. reesei (viride)	Reese (1956), Mandels and Reese (1960), Berghem et al. (1976), Ghose (1977), Nisizawa et al. (1978), Shoemaker and Brown (1978a), Sternberg and Mandels (1979), Ryu and Mandels (1980), White and Brown (1981), Montenecourt et al. (1981), Fägerstam (1981), Bhikhabhai et al. (1985)
Trichurus spiralis	Mishra et al. (1981)
Verticillium albo-atrum	Gupta and Heale (1971), Heale and Gupta (1971)
Volvariella volvacea	Chang and Steinkraus (1982)

[a] *Chrysosporium lignorum* and *Sporotrichum pulverulentum* are now named *Phanerochaete chrysosporium*.

Basidiomycetes also degrade crystalline cellulose. However, these fungi do not seem to produce exo-1,4-β-glucanases (Ljungdahl and Eriksson 1985).

2.3 Cellulolytic Enzymes

2.3.1 The Old Concept

Anyone who has read the early literature on cellulolytic enzymes must have noticed the large varieties existing in the proposed composition of the cellulase systems and the mechanism of action of the cellulolytic enzymes. It was obvious at an early stage of research that some cellulolytic microorganisms, particularly fungi, can degrade cellulose only after some kind of modification of the substrate has occurred. However, some organisms have the ability to degrade cellulose, such as cotton fibers, in its native form. By definition only the latter organisms have been considered as truly cellulolytic.

In 1950 Reese et al. (1950) introduced their now classic "C_1, C_x" hypothesis. It was postulated that crystalline substrates were first rendered susceptible to hydrolysis by the C_1-component. This component was suggested to be a nonhydrolytic chain-separating enzyme. The separation of the cellulose chains was suggested to take place by a splitting of the hydrogen bonds. Cellulose modified in this way is then hydrolyzed by the C_x-enzyme fraction and by the β-glucosidase. It was also suggested that only "truly cellulolytic microorganisms" are equipped with both C_1 and C_x enzymes, while microorganisms only able to hydrolyze amorphous or modified cellulose lack the C_1 enzyme. Mandels and Reese (1964) gave the C_1-C_x complex the following schematic description:

Cellulose	→	Reactive cellulose	→	Cellobiose	→	Glucose
	C_1		C_x		β-glucosidase	
	?		hydrolytic		hydrolytic	

The above scheme was the working hypothesis for the enzyme mechanisms involved in cellulose degradation for a long time. However, the as yet undefined nature of the C_1 enzyme was of course unsatisfactory. Therefore, for a decade almost all of the cellulase research was focused on unraveling the nature of the C_1 enzyme.

King (1965) found that, although glucose was released from cellulose by a crude enzyme mixture, the amount of glucose decreased as the purification of the so-called C_1 component proceeded. Finally, the major product turned out to be cellobiose. King recognized the speculation concerning the nature of its mechanism of action but without entering the discussion of the relationship between C_1 as a swelling factor, as "hydrogen bondase", or other suggested activities, he simply noted the fact that cellobiose as a product of cellulose hydrolysis increased with an increasing purity of the enzyme.

At the symposium "Cellulases and their applications" in Atlantic City 1968, the true nature of the C_1-component was finally solved. Here, King and Vessal (1969) stated that the C_1-component is required for the attack on crystalline cellulose. The C_1-enzyme has thus been considered by some workers to carry out an hydrolytic attack on crystalline cellulose thereby producing cellobiose (King 1965), and by others to be an enzyme of undefined nature converting crystalline cellulose to a form degradable by the C_x enzymes (endoglucanases). Selby (1969) at this symposium demonstrated that the powerful synergistic action on cotton previously found only between the C_1 and C_x components of *Trichoderma viride* was also displayed by the same enzymes from *Penicillium funiculosum*. Selby and co-workers had also demonstrated a cross-synergism between the C_1 and C_x enzymes from the two fungi. However, in his attempt to explain the action of the C_1 enzyme, Selby was still thinking of the C_1-enzyme mechanism in terms of a "hydrogen bondase enzyme".

At the same symposium, Eriksson (1969) presented a solution to the C_1 mystery. Studies of cellulases, mainly CMC-ases (carboxymethyl cellulases) from culture solutions of the white-rot fungus *Chrysosporium lignorum* (later *Sporotrichum pulverulentum/Phanerochaete chrysosporium*) had given the following results: A concentrated cell-free culture solution was found to hydrolyze cotton fibers at a reasonable rate. Fractionation of this culture solution on a Sephadex G-75 column gave three CMC-ase peaks. Hydrolysis of the cotton fibers was much less if such fibers were incubated with the individual CMC-ase peaks, the number of enzyme units being the same in both the unfractionated and the fractionated cases.

Fractionation of the *C. lignorum* culture solution also gave rise to a protein peak with no CMC-ase activity. This peak was thought to contain the C_1-enzyme activity which also turned out to be correct (Fig. 2.8). When a cotton fiber suspension was incubated with a mixture of one of the CMC-ase activity peaks and the same amount of protein from the unknown peak, the rate of hydrolysis was increased tenfold. After incubation of a solution of cellohexaose with the unknown protein the hydrolysis product was cellobiose. These observations made Eriksson (1969) suggest that the mechanism for degradation of crystalline cellulose proposed by Mandels and Reese (1964) was reversed, i.e., the C_x enzyme, showing activity toward CM-cellulose and attacking the cellulose chain at random over the chain, comes in first and opens up the 1,4-β-glucosidic bonds, giving rise to both reducing and nonreducing end groups. The C_1 enzyme then comes in to split off cellobiose units from the nonreducing end. Such a mechanism would also explain the synergistic action between the two enzymes. When a randomly acting cellulase (CMC-ase, C_x) hydrolyzes a 1,4-β-glucosidic linkage the chances that the broken bond stays

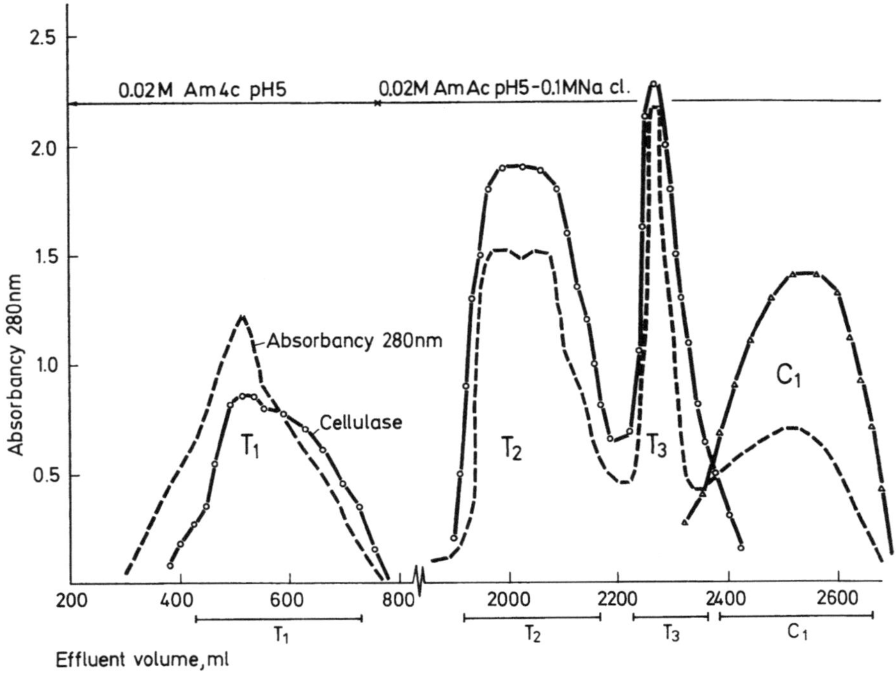

Fig. 2.8. Fractionation of culture solution of *C. lignorum* (*S. pulverulentum/P. chrysosporium*) on a Sephadex anion exchanger A-50. (– – –) Absorbancy 280 nm; (○) CMC-ase activity (endoglucanase); (△) C_1-activity (exoglucanase) (Eriksson and Pettersson 1972)

broken are, due to steric factors, not very good. Interchain hydrogen bonds will probably put the glucose units in a position that allows reformation of the broken bond. In the presence of exoenzyme (C_1), this will probably not happen.

The suggestion by Eriksson (1969) that the nature of the C_1 enzyme was that of an exoenzyme was in a few years confirmed to be valid by several researchers. The same enzyme mechanism for degradation of crystalline cellulose was also shown to be used by a large number of fungi, i.e., *C. lignorum* (Eriksson and Pettersson 1972), *Trichoderma koningii* (Halliwell et al. 1972, Wood and McCrae 1972), *Trichoderma* (*viride*) *reseii* (Berghem and Pettersson 1973, Emert et al. 1974), *Fusarium solani* (Wood and McCrae 1977a), and *Sclerotium rolfsii* (Patil and Sadana 1984).

In spite of the fact that the $C_1 - C_x$ hypothesis did not hold true, it was an important conclusion of the work by Reese et. al. (1950) that crystalline cellulose is hydrolyzed by most fungi via a three-step enzyme reaction sequence. The attack on crystalline cellulose is initiated by the endo-/exoglucanase reactions as described above, followed by an attack on the released water-soluble oligosaccharides by the third hydrolytic enzyme, the 1,4-β-glucosidase, converting these cellodextrins to glucose. It should be emphasized that the terms C_1 and C_x should be avoided since these terms will cause confusion, particularly since their reaction sequences are opposite to what was first suggested.

2.3.2 Hydrolytic Enzymes

The cellulolytic enzymes have been defined as the enzymes hydrolyzing cellulose, thereby ultimately yielding soluble sugars small enough to pass through the microbial cell walls (Eriksson 1967). Today, a modification of this definition seems necessary since, in addition to the hydrolytic enzymes, oxidative enzymes also participate in cellulose degradation (Eriksson and Wood 1985).

A large number of microorganisms grow on cellulose but relatively few are able to produce the complete enzyme system necessary to hydrolyze crystalline cellulose. However, several organisms produce enzymes that degrade amorphous cellulose or water-soluble derivatives (Mandels and Weber 1969).

A complete set of hydrolytic (cellulolytic) enzymes is minimally composed of the following enzymes:

1. Endo-1,4-β-glucanases (EC 3.2.1.4) which over the cellulose chain randomly attack and split $\beta(1\rightarrow4)$-glucosidic linkages.
2. Exo-1,4-β-glucanase (EC 3.2.1.91) that splits off either cellobiose or glucose from the nonreducing end of the cellulose.
3. 1,4-β-Glucosidase (EC 3.2.1.21) hydrolyzing cellobiose and other water-soluble cellodextrins to glucose.

2.3.3 Oxidative Enzymes

In addition to the hydrolytic enzymes described above, oxidative enzymes are also involved in cellulose degradation. Thus, the fungus *S. pulverulentum* (*P. chrysosporium*) has two different pathways to oxidize cellobiose. The enzymes involved in these reactions can be described as follows:

1. Cellobiose: quinone oxidoreductase (cellobiose dehydrogenase) reduces quinones and phenoxy radicals in the presence of cellobiose which is oxidized to cellobiono-δ-lactone (Westermark and Eriksson 1974a, b).
2. Cellobiose oxidase oxidizes cellobiose and higher cellodextrins to their corresponding-onic acids using molecular oxygen (Ayers et al. 1978).

2.3.4 Phosphorolytic Enzymes

Some aerobic and anaerobic bacteria produce another type of enzymes for cellulose degradation, namely, phosphorylases (Ljungdahl and Eriksson 1985).

2.4 Assay of Enzymes Participating in Cellulose Degradation

A wide variety of substrates have been used for the assay of cellulase activity. If these assays are restricted to the hydrolytic enzymes, the most commonly used substrates have been native cellulose, regenerated cellulose, and soluble cellulose derivatives. Methods and substrates used in these assays are listed in Table 2.2.

For the assay of cellulase activity, as for most other enzyme activities, it is desirable to use a method in which the number of enzyme units can be directly determined. A number of quantities are defined in terms of units of enzymes, such as the concentration of an enzyme, normally given as units/ml, the specific activity, defined as units/mg protein, and the molecular activity defined as units/μmol enzyme.

The determination of activity units does not present difficulties for enzymes where the change in the substrate can be expressed in absolute terms. However, for enzymes whose activities are not measured in terms of a chemical reaction but in terms of some physical change, such as loss of weight of an insoluble substrate, reduction in mechanical properties, or a decrease of viscosity, complications arise and it is not possible to express their activity in the units mentioned above. These difficulties are particularly apparent if a mixture of enzymes is required to give rise to a product, as is the case for degradation of crystalline cellulose. The situation is further complicated by the fact that a

Table 2.2. Assay of hydrolytic enzymes involved in the degradation of cellulose and cellodextrins

Enzymes	Substrate	Product or measured effect
Mixture of cellulases	Insoluble celluloses Dyed celluloses Cellulose agar [U-^{14}C]-cellulose	Loss in weight of insoluble substrates Decrease in mechanical properties of fibers or films Change in turbidity of cellulose suspensions Release of reducing sugars Release of dye Clarification of agar Water-soluble ^{14}C-products
Endo-1,4-β-glucanases	Cellulose derivatives Cellodextrins	Decrease in viscosity Release of reducing sugars
Exo-1,4-β-glucanases	Amorphous cellulose Cellodextrins p-Nitro-phenyl-β-cellobioside p-Nitro-phenyl-β-lactoside Carboxymethyl substituted cellodextrins	Reducing sugars Cellobiose p-Nitro-phenol
1,4-β-glucosidase	p-Nitro-phenyl-β-D-glucoside Cellobiose Cellodextrins	Reducing sugars Glucose P-Nitrophenol

homogeneous product is not generated but rather a mixture of monomers and various oligomers. All these difficulties are encountered if the real substrate, i.e., crystalline cellulose, is to be used as substrate for the three hydrolytic enzymes necessary to completely convert this type of cellulose to glucose.

Due to the lack of specific substrates and lack of standardization of activity determinations, it has been difficult to compare cellulase production in one fungal or bacterial strain with another, particularly if such efforts were undertaken in different laboratories. Attempts to standardize the measurements have been made among scientists, but they have so far been fruitless. An attempt to coordinate such activity determinations was recently made by the Commission on Biotechnology IUPAC (Ghose 1987). However, the recommendations (Commission of Biotechnology, IUPAC) consisted only of a list of the most commonly used methods, and no real standardizations or improvements were accomplished.

The most commonly used assay of complete cellulase activity is based on saccharification of filter paper (Mandels and Weber 1969, Mandels et al. 1976, Montenecourt et al. 1978). The original filter paper activity was defined by Mandels et al. (1976) as the amount of reducing sugars released in 1 h when using 50 mg of Whatman No. 1 filter paper under standardized conditions. This technique has been further developed by Montenecourt et al. (1978), who used filter paper and bioassay disks in place of the 50 mg filter paper strips. The assays are carried out at 50 °C in a stationary water bath for 1 h. Free reducing sugars are quantified by the dinitrosalicylic (DNS) acid method by Miller (1959) and the results are expressed as glucose equivalents. It is evident that these methods, which register only the number of reducing end groups, are very dependent upon the β-glucosidase levels in the cultures. This was also pointed out by Bailey (1981) and by Joglekar et al. (1983a) who demonstrated that the β-glucosidase levels have a marked effect on filter paper hydrolyzing activity even during short-time incubations.

To obtain a better appreciation of the composition of a mixture of hydrolytic cellulases and the effectiveness of this mixture, the following amendments to the measurement techniques can be recommended:

1. Measure the number of reducing endgroups by the DNS method (Miller 1959) and total sugar released by the orcinol method (Vasseur 1948). The ratio between the values obtained with the orcinol method and the DNS method, respectively, will be an approximate average DP of the released sugars.
2. Measure reducing endgroups with the DNS method, add extracellular β-glucosidase and continue the hydrolysis until a final (steady) result is obtained.
3. Fractionate and determine the released sugars by HPLC.
4. Use the cellobiose:quinone oxidoreductase method (Kelleher et al. 1987) for monitoring saccharification. Only cellobiose and higher dextrins are registered by this method.
5. Measure the amount of glucose by the glucose-oxidase method.

Better still than to use filter paper as substrate would be to use a standardized Avicel-cellulose preparation. Such a preparation could be standardized by screening the cellulose particles with both a 50 and a 100 mesh sieve. A screened Avicel preparation kept suspended by magnetic stirring during the enzymatic hydrolysis would offer a better substrate than filter paper. If an international agreement could be reached to use a screened Avicel for comparison of cellulase preparations produced in different laboratories, much would be gained.

Here a word of caution is needed; it does not necessarily mean that a cellulase preparation, more effective toward pure cellulose than another enzyme preparation, is also more effective toward the lignocellulosic substrate chosen for saccharification in a certain laboratory or factory. To test the efficiency of an enzyme preparation meant for a certain substrate there is no substitute for the waste lignocellulosic material to be used in practice.

The release of a dye from a dyed insoluble cellulosic substrate is also a convenient way to measure the hydrolyzing effect of a mixture of hydrolytic cellulases. Dyed filter paper (Mandels and Weber 1969) and dyed Avicel (Leisola and Linko 1977, Ng and Zeikus 1981) have been used as substrates for these purposes. The method to use dyed celluloses is also suitable for screening for cellulase production and can be applied to automatic analysis of cellulase activity during fermentation (Leisola and Kauppinen 1978).

An alternative to using dyed solid cellulosic substrates for assay of cellulase activity is to use a ^{14}C uniformly labeled cellulose (Vardanis and Finkelman 1981). As with dyed solids used as a substrate, it gives no information on the composition of the mixture of glucose and cellodextrins released from the solid substrate. However, contrary to the case with the dyed celluloses, degradation of ^{14}C-labeled cellulose is easy to quantify, i.e., the amount of ^{14}C set free in solution can easily be related to the total amount of ^{14}C in the solid substrate.

An immunoelectrophoretic method for the detection of cellulases was developed by Nummi et al. (1980). In their studies of *Trichoderma reesei* cellulases it was shown that the antiserum produced reacted with all *T. reesei* proteins, and by using amorphous cellulose as substrate it was possible to estimate the active cellulases.

2.4.1 Endo-1,4-β-Glucanase Activity

Endo-1,4-β-glucanases act randomly on the cellulose chains and open up nonreducing ends for a continued action of exo-enzymes (Eriksson and Wood 1985, Ljungdahl and Eriksson 1985). Water-soluble cellulose derivates are convenient substrates for the determination of endoglucanase activity since such substrates are not hydrolyzed by exoglucanases (Pettersson 1975).

One unit of endoglucanase activity is usually defined as the amount of enzymes hydrolyzing 1,4-β-glucosidic linkages at an initial rate of 1 µM equiv/min. Determination of the rate of formation of reducing groups will directly give such information. Therefore, measuring the release of reducing sugars from cellulose derivatives is probably the most commonly used assay technique of endoglucanase activity. However, too many drawbacks are involved in this method, at least if the rule set by the Commission on Enzymes is obeyed. This rule states that wherever possible enzyme assays should be based on measurements of initial rates of reactions. This cannot easily be done if one relies upon measurements of the increase of reducing endgroups. Such measurements can be carried out only at the leveling-off stage of the reaction, where complications such as competitive inhibition and transglucosidation reactions occur. A viscometric assay method should be particularly suitable for avoiding such drawbacks since a drop in viscosity is encountered as soon as the first bond in the polymer is broken.

Almin and Eriksson (1967, 1968) and Almin et al. (1967, 1975) have developed a method by which viscosity changes of carboxymethyl cellulose (CMC) solution can be recalculated to absolute enzyme units. This has turned out to be a valuable tool in cellulase studies, and developments along similar lines have later been done by others (Hulme 1971, Demeester et al. 1977). The viscometric method is, however, somewhat difficult to use since it demands a thorough analysis and knowledge of the properties of the cellulose derivatives used as substrate before calculations of absolute enzyme units can be carried out.

Hydroxyethyl cellulose (HEC) has also been used successfully as a substrate for endoglucanases (Child et al. 1973, Demeester et al. 1977). In some respects it can be considered as a preferred substrate over CMC since HEC is a neutral compound and is thus not affected by changes in the pH value or the ionic strength of the solvent. However, HEC is mainly used for studies of released reducing sugars, since it suffers from the drawback that the substituents may be of different size due to polymerization. This makes a mathematical development of a formula that allows the calculation of the number of β-1,4-glucosidic bonds broken per unit time more difficult.

Eriksson and Pettersson (1973, 1975a) developed a zymogram technique using viscous solutions of CMC for the detection of endoglucanases in polyacrylamide gels used for isoelectric focusing. The CMC solution is sprayed over the gels. Reducing sugars are produced where the endoglucanases are located. These sugars are soaked into an overlaid paper and detected by spraying the paper with a reagent for reducing sugars. Saddler and Khan (1981) incorporated CMC in the gels and detected the endoglucanase positions by staining the gels with dinitrosalicylate. Other staining techniques to detect endoglucanases in electrophoretic gels have been developed by McHale and Coughlan (1981b) and by Beguin (1983).

2.4.2 Exo-1,4-β-Glucanase Activity

No substrate is available for a direct, specific measurement of exoglucanase activity in a mixture of cellulolytic enzymes. Therefore, direct measurements of exoglucanase activity can only be made with preparations lacking other cellulolytic enzymes or with purified enzymes. For such measurements amorphous cellulose, phosphoric acid swollen cellulose, or cellodextrins may be used (Mandels 1982). The most common exo-1,4-β-glucanases remove dimers (cellobiose) from the nonreducing end of the cellulose chain. These enzymes are not active against carboxymethyl cellulose since their chain-cutting activity will stop as soon as they hit the first substituent.

Measurement of exoglucanase activity in a mixture of cellulolytic enzymes is more difficult. However, the methods mentioned below have been used for this purpose.

Eriksson and Johnsrud (1982) used amorphous cellulose as substrate and recommended the following approach. In a mixture of enzymes the endoglucanase activity is first determined. From a standard curve made up with known numbers of units of purified endoglucanases toward the same amorphous cellulose, the amounts of reducing sugars produced by the same amount of pure endoglucanase is determined. The amount of reducing sugars obtained with the enzyme mixture is then measured. The amount of sugar obtained from the endoglucanase activity is deducted from the amount of sugar obtained with the enzyme mixture. If the β-glucosidase activity in the mixture offers a problem, this activity can be inhibited by addition of δ-glucono-lactone. Crystalline cellulose cannot be used in this test since the synergistic effects obtained from the endo- and exoglucanases will obscure the results.

Another technique for estimation of exoglucanase activity, i.e., cellobiohydrolase activity, has been suggested by Deshpande et al. (1984). These authors used the heterobiosides p-nitrophenyl-β-D-cellobioside (PNPC) or p-nitrophenyl-β-D-lactoside (PNPL) as selective substrates for the measurement of exoglucanase activity. The exoglucanase, splitting off cellobiose units from the nonreducing end of the cellulose chains, specifically acts on the agluconic bond (between the p-nitrophenyl and the disaccharide moiety) and not on the holosidic bond (between the two glucose units of cellobiose). The interfering effect of β-glucosidase, which acts on both the agluconic and the holosidic bond, is as mentioned above overcome by the addition of δ-glucono-lactone, which specifically inhibits the activity of this enzyme. The interference of endoglucanases, which also act on both agluconic and holosidic bonds can be compensated for by prior standardization of the assay procedure with a purified endoglucanase from the mixture of cellulases studied. The activities of the three different types of hydrolytic enzymes toward the heterobiosides used are presented in Table 2.3. The method is laborious and probably not practical unless long-time work is planned with the same enzyme mixture. Only then is it worthwhile to purify the endoglucanases and to prepare a standard curve for the action on the substrate by this enzyme.

Another procedure to assay for exoglucanases, i.e., cellobiohydrolase activity, has recently been suggested (Kelleher et al. 1987). The method takes advantage of the enzyme, cellobiose : quinone oxidoreductase, which in the presence of cellobiose or higher cellodextrins reduces quinones. This enzyme has earlier been effectively used to monitor cellulolysis (Canevascini 1985).

Table 2.3. Activities of the various purified enzymes toward heterobiosides[a] (Deshpande et al. 1984)

Enzyme[c]	pNPG Agluconic	Specific activity (IU mg^{-1})			
		pNPC[b]		pNPL[b]	
		Agluconic	Holosidic	Agluconic	Heterosidic
Exoglucanases					
Exo-1	0	5.0	0	4.0	0
Exo-2	0.008	1.5	0	0.5	0
Exo-3	0.12	10.9	0	1.9	0
Endoglucanases					
Endo-1	0.005	1.3	1.35	1.3	0
Endo-2	0.09	1.26	0.4	1.3	0
β-Glucosidase	90.8	3.3	3.7	0	0

[a] Heterobiosides: pNPG, p-nitrophenyl-β-D-glucopyranoside; pNPC, p-nitrophenyl-β-D-cellobioside; pNPL, p-nitrophenyl-β-D-lactoside.
[b] Action on the aglugonic bond is measured by the release of p-nitrophenol, and the action on holosidic and heterosidic bonds, by estimating the glucose produced.
[c] Exo-1 and 2 are from *Trichoderma reesei*, Exo-3 from *Sporotrichum pulverulentum*; Endo-1 from *T. reesei* and Endo-2 from *S. pulverulentum*, β-glucosidase from *T. reesei*.

Exo-1,4-β-glucanases removing glucose from the nonreducing end of the cellulose chain have also been identified but have been less studied than the cellobiohydrolases. Rabinowich et al. (1982) have suggested the use of carboxymethyl-substituted cellodextrins for the selective determination of this exoglucanase activity.

2.4.3 1,4-β-Glucosidase Activity

The 1,4-β-glucosidase enzymes hydrolyze both cellobiose and higher cellodextrins to glucose. This is the only type of hydrolytic enzymes participating in cellulose degradation for which low-molecular weight specific substrates are available for estimation of the activity, also in mixtures of cellulolytic enzymes. The most commonly used methods for assaying β-glucosidases include the measurement of the release of glucose from cellobiose or other, higher cellodextrins (Selby and Maitland 1967, Sternberg et al. 1977) and to use β-glucosides with a chromophore such as p-nitrophenyl-β-D-glucopyranoside (Berghem and Pettersson 1973). These two enzyme activities often coincide, although an enzyme splitting cellobiose without action on other β-glucosides has been found in the rot fungus *Stereum sanguinolentum* (Bucht and Eriksson 1969).

The most specific method used to study the production of glucose from cellobiose or other cellodextrins is the glucose oxidase/peroxidase method

Table 2.4. Assay of oxidative enzymes involved in the degradation of cellulose and cellodextrins

Enzymes	Substrate	Product or measured effect
Cellobiose: quinone oxidoreductase (cellobiose dehydrogenase)	Reducible quinones such as 2,6-dichlorophenol-indophenol (DCIP), 3-methoxy-5-tert-butyl-benzoquinone Cellobiose	Decrease in absorbance at a specific wavelength
Cellobiose oxidase	Cellobiose Cellodextrins Higher cellodextrins Lactose	Onic acids

(Bergmeyer and Bernt 1963). The commonly used procedure for determination of reducing endgroups, the dinitrosalicylate method (DNS) (Miller 1959, Nelson 1944), is also used to estimate glucose released from cellobiose or other cellodextrins. Since the release of glucose, and thus the number of bonds broken per unit time can be easily determined, it is easy to express β-glucosidase activity in absolute units.

β-Glucosidase activity is easily localized in gels by soaking the gels in a solution containing p-nitrophenyl-β-D-glucoside. After washing, the gels are immersed in an alkaline buffer where the yellow color of the p-nitrophenolate ion corresponding to the location of the enzyme will become apparent.

2.4.4 Oxidative Enzymes

It was suggested many years ago that also oxidative processes are involved in cellulose degradation. Those concepts were, however, never clarified or accepted until Eriksson et al. (1974), using a cell-free fungal culture solution to degrade cotton, found that the degradation was considerably enhanced in the presence of oxygen compared to the degradation obtained in nitrogen atmosphere. The oxidative enzymes involved in cellulose degradation by the white-rot fungus *Sporotrichum pulverulentum* (*Phanerochaete chrysosporium*) were later identified as cellobiose dehydrogenase by Westermark and Eriksson (1974a, b, 1975) and as cellobiose oxidase (Ayers et al. 1978, Ayers and Eriksson 1982). The enzyme cellobiose dehydrogenase has now been found in many fungi (Westermark and Eriksson 1988). The assay techniques used for the oxidative enzymes involved in cellulose degradation are presented in Table 2.4.

2.4.5 Phosphorolytic Enzymes

Some aerobic and anaerobic bacteria employ an additional category of enzymes for cellulose degradation, i.e., phosphorylases. Two types have been distinguished, namely, those specific for cellobiose (Sasaki et al. 1983) and those utilizing higher cellodextrins for the formation of α-glucosyl-phosphate (Schafer and King 1965). The substrates for these enzymes are cellobiose or higher cellodextrins. The products of enzymatic activity are glucose-1-P and glucose (Ljungdahl and Eriksson 1985).

2.5 Cellulose Degradation by White-Rot, Brown-Rot, Soft-Rot, and Anaerobic Fungi

2.5.1 White-Rot Fungi

The white-rot group of fungi is rather heterogeneous. These fungi have in common the capacity to degrade lignin as well as the other wood components. They also have in common the extracellular production of enzymes oxidizing phenolic compounds related to lignin. This ability has therefore been utilized for the identification of white-rot fungi (Bavendamm 1928, Davidson et al. 1938). The relative amounts of lignin and polysaccharides degraded and utilized by these fungi vary as does the order of preferential attack on these wood components (Eriksson 1981 c). However, the wood polysaccharides are degraded during primary metabolism of the fungus, while lignin metabolism occurs only during secondary metabolism (Keyser et al. 1978, Kirk et al. 1978a, b). The primary growth phase prevails until the fungal environment becomes limited for an essential nutrient, such as nitrogen, carbon, sulfate or phosphate.

The white-rot fungus most commonly used in studies on enzymes involved in cellulose degradation, their production, and their regulation is *Sporotrichum pulverulentum* (*Phanerochaete chrysosporium*). The mechanisms of degradation are presented in Fig. 2.9.

The fungus was first isolated from wood chip piles and was given the name *Chrysosporium lignorum* (Bergman and Nilsson 1966). This name was later changed to *Sporotrichum pulverulentum* for its imperfect stage (von Hofsten and von Hofsten 1974). The third name, *Phanerochaete chrysosporium* has been assigned to it based on its perfect state (Burdsall and Eslyn 1974). The fungus is characterized as thermotolerant and has an optimum growth rate around 38 – 39 °C. Details for production of cellulolytic enzymes, concentration of culture solutions, and separation of the cellulases are given by Eriksson and Pettersson (1975a). Results from research carried out mainly during the 1960's and the 1970's revealed that *S. pulverulentum* produces the following pattern of hydrolytic enzymes degrading cellulose:

1. Five endo-1,4-β-glucanases (EC 3.2.1.4) attacking the cellulose chain at random, thereby splitting the 1,4-β-glucosidic linkages (Eriksson and Pettersson 1975a).
2. One exo-1,4-β-glucanase (EC 3.2.1.91) releasing either cellobiose or glucose from the nonreducing end of the cellulose chain (Eriksson and Pettersson 1975b).

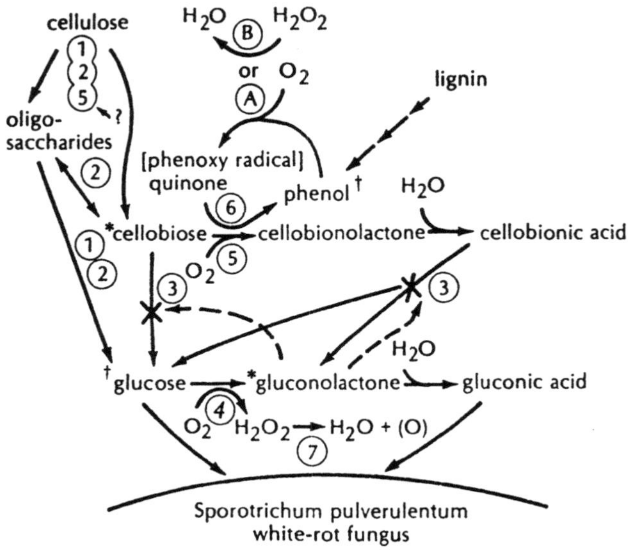

Fig. 2.9. Enzyme mechanisms involved in degradation of cellulose by the white-rot fungus *S. pulverulentum*. *1* Endo-1,4-β-glucanases; *2* exo-1,4-β-glucanases; *3* 1,4-β-glucosidases; *4* glucose oxidase; *5* cellobiose oxidase; *6* cellobiose: quinone oxidoreductase; *7* catalase (Eriksson 1978)

3. Two 1,4-β-glucosidases (EC 3.2.1.21) hydrolyzing cellobiose and water-soluble cellodextrins to glucose (Deshpande et al. 1978).

All these enzymes have been separated, purified, and characterized physicochemically as well as functionally (Fig. 2.10; Eriksson and Rzedowski 1969a, b, Eriksson and Pettersson 1975a, b, Streamer et al. 1975, Deshpande et al. 1978). All purifications were carried out on a quantitative basis, and the purity of the enzymes was tested in several ways. After purification, each enzyme yielded a single protein band on analytical polyacrylamide gel electrophoresis and on dodecyl sulfate gels.

From the purification data obtained it was possible to calculate that the ratio between the five endoglucanases produced by the fungus was approximately 4:1:1:1:1. The quantitative calculations were possible by using the viscometric method developed by Almin, Eriksson, and co-workers (Almin et al. 1967, 1975, Almin and Eriksson 1967, 1968) which allows calculation in absolute units of endoglucanase activities. The separation and purification work was also facilitated by the development of a zymogram technique for detection of polysaccharide degrading enzymes after isoelectric focusing on flat bed polyacrylamide gels (Eriksson and Pettersson 1973). This technique was particularly useful for identification of the individual endoglucanases.

Small but significant differences in the amino acid composition of the five different endoglucanases were found (Table 2.5). All but one of these enzymes are glycoproteins, and the carbohydrate content varies between 0 and 10.5%. The carbohydrate composition is also given in Table 2.5. The molecular

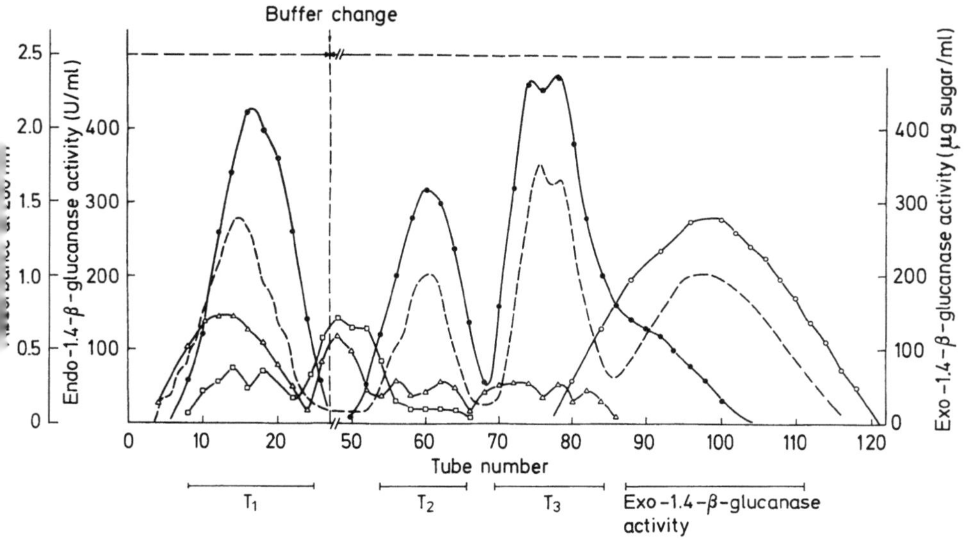

Fig. 2.10. Distribution of protein, endo-1,4-β-glucanases, exo-1,4-β-glucanase, xylanase, and β-glucosidase activities after chromatography of a concentrated culture solution from *Sporotrichum pulverulentum* on DEAE-Sephadex A-50. The buffer was changed from 0.02 to 0.1 M ammonium acetate pH 5 at the point indicated. The fraction volume was 20 ml. (– – –) Absorbance at 280 nm; (●) endo-1,4-β-glucanase; (○) exo-1,4-β-glucanase; (△) xylanase; (□) β-glucosidase (Eriksson and Pettersson 1975a)

weights, determined by ultracentrifugation and calculated on the basis of the amino acid composition and the carbohydrate content, vary between 28 000 and 37 000.

The endoglucanases were also characterized with respect to their molecular activities, A_m and the Michaelis-Menten constants K_m (Almin et al. 1975). The enzymes differ in these values. The A_m and K_m calculations were carried out with carboxymethyl cellulose as substrate. The A_m-values seem to be more valuable as a distinguishing parameter between the enzymes than the K_m values. The endoglucanases were also found to vary in their substrate preferences. The most striking difference between them is, however, their transglucosylation activities (Streamer et al. 1975).

An exo-1,4-β-glucanase was also isolated from culture solution of *S. pulverulentum* (*P. chrysosporium*) grown on cellulose powder as sole carbon source. This enzyme has been extensively purified and characterized with respect to some physicochemical properties (Eriksson and Pettersson 1975b). The enzyme purification was carried out in five steps, and the purified enzyme was found to be homogeneous by analytical polyacrylamide electrophoresis on dodecyl sulfate gels and by analytical isoelectric focusing. A single symmetrical peak was obtained using the free-zone electrophoresis method. The purification factor was about 15 and the yield of enzyme activity around 7%. The exoglucanase was isoelectric at pH 4.3.

Table 2.5. Amino acid and carbohydrate compositions of the five endo-1,4-β-glucanase enzymes from culture solution of the fungus *S. pulverulentum* (Eriksson and Pettersson 1975a)

Amino acid	Number of residues in				
	T_1	T_{2a}	T_{2b}	T_{3a}	T_{3b}
Lysine	5	7	5	7	7
Histidine	4	5	2	4	5
Arginine	5	6	2	6	5
Aspartic acid	29	36	25	36	37
Threonine	27	33	28	35	32
Serine	31	42	32	39	34
Glutamic acid	22	29	21	29	31
Proline	21	26	16	24	23
Glycine	28	36	31	35	39
Alanine	27	35	20	31	32
Half-cysteine	1	2	3	–	–
Valine	20	25	16	18	18
Methionine	1	2	1	6	5
Isoleucine	11	15	8	18	19
Leucine	21	27	12	23	25
Tyrosine	10	11	15	10	10
Phenylalanine	11	13	9	17	19
Tryptophan	4	5	6	4	4
Σ	278	355	251	342	345
Monosaccharides	21 [a]	0	14 [b]	11 [c]	5 [c]
Σ	299	355	265	353	350
Carbohydrate, %	10.5	0	7.8	4.7	2.2
Molecular weights calculated from amino acid composition (closest integer values)	32400	36800	28500	37600	37000

[a] 2 glucose and 19 mannose per enzyme molecule; number of determinations = 6.
[b] 5 mannose, 7 galactose, 1 glucose and 1 arabinose per enzyme molecule; number of determinations = 2.
[c] Carbohydrate composition has not been determined with sufficient accuracy.

The molecular weight, 48 600, was calculated on the basis of knowledge of partial specific volume, ultracentrifugation data and the amino acid composition. The enzyme contains no carbohydrate.

A functional characterization of the exoglucanase has been presented by Streamer et al. (1975). It was found that the purified enzyme does not cause any decrease in viscosity of carboxymethyl cellulose solutions. With cellodextrins as substrates, the products, glucose and cellobiose are released in the α-configuration. With dewaxed cotton and Avicel as substrates a strong synergistic effect between the five endoglucanases and the exo-glucanase was observed. However, if Walseth cellulose (Walseth 1952) – an acid swollen cellulose – was used as substrate, no synergism was observed. The results obtained at that time strongly supported the hypothesis that the endoglucanases act randomly over the cellulose chain, producing nonreducing termini on which the exoenzyme acts.

It has been found in recent studies that there is a homology between the exoglucanase from *S. pulverulentum* and cellobiohydrolase I (CBH I) from *Trichoderma reesei* (Fägerstam et al. 1984). In these studies it also has been found that an additional exoglucanase is produced by *S. pulverulentum*. This enzyme has strong similarities with cellobiohydrolase II (CBH II) from *T. reesei* (Fägerstam and Pettersson 1980, G. Pettersson, personal communication).

Enzymes of the β-glucosidase type hydrolyze compounds containing β-D-glucoside linkages by splitting off the terminal β-D-glucose residue. *S. pulverulentum* produces two different 1,4-β-glucosidases. Cellobiose or cellulose are necessary substrates to cause induction of the enzymes. However, with cellobiose as the sole carbon source, only cell wall-bound enzymes are produced. For these enzymes to appear in the culture solution cellulose seems to be a necessary carbon source.

The molecular weights of the enzymes, determined by dodecyl sulfate polyacrylamide gel electrophoresis, vary between 165 000 and 182 000 (Deshpande et al. 1978). These findings are in line with the suggestion that fungal extracellular enzymes, cellulases, mannanases, and xylanases, i.e., those capable of hydrolyzing insoluble high-molecular weight polysaccharides, are smaller molecules than the corresponding β-glucosidases hydrolyzing water-soluble oligosaccharides (Ahlgren and Eriksson 1967).

The kinetic constants K_m and K_i have been determined for both the β-glucosidase in the culture solution and for the cell-bound β-glucosidase activity. K_m for cellobiose was for both enzyme types higher than the corresponding K_m obtained for p-nitro-phenyl-β-glucoside. K_m/K_i of the free enzymes for gluconolactone is approximately 2500 to 13 000 with cellobiose as substrate. The inhibition of β-glucosidases by gluconolactone is an important factor in the regulation of cellulose hydrolysis. This lactone is produced both by oxidation of glucose by glucose-1-oxidase and by splitting cellobionic acid, formed by the two oxidative enzymes cellobiose dehydrogenase (quinone) and cellobiose oxidase, into glucose and gluconic acid (gluconolactone).

In 1974, Westermark and Eriksson (1974a, b, 1975) discovered the enzyme cellobiose dehydrogenase (quinone) in culture filtrates of *Polyporus versicolor* and *S. pulverulentum* (*Phanerochaete chrysosporium*). The enzyme (EC 1.1.5.1) contains an FAD-group and catalyzes the oxidation of cellobiose and higher oligosaccharides up to cellopentose to their corresponding glyconic acid. Lignin degradation products, such as quinones and probably also phenoxy radicals, act as electron acceptors. The enzyme can be considered as a link between the degradation of lignin and cellulose in white-rot fungi (Fig. 2.9; Eriksson 1978). For cellulose degradation, the enzyme seems to be particularly important in the regulation of the cellobiose concentration, cellobiose being an important competitive inhibitor of the endo- and exoglucanases. In lignin degradation, as is demonstrated in Chap. 4, cellobiose dehydrogenase seems to participate in the reduction of phenoxy radicals, and thus prevents repolymerization of the lignin degradation products formed by oxidative enzymes. It also reduces quinones formed by phenoloxidases. The quinone reduction

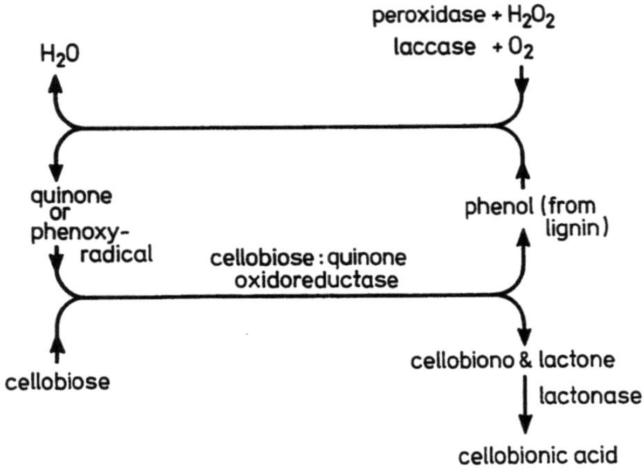

Fig. 2.11. Proposed reaction scheme for the enzyme cellobiose : quinone oxidoreductase (cellobiose dehydrogenase) (Westermark and Eriksson 1974b)

Table 2.6. Degradation of cotton cellulose by enzymes from *S. pulverulentum* (Eriksson et al. 1974)

Tube number	Enzyme preparation	Cellulose degradation weight loss (%)
1	Concentrated culture solution	51.1 (Oxygen atmosphere)
2	Concentrated culture solution	21.5 (Nitrogen atmosphere)
3	Mixture of endo- and exoglucanases	20.0
4	Endo-1,4-β-glucanase	0.0
5	Exo-1,4-β-glucanase	0.0

seems to be an important feature of cellobiose dehydrogenase since quinones have been shown to be very toxic to the organism and inhibitory to its enzymes (Buswell et al. 1979b). The enzyme cellobiose dehydrogenase has subsequently been found in several other white-rot fungi as well as in species of Ascomycetes and Fungi Imperfecti (Westermark and Eriksson 1988). The reaction mechanisms of the enzyme are presented in Fig. 2.11.

In addition to the enzyme cellobiose dehydrogenase, another oxidative enzyme of importance for cellulose degradation was discovered by Eriksson et al. (1974). They demonstrated that in vitro degradation of cotton cellulose in cell-free culture solution of *S. pulverulentum* occurs twice as fast in an O_2 compared to N_2 atmosphere (Table 2.6). The enzyme was later purified and characterized by Ayers et al. (1978) as cellobiose oxidase. It oxidizes cellobiose and higher cellodextrins to their corresponding glyconic acids using molecular oxygen. The enzyme is a heme protein but also contains an FAD-group. The final deposition of the second oxygen atom consumed in this reaction has not been determined. However, the superoxide anion has been detected (Ayers and

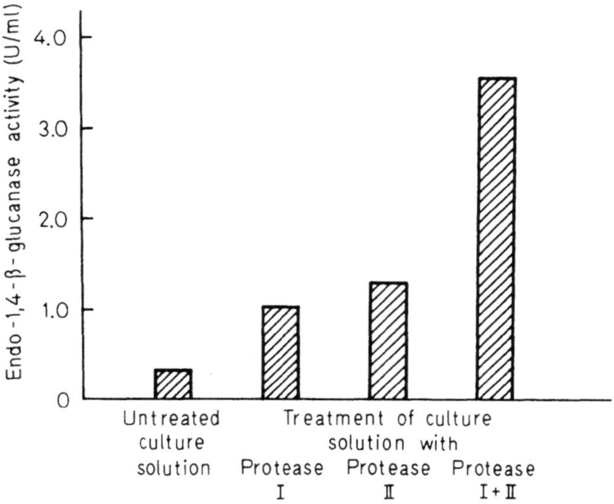

Fig. 2.12. Activation of endo-1,4-β-glucanases from *S. pulverulentum* after treatment with protease *I* and protease *II* (Eriksson and Pettersson 1982)

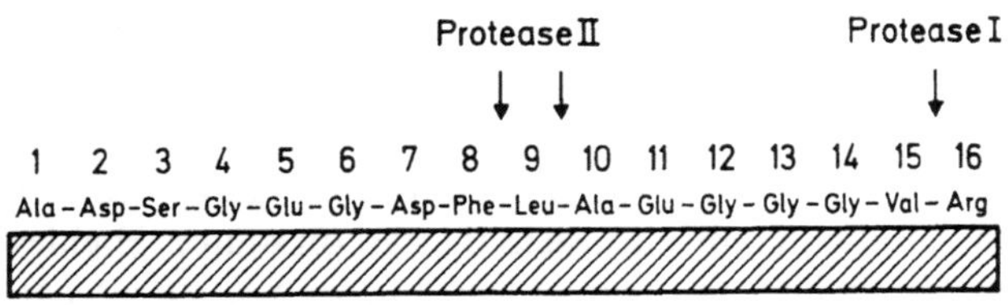

Fig. 2.13. Point of attack on human fibrinopeptide A by protease *I* and protease *II* (Eriksson and Pettersson 1982)

Eriksson 1982). Assignment of an EC number and final name of the enzyme awaits clarification of the reaction products.

Two acidic proteases have been demonstrated to be of importance for degradation of cellulose by *S. pulverulentum* (*P. chrysosporium*) since the proteases increase the activity of the endoglucanases. The two proteases have been purified and partly characterized (Eriksson and Pettersson 1982). If a culture solution of *S. pulverulentum*, grown on cellulose and thus containing endoglucanases, is treated with the individual proteases or a mixture of both, the endoglucanase activity in the solution increases. Treatment with both proteases gives a tenfold increase of the endoglucanase activity, particularly if the treated culture solution is harvested at an early stage of cultivation (Fig. 2.12). The specificities of protease I and II were investigated using human fibrinopeptide A as substrate. Protease I splits off arginine in the C-terminal position, while protease II splits at the carboxyl side of both Phe-8 and Leu-9 in fibrinopeptide A (Fig. 2.13). The possible explanation of the effect of the proteases is that the endoglucanases are activated or that the proteases destroy an endoglucanase inhibitor of protein origin. The *S. pulverulentum* proteases can also increase the activity of endoglucanases other than its own. It was thus demonstrated that the proteases from *S. pulverulentum* can double the endoglucanase activity of certain commercial cellulases (Eriksson and Pettersson 1982).

2.5.2 Brown-Rot Fungi

In wood decayed by brown-rot fungi the hemicellulose fraction is virtually absent and the degree of polymerization (DP) of the cellulose is drastically reduced (Cowling 1961, Rypácek 1975). The attack on the lignin, however, seems to be limited to a removal of part of the methoxyl groups (Ander and Eriksson 1978, Crawford 1981 b).

Brown-rot fungi, which rapidly depolymerize cellulose during early stages of wood decay, seem to utilize mechanisms for cellulose degradation different from those operating in *S. pulverulentum* (*P. chrysosporium*) and *T. reesei*. Brown-rot fungi produce endo-1,4-β-glucanases but no reports exist on their production of exo-1,4-β-glucanases (Highley 1973). Thus, brown-rot fungi do not seem to degrade crystalline cellulose by a synergistic action between endo- and exoglucanases as is the case in *T. reesei* and in *S. pulverulentum*. While brown-rot fungi depolymerize cellulose rapidly during early stages of wood decay, white-rot fungi depolymerize cellulose more slowly and utilize the degradation products simultaneously. There are also striking differences in the induction and repression of cellulases in brown- and white-rot fungi. Brown-rot fungi produce abundant amounts of cellulases with only glucose as the carbon source (Highley 1973). This is not the case with white-rot fungi, at least not with *S. pulverulentum*, where cellulases are induced by cellobiose and repressed by small amounts of glucose in the culture medium (Eriksson and Hamp 1978).

Brown-rot fungi apparently lack the synergistic endo-exo-glucanase cooperation to degrade crystalline cellulose. However, no other enzyme systems are known to substitute for these effects. Other explanations must therefore be sought to clarify the mechanism of cellulose degradation by this group of fungi. Koenigs was probably the first to suggest that brown-rot fungi oxidize cellulose (Koenigs 1974 a). He found that brown-rot fungi are more powerful producers of H_2O_2 than white-rot fungi (Koenigs 1974 b), and he therefore suggested that the initial attack on crystalline cellulose by brown-rot fungi is via an H_2O_2/Fe^{2+} system (Koenigs 1974 a).

The eventual involvement of H_2O_2 in degradation of cellulose by brown-rot fungi has been thoroughly studied by Highley (Highley 1980, 1982 a, 1987 b, Highley and Murmanis 1985). In the first of these papers (Highley 1980), evidence was obtained that H_2O_2 might be involved in cellulose degradation by *Poria placenta* since a decreased cellulose degradation was the result of the addition of $\cdot OH$ and H_2O_2 quenching agents. However, in a later work (Highley 1982 a) it was found that only one brown-rot fungus of six studied produced significant amounts of extracellular H_2O_2. However, several white-rot fungi produce H_2O_2 extracellularly. Peroxidase or catalase that might destroy H_2O_2 were not found in filtrates of the brown-rot cultures, nor did culture filtrates from brown-rot fungi incubated with small quantities of H_2O_2 destroy it. Furthermore, in the most studied brown-rot fungus *P. placenta*, compounds affecting the $H_2O_2/\cdot OH$-system did not affect degradation of wood by this organism. Thus, these results strongly question involvement of extracellularly produced H_2O_2 in cellulose depolymerization by brown-rot fungi. In a later work, to gain a better understanding of the role of H_2O_2 in wood degradation by *P. placenta* and the white-rot fungus *Coriolus versicolor*, Highley and Murmanis (1985) used diaminobenzidine (DAB) to cytochemically localize H_2O_2 during degradation of hemlock wood by these two fungi. No H_2O_2 was detected with DAB inside or outside the cells of *P. placenta*. However, with *C. versicolor* large amounts of H_2O_2 were found predominantly in the periplasmic space, the cytoplasm, and hyphal sheath. In a recent paper Highley (1987b) describes how low concentrations of nitrogen and carbohydrates stimulate H_2O_2 production in brown-rot fungi. However, similar conditions had

Table 2.7. Effect of carbohydrate source and concentration on H_2O_2[a] formation by white- and brown-rot fungi (Highley 1987b)

Fungus	Coloration of medium[b]/Carbohydrate source						
	None	0.1% Cellobiose	0.1% Glucose	1.0% Glucose	0.1% Mannose	0.1% Xylose	0.2% Malt extract
Brown-rot fungi							
L. arizonia	0	0	0	0	0	0	0
L. olviascens	0	2	0	0	0	3	0
C. puteana	4	4	4	0	4	4	4
G. trabeum	0	0	0	0	0	0	0
L. lepideus	2	2	4	0	1	1	1
P. carbonica	4	4	3	0	4	4	4
P. xantha	4	4	3	0	4	0	1
A. serialis	4	4	4	2	4	4	4
F. vailliantii	1	0	0	0	0	0	0
S. incrassata	0	0	2	0	4	2	2
L. sulfureus	2	2	3	3	3	3	2
F. meliae	0	0	0	0	2	1	0
P. placenta	0	3	3	0	3	3	3
White-rot fungi							
P. chrysosporium (ME-446)	0	1	0	0	0	0	0
P. chrysosporium (ME-461)	3	3	3	3	3	3	3
G. applanatum (MAD-708)	0	0	0	0	0	0	0
G. applanatum (7823-S)	2	4	4	4	4	4	4
X. frustulatus	0	1	0	0	0	0	0
C. versicolor	4	4	4	4	4	3	4

[a] H_2O_2 detected with ABTS.
[b] O = no coloration; 1 = very slight coloration; 2 = slight coloration; 3 = moderate coloration; 4 = heavy coloration.

little effect on H_2O_2 production by white-rot fungi. In Table 2.7 the effect of different carbohydrates on H_2O_2 formation by brown- and white-rot fungi is presented.

In his thesis, Cobb (1982) was able to demonstrate clearly that a nonenzymatic cellulose decay mechanism exists in brown-rot fungi. Using an ultrafiltration membrane which was placed between the fungal culture and the cellulose substrate, and which did not allow passage of enzymes, it was demonstrated that the fungal culture grown on hemicellulose monosaccharides decomposed the cellulose substrate more than a fungal culture grown on ball-milled pine.

Several other features, unique to brown-rot fungi in their cellulose degradation, deserve to be mentioned. It was thus found by Highley (1977) that the brown-rot fungus *P. placenta* was unable to utilize cellulose unless it was placed in contact with wood. Nilsson (1974a) found no morphological changes in cotton fibers placed between wood pieces inoculated with a brown-rot fungus. However, no tensile strength was left in the cotton fibers, thus indicating an attack by the fungus. All these studies suggest that when cultured on wood, brown-rot fungi produce a factor able to diffuse into the pure cellulose and cause depolymerization. Highley (1977) also found that when other carbohydrate supplements such as cellobiose, glucose, or mannose were added to cellulose, this resulted in a depolymerization of cellulose by *P. placenta*. Taking all this information into account, it seems as if a compound present in wood or low molecular weight sugars such as glucose or mannose, must be present in order to induce the brown-rotters to begin their cellulose degradation. Low-molecular weight sugars serve as substrates for oxidizing enzymes whereby H_2O_2 is produced.

The conclusion from the above work must be that a nonenzymatic mechanism is involved in cellulose degradation by brown-rot fungi. However, most brown-rot fungi exhibit sparse activity on pure cellulose despite their rapid degradation of cellulose in wood. Cell-free enzyme preparations from brown-rot fungi are also inactive on insoluble cellulose but active on soluble cellulose such as cellulose derivatives. This indicates that brown-rot fungi display endoglucanase activity, a fact that has been demonstrated by several workers. However, the endoglucanase from one brown-rot fungus, namely, *P. placenta*, seems to exhibit some unusual properties. It was thus demonstrated by Highley and co-workers (Wolter et al. 1980, Highley et al. 1981, Highley and Wolter 1982) that the endoglucanase from *P. placenta*, is a multifunctional enzyme, active on both polysaccharides and glucosides. The enzyme hydrolyzes xylan, mannan, and carboxymethylcellulose in addition to several glucosides (Table 2.8). Liquid column chromatography separations using Sepharose 6-B, Sephadex 100, Sephadex 200, hydroxylapatite, DEAE-Bio-Gel A, Ultrogel AcA 34 as well as other standard procedures were unsuccessful in separating the various enzyme activities presented in Table 2.8. The enzyme complex has a molecular weight of 185 000. The carbohydrate degrading complex from *P. placenta* seems to be composed of an aggregate of polypeptides, which under the right conditions break into small subunits. The ability of the enzyme complex

Table 2.8. Polysaccharidase and glycosidase activities in crude, collodion membrane (75000 MW) concentrated and thin-layer isoelectric focusing (TLIEF) purified culture solution of *Poria placenta* (Wolter et al. 1980)

Substrate	Enzyme	Specific enzyme activity ($U\ mg^{-1}$ protein)		
		Crude extract dialyzed	Collodium concentrate	TLIEF eluate
Xylan	Xylanase	4.4	7.3	3.5
Glucomannan	Mannanase	2.5	6.0	1.8
Carboxymethylcellulose	Cellulase	1.4	2.1	0.8
β-D-Glucoside	β-D-Glucosidase	7.2	13.1	0.26
α-D-Glucoside	α-D-Glucosidase	1.3	5.8	0.095
β-D-Galactoside	β-D-Galactosidase	18.7	61.9	29.6
α-D-Galactoside	β-D-Galactosidase	25.7	61.9	35.1
β-D-Xyloside	β-D-Xylosidase	1.9	7.6	0.17

to degrade a large number of different carbohydrate substrates distinguishes it from other reported carbohydrases. However, few carbohydrate degrading enzymes have been purified from brown-rot fungi. Those that have are not very substrate specific. Thus a purified β-glucosidase from the brown-rot fungus *Lenzites trabea* also hydrolyzes β-xylosides (Herr et al. 1978a). The purified xylanase from the brown-rot fungus *Tyromyces palustris* exhibits CMCase activity but no glycosidase activity. The purified endoglycanase from *Polyporus schweinitzii* also showed both mannanase and xylanase activities (Keilich et al. 1969). The purified multifunctional enzyme from *P. placenta* appears to be even more nonspecific than the above mentioned enzymes since it can degrade both polysaccharides and glycosides.

In addition to the above described multifunctional properties of the *P. placenta* enzyme, it also exhibits other unusual traits. Thus, the synthesis of the complex carbohydrolase apparently is not controlled by catabolite repression since it is produced constitutively on glucose. Also, the carbohydrolase complex is isoelectric at a very low pH-value, 1.8, and the pH optimum is very low, between pH 3 and 4. It is also a very thermostable enzyme complex with the exception of the β-glucosidase activity. Some other activities were maintained after incubation at 70 °C for 1 h. Endoglucanase and xylanase activities were high over a broad pH range (Wolter et al. 1980, Highley et al. 1981, Highley and Wolter 1982).

From the above description of cellulose degradation by brown-rot fungi, it is obvious that much more work is needed until the exact mechanisms can be elucidated.

2.5.3 Soft-Rot Fungi

Soft-rot fungi have the capability to deplete both polysaccharides and lignin in wood. However, the polysaccharides are the main targets for this group of fungi while lignin is largely undigested. Patterns of the attack on wood by soft-rot fungi are presented in Table 1.5.

Numerous studies of cellulose degradation by soft-rot fungi have been carried out. Detailed descriptions of at least early studies can be found in most of the existing reviews on cellulases (Goksöyr and Eriksen 1980, Lee and Fan 1980, Mandels 1982, Enari 1983, Coughlan 1985, Eriksson and Wood 1985, Ljungdahl and Eriksson 1985, Fan et al. 1987). The ability of soft-rot fungi to produce all necessary extracellular enzymes for degradation of crystalline, highly-ordered cellulose varies. The most well-known of those producing a complete setup of enzymes is *Trichoderma viride* (Nisizawa et al. 1972, Ogawa and Toyama 1972), *T. reesei* (descendants of formerly *T. viride* QM6a and mutants of this strain are named *T. reesei* in honor of Elwyn T. Reese) (Mandels and Reese 1964, Gritzali and Brown 1979, Fägerstam and Pettersson 1980, Bhikhabhai et al. 1984). *Penicillium funiculosum* (Wood et al. 1980), *T. koningii* and *Fusarium solani* (Wood and Phillips 1969) have been used extensively for studies on the mechanisms of cellulase action (Mandels and Reese 1964, Wood 1971, Wood and Phillips 1969, Nisizawa et al. 1972, Ogawa and Toyama 1972, Wood and McCrae 1972, 1977a, 1978, Berghem and Pettersson 1973, Halliwell and Griffin 1973, Berghem et al. 1976) in this group of fungi. Fractionation studies on these cellulases are numerous and well accounted for in the extensive review literature (Goksöyr and Eriksen 1980, Lee and Fan 1980, Mandels 1982, Enari 1983, Coughlan 1985, Eriksson and Wood 1985, Ljungdahl and Eriksson 1985, Fan et al. 1987).

Characterization of these cellulases has shown that they are in many respects similar to the cellulases of white-rot fungi. The culture solutions contain endoglucanases, exoglucanases (normally cellobiohydrolases) and 1,4-β-glucosidases. In most other studied cell-free culture filtrates of soft-rot fungi, appreciable amounts of endoglucanases, variable amounts of β-glucosidases, but small amounts of exoglucanases have been found. Fungi producing an incomplete array of enzymes are capable of degrading only the more disordered, amorphous celluloses. Why only some soft-rot fungi produce a "complete enzyme complex" while others do not is not clear. However, it may be very important to understand the mechanisms of cellulose degradation by these fungi, since probably more than one mechanism of action operates in soft-rot fungi (Wood 1983).

The filamentous fungus *T. reesei* is one of the most important cellulose degrading organisms known (Mandels 1982, Montenecourt 1983, Eriksson and Wood 1985). It excretes large amounts of different cellulases, which combined, solubilize crystalline cellulose efficiently. The number of enzymes excreted and their individual roles in the solubilization of cellulose are still not clear. The picture is complicated due to the use of different strains and different cultivation conditions, resulting in variable expressions of secreted proteins. The existence of structural heterogeneity among several of the cellulases further complicates the picture. An evaluation of the rich flora of published papers on *Trichoderma* cellulases gives the following picture of the composition of the three different types of enzymes involved in cellulose saccharification. This picture has been reasonably clear only during the last couple of years:

1. Three endo-1,4-β-glucanases (EC 3.2.1.4), EGI-III (Ståhlberg et al. 1988).
2. Two enzymes characterized mainly as exo-1,4-β-glucanases (EC 3.2.1.91) (cellobiohydrolases, CBH-I and CBH-II) (Enari and Niku-Paavola 1987, Teeri et al. 1987).
3. 1,4-β-glucosidases (EC 3.2.1.21) (Enari et al. 1981, Enari and Niku-Paavola 1987).

Particularly two of the endoglucanases and the two enzymes characterized mainly as cellobiohydrolases are well characterized both physicochemically and functionally. For the two endoglucanases and the cellobiohydrolases, their corresponding genes have also been isolated and their complete nucleotide sequences determined, EG-I (Penttilä et al. 1986) and EG-III (Saloheimo et al. 1988), CBH-I (Shoemaker et al. 1983a, Teeri et al. 1983, Fägerstam et al. 1984), and CBH-II (Teeri et al. 1987).

The major endoglucanase component, EG-I, (named endo-II by Håkansson et al. 1979) represents approximately 5–10% of the secreted protein in *T. reesei* QM 9414 cultures. This enzyme has been purified and characterized in several laboratories (Håkansson et al. 1979, Shoemaker et al. 1983b, Bhikhabhai et al. 1984, Henrissat et al. 1985, Niku-Paavola et al. 1985).

The M_r of the EG-I glycoprotein has been reported to be 55 000 (Bhikhabhai et al. 1984), 54 000 (Shoemaker et al. 1983b), 50 000 (Henrissat et al. 1985), and 43 000 (Niku-Paavola et al. 1985). The N-terminal sequence of EG-I has been determined (Shoemaker et al. 1983b, Bhikhabhai and Pettersson 1984) and also some sequences of internal peptides. The results obtained point to a considerable sequence homology between EG-I and CBH-I in spite of the clear differences in substrate specificity. Shoemaker et al. (1983b) found, on the other hand, that their tryptic maps are dissimilar. With the goal of clarifying the relationship between the CBH and EG classes of cellulases and to describe their roles in cellulose hydrolysis, Penttilä et al. (1986) isolated the chromosomal egl-1 gene and its full-length cDNA copy. These isolates were compared with the cbh-I gene sequence to clarify the relationship between the EG and CBH classes of cellulases. The deduced 437 amino acids (aa) long EG-I protein with a 22-aa long signal peptide is to 45% identical in the aa sequence with CBH-I. The best conserved region is found at the C-terminus and it shows about 70% homology. These data suggest that the two enzymes originate from a common ancestor by gene duplication. In spite of this the intron positions have not been conserved in these genes which both contain two short intrones. It is presumed that the deduced EG-I sequence contains six N-glycosylation sites and a presumed *O*-glycosylated region appears near the C-terminus. This region is very similar to that at the C-terminus of CBH-I. A comparison of the amino acid sequences between EG-I and the

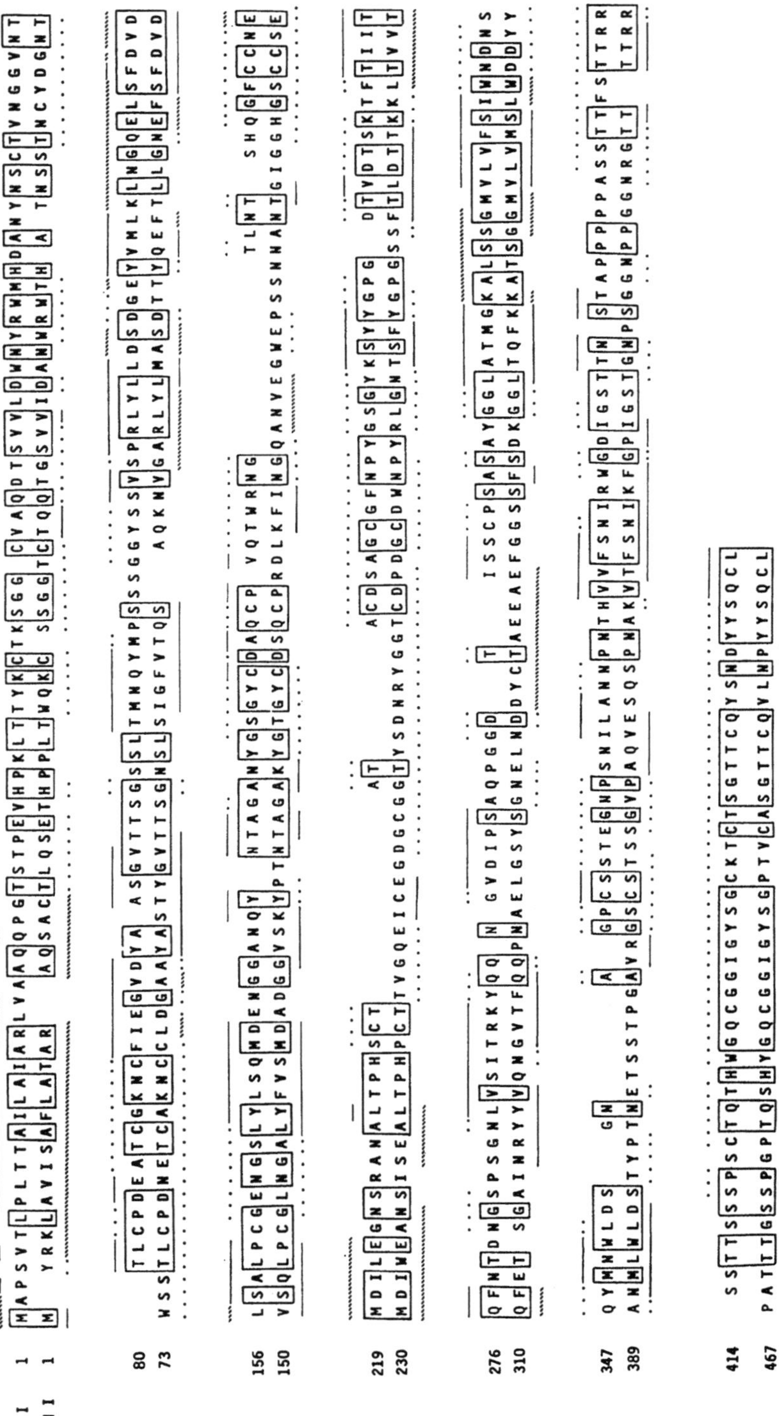

Fig. 2.14. Homology of EG-I and CBH-I aa sequences (Penttilä et al. 1986)

CBH-I suggests that the evolutionary divergences have involved four separate 10–20 amino acid deletions from the ancestral protein. The homology of EG-I and CBH-I aa sequences is given in Fig. 2.14.

While the number of cellobiohydrolases produced by *T. reesei* (CBH-I and CBH-II) is well defined, the number of endoglucanases produced is less clear. Depending on growth conditions and methods of isolation, different groups have identified a range of endoglucanases from *T. reesei* culture filtrates (Shoemaker and Brown 1978a, Bhikhabhai et al. 1984, Odegaard et al. 1984, Sheir-Neiss and Montenecourt 1984). As described above, the gene for EG-I has been sequenced (Penttilä et al. 1986, van Arsdell et al. 1987), and sections of the protein have also been sequenced (Bhikhabhai and Pettersson 1984).

A new endoglucanase from *T. reesei* EG-III has recently been purified, and its catalytic properties studied (Saloheimo et al. 1988). EG-III is a 49.8 kDa protein. As almost all *Trichoderma* cellulases characterized so far, it is a glycoprotein (Bhikhabhai et al. 1984, Beldmann et al. 1985). The total sugar content of the enzyme was estimated to be an average of 47 mannose equivalents per protein molecule. This corresponds to 15% w/w of the total mass, 49.8 kDa.

A remarkable sequence homology can be detected between the N-terminus of EG-III and the N- or C-terminal regions of three other well-characterized *T. reesei* cellulases (Shoemaker et al. 1983a, Penttilä et al. 1986, Teeri et al. 1987; Fig. 2.15). Two homologous regions named A and B, of about 35 amino acids, are present at the N-terminus of EG-III and CBH-II and in the C-terminus of CBH-I and EG-I. However, outside the relatively short terminal A and

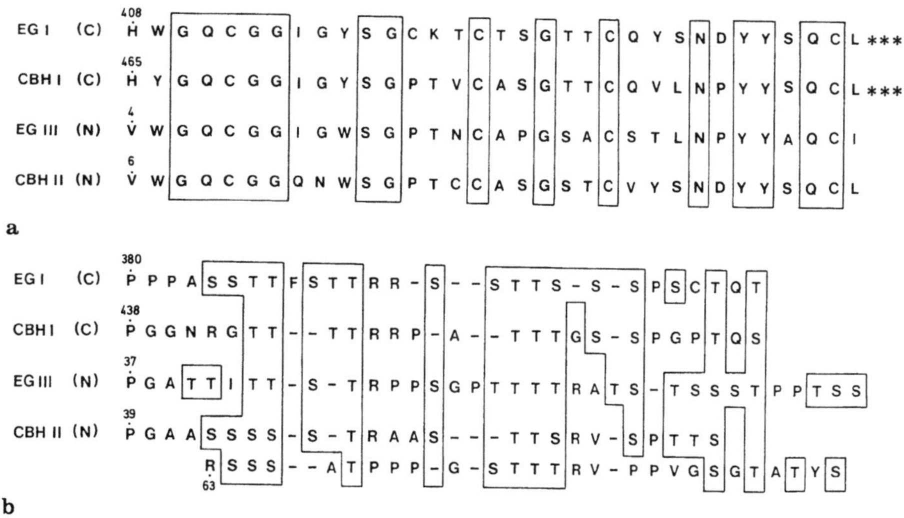

Fig. 2.15. Amino acid sequence homology between four different cellulases of *T. reesei* (**a**). The conserved terminal domain blocks A. The residues conserved in all four proteins are *boxed*. The *three asterisks* indicate a stop codon. **b** The putative O-glycosylated domains blocks B and B′ (Fig. 2.16). All Ser and Thr residues are *boxed* (Teeri et al. 1987)

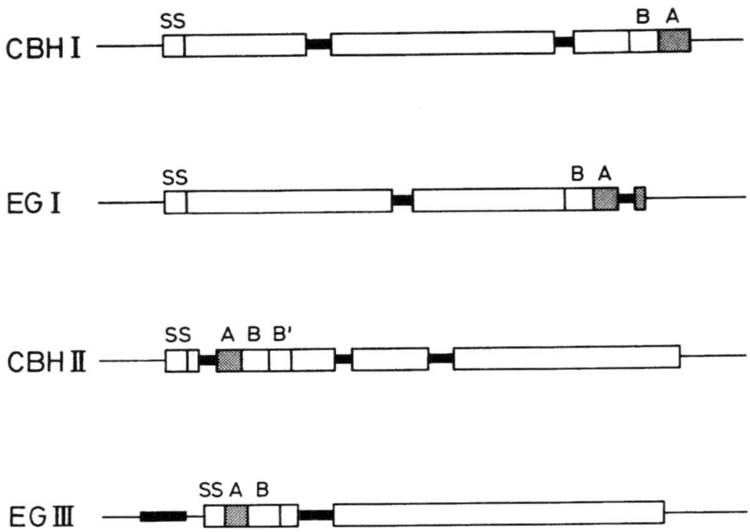

Fig. 2.16. Structural organization of the *T. reesei* genes coding for cellulolytic enzymes. Exons are indicated by *open* or *filled boxes* and introns by *solid bars*. The homologous regions common to all the four cellulases are denoted by *A*, *B*, and *B'*. The signal sequence regions are denoted by *SS* above the boxes. Sequence data are from: *CBH-I* (Shoemaker et al. 1983a), *EG-I* (Penttilä et al. 1986), *CBH-II* (Teeri et al. 1987), *EG-III* (Teeri et al. unpublished, Teeri et al. 1987)

B blocks, EG-III shows very little, if any, homology to the other *T. reesei* cellulases. The structural organization of the *T. reesei* genes coding for the four cellulolytic enzymes is outlined schematically in Fig. 2.16.

The homology of block A of EG-III is approximately 70% to the same regions in the other cellulases. The homology in the B-blocks is 50% – 60%. The striking similarity of these regions in the four proteins suggest that they most likely have a functional significance. These A-B-blocks fold to a structure which can be removed from both CBH-I and CBH-II by limited proteolyses. The core enzymes, remaining after the proteolysis, have a significantly decreased affinity and activity toward microcrystalline cellulose. However, the activity on soluble substrates is retained (van Tilbeurgh et al. 1986, Tomme et al. 1988). This type of domain structure is found in all four of the sequenced *Trichoderma* cellulases. An analogous overall structure is found also in the cellulases of *Cellulomonas fimi* and *Clostridium thermocellum* (Joliff et al. 1986, Warren et al. 1987).

It is possible that proteolytic removal of terminal homologous regions could serve as an in vivo mechanism to modify the properties of cellulases during the course of hydrolysis. The long-chain substrates are gradually changed to shorter, more accessible and water-soluble substrates. The basidiomycete, *Schizophyllum commune*, has been shown to produce two forms of endoglucanases, EG-I and EG-II (Paice et al. 1984). EG-II is proposed to be created from EG-I by a proteolytic cleavage at the N-terminus. The *T. reesei*

EG-III sequence shows much homology with the *S. commune* EG-I protein (Yaguchi, personal communications). However, the *S. commune* endoglucanases do not have the amino acid blocks A and B common to all *T. reesei* cellulases sequenced so far (Teeri et al. 1987).

The idea of proteolytic modification of the *T. reesei* cellulases is supported by the amino acid composition data of Shoemaker and Brown (1978b) who isolated both an EG-III-like enzyme and what they presumed to be a proteolytic cleavage product. This contained less carbohydrate, and its amino acid composition is well explained if removal of the terminal homologous sequences A and B are presumed.

The theory that proteolytic activity in the culture filtrates contributes to the polydispersity of endoglucanases has recently been proved to be true by Ståhlberg et al. (1988). From the culture filtrate of *T. reesei* they isolated a novel endo-glucanase (38 kDa) which was shown to be derived from EG-III (50 kDa). Through proteolysis EG-III has lost an N-terminal domain of 61 amino acids, corresponding to the A-region and most of the B-region common to the four most studied *T. reesei* cellulases. These regions contain 80% of the carbohydrate present in intact EG-III. Two endoglucanases with molecular masses of 37 and 49,5 kDa have been isolated and characterized earlier by Gritzali and Brown (1979). These two enzymes had similar amino acid compositions, and it seems likely that they isolated the EG-III and EG-III core proteins. The core protein, designated EG-III core, is fully active on soluble substrates such as carboxymethyl cellulose, while both the activity against and adsorption on microcrystalline cellulose (Avicel) is markedly decreased. Sedimentation velocity experiments revealed that the loss of the A-B-region produces a core protein that is considerably more compact than the native enzyme, suggesting that the A-B-domain constitutes a protruding part, which largely contributes to the pronounced asymmetry of the intact EG-III enzyme. An asymmetric shape has also been demonstrated for the major cellobiohydrolase, CBH-I (Schmuck et al. 1986). It is likely that the protruding A-B-region is responsible for the asymmetrical shape also in this protein. Based on the analysis of amino acid sequence data and on the low-angle X-ray scattering studies of Schmuck et al. (1986), a hypothetical model of a cellulolytic enzyme was contructed by Knowles et al. (1987; Fig. 2.17). It is suggested that cellulases consist of a catalytic domain probably linked via a flexible "hinge" region to a "tail" domain that is most likely responsible for disrupting the structure of the semicrystalline substrate.

It seems clear that the A-B-region increases the binding affinity to cellulose, thereby increasing the activity toward insoluble substrates. It has been demonstrated earlier by van Tilbeurgh et al. (1986) and Tomme et al. (1988) that cellobiohydrolyses from *T. reesei* have a special adsorption domain binding the protein to cellulose and thus increasing the activity toward the substrate. The results obtained by Ståhlberg et al. (1988) show that this is also the case with EG-III. It thus seems possible that all *Trichoderma* cellulases may have the same functional organization with a domain specialized in binding to insoluble cellulose.

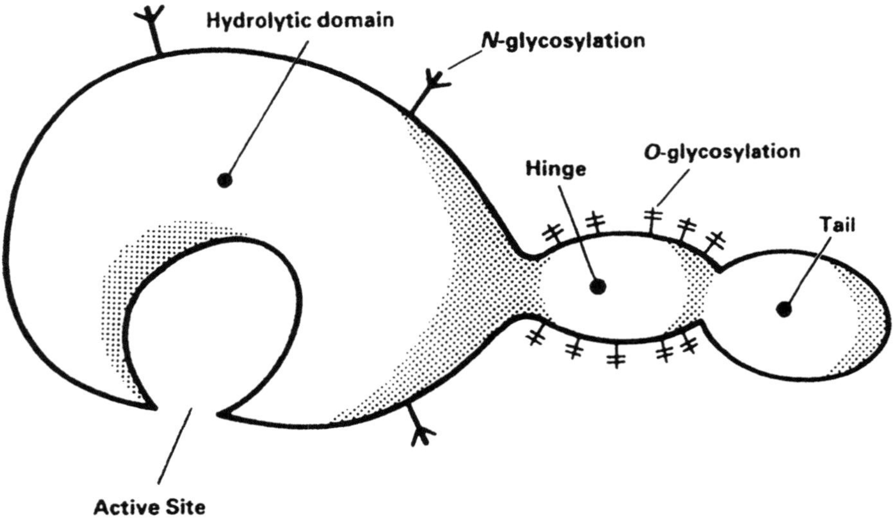

Fig. 2.17. A hypothetical model of a cellulolytic enzyme. This model is based on the analysis of amino acid sequence data and on low-angle X-ray diffraction studies (Knowles et al. 1987)

The importance of the binding of cellulases to cellulose is clearly demonstrated in saccharification experiments. In saccharification of steam-exploded wheat straw approximately 40% of the added enzymes are free in solution after 24 h of hydrolysis, while 60% of the enzymes are adsorbed on the solid residue (Deshpande and Eriksson 1984, Vallander and Eriksson 1987). It seems (Vallander and Eriksson, to be published) as if the enzymes binding to the solid residue are much more efficient in hydrolyzing the solid substrate, and therefore are more important to recirculate and reutilize than the enzymes free in solution. The technique to also reutilize the adsorbed enzymes has been developed by Vallander and Eriksson (Swedish patent application, 1987). It involves mixing of new substrate with the enzyme-rich residue from earlier hydrolysis. By recirculating the enzyme-rich residue in a specific way, a very high percentage of the enzymes can be reutilized. Recirculation of enzymes in solution only, gives a too low degree of reutilization to give an economically feasible process.

While the binding domain is composed of two clearly distinguishable blocks, A and B, the main part of the enzyme is the core protein containing the active site. The specificity of EG-III has been studied using cellodextrins, containing 2–5 glucose units, and their chromophoric glycosides as shown in Fig. 2.18. The cleavage sites of the underived cellodextrins, cellotriose, and cellopentose could not be unequivocally determined in contrast to the chromophoric substrates. The K_m values of the 4-methylumbelliferyl derivatives were in the 10–60 μM range, while those for EG-I toward the same substrates were in the mM range. The product patterns for EG-I were of course different from those of EG-III (van Tilbeurgh, unpublished results). The turn-

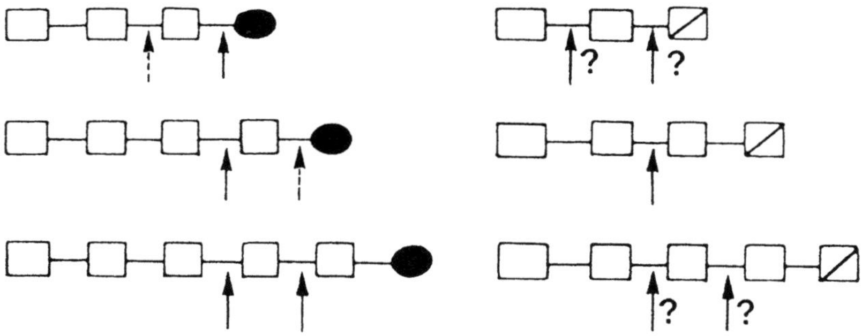

Fig. 2.18. Substrate specificity of EG-III against 4-methylumbelliferyl-derived and underived cellodextrins. The cutting points of cellotriose and cellopentaose could not be unequivocally determined. Symbols: □, *β*-glucopyranosyl (-1,4); ●, 4-methylumbelliferyl; ▨, *β*-glucopyranose (reducing end); *arrows*, main hydrolysis sites; *dashed arrows*, secondary hydrolysis sites (Saloheimo et al. 1988)

over numbers for EG-III, 10–200/min, are 50–200 times lower than for EG-I. EG-III does not hydrolyze cellobiosides and lactosides in contrast to both EG-I and CBH-I. According to van Tilbeurgh et al. (1985), the specificity of the EG-III therefore shows some similarity to that of CBH-II. This should be considered in view of the fact that both EG-III and CBH-II have their homologous domains at the N-termini in contrast to CBH-I and EG-I, which have similar sequences at the C-termini. In contrast to EG-I, no transferase activity was observed for EG-III (van Tilbeurgh, unpublished results). Differences in transferase activities among the five endoglucanases isolated from *Sporotrichum pulverulentum* were also demonstrated by Streamer et al. (1975). Three of the five endoglucanases exhibited transferase activity, although of different capacity, while two of the endoglucanases and the exoglucanase did not show any such activity.

In addition to EG-I and EG-III, another endoglucanase with a low molecular mass of 20 kDa has also been isolated from *T. reesei* culture solution (Håkansson et al. 1978). The purification of the enzyme was carried out by a two-step procedure of gel filtration and ion exchange chromatography. The enzyme is isoelectric at pH 7.5 and the purified enzyme was not a glycoprotein. A molecular weight of 20 200 was determined for the endoglucanase by chromatography in 6 M guanidine HCL. This value is in good agreement with that obtained by sedimentation equilibrium centrifugation 20 400.

In addition to the above described endoglucanases, *T. reesei* produces two different enzymes first classified as exoenzymes, cellobiohydrolases (EC 3.2.1.91). However, later more and more of properties generally assigned to endoglucanases have been found to be exhibited also by these enzymes. These activities were shown to be displayed by at least two structurally unrelated enzymes (Fägerstam and Pettersson 1979, 1980, Teeri et al. 1987). CBH-II was first depicted as an enzyme immunologically distinct from CBH-I, but also producing cellobiose from cellulose.

Under optimal culturing conditions, the most abundant cellobiohydrolase, CBH-I, constitutes about 60% of the protein released into the culture medium (Montenecourt et al. 1980). However, the specific activity of CBH-II, as measured by the formation of the reducing sugars from Avicel, is about twice as high as that of CBH-I (Pettersson et al. 1981, Henrissat et al. 1985). CBH-I has been extensively studied with respect to its structural properties. The gene coding for this enzyme was isolated and characterized by Teeri et al. (1983) and Shoemaker et al. (1983a). The enzyme is a glycoprotein consisting of a single polypeptide chain (Berghem et al. 1975, Gum and Brown 1976). Except for the heavily glycosylated regions, the CBH-I protein (65 kDa) has been sequenced to near completion (Fägerstam et al. 1984).

The two cellobiohydrolases, CBH-I and CBH-II, lack any apparent homology in their amino acid sequences (Fägerstam and Pettersson 1979, Gritzali and Brown 1979, Chanzy et al. 1983, van Tilbeurgh et al. 1984, Niku-Paavola et al. 1986) except for a conserved region present at their C- and N-termini, respectively (Figs. 2.15 and 2.16; Teeri et al. 1987). Limited action of papain on the two cellobiohydrolases from *T. reesei* (CBH-I, 65 kDa, and CBH-II, 58 kDa) has led to the isolation of the respective core fragments (56 and 45 kDa). These core fragments were fully active on low molecular mass water-soluble substrates but have a strongly reduced activity (10 and 50%, respectively, of the initial value) on microcrystalline cellulose (Avicel).

The specific activities of both the intact enzymes and their cores on crystalline and amorphous cellulose differentiate the CBH-I and CBH-II in terms of adsorption and catalytic properties. Distinct functions can be attributed to the terminal peptides. In intact CBH-II, the N-terminal region, containing the A-block (Fig. 2.16) is involved in the binding of the enzyme both to amorphous and crystalline cellulose. However, the homologous C-terminal peptide in CBH-I only affects the interaction with microcrystalline cellulose. It is now known that the A-block is responsible for the binding to cellulose and also that this block shows a very strong homology in all *Trichoderma* cellulases. This is also the case with the cellulases from *S. pulverulentum* (G. Pettersson, personal communications). The fact that CBH-I binds preferentially to crystalline regions while CBH-II binds both to crystalline and amorphous regions is therefore most likely a reflection of differences in the active surfaces of the enzymes.

Although CBH-I and CBH-II from *T. reesei* QM9414 show common features, such as binding to the same affinity column (van Tilbeurgh et al. 1984) and high activity toward microcrystalline cellulose (Fägerstam and Pettersson 1980), the modes of action of the two enzymes and the configuration of their active sites seem to differ considerably. Whereas CBH-I can be eluted from a biospecific adsorbant by either cellobiose or lactose, CBH-II can be released only by cellobiose (van Tilbeurgh et al. 1984). The reaction patterns toward 4-methylumbellyferyl-β-D-glucosides, MeUmb(Glc)$_4$ and MeUmb(Glc)$_5$, are totally different for the two enzymes. With MeUmb(Glc)$_5$ as substrate no MeUmb(Glc)$_3$ can be observed as a reaction product from CBH-I hydrolysis (van Tilbeurgh et al. 1982) while it is the main product from CBH-II hydrolysis.

Features that contradict the characterization of CBH-I as a cellobiohydrolase (EC 3.2.1.91) are the following: A lack of specificity of CBH-I for the terminal glycosidic bonds in the hydrolysis of soluble cellodextrins (Gritzali and Brown 1979, van Tilbeurgh et al. 1982). CBH-I was shown by collodial gold labeling to bind randomly in an endo-type fashion to *Valonia* cellulose fibers (Chanzy et al. 1983, 1984). CBH-I alone hydrolyzes both cotton (Nummi et al. 1983) and *Valonia* cellulose (Chanzy et al. 1983) to water-soluble products, indicating that some endo-type activity is also included in its behavior (Enari and Niku-Paavola 1987).

Concerning CBH-II, activity on CMC, hydroxyethyl cellulose (HEC), or barley β-glucan has recently been obtained with purified CBH-II in contrast to earlier findings (Niku-Paavola et al. 1986, Biely and Markovic 1988). These results even led Kyriacou et al. (1987) and Biely and Marcovic (1988) to characterize CBH-II as a specific endoglucanase.

To reach a final conclusion regarding the functions of CBH-I and -II further results are needed. However, it deserves to be pointed out that because of the strong synergistic interactions between the different types of cellulases, characterization studies require pure enzymes since even traces of activities other than those expected can obscure the results.

The least studied of the enzymes involved in cellulose hydrolysis by *T. viride* (*T. reesei*) are the β-glucosidases. The reason may be that β-glucosidase activity in culture solutions of *T. reesei* represents only about 1% of the total protein secreted, while the main parts consist of cellobiohydrolase and endoglucanase (Enari et al. 1981). During the early stages of growth, the enzyme is mycelium-bound but is released into the culture medium after lysis of the mycelium, which occurs in the later stages of growth (Berg and Pettersson 1977). Kubicek (1981) found that the *T. reesei* cellulases and β-glucosidases are differently bound to the cell wall. Treatment of intact cell walls with an α-mannanase or trypsin released the endoglucanases, while treatment with laminarinase or chitinase released the β-glucosidases. Both enzymes were also released during autolysis in phosphate buffer. The results suggest that cleavage of chemical bonds of certain cell wall polymers of *T. reesei* is necessary until endoglucanases and β-glucosidases appear in the culture solution. Since β-glucan and chitin form the inner rigid structure of fungal cell walls, both results indicate that β-glucosidase is located within the intramural space and that some autolytic deformation or destruction of specific wall components is of great importance for the excretion of the β-glucosidase. Using ferritin-conjugated antibodies as an electron microscopic marker, Sprey (1986) found that *T. reesei* QM 9414 β-glucosidase was localized within the cell walls. At different stages of cell wall degradation, obtained with cell wall lysing culture filtrate from *Micromonospora calcea*, β-glucosidase was mainly detected within the outer, fibrous exopolysaccharide layer and the outer face of the plasma membrane.

Enari et al. (1981) purified the β-glucosidases of *T. reesei* and *Aspergillus niger* and studied their properties to obtain information necessary for the design of a technical process based on enzymatic hydrolysis of various cellulosic materials. The purified *β*-glucosidase component of *T. reesei* showed an apparent molecular weight of 35 000 on gel chromatography. The molecular weight calculated from SDS electrophoresis of the purified *Trichoderma β*-glucosidase was 73 000. The discrepancy is most likely due to the glycoprotein nature of the enzyme (0.10 mg sugar/mg protein). Glycoproteins are known to behave abnormally in SDS electrophoresis. Similar results have been reported for *T. viride β*-glucosidase. Using gel chromatography, Berghem and Pettersson (1974) estimated the molecular weight at 47 000, but Gong et al. (1977), using SDS electrophoresis, found it to be 76 000. In addition to the purified *β*-glucosidase from *T. reesei*, Enari et al. (1981) also observed another *β*-glucosidase component with a molecular weight of 130 000, but this component represented only 2% of the total *β*-glucosidase activity. Inglin et al. (1980) isolated an intracellular *β*-glucosidase from *T. reesei*. The enzyme had a molecular mass of 98 000, estimated by Sephadex G-150 chromatography. Its optimal activity was at pH 6.5 and the K_m values for sophorose and cellobiose were 6.7 and 3.3 mM, respectively.

Aspergillus species seem to produce *β*-glucosidases of higher molecular weight than does *T. reesei*. Thus, the purified *β*-glucosidase from *A. niger* has a molecular weight of 150 000 and consists of two polypeptide chains which can be separated by SDS treatment (Enari et al. 1981). *A. oryzae* has been reported to form an enzyme with a molecular weight of 218 000, which also consists of two polypeptide chains (Mega and Matsushima 1979).

For the enzymatic conversion of cellulose to glucose on a technical scale it is necessary to have all cellulolytic components at their optimal levels. Since the *β*-glucosidase activity in *T. reesei* cellulase preparations is suboptimal for the saccharification process (Sternberg 1976) it is essential to find a way of supplying additional *β*-glucosidase to such reactions. Many attempts have therefore been made to find alternative sources of *β*-glucosidase and also to maximize its production. Sternberg et al. (1977) found the black *Aspergillii* (*A. niger* and *A. phoenicis*) to be superior producers of *β*-glucosidase and presented a method for production of this enzyme. Thus, when *Trichoderma* cellulase preparations were supplemented with *β*-glucosidase from *Aspergillus* during practical saccharifications, glucose is the predominant product and the rate of saccharification is significantly increased (Vallander and Eriksson 1987).

Other fungal *β*-glucosidases, particularly those from thermophilic organisms, have been studied for saccharification purposes. Thermostable enzymes would be particularly useful if they could be produced in large quantities. The *β*-glucosidase from *Thermoascus aurantiacus* has pH and temperature optima of 5 and 70 °C, respectively (Tong et al. 1980). From *Sporotrichum thermophile* Meyer and Canevascini (1981) isolated two distinctly different *β*-glucosidases. One, with M_r 440 000, had only aryl-*β*-glucosidase activity, while the other, with M_r 40 000, had cellobiase activity with only small activity toward aryl-*β*-glucosides.

Interesting kinetics in this context have also been reported for *β*-glucosidase from *Botryodiplodia theobromae* (Umezurike 1971, 1975, 1979). Aging of this enzyme was found to lead to its dissociation from molecular weights of

350 000–380 000 to 45 000–47 000. Electron micrographs showed that the high molecular weight form of the enzyme was an octamer.

In the white-rot fungus *Sporotrichum pulverulentum*, two oxidizing enzymes, cellobiose: quinone oxidoreductase (Westermark and Eriksson 1974a, b, 1975) and cellobiose oxidase (Ayers et al. 1978) were found to participate in cellulose degradation. No oxidative enzyme has so far been shown to be involved in cellulose degradation by *T. reesei*. However, acidic carbohydrates were produced during enzymatic hydrolysis of cellulose by unpurified *T. reesei* cellulase complex. The acids were identified by means of gas chromatography/mass spectrometry. Three monomeric acids, gluconic, arabinonic, and erythronic acids as well as three aldobionic acids, glucosylgluconic (cellobionic), glucosylarabinonic, and glycosylerythronic acids and their lactones were found as degradation products. The formation of aldonic acids was continuous and amounted to about 10% of reducing sugars under conditions where degradation products were removed by ultrafiltration (Vaheri 1982a). This led Vaheri (1982b) to assume that an oxidative reaction is involved also in the degradation of crystalline cellulose by *T. reesei*. In the early stages of the breakdown of native cellulose, the appearance of short, insoluble fibers from cotton catalyzed by unpurified filtrate of *T. reesei* was demonstrated. Oxygen was shown to be necessary for this formation of short fibers, and no loss of short segments was detected under strictly anaerobic conditions. For this to take place, the oxygen concentration had to be kept low. Under high oxygen concentrations, short fiber formation did not occur. Vaheri (1982b) therefore suggested that, in the initial stage of breakdown of cotton, oxidative reactions create changes in the cellulose. These changes are not reflected in the formation of soluble products but make the substrate more accessible to attack of endoglucanases.

Halliwell and co-workers (Halliwell and Vincent 1981) have in several papers presented results from short fiber formation by *Trichoderma koningii* from native cellulose substrate. These authors found that the breakdown of cellulose fibers to short segments depends on two activities, i.e., that of an endoglucanase and that of a less stable enzyme named component C_2. The purified endoglucanase alone was shown to have no specific activity toward insoluble cellulose, while component C_2 degraded cellulose weakly to short fibers. The components together showed strong synergistic effects. The nature of the C_2 component has not yet been elucidated.

2.5.4 Anaerobic Fungi

Until recently, the complex microbial population of the rumen was believed to be composed of bacteria and protozoa only. However, a large population of fungi possessing the exceptional property of being obligate anaerobes have now been demonstrated to also be part of this population (Orpin 1975, 1977, Bauchop 1979, 1981). These anaerobic rumen-inhabiting fungi are of course not directly involved in the degradation of cellulose in wood. However, they do degrade cellulose to glucose. The intention with this chapter has been to incisively clarify and present existing information concerning the enzyme mechanisms involved in degradation of cellulose by all the different categories of fungi and bacteria.

The fermentation of cellulose by a rumen anaerobic fungus resulted in the formation of acetate, carbon dioxide, formate, ethanol, lactate, and hydrogen (Bauchop and Mountfort 1981). However, in coculture with a rumen methanogenic bacterium the major products with a stable coculture were acetate, carbon dioxide, and methane with low or very low production of hydrogen, formate, or lactate. These results show that the methanogenes utilize electrons produced in the fermentation of cellulose by the fungus to form methane and also that the acetate is converted to carbon dioxide and methane by *Methanosarcina barkeri*. Cellulolytic anaerobic fungi may therefore play an important role in cellulose degradation at least in the rumen.

The cellulolytic enzyme system in one of these rumen anaerobic fungi, *Neocalimastixs frontalis*, has recently been studied. Pearce and Bauchop (1985) found that with Avicel as substrate the major digestion product obtained with *N. frontalis* from cellulose was glucose and not cellobiose. This result suggests that excess β-glucosidase is produced relative to cellulase by *N. frontalis*. This contrasts with the situation in *Trichoderma viride* (*T. reesei*), which secretes only small quantities of β-glucosidase (Mandels 1982).

Wood et al. (1986) reported the first isolation from a rumen anaerobic fungus, *N. frontalis*, of an extracellular enzyme system with the capacity for extensive solubilization of cotton fibers. The fungus was isolated from the rumen of a sheep. The cellulase was found to be severalfold more active in solubilizing the cotton fiber per unit of endo-1,4-β-glucanase than the cellulase of the aerobic fungus *Trichoderma reesei* mutant strain C-30, which is one of the most active cellulases isolated so far. The cellulase obtained from *N. frontalis* strain RK 21, grown on filter paper as carbon source, contained both endoglucanase and β-glucosidase activities. Under the assay conditions used, the enzyme sample containing 0.37 IU of endoglucanase and 0.39 IU of β-glucosidase solubilized cotton to an extent of 16% in 72 h. However, when growing in coculture with the methanogenic bacterium *Methanobrevibacter smithii*, which removes the hydrogen produced during fermentation, the same amounts of endoglucanase and β-glucosidase activities now accomplished 98% solubilization of the cotton cellulose in the same time.

The results obtained suggest that the cellulase from *N. frontalis* RK 21 contains a factor that acts in concert with the endoglucanase and β-glucosidase enzymes to solubilize the crystalline cellulose. The nature of this factor is presently under investigation by Wood and co-workers (Wood et al. 1986). In the case of the aerobic fungi *S. pulverulentum*, *T. reesei*, *Penicillium funiculosum*, *Fusarium solani* and *Trichoderma koningii* the factor that acts in synergism with the endoglucanase and β-glucosidase enzymes has been demonstrated to be the enzyme cellobiohydrolase (Eriksson and Wood 1985).

The cellulases obtained from *N. frontalis* RK 21 and its methanogen coculture were demonstrated to be considerably better than the cellulase of

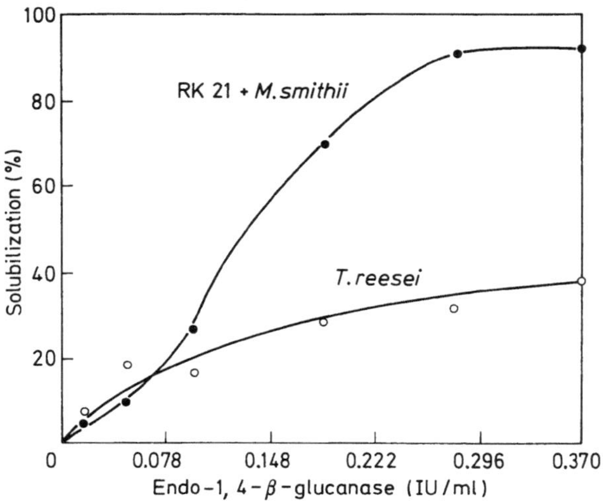

Fig. 2.19. Comparison of the cellulase (cotton-solubilizing) activity of *T. reesei* C-30 and *N. frontalis* RK21. Assays, which were carried out at 37 °C (RK21) and 50 °C (*T. reesei*) for 72 h, were equivalent in terms of endo-1,4-β-glucanase and β-glucosidase activities; *T. reesei* cellulase was supplemented with β-glucosidase from the fungus *F. solani*. ●, Cellulase from RK21/methanogen co-culture; ○, *T. reesei* cellulase (Wood et al. 1986)

T. reesei for solubilizing crystalline cellulose per unit of endoglucanase. This is demonstrated in Fig. 2.19. The existence of a cellulolytic system that can effectively solubilize both amorphous and highly crystalline cellulose will be of considerable importance, not only for a better understanding of cellulose degradation in the rumen but particularly for a biotechnical saccharification of lignocellulosic materials.

2.6 Cellulose Degradation by Bacteria

Degradation of woody material is mainly a task for fungi. This depends in part upon the organization of their hypha, which gives these organisms a penetration capacity. The ability of bacteria to degrade wood is presented in Sect. 1.5. Normally, bacteria have a limited capability to degrade lignified wood cell walls. However, degradation of lignocellulosic materials in nature does not normally take place with one organism in monoculture but rather with a consortium of organisms, where bacteria participate, mainly in degradation of polysaccharides.

Degradation of cellulose by bacterial systems occurs both aerobically and anaerobically. It is only recently that the mechanisms for bacterial cellulose degradation have been seriously investigated. The cellulase systems of bacteria are not directly comparable to those of fungi. Bacterial degradation of cellulose seems to take place by a cluster of enzymes rather than by individual endo- and exo-1,4-β-glucanases as is the case in fungi. Furthermore, many bacteria lack 1,4-β-glucosidases (cellobiases); they rather employ phosphorylases for the degradation of cellobiose (Ljungdahl and Eriksson 1985).

2.6.1 Aerobic Bacteria

Due to the abundance of cellulose, its biological degradation has been of paramount importance in the generation of energy for living systems. The presence in soil of aerobic cellulolytic bacteria has long been recognized. In Table 2.9 aerobic bacteria with cellulolytic activity are listed. It is apparent from what has already been said in this chapter that fungal degradation of cellulose has been studied most intensively. Prokaryotic systems have only recently gained a real interest, mainly due to their potentials for various types of biotechnology based on lignocellulosic materials.

Cellulomonas Species. The genus *Cellulomonas* constitutes a group of gram-variable nonspore-forming mesophilic rods, and can grow either aerobically or anaerobically. Biochemical, serological, and DNA–DNA homology studies revealed that there are at least eight distinct, validly described *Cellulomonas* species: *C. biazotea*, *C. flavigena*, *C. cellasea*, *C. fimi*, *C. gelida*, *C. cartae*, *C. uda*, and *C. turbata* (Ljungdahl and Eriksson 1985). In addition, several unclassified *Cellulomonas* isolates are available and have been studied for their cellulolytic activities.

Table 2.9. Aerobic bacteria with cellulolytic activity (Ljungdahl and Eriksson 1985)

Cellulomonas flavigena	*'Cellvibrio viridus'*
C. biazotea	*'C. flavescens'*
C. cellasea	*'C. ochraceus'*
C. fimi	*'C. fulvus'*
C. gelida	*'C. vulgaris'*
C. cartae	*'C. gilvus'*
C. uda	*Cytophaga hutchinsonii*
c. turbata	*C. aurantiaca*
Bacillus brevis	*C. rubra*
B. firmus	*C. tenuissima*
B. lichenformis	*C. winogradskii*
B. pumilus	*C. krzemieniewskae*
B. subtilis	*Herpetosiphon geysericolus*
B. polymyxa	*Sporocytophaga myxococcoides*
B. cereus	*Streptomyces flavogriseus*
Serrata marcescens	'*Thermoactinomyces* sp.'
'Pseudomonas fluorescens var. *cellulosa'*	Thermomonospora curvata

[a] The bacteria within prime signs are not validly classified.

Cellulases of the *Cellulomonas* species appear to be tightly bound to their cellulose substrates. Extracellular cellulases produced by the *Cellulomonas* isolate IIbc (Béguin et al. 1977) were found mainly to be bound to the cellulose in the growth phase and to appear in the culture solution only in the stationary phase. In a paper by Béguin and Eisen (1978) the purification of the extracellular cellulases from cultures of *Cellulomonas* IIbc was described. One of the enzymes was found in the culture solution, and the two others were bound to the cellulose used as carbon source. The free enzyme and one of the cellulose-bound enzymes bind to Sephadex. The two cellulose-bound enzymes are glycosylated. All three enzymes behave as endocellulases toward soluble CMC and have little activity on cellulose powder. The three cellulases are therefore thought not to represent the whole enzyme system used by *Cellulomonas* to degrade cellulose. Particularly their low activity toward crystalline cellulose suggests that additional enzyme components are necessary for digestion of such a substrate. Analyses under nondenaturing polyacrylamide gel electrophoresis conditions show that at least two other cellulases, active toward CMC, are present in the culture solution.

Subsequent investigations with another strain, *C. fimi*, have confirmed these findings. Thus, Langsford et al. (1984) reported that supernatants from cultures of *C. fimi* contain up to ten components with CMC-ase activity as determined by nondenaturing polyacrylamide gel electrophoresis (PAGE). It was found that some of the active components were glycosylated. The number of active components, as determined by PAGE, varied with the cellulosic substrate used as carbon source and with the age of the culture. However, the enzyme profile is not as complex as it might seem. The variations were demonstrated to be a consequence of proteolysis and also of a reduction in the glycosylation of some of the components. The number of enzymes seems to be no more than three or four, at least two of which are glycosylated. The glycosylated enzymes bind very strongly to Avicel and are stabilized by this substrate. Enzymes, free in the supernatant, are unstable and are altered by proteolysis and deglycosylation. This leads to a diversity of products still enzymatically active but reduced in their substrate affinity. Three cellulases have also been described for a different *C. fimi* strain. Two of these are glycosylated and bind strongly to cellulose (Béguin and Eisen 1978). The cellulase system from *C. fluorescence* var. *cellulosa* also contains three cellulases, and these are all glycosylated (Yamane et al. 1970).

It thus seems as if posttranslational modifications are contributing factors to the multiplicity of cellulolytic enzymes in culture solutions of *C. fimi*. Studies have been focused on the two tightly substrate-bound enzymes designated CB 1 and CB 2 (Gilkes et al. 1984). Both components are glycoproteins in which mannose appears to be the sole glycosyl component (Béguin et al. 1987). The two enzymes have similar molecular weights ($M_r = 58\,000$ and $56\,000$, respectively). However, there are significant differences in their modes of action and substrate specificities. Both enzymes hydrolyze CMC but with very different specific activities (370 and 85 units/mg protein for CB 1 and CB 2, respectively). These properties led Gilkes et al. (1984) to characterize CB 1 as

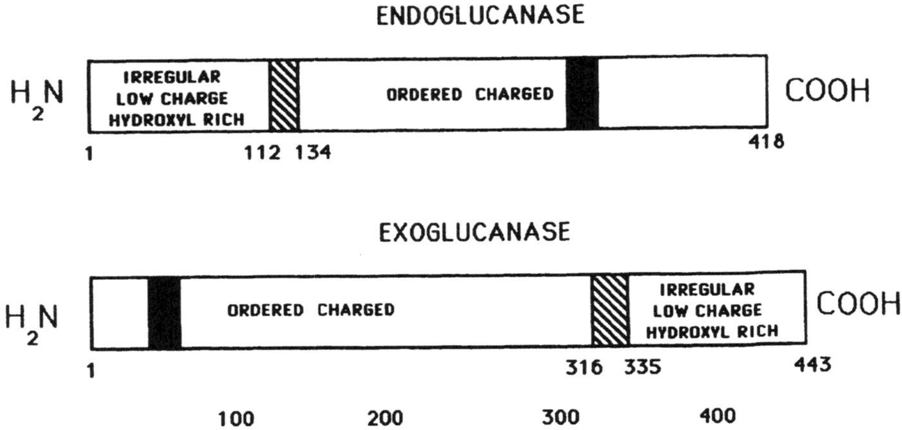

Fig. 2.20. General organization of domains within *C. fimi* endoglucanase CB-I and CB-II. The *hatched areas* designate potential active sites. The Pro-Thr region is indicated by a *solid bar*. Numbering of amino acids starts at the amino terminus of the mature protein (Warren et al. 1987)

an endoglucanase and CB 2 as an exoglucanase. Both enzymes hydrolyze cellodextrins in the range cellotriose to cellohexaose. However, the hydrolysis rate is more than 1000-fold higher for CB 1 than for CB 2. Preference for the penultimate bond from the nonreducing end for both enzymes is observed. However, with cellohexaose cleavage at the central linkage also occurs. Polarimetric and NMR spectroscopic investigations indicate that hydrolysis of CB 1 proceeds with inversion of configuration, while CB 2 retains configuration (Withers et al. 1986). Degradation of cellodextrins by fungal cellulases leads to inversion of configuration only with exoglucanases (Streamer et al. 1975), but here it is obtained with CB 1 designated as an endoglucanase.

Comprehensive cloning of *C. fimi* cellulase genes have been carried out particularly by Miller and co-workers (Béguin et al. 1987). Studies of the organization of different domains within the *C. fimi* endoglucanase CB 1 and exoglucanase CB 2 also have been carried out (Fig. 2.20). A sequence of 41 amino acids was proposed to be a signal peptide of the exoglucanase to direct the secretion of this enzyme. The hypothesis was demonstrated by gene fusion experiments in *Escherichia coli* (O'Neill et al. 1986).

Analyses of the purified, presumably unglycosylated, *E. coli*, recombinant proteins have revealed no significant differences in the kinetic properties of these enzymes and their *C. fimi* counterparts (Béguin et al. 1987).

The hydropathy profiles, their secondary structures, and sequence conservation between the CB 1 and CB 2 enzymes predicted from the nucleotide sequence of the cex and cenA genes were analyzed (Warren et al. 1987). Both enzymes seem to contain three domains. There is a considerable sequence conservation in two of the domains. It can be seen in Fig. 2.20 that the domains occur in different order in the two enzymes. There is very little amino acid conservation in the ordered domains but both contain the sequence Glu-Xaa$_7$-Asn-

Xaa_6-Thr. These sequences appear at widely separated sites in the two enzymes. Related sequences have been found in *T. reesei* CBH-I, in an endoglucanase from *Schizophyllum commune*, and in a xylanase from *Cryptococcus albidus* (Fägerstam et al. 1984, Paice et al. 1984, Morosoli et al. 1986b). These sequences have been proposed as the active sites of these enzymes on the basis of their homology with the active site of lysozyme (Paice et al. 1984, Yaguchi et al. 1983).

Bacillus Species. The genus *Bacillus* comprises a large number of spore-forming, aerobic, or facultative anaerobic species. The variety of exo-enzymes synthesized by the *Bacilli* has been summarized by Priest (1977). In this extensive review, *B. brevis*, *B. firmus*, *B. polymyxa*, *B. pumilus*, and *B. subtilis* are listed as being producers of cellulases. Other cellulase-producing bacilli are *B. licheniformis* (Knösel 1971) and *B. cereus* (Thayer 1978).

The species mentioned above are all able to hydrolyze CMC with cellobiose as the main product. The endoglucanase formation in *B. polymyxa* is dependent on the carbon source in the growth medium. Thus, the cellulase was produced when starch, glucose, cellulose, cellobiose, CMC, xylan, maltose, and sucrose were added to a basal salts-peptone medium (Greaves 1971b). Similarly, Thayer (1978) found that glucose and cellobiose stimulated enzyme formation by *B. cereus*, and Robson and Chambliss (1984), working with *B. subtilis* DLG, demonstrated an enhanced production of CMC-ase in the presence of sucrose, maltose, glucose, cellobiose, or lactose in the medium. However, the isolate DLG did not degrade crystalline forms of cellulose.

There are relatively few reports of *Bacilli* producing endoglucanases. However, two relatively well-characterized examples, the alkalophilic *Bacillus* sp. strain N-4 (Sashihara et al. 1984) and *B. subtilis* DLG (Robson and Chambliss 1984, 1986, 1987) produce such enzymes. In *B. subtilis* DLG, endoglucanase production begins at the onset of the stationary phase (Robson and Chambliss 1984). Glucose does not cause repression of the cellulase activity, as is the case in most other cellulolytic systems (cf. Sect. 2.7). The possibility even exists that endoglucanase in *B. subtilis* DLG is induced by glucose. In a study by Robson and Chambliss (1987), in vivo endoglucanase mRNA was observed only when the bacterium was grown in a glucose-containing medium; almost no gene-specific mRNA was observed in the absence of glucose. Thayer (1978) also reported increased endoglucanase activity in *B. cereus* in the presence of glucose. However, other investigations are necessary to evaluate whether glucose itself acts as an inducer of endoglucanase synthesis or if the stimulatory effect on endoglucanase production by glucose is more indirect.

The endoglucanase character of the *B. subtilis* DLG glucanase was confirmed by measuring the ability of the enzyme to reduce the viscosity of a CMC solution. *T. reesei*, EG-I showed the same specific activity as *B. subtilis* DLG endoglucanase in such an assay. However, the bacterial enzyme displayed almost twice the specific activity of *T. reesei* EG-I if trinitrophenyl CMC was used as substrate. This illustrates the risk of using only one assay system for comparison of specific activities of two cellulases.

Robson and Chambliss (1987) also evaluated the DNA sequence of the *B. subtilis* DLG endoglucanase gene. Like many other genes coding for a secreted enzyme, the *B. subtilis* endoglucanase codes for a signal sequence at the N-terminal end of the enzyme. Such a signal is believed to be necessary for an efficient export of a protein in most organisms, and is usually composed of a short sequence of charged amino acids followed by a longer stretch of strongly hydrophobic amino acids (Perlman and Halvorson 1983). These DNA sequence studies also revealed that the endoglucanase gene of *B. subtilis* has a coding capacity for approximately a 55000 dalton protein. However, the isolated mature extracellular endoglucanase is only 35200 daltons. This result was not unexpected since in earlier studies a 51500 dalton endoglucanase had been observed in intracellular extracts of *E. coli* and *B. subtilis* transformants carrying the endoglucanase gene on a high-copy number plasmid (Robson and Chambliss 1986).

The apparent ca. 16500 dalton difference between the extracellular endoglucanase and its precursor form minus the signal sequence is not fully revealed but is most likely due to posttranslational modifications. Apparently the endoglucanase precursor undergoes signal sequence processing at the amino terminus at some point before or during its secretion. The site at which this processing occurs has been located between two alanine residues following the hydrophobic region. This seems to correlate with the accumulated data on signal sequence cleavage sites in members of the family Bacillaceae and other organisms (Murphy et al. 1984).

Pseudomonas Species. The genus *Pseudomonas* is a large group of rod-shaped, nonspore-forming, gram-negative, aerobic bacteria. The genus has been incisively described by Palleroni (1984). Cellulolytic *Pseudomonas* species have been studied by Ueda et al. (1952). The best-studied strain concerning its cellulase production was isolated from soil and characterized as *Pseudomonas fluorescens* var. *cellulosa.* Bacteria placed in the genus "*Cellvibrio*", which are all cellulolytic, were suggested by Doudoroff and Palleroni (1974) to perhaps belong to the genus *Pseudomonas.* However, the classification of the cellulolytic *Pseudomonas* and *Cellvibrio* species is still obscure and remains to be clarified.

The cellulolytic enzyme system of the most-studied organism, *P. fluorescęnce* var. *cellulosa,* seems to be quite complex. Yamane et al. (1970) identified two extracellular (A and B) and one cell-bound (C) endoglucanase component of *P. fluorescence* var. *cellulosa,* which were purified and characterized. All three of these cellulases hydrolyzed a variety of substances, including crystalline and amorphous cellulose and cellodextrins. The mode of action of these cellulases toward CMC appears to be similar to that shown by fungal cellulases of this type. However, the fact that crystalline cellulose also is degraded by the three enzymes suggests that a certain exoglucanase activity can be ascribed to at least one of them.

Concerning their substrate specificities, certain differences were noticed. Thus, cellulase A and C hydrolyze soluble substrates, including CMC, more easily than cellulase B. However, the three cellulases showed activity of a similar order of magnitude toward insoluble substrates. The hydrolysis patterns of the three cellulases toward nonreduced and reduced cellodextrins are presented in Table 2.10. It can be seen that G3 and G4H were hydrolyzed only by cellulase C. The rate of hydrolysis of the cellodextrins increased roughly in parallel to their chain lengths. Generally, nonreduced cellodextrins were hydrolyzed more easily than their corresponding reduced derivatives. However, the effect of a terminal sorbitol group was less on cellulase C than on the other two. Cellulases A and B are unique in the respect that they do not hydrolyze G3. All three *Pseudomonas* cellulases contain a considerable amount of carbohydrates, mainly glucose and galactose.

In a later work by Yoshikawa et al. (1974) it was shown that the enzyme components A and B are not homogeneous. Five extracellular cellulases (peaks I–V) differing in molecular mass were isolated from a bacterial culture. The number of components increased with increasing time of cultivation. The relationship between the electrophoretically separable components (A and B) and the five components (peak I–V) of different molecular mass was investigated. It was found that peaks I and V were composed almost exclusively of B and A, respectively, whereas the other peaks consisted of different ratios of A and B. It was also found that the ratio of A to B increased with time of bacterial growth. Furthermore, the smaller molecular mass components increased with age of the culture, while the larger molecular mass components decreased. This suggests that a large molecular mass component is initially released from the bacterial cells into the culture medium, which subsequently is modified to form multiple cellulase components of smaller molecular masses.

In a subsequent work by Yoshikawa et al. (1975), peak I was found to have a molecular weight of more than 200 000. The purified component was composed of only component B. To study whether proteolytic activity converts the large molecular mass component I to components of smaller molecular mass, pronase was used. Peak I was found to be liable to limited proteolysis by this enzyme and was readily converted into peaks IV and V, however, without apparent production of peaks II and III.

The cellulases of *P. fluorescence* var. *cellulosa* were also treated with a β-glucosidase preparation from *Trichoderma viride* to examine the possibility of removal of the carbohydrate moieties. Small quantities of reducing substances were liberated. A conversion of component B to A by treatment with β-glucosidase was found in all peak fractions. The conclusion was that both β-glucosidases and proteases contribute to the multiplicity of cellulases formed by the bacterium. A parallel can be drawn with the results obtained by Eriksson and Pettersson (1982), who isolated and characterized two acidic proteases from *Sporotrichum pulverulentum*. These proteases were demonstrated to activate the endoglucanases from the same fungus.

Studies of the cellulases from *P. fluorescence* var. *cellulosa* were also carried out by Ramasamy and Verachtert (1980). They particularly investigated the localization of the cellulase components and observed that cell-bound activity was obtained only when the bacterium was grown on cellobiose.

Table 2.10. Relative rate of hydrolysis of the bonds of several nonreduced and reduced cellodextrins (Yamane et al. 1970)

Substrates		Bond attacked preferentially[a]		
		Cellulase A	Cellulase B	Cellulase C
G3	1 2 ○ – ○ – ○	–[b]	–[b]	2 or 1
G4	1 2 3 ○ – ○ – ○ – ○	2>1 or 3	2>1 or 3	2>1 or 3
G4H	1 2 3 ○ – ○ – ○ – ●	–[b]	–[b]	1>2>>>3
G5	1 2 3 4 ○ – ○ – ○ – ○ – ○	2 or 3>>>1 or 4	2 or 3>>>1 or 4	2 or 3>>>1 or 4
G5H	1 2 3 4 ○ – ○ – ○ – ○ – ●	2>3>>>1 or 4	2>3>>>1 or 4	2>>>1, 3 or 4
G6	1 2 3 4 5 ○ – ○ – ○ – ○ – ○ – ○	3≧4 or 2>>>1 or 5	3≧4 or 2>>>1 or 5	2 or 4>3>>>1 or 5
G6H	1 2 3 4 5 ○ – ○ – ○ – ○ – ○ – ●	3>>4=2>>>1 or 5	3>>4>2>>>1 or 5	3=4>>>1, 2 or 5

[a] Numbers from 1 to 5 indicate the position of bonds from the nonreducing end. The signs >, >> and >>> represent the relative rates of bond hydrolysis in the factors of 1.5–3 times, 5–10 times, and in extremely high, respectively, as compared with the rate to the next mentioned.
[b] Not hydrolyzed.

Table 2.11. Activity of *Pseudomonas* endoglucanases in *Escherichia coli*[a] (Gilbert et al. 1987)

Plasmid	Addition to growth medium	Endoglucanase (U mg^{-1} protein)	β-Glucosidase (U mg^{-1} protein)	Exoglucanase (U mg^{-1} protein)
pJHH1	–	122	0.12	0
pJHH1	Glucose	12	N.D.	N.D.
pJHH2	–	114	0.15	0
pJHH3	–	71	0.1	0
pJHH3	Glucose	5	N.D.	N.D.
pJHH4	–	66	0.12	N.D.
pGJH2	–	51	0.1	0
pGJH2	Glucose	53	N.D.	N.D.
pGJH3	–	56	0.11	0
pJHH8	–	52	N.D.	N.D.
JHH8	Glucose	47	N.D.	N.D.
pJHH9	–	0	N.D.	N.D.

[a] Plasmids were inserted into *E. coli* JM83. N.D. indicates no determinations. Glucose was added to the culture medium at a final concentration of 0.2%.

Gilbert et al. (1987) have studied the molecular biology of the cellulase complex from *P. fluorescence ssp cellulosa.* While Wolff et al. (1986) could isolate only one endoglucanase gene from *P. fluorescence ssp cellulosa*, suggesting that the multiplicity of endoglucanases from these species obtained in earlier studies (Yamane et al. 1970) is a result of posttranslational modification of a single gene product, Gilbert et al. (1987) isolated four distinct endoglucanase genes that showed no homology with one another.

Expression of genes coding for polysaccharases is often repressed by addition of glucose to the growth medium. The regulation of the *P. fluorescence ssp cellulosa* genes, coding for endoglucanases, cloned in *Escherichia coli*, was therefore investigated. Cells were cultured with and without glucose, and the endoglucanase activity was measured. Two of the genes encoded by the plasmids were repressed by glucose, while the other two were not (Table 2.11). The conclusion drawn by Gilbert et al. (1987) was that the regulatory region of the gene encoded by the different plasmids must contain a recognition sequence for the catabolite activator protein. Furthermore, the recognition sequence is conserved in both bacteria.

The endoglucanase gene eglX from *P. fluorescence* var. *cellulosa* has also been cloned and expressed in the bacterium *Zymomonas mobilis* (Lejeune et al. 1988). The enzyme was intracellular in its new host. It was produced during the growth phase and was not repressed by glucose.

Cellulose degradation by "*Cellvibrio*" species will be discussed in the *Pseudomonas* section. Particularly the cellulases of *Cellvibrio gilvus* have been studied rather extensively. The results up to 1969 were summarized by King and Vessal (1969). *C. gilvus*, which grows well on cellobiose, other cellodextrins, and cellulose but less well on glucose has, similar to *P. fluorescence* var. *cellulosa*, a complex extracellular cellulolytic system. Degradation products of cellulose are chiefly cellobiose and cellotriose. The cellulase system in *C. gilvus* is constitutive and contains at least four electrophoretically, distinctly different glucanases (Storvick et al. 1963). Cellobiose and higher cellodextrins are metabolized by *C. gilvus*, and the cell yield per glucose equivalent increases with the DP of the cellodextrins. This indicates that the energy of the glucosidic bonds is preserved, and that cleavage of these bonds is by phosphorolysis (Alexander 1968). The cellobiose phosphorylase from *C. gilvus* has also been purified by Sasaki et al. (1983). This enzyme does not cleave cellotriose or higher cellodextrins. King and Vessal (1969) therefore suggested that transglucosylation reactions were involved in the metabolism of cellotriose and higher cellodextrins to yield cellobiose, which subsequently is cleaved by cellobiose phosphorylase.

Other "*Cellvibrio*" strains have also been examined for their cellulolytic activities. Thus Berg et al. (1972) studied the cellulase system of *C. fulvus*. In contrast to *C. gilvus*, the cellulases of *C. fulvus* are repressed by glucose. As for *P. fluorescence* var. *cellulosa*, the location of the cellulase in *C. fulvus* seems to depend on the carbon source for growth (Berg 1975). With glucose and cellobiose, the endoglucanase activity is cell-bound during the active growth phase, whereas with cellulose as substrate it is both free in the culture solution and partly bound to the cellulose. Similarly, when *C. vulgaris* grows on cellulose, it produces a truly extracellular endoglucanase (Oberkotter and Rosenberg 1978).

Recently, Wynne and Pemberton (1986) isolated a gram-negative soil bacterium capable of rapid degradation of crystalline cellulose. Subsequent studies have shown that this bacterium belongs to the genus "*Cellvibrio*", and it has been named *C. mixtus* (Blackall et al. 1985). The bacterium hydrolyzes a wide variety of polysaccharides, including CMC, xylan, chitin, starch, and pectin. Using cosmid cloning, the genes for the degradation of CMC, chitin, starch, pectin, esculin, and cellobiose were cloned and expressed by *E. coli*. Microcrystalline cellulose hydrolysis was confirmed by complementation of Cel^- mutants of *C. mixtus* by one 39.6-kb cloned fragment of a plasmid. The results suggest that the cloned fragment contains some if not all the genes responsible for coding of the enzymes degrading microcrystalline cellulose. Within this fragment also the genes for amylase, chitinase, β-glucosidase, and CMC-ase were clustered in a 9.8-kb segment. This clustering of genes associated with the degradation of a range of polysaccharides conforms with a well-established pattern for gram-negative bacteria. Functionally related genes, such as those for photosynthesis in *Rhodopseudomonas sphaeroides*, nitrogen fixation in *Klebsiella species*, and degradation of a wide variety of aromatic and aliphatic compounds by members of the genus *Pseudomonas* are clustered on the main chromosome (Wynne and Pemberton 1986). The cloning of a cluster of polysaccharide genes has not been reported earlier, although individual cellulase genes have been cloned from a variety of microorganisms (cf. *Cellulomonas* species).

Cytophaga Group. The *Cytophaga* group of bacteria has been covered in an excellent review by Christensen (1977). *Sporocythophaga* and *Cytophaga* species are gliding bacteria which grow in close contact with cellulose fibrils, which are their natural substrate. They inhabit soil, sea, and fresh water and degrade a range of rather resistant polysaccharides, including cellulose. The genus *Cythophaga* includes the following known cellulolytic species: *C. hutchinsonii*, *C. aurantiaca*, *C. rubra*, *C. tenuissima*, and *C. winogradskii*. The genus *Sporocytophaga* has one well-characterized organism, i.e., *S. myxococcoides*. The cellulolytic activity of *S. myxococcoides* has been studied by several authors, and the cellulases have been isolated and characterized by Osmundsvåg and Goksöyr (1975).

Berg et al. (1972) found that *S. myxococcoides* grows in close contact with cellulose fibers. This close contact seems necessary for degradation to take place, although an endoglucanase can be recovered from the cell-free medium. However, this enzyme alone does not hydrolyze crystalline cellulose such as Avicel. During growth on cellulose, *S. myxococcoides* produces a slimy polysaccharide.

Carpentier and Robic (1974) later found in *S. myxococcoides* an exoglucanase yielding glucose with cellobiose and higher dextrins as substrates. Osmundsvåg and Goksöyr (1975), however, found only endoglucanase activity to be produced by *S. myxococcoides*. The major proportion of this endoglucanase activity was in the growth medium (about 90%). Two major cellulase components, cellulases I and II were present in the culture medium. Cellulase II occurred in the greatest amount. It had a molecular weight of 52 000, a pI value of 4.75, and a relatively broad pI optimum (5.5 – 7.5). Cellulase I had a molecular weight of 46 000, a pI value of 7.5, and a pH optimum of 6.5 – 7.5. The two cellulases were typically random acting endoglucanases, causing a rapid decrease in viscosity of CMC solutions compared with the rate of formation of reducing end groups. No effect could be found on native cellulose with purified enzymes, concentrated cell-free supernatant, extracts from broken cells, or with supernatants from Triton X-100-treated cells.

In a more recent study, Vance et al. (1980) investigated the spectrum of extracellular cellulases produced by *S. myxococcoides*, and the possibility to increase the yield of these enzymes. Contrary to the findings of Berg et al. (1972) and to Osmundsvåg and Goksöyr (1975), Vance et al. (1980) found that *S. myxococcoides* produces extracellular enzymes that degrade Avicel. This activity was detectable by measuring the soluble carbohydrate released.

Actinomycetes Group. Actinomycetes are a heterogeneous group of gram-positive bacteria of which terrigenous saprophytes are the most common forms. Most *Actinomycetes* develop branching hyphae similar to those of filamentous fungi. Examples of bacteria belonging to the *Actinomycetes* are found among the genera *Thermomonspora*, *Thermoactinomyces*, and *Streptomyces*. A detailed discussion of *Actinomycetes* and their importance in soil, compost, and fodder has been given by Lacey (1973). A comprehensive and more recent review of lignocellulose degrading *Actinomycetes* has been given by McCarthy (1987).

Table 2.12. Temperature and pH relationships of actinomycetes cellulases (McCarthy 1987)

Species	Enzyme activity[a]	Optima for		Thermostability
		Production	Activity	
Stm. lividans	Endoglucanase	34 – 37 °C	55 °C	30 min at 60 °C
	β-Glucosidase	31 °C	40 °C, pH 6.5	< 10 min at 40 °C
Stm. flavogriseus	Endoglucanase		50 °C, pH 6 – 7.5	2 h at 40 °C
	Cellulase	30 °C, pH 7	40 – 45 °C, pH 5.3 – 6	2 h at 40 °C
	β-Glucosidase		40 °C, pH 6.5 – 7.5	< 10 min at 40 °C
Stm. albogriseus	Endoglucanase	25 – 35 °C	45 – 55 °C	> 24 h at 50 °C
Stm. nitrosporeus		pH 6.7 – 7.2	pH 5 – 6	
Mim. melanospora	Cellulase			< 24 h at 50 °C
Thm. curvata	Endoglucanase	45 – 55 °C, pH 8.0	65 °C, pH 6 – 6.5	> 60 min at 70 °C
	Cellulase			
Thermomonospora sp.	Endoglucanase		70 °C, pH 6	24 h at 65 °C
	Cellulase	55 °C, pH 7.4	65 °C, pH 7	4 h at 65 °C
	β-Glucosidase		55 °C, pH 6.5	< 1 h at 55 °C

[a] Endoglucanase is a measure of activity against carboxymethylcellulose; cellulase is a measure of activity against Avicel or filter paper; β-glucosidase is a measure of activity against cellobiose or nitrophenol β-D-glucoside.

Actinomycetes form an important part of the microbial community responsible for lignocellulose degradation and nutrient recycling of natural substrates. The characterization of the cellulolytic activities of *Actinomycetes* has understandably received most attention due to an abundance of cellulose in plant biomass. Generally, *Actinomycetes* cellulases are inducible extracellular enzymes, which appear to attack cellulose in a way similar to fungal cellulases. Cellulolytic activity has been reported in a number of *Actinomycetes* but it has been further characterized in relatively few (Table 2.12). The thermophilic species *T. curvata* and *T. fusca* and the mesophilic *Streptomycetes flavogriseus*, are the most studied. Stutzenberger (1972a, b) initiated studies of cellulolytic activity of *T. curvata*. It was found that *T. curvata* produces cellulases active against both cotton fibers and CMC. The cellulase activity could be separated into two fractions and it was found that the sole product of purified cellulase activity on ground cotton fibers appeared to be cellobiose. In a later study by Stutzenberger and Kahler (1986), it was found that glucose could not repress cellulase biosynthesis in *T. curvata* in the presence of cellobiose due to the strong preference for uptake of the disaccharide even by glucose-grown cells. Stutzenberger and Lupo (1986) found that *T. curvata* excretes a cellulase complex capable of saccharifying crystalline cellulose. The saccharifying capacity of this cell-free culture solution was rapidly inactivated by heating at acidic pH values, while at alkaline pH values significant activation of the cellulolytic complex was obtained by heat treatment. This pH-dependent thermal activation occurred in the major endoglucanase, the only purified cellulase component of the cellulase complex (Fig. 2.21). A thermal activation step may thus provide a convenient method for increasing the cellulolytic activity of *T. curvata* enzymes.

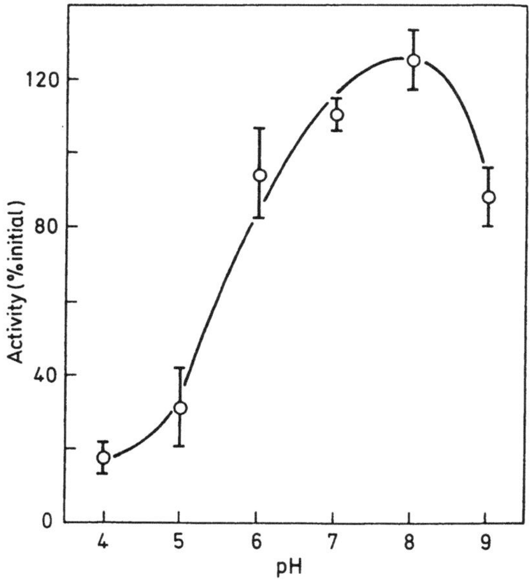

Fig. 2.21. pH-dependent thermal activation/inactivation of crystalline cellulose-saccharifying enzymes in culture fluid of *T. curvata*. *Points* show averages and standard *error bars* for the cumulative data of three experiments in which fluid was adjusted to pH values from 4 to 9, heated for 30 min at 65 °C, then assayed for residual activity (Stutzenberger and Lupo 1986)

In the *T. fusca* strain, YX, Hägerdal et al. (1978) identified three extracellular cellulases with activities toward both CMC and Avicel. Based on these activities it was not possible to discriminate between endo- and exoglucanase activities. In a later study by Hägerdal et al. (1979), little β-glucosidase activity was found in the culture filtrate. The major part was associated with the culture solids. Mandels (1975) suggested that the culture filtrate of *Thermoactinomyces* sp. has an incomplete cellulolytic enzyme system. It was suggested that the culture filtrate lacked the exoglucanase component, while the study by Hägerdal et al. (1979) showed that the β-glucosidase (cellobiase) activity is lacking. According to these authors, it seems unlikely that the exo-glucanase component is missing in the *Thermoactinomyces* culture filtrate since its specific activity toward cotton is two to six times higher than that of the filtrate from *T. viride* QM 9414 (Mandels 1975).

Two *T. fusca* YX strain endoglucanases were later isolated in nearly pure form by Calza et al. (1985). These two enzymes account for at least 90% of the total extracellular CMC-ase activity. However, at least one additional endoglucanase component is present in the culture solution. The two enzymes differ in their molecular weights, substrate specificities, and carbohydrate contents and are immunologically, distinctly different proteins. The molecular weights are 94 000 and 46 000, respectively. The specific activities toward CMC of the two enzymes differ widely. The most active enzyme (94 000) has a specific activity higher than 700 IU/mg while the less active (46 000) has a specific activity of 77 IU/mg. The high molecular mass enzyme contains less than 1% carbohydrate and is isoelectric at pH 3.5, whereas the low molecular mass enzyme is a glycoprotein containing 25% carbohydrate and has its isoelectric point at pH 4.5. It was concluded by Calza et al. (1985), however, that other components are required for filter paper hydrolysis, since the specific activity of the starting material is three times higher than that of any purified enzyme. It therefore seems likely that not all of the cellulolytic components from *T. fusca* are identified. The gene for the larger endoglucanase has now been cloned in *E. coli* (Collmer and Wilson 1983). Finally, it should be pointed out that another thermophilic actinomycete, *Microbispora bispora*, produces both endo- and exoglucanases. The latter has been positively identified as a cellobiohydrolase by Bartley et al. (1984).

The mesophilic actinomycete, *Streptomyces flavogriseus*, has been demonstrated by MacKenzie et al. (1984) to produce high levels of extracellular enzymes capable of hydrolyzing cellulose and xylan. One such enzyme, an exoglucanase, was purified to homogeneity. The purified enzyme has a molecular mass of approximately 45 000 and an isoelectric point of 4.15. The enzyme showed only negligiable activity toward CMC but was able to extensively hydrolyze acid-swollen cellulose. The main product of enzyme action is cellobiose. During the purification of the exoglucanase, several distinct endoglucanase fractions were identified. It was concluded that *S. flavogriseus* seems to more closely resemble their fungal counterparts than other bacterial cellulose degraders. This conclusion was also reached in a more recent paper by MacKenzie (1986). MacKenzie et al. (1987) examined the production of cellulases, xylanases, and other xylan-degrading enzymes produced by *S. flavogriseus* and *S. olivochromogenes* when grown on various substrates. With cellulose as substrate, the culture solution contained primarily cellulase activities, while culture solutions from xylan-grown cultures possessed high levels of both cellulase and xylanase activities. This indicates that the enzymes required for hydrolysis of cellulose and xylan are under separate control in *S. flavogriseus* and *S. olivochromogenes.*

2.6.2 Anaerobic Bacteria

Studies of cellulose degradation by anaerobic bacteria have recently accelerated enormously. The interest emanates from the possibility to use them in industrial fermentations, particularly for a direct conversion of both cellulose and hemicelluloses into ethanol. Anaerobic bacteria with cellulolytic activity are listed in Table 2.13. The cellulolytic system in anaerobic bacteria seems to be similar in so many respects that it is possible to make some generalizations. Thus, the production of a multicomponent cellulolytic complex seems to be a common denominator. Lamed et al. (1983a) gave this multicomponent cellulolytic complex, produced by *Clostridium thermocellum*, the term "Cellulosome". The enzymes involved in cellulose degradation by *C. thermocellum* have been studied more thoroughly than those of any other anaerobic bacterium, and the mechanisms for cellulose degradation by anaerobic bacteria will therefore be focused here on those employed by *C. thermocellum.*

C. thermocellum is an anaerobic thermophilic spore-forming bacterium. It was first described by Viljoen et al. (1926), but a pure culture giving reproducible results was first obtained by McBee (1948). *C. thermocellum* is extremely efficient in solubilizing crystalline cellulose.

It is generally recognized today that bacteria attacking insoluble materials maintain a physical contact with the substrate (Wiegel and Dykstra 1984). Thus, evidence indicates that the anaerobic cellulolytic bacteria attach themselves to cellulose to effect its degradation (Marshall 1980). There are several mechanisms by which bacterial cells bind to substrate surfaces. Ljungdahl et al. (1983) discovered that the yellow substance formed by *C. thermocellum* during growth on cellulose is an affinity substance that binds both the bacteria and their cellulolytic enzymes to cellulose. The yellow coloring of cellulose by *C. thermocellum* had been noticed earlier by many investigators (Viljoen et al.

Table 2.13. Anaerobic bacteria with cellulolytic activity (Ljungdahl and Eriksson 1985)

Acetivibrio cellulolyticus	*Butyrivibrio fibrisolvens*
Clostridium cellobioparum	*Ruminococcus albus*
C. papyrosolvens	*R. flavefaciens*
C. stercorarium	*Eubacterium cellusolvens*
C. acetobutylicum	*Micromonospora ruminantium*
Bacteroides succinogenes	*M. propionici*

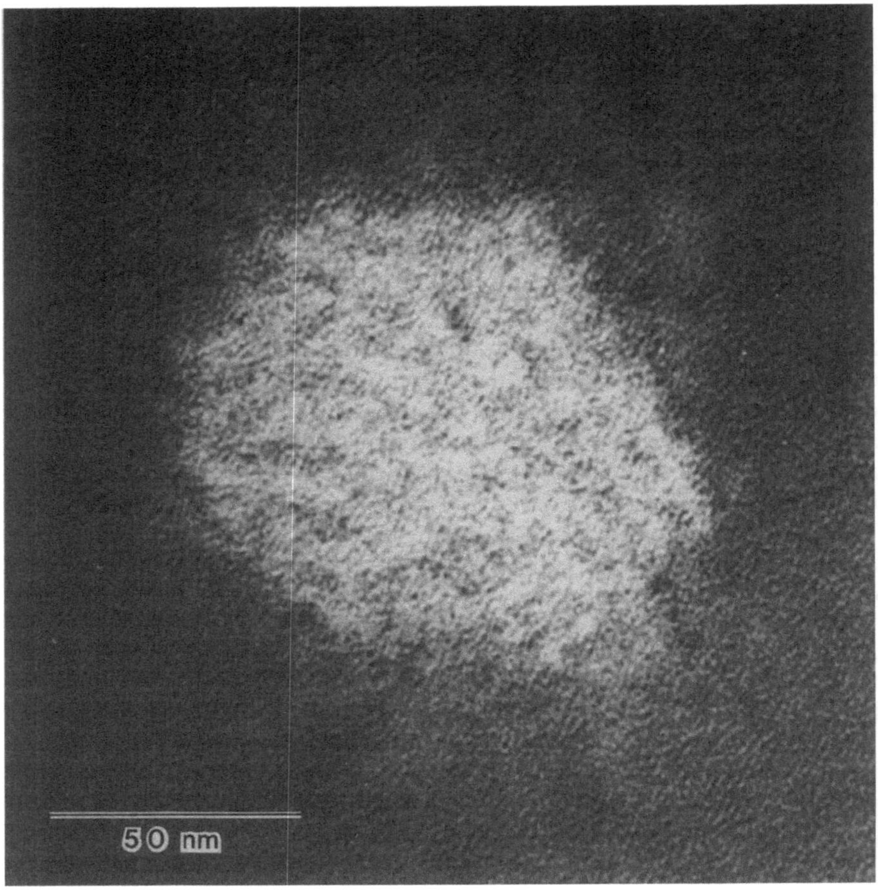

Fig. 2.22. Transmission electron micrograph of the larger, originally bound cellulose complex of *C. thermocellum* (Coughlan et al. 1985)

1926; McBee 1948; Ng et al. 1977). The appearance of cellulase activity in the culture solution is preceded by the secretion of the yellow affinity substance (YAS). Aggregates of the cellulosomes mentioned earlier (Lamed et al. 1983 a), i.e., polycellulosomes, located on the cell surface of the bacteria, are also involved in the attachment of the cell to the cellulosic substrate (Lamed et al. 1983 b, Bayer et al. 1985, Bayer and Lamed 1986, Mayer et al. 1987). It was shown by Bayer et al. (1983) that the adherence of the bacterial cell via the polycellulosomes to the substrate is specific for cellulose.

Several electron microscopy studies of the cellulase complex of *C. thermocellum* have been undertaken. Lamed et al. (1983 a, b) reported repeating globular complexes of the dimension 18 nm. However, larger aggregation products were also seen, particularly in the cell-associated cellulosome. Polycellulosomes isolated from residual cellulose in cultures of *C. thermocellum* strain

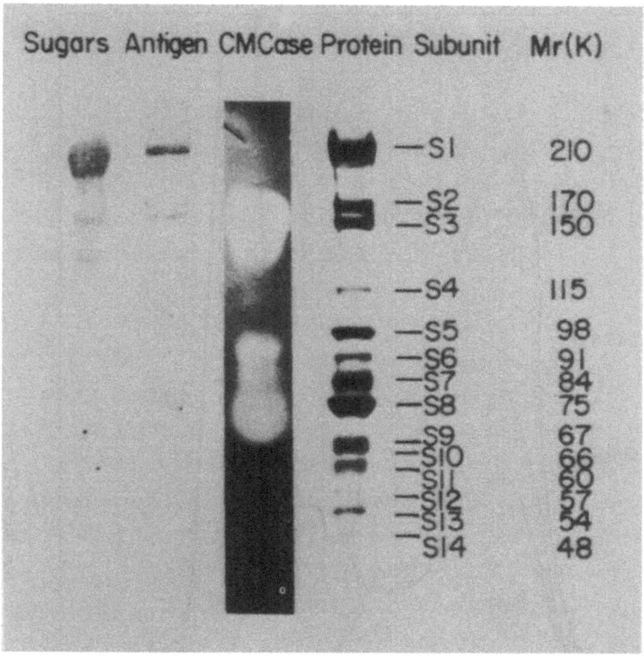

Fig. 2.23. Subunit composition and biochemical characterization of the cellulosome isolated from cellulose-grown *C. thermocellum* YS. Following SDS-PAGE (6% gels), protein content was determined by Coomassie Blue staining, endoglucanase activity by CMC overlay, anti-"cellulosome" specificity by immunoblotting, and sugar content by the enzyme hydrazide technique (Lamed and Bayer 1988)

JW 20 consist of two types of complexes (Hon-Nami et al. 1985). The larger of these complexes, termed the originally bound large (OBL) complex, was estimated to have a particle mass of 100×10^6 daltons, while the smaller, termed originally bound small (OBS) complex, was estimated to have a particle mass of 4.5×10^6 daltons (Coughlan et al. 1985). A major proportion of the complexes was of a larger size. Such a complex is visualized in Fig. 2.22.

Lamed and Bayer (1988), in an overview article, summarized the properties of the cellulosome from *C. thermocellum*. The subunit composition and biochemical characterization of the cellulosome isolated from cellulose grown *C. thermocellum* strain YS are presented in Fig. 2.23. By SDS treatment, 14 different polypeptides ranging in M_r from 48 000 to 210 000 can be distinguished by electrophoresis.

In a recent paper by Wu and Demain (1988) purification and characterization of the extracellular cellulase components of *C. thermocellum* based on their activities toward crystalline cellulose were described. Attempts were also made to determine the synergistic behavior between the different cellulase components. The approach was the following: A concentrated cotton-grown culture filtrate was fractionated on a Sepharose 2B column. As can be seen in

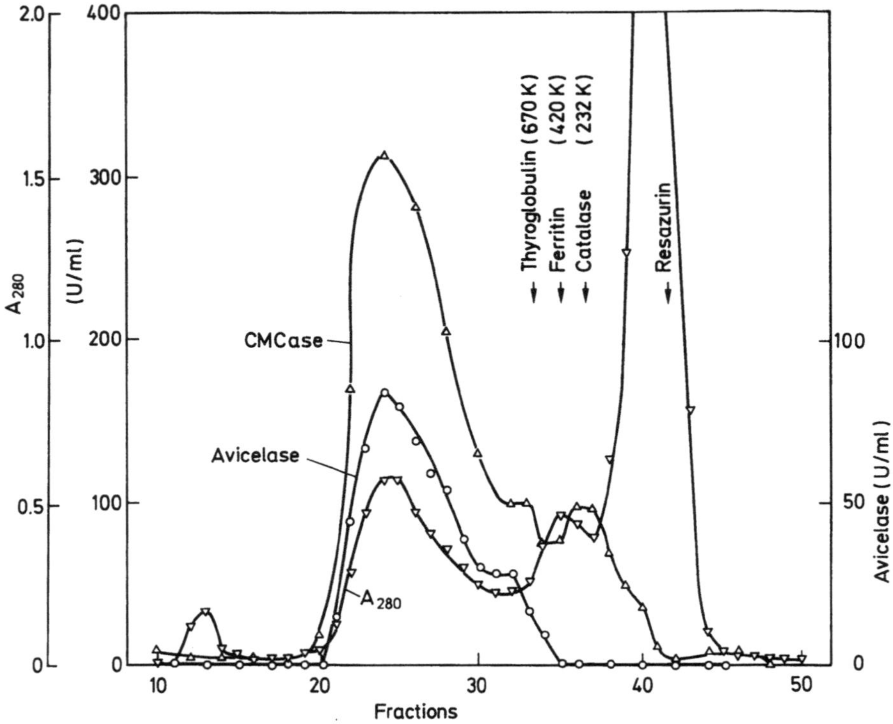

Fig. 2.24. Sepharose-2B chromatography of concentrated culture solution of *C. thermocellum* grown on cotton cellulose (Wu and Demain 1988)

Fig. 2.24, most of the so-called Avicelase and CMC-ase activities are found in the second peak of absorbancy. This peak eluted before thyroglobulin (MW 670 000). The aggregate was dissociated by sodium dodocyl sulfate (SDS) under nondenaturing conditions. The resulting protein sample was fractionated on an Ultrogel AcA 34 column. As can be seen in Fig. 2.25, peak II from Fig. 2.24 was resolved into four peaks. Without removal of the detergents, the individual fractions from the Ultrogel AcA 34 column were examined for Avicelase activity. No such activity was detected. When the detergent was removed, fraction 17 displayed low but noticeable activity toward Avicel. However, if fractions 17, 21, 27, and 43 were combined and detergents removed extensive hydrolysis of Avicel was observed. These results suggest that degradation of crystalline cellulose involves more than one protein subunit of the aggregate.

The four fractions from Ultrogel AcA 34 (Fig. 2.25) were studied by SDS gel electrophoresis. It was shown that peak I contained a mixture of high molecular mass proteins ($M_r = 82\,000-250\,000$). These protein species were designated S_L. Peak II, on the other hand, contained a mixture of low-molecular mass proteins designated S_S. Similar protein species were also found in peak III while peak IV had little or no protein.

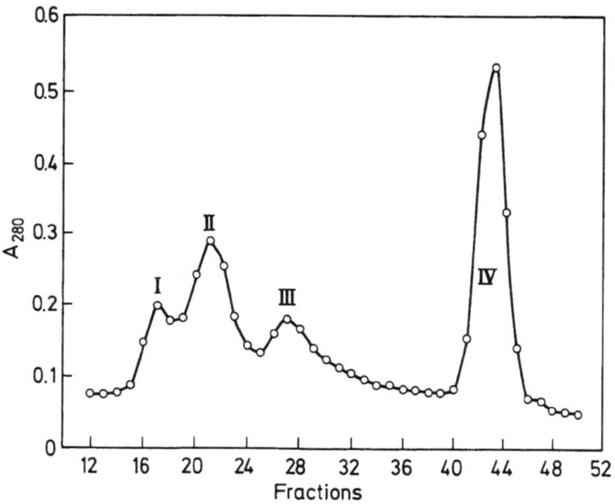

Fig. 2.25. Ultrogel AcA 34 chromatography of peak II from the Sepharose-2B column (Fig. 2.24) after treatment with SDS (Wu and Demain 1988)

Further investigations of Avicel degradation showed that a significant degradation was obtained by the combination of only two fractions. The results presented by Wu and Demain (1988) indicate that S_S and S_L were both necessary and sufficient for the hydrolysis of Avicel. A significant amount of reducing sugars was released from Avicel only when both S_L and S_S were combined. HPLC analysis indicated that cellobiose is the major product of hydrolysis. When cellobiose was added to the reaction mixture, Avicel degradation by S_S and S_L was dramatically decreased. Thus, the reconstituted Avicelase, like the crude enzyme mixture (Johnson et al. 1982), is sensitive to cellobiose inhibition.

Lamed and Bayer (1988) have compared the properties of the cellulosome with previously reported characteristics of crude cellulase preparations (Johnson et al. 1982). The cellulolytic activity was found to be enhanced by both CA^{2+} ions and thiols. A ninefold increase was obtained for the activity toward microcrystalline cellulose and a threefold increase for activity toward amorphous cellulose, while no increase of CMC-ase activity could be observed. However, cellobiose inhibited the detected enhancement in both cases. Wu and Demain (1988) studied the effect of Ca^{2+} on reconstituted Avicelase activity, i.e., a combination of the S_L and the S_S fractions. It was found that the amount of cellobiose produced from Avicel was significantly reduced in the absence of Ca^{2+}, namely 44% of that produced in the presence of Ca^{2+}. When EDTA (12 mM) was added to the reaction mixture in the absence of Ca^{2+}, no significant amount of cellobiose was detected.

In both white-rot (cf. Sect. 2.5.1) and soft-rot fungi (cf. Sect. 2.5.3), the accepted model for degradation of crystalline cellulose involves the synergism between an endo- and an exoglucanase, or between enzymes that at least partly

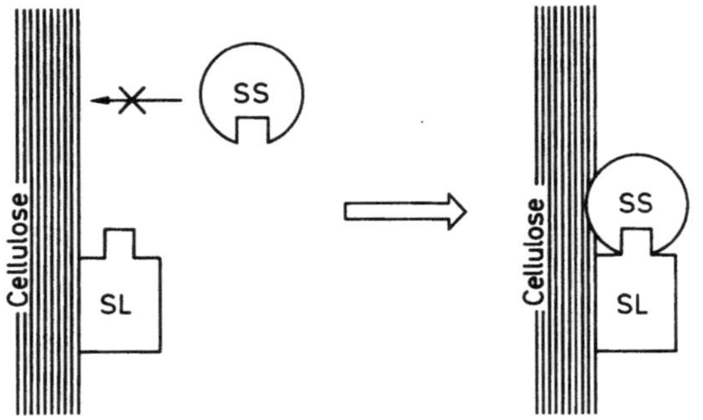

Fig. 2.26. Model of the anchoring of S_S by S_L to the cellulose surface (Wu and Demain 1988)

exhibit both of these activities (cf. Sect. 2.5.3). This model also seems to explain the synergism between the cellulase components of *C. stercorarium*, the only case of reported synergism between identified endo- and exoglucanase components in anaerobic bacterial cellulase systems (Creuzet et al. 1983). However, an exo-cellobiohydrolase has been found in the aerobic bacterium *Microbispora bispora* (Bartley et al. 1984). The *M. bispora* system is also similar to those of white- and soft-rot fungi since extracellular endoglucanases and cellobiohydrolases act in synergism to hydrolyze crystalline cellulose. However, the stable aggregation of the cellulase components S_L and S_S seems to be necessary for the function of the *C. thermocellum* cellulase system and has not been reported to occur in any other cellulolytic organism. Wu and Demain (1988) therefore suggest that its seems proper to consider the cooperative action of S_L and S_S as some kind of a subunit interaction. The cellulase aggregate of the cellulosome has a high affinity for cellulose. However, S_S seems to lack this affinity unless it is mixed with S_L. Wu and Demain (1988) therefore suggest that S_L appears to function as an anchor for S_S on the cellulose surface. Their model is outlined in Fig. 2.26.

It is noteworthy in young cultures of *C. thermocellum* that the extracellular cellulolytic enzyme system, the "cellulosome", is tightly bound to the substrate. As mentioned above, this binding is facilitated by the low molecular mass yellow affinity substance produced during the early stages of growth (Ljungdahl et al. 1983, Lamed et al. 1985). When the cellulose is consumed, the bound enzymes are released to the culture fluid. Several endo-1,4-β-glucanases with M_r values from 56 000 to 125 000 have been purified from this organism (Ait et al. 1979b, Ng and Zeikus 1981, Petré et al. 1981, Cornet et al. 1983). It seems likely that at least some of the above mentioned endoglucanases may be constituents of the cellulosome.

Since cellobiose and cellodextrins are the products of the concerted action of the *C. thermocellum* cellulosome and subunits, the question arises as to how these products are further metabolized. In most fungi, the cellobiose is hydrolyzed by a β-glucosidase, cellobiase. Such an enzyme also has been isolated from *C. thermocellum* (Ait et al. 1979a). The M_r of this enzyme is 50 000, and it has the isoelectric point at pH 4.68. The β-glucosidase of *C. thermocellum* was shown to be very heat-stable. However, the affinity of the en-

Table 2.14. Summary of the properties of the cellulolytic complexes of anaerobic bacteria (Coughlan and Ljungdahl 1988)

Organism	Surface location	Polycellulosome		Cellulosome		Subunits		Size range (K)	Calcium required	DTT required
		Diam. (nm)	Mass (million daltons)	Diam. (nm)	Mass (million daltons)	Number by SDS	Number by TEM			
A. cellulolyticus	–[a]	–	–	–	–	–	–	20–210	+	+
B. succinogenes	–	60	–	–	4.5	20+	–	20–200	–	+
C. thermocellum JW20	+	60	50–80	16–18	2.0–2.5	20+	35	45–200	+	+
C. thermocellum YM4	–	–	–	23–30	3.5	–	45–50	45–200	+	+
C. thermocellum YS	+	–	–	18	2.1	20+	–	45–210	+	+
C. thermocellum ATCC 27409	–	–	–	–	6.5	–	–	–	+	+
R. albus	+	50–60 200–300	–	–	>1.5	–	–	–200	–	+
R. flavefaciens	+	–	–	–	>3.0	–	–	–	+	+

[a] – Not studied.

zyme for cellobiose and PNPG was very low with a K_m value of 2×10^{-3} M for PNPG and 8.3×10^{-2} M for cellobiose. It therefore seems unlikely that this β-glucosidase functions as a cellobiase. It is posible that the β-glucosidase has an unspecific action and may participate in hydrolyzing saccharides having other types of β-structure linkages. The cellobiose and cellodextrins produced may instead enter the cell where they are metabolized by specific cellodextrin and cellobiose phosphorylases to yield glucose 1-phosphate and free glucose. As was demonstrated by Alexander (1968) and Sheth and Alexander (1969), formation of glucose 1-phosphate is an advantage for an organism from an energy point of view, since the energy of the glucosidic bond is then preserved. Glucose 1-phosphate and glucose are both fermented by *Clostridia* via the Emden-Meyerhof pathway.

In addition to the cellulolytic system of *C. thermocellum*, several other anaerobic bacteria have been studied with respect to their mechanisms for cellulose degradation. The properties of the cellulase complexes produced by *Acetivibrio cellulolyticus* (Saddler and Khan 1981, MacKenzie et al. 1987), *Ruminococcus flavefaciens* (Pettipher and Latham 1979, Latham et al. 1978), *Ruminococcus albus* (Wood and Wilson 1984, Stack and Cotta 1986), and *Bacteroides succinogenes* (Groleau and Forsberg 1983) are summarized in Table 2.14. The polypeptide of 200–210 K which in *C. thermocellum* is responsible for the attachment of the enzyme complex to the cell and to the cellulose substrate is produced also by each of the organisms presented in Table 2.14.

It has become increasingly apparent from recent studies that the enzyme mechanisms of cellulose degradation by *C. thermocellum* (Lamed and Bayer 1988, Wu and Demain 1988) are fundamentally different from those of the well-characterized fungal systems (cf. Sects. 2.5.1 and 2.5.3). The mechanisms for degradation of cellulose by *A. cellulolyticus* and *C. thermocellum* have many things in common. Thus, *A. cellulolyticus* also produces a yellow substance like *C. thermocellum* during active growth on cellulose. In addition Ca^{2+} and a reducing agent such as dithiothreitol are required for maximum hydrolysis of Avicel, as is the case with *C. thermocellum*. Another bacterium, the aerobic *Cellvibrio mixtus* (Wynne and Pemberton 1986), also secretes a water-insoluble yellow pigment when growing on cellulose.

The enzyme complex produced by *R. albus* is formed only in the presence of 3-phenylpropanoic acid (PPA) in the medium. Rumen fluid is also a source of PPA (Stack and Cotta 1986). PPA seems to be involved in the formation of the capsule surrounding the cell that contains the cellulase complex (Stack and Hungate 1984).

From what has been said here it can be concluded that the mechanisms and biochemistry of cellulose degradation by anaerobic bacteria and fungal systems are fundamentally different.

2.7 Regulation of Cellulase Synthesis

Production of cellulases is generally assumed to be subject to genetic regulation. Cellulases are inducible enzymes, the best inducers being cellulose, cellobiose, sophorose (2-0-β-D-glucopyranosyl-D-glucose), and lactose. Synthetic compounds such as palmitate and acetate esters of disaccharides (Reese et al. 1969) and thiocellobiose (Rho et al. 1982) have also been shown to function as inducers of cellulases.

The highest cellulase activities are normally found with cellulose as carbon source. In virtually all microorganisms examined, the synthesis of cellulases is induced by the presence of cellulose and repressed in the presence of glucose. If cellulase synthesis, as it seems, is regulated by an induction mechanism, the obvious difficulty is that cellulose itself cannot act as an inducer since it is insoluble in water and thus cannot enter the microbial cell. The most generally accepted view of the induction process is that the organisms in question produce a basic level or a constitutive amount of cellulase. This cellulase gives rise to soluble hydrolysis products of cellulose, which enter the microbial cells and function as inducers.

The search for inducers of this kind is long-standing and the best suggested so far is probably cellobiose (Mandels and Reese 1960). Cellobiose is a product of cellulase action and fungal growth on cellobiose results in cellulase production (Vaheri et al. 1979b). Cellobiose has also been found to be an inducer of cellulases in *Sporotrichum pulverulentum* (Eriksson and Hamp 1978), *Sporotrichum thermophile* (Canevascini et al. 1979), and *Neurospora crassa* (Eberhart et al. 1977). In Fig. 2.27 the inductive effect of cellobiose is demonstrated. Pellets of *S. pulverulentum* grown on glucose were washed and starved and were suspended in a CMC solution of known viscosity. It is clear from Fig. 2.27 that addition of cellobiose to the CMC medium gives rise to a faster drop in viscosity compared to when no cellobiose is added. The drop in viscosity is accomplished by induced endoglucanases. However, cellobiose does not

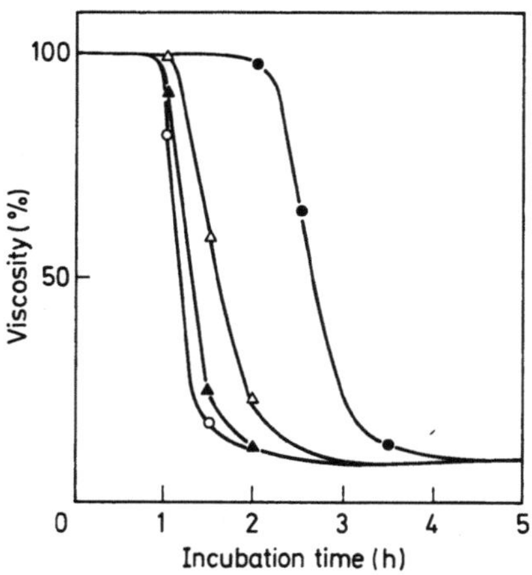

Fig. 2.27. Influence of cellobiose upon the induction time of endo-1,4-β-glucanase in *S. pulverulentum* in the CM-cellulose medium. Mycelium concentration 0.47 g l^{-1}. Initial concentration of cellobiose: (●) 0 mg l^{-1}; (△) 1 mg l^{-1}; (▲) 10 mg l^{-1}; (○) 50 mg l^{-1} (Eriksson and Hamp 1978)

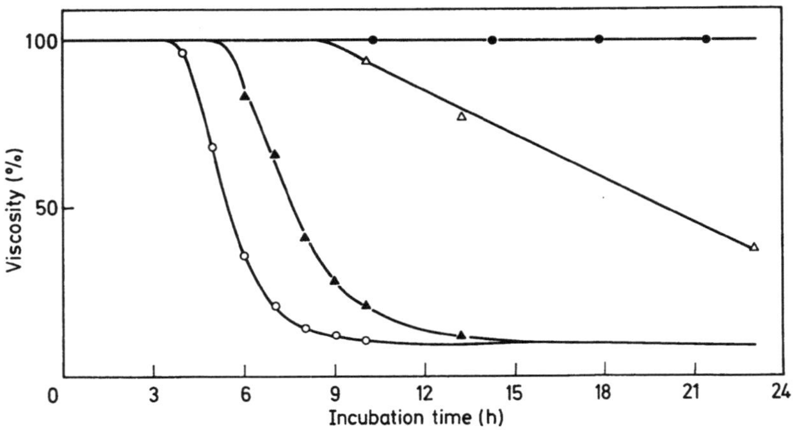

Fig. 2.28. Influence of sophorose upon the induction time of endo-1,4-β-glucanase in *Trichoderma viride* in the CM-cellulose medium. Mycelium concentration 0.17 g l^{-1}. Initial concentration of sophorose: (●) 0 mg l^{-1}; (△) 1 mg l^{-1}; (▲) 10 mg l^{-1}; (○) 50 mg l^{-1} (Eriksson and Hamp 1978)

seem to be an inducer of cellulase in *Trichoderma*, at least not with washed mycelium (Eriksson and Hamp 1978, Sternberg and Mandels 1979). A seemingly specific cellulase inducer was discovered by Mandels et al. (1962), namely, the disaccharide sophorose, which is a powerful inducer of both endo- and exocellulase synthesis in *T. reesei* (Gritzali and Brown 1979, Sternberg and Mandels 1979). Sophorose is also an inducer of cellulase synthesis in *S. pulverulentum* (Eriksson and Hamp 1978). However, it is an efficient inducer in only a limited number of cellulolytic fungi and not even all *T. viride* strains respond to it (Mandels et al. 1962, Mandels 1982).

During growth both on cellulose and on cellobiose, products of glucosyl transfer are found in the medium, one of these being sophorose. These glucosyl transfer products are reported to be better inducers than cellobiose (Mandels and Reese 1960, Reese and Maguire 1971). Fungal β-glucosidase is a transferase and it has been shown in vitro that this enzyme in the presence of cellobiose and glucose can form all the β-linked disaccharides, including sophorose (Crook and Stone 1957). If such a transferase activity takes place in vivo, sophorose might be a naturally occurring inducer of cellulase. It was in fact found by Vaheri et al. (1979a) that when cellulose is hydrolyzed with *T. reesei* a number of reversion disaccharides including sophorose were identified. Such a system would have some similarities to β-galactosidase induction in *Escherichia coli*, in which allolactose, the natural inducer, is formed by the transferase activity of β-galactosidase (Jobe and Bourgeois 1972).

There seems to be a considerable difference between *S. pulverulentum* and *T. viride* in regulating the production of cellulase. While *S. pulverulentum* endoglucanase activity is induced by cellobiose in as low concentrations as 1 mg/l (cf. Fig. 2.27), cellobiose at the same or even higher concentrations does

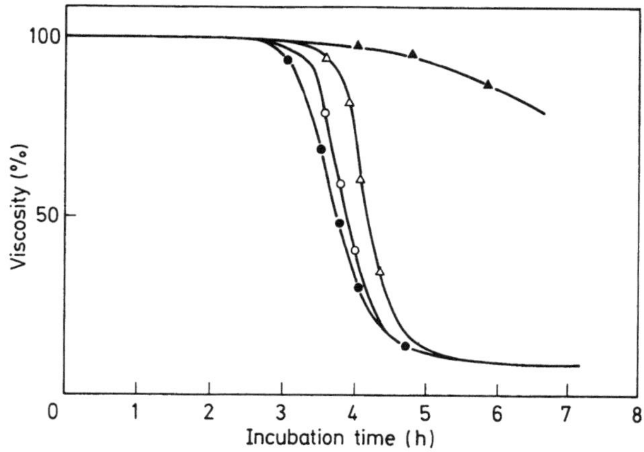

Fig. 2.29. Catabolite repression by glucose of endo-1,4-β-glucanase production in *S. pulverulentum* in the CM-cellulose medium. Mycelium concentration 0.09 g l^{-1}. Initial concentration of glucose: (●) 0 mg l^{-1}; (○) 50 mg l^{-1}; (△) 100 mg l^{-1}; (▲) 1000 mg l^{-1} (Eriksson and Hamp 1978)

not induce endoglucanase activity in *T. viride* QM6a (Eriksson and Hamp 1978). However, in *T. viride* sophorose causes induction of endoglucanases at a concentration of 1 mg/ml (Fig. 2.28). A study of endoglucanase production in the two fungi (Eriksson and Hamp 1978) also demonstrated several other differences. Thus, in a solution of CMC alone, endoglucanase production can be registered in *S. pulverulentum* but not in the *T. viride* QM6a strain (cf. Figs. 2.27 and 2.28). The conclusion drawn from this experiment was that there is a "basal synthesis level" of endoglucanases in *S. pulverulentum*, while this does not seem to be the case with washed mycelium of *T. viride* QM6a.

Nisizawa et al. (1971) found that in *T. viride* sophorose induction is very sensitive to the concentration of the inducer. While sophorose is a very efficient inducer in low concentrations, in high concentrations it inhibits production of cellulolytic enzymes. This has been explained by assuming that sophorose is enzymatically hydrolyzed to glucose, which inhibits cellulase synthesis. In a subsequent paper by Nisizawa et al. (1972), the repressive effect of glucose was clearly demonstrated. The formation of cellulase and aryl-β-glucosidase was in *T. viride* strongly repressed by 10 mM glucose and somewhat repressed by 1 mM glucose. In *S. pulverulentum*, catabolite repression of endoglucanase production is obtained at glucose concentrations as low as 0.3 mM (Fig. 2.29).

Nisizawa et al. (1972) suggested that catabolite repression in *Trichoderma* occurs at the translational level rather than at the transcriptional, based on results of experiments using actinomycin D and puromycin. In experiments by Montenecourt et al. (1981) it was found that during the period of endoglucanase induction, following exposure to sophorose, no particular differences in intracellular cAMP levels could be observed between induced and

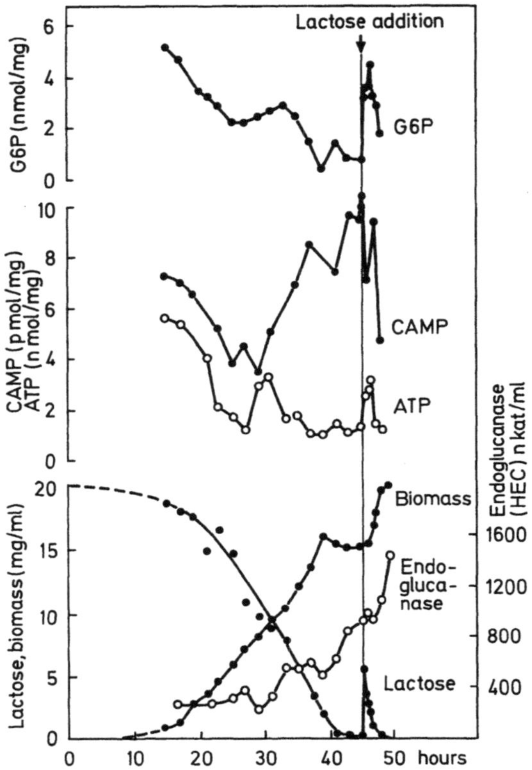

Fig. 2.30. Time course of cultivation of *Trichoderma reesei* on lactose as the sole source of carbon. *Arrow* and *dotted line* indicate the time of addition of lactose into the fermenter (Farkas et al. 1987)

noninduced cultures. Thus, in *Trichoderma*, synthesis of endoglucanases seems to be unrelated to the intracellular cAMP levels. This result is consistent with that of Nisizawa et al. (1972) that catabolite repression of inductive cellulase formation in *Trichoderma* occurs primarily at the level of translation.

Farkas et al. (1987) investigated the intracellular level of some key metabolites at the time of cellulase induction in a *T. reesei* mutant. The results are shown in Fig. 2.30; they indicate that the metabolic state of the cells is probably an important factor influencing the induction of cellulases. The period of cellulase induction was preceded by the transient peak of ATP concentration and accompanied by an increased cAMP level. When the substrate, lactose, was added, endoglucanase secretion was briefly stopped, ATP and glucose 6-phosphate concentration increased while the levels of cAMP decreased. However, the added lactose was rapidly consumed, and after 2 h endoglucanase secretion again resumed. The increased ATP level prior to cellulase induction may be due to a decrease in catabolite repression of enzymes involved in the respiratory pathway (Polakis et al. 1965, Cochet et al. 1984). The increased levels of cAMP at the time of cellulase synthesis indicate that it may play a role in modulating catabolite repression, although this was not the conclusion drawn by Montenecourt et al. (1981).

When the organism in question is grown on glucose, no or very low activity is registered, at least as long as the glucose concentration is not growth-limiting. However, Hulme and Stranks (1970) found in studies with *Myrothecium verrucaria* that, when glucose was exhausted, the endogenous metabolism of the organism could support the production of cellulases in the absence of cellulose. The conclusion drawn was that cellulases of *M. verrucaria* were produced in response to a strict regulation of metabolism rather than to the inductive influence of specifically added chemicals. In this way cellulases also seem to be constitutive enzymes, with production repressed by an abundance of any carbon source that can be rapidly utilized. However, it is not known whether this so-called constitutive synthesis is the "basal synthesis" referred to above in *S. pulverulentum*.

The production of cellulases in fungi can also be hampered by factors other than catabolite repression. Thus, Varadi (1972) demonstrated that a wide variety of phenols can repress the production of both cellulases and xylanases in *Chaetomium globosum* and *Schizophyllum commune*. The production of these enzymes is considerably repressed at concentrations of less than 1 mM of vanillic acid, vanillyl alcohol, and vanillin. It was shown by Ander and Eriksson (1976) that the production of endoglucanases was drastically repressed in a phenol oxidase-less mutant (Phe 3) of *S. pulverulentum* in the presence of kraft lignin and phenols at a concentration of 10^{-3} M. However, both in the wild-type (WT) and in a phenol oxidase-positive revertant (Rev 9), obtained by UV-irradiation of the Phe 3, the enzymes were produced without significant repression by phenols. Furthermore, if a highly purified phenol oxidase preparation was added to the growth medium of Phe 3, the endoglucanase production increased to normal also in the presence of phenols. These results indicate that because of the absence of phenol-oxidizing enzymes kraft lignin and phenols decrease endoglucanase synthesis in Phe 3.

In a recent paper by Müller et al. (1988), it was demonstrated that cellulase production in most white-rot fungi is stimulated by lignin-related phenolics. An explanation of this phenomenon also is offered by these authors, who suggest that stimulation of cellulase activity is due to an indirect rather than a direct induction by the phenols themselves. By definition, all white-rot fungi produce phenoloxidases that oxidize phenolic compounds to their corresponding quinones. These quinones are then reduced to phenols by the enzyme cellobiose: quinone oxidoreductase (Westermark and Eriksson 1974a, b). In this reaction, cellobiose is oxidized to cellobiono-lactone (cellobionic acid). Cellobionic acid and, especially its lactone, are powerful inducers of cellulases (Bruchmann 1978, Bruchmann et al. 1978). Figure 2.31 shows a possible mechanism for the inducing effect of lignin-related phenolics on cellulase production by white-rot fungi according to Müller et al. (1988).

The activity of cellulose-hydrolyzing enzymes in culture filtrates of fungi is dependent not only on mechanisms regulating their biosynthesis, but also on the presence of specific inhibitors regulating the activity of the enzymes themselves. One such inhibitor is gluconolactone, produced either by oxidation of glucose by the enzyme glucose oxidase or by hydrolysis of cellobionolactone

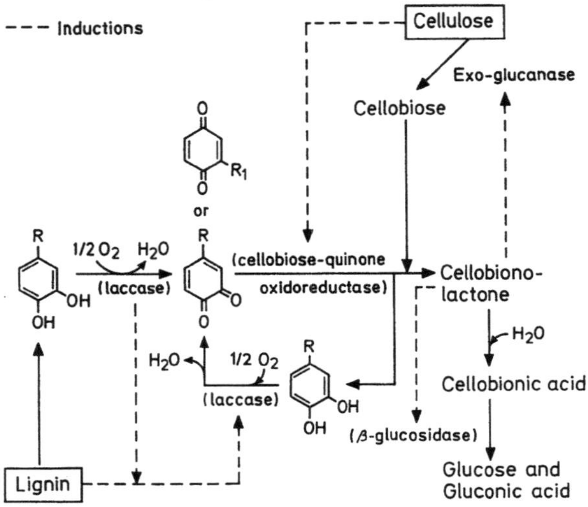

Fig. 2.31. Possible regulation mechanism of cellulase induction in *Trametes versicolor* in the presence of phenols (Müller et al. 1988)

(cellobionic acid). This lactone (acid) is formed by oxidation of cellobiose by either cellobiose oxidase or cellobiose: quinone oxidoreductase (Eriksson 1981 a, c). The importance of gluconolactone for the regulation of 1,4-β-glucosidases from *S. pulverulentum* has been studied by Deshpande et al. (1978). Cell-bound as well as extracellular 1,4-β-glucosidase activity appear in *S. pulverulentum*. The K_1 values for gluconolactone as a competitive inhibitor of the two extracellular β-glucosidases isolated from *S. pulverulentum* culture solution were 3.5×10^{-7} and 15×10^{-7}M, respectively.

However, not only inhibition mechanisms but also activation mechanisms seem to function in culture solutions of *S. pulverulentum*. Thus, two acidic proteases have been purified and demonstrated to increase the activity of at least the endoglucanases, as shown in Fig. 2.12. Particularly if the culture solution is harvested at an early stage of cultivation, protease treatment can produce up to a tenfold increase of the endoglucanase activity. An investigation of what the two acidic proteases actually do to the endoglucanase molecules is now under way (G. Pettersson and K.-E. Eriksson).

2.8 Synergistic Mechanisms Involved in Cellulose Degradation

It is now a well-established fact that degradation of crystalline cellulose, at least by white- and soft-rot fungi, is carried out by a multicomponent enzyme complex where the individual components interact in a synergistic way to convert the solid substrate into glucose. When isolated and purified to homogene-

Table 2.15. Cellulase (cotton-solubilizing) activity of endoglucanases, cellobiohydrolase, and β-glucosidase of *T. koningii* and *F. solani* when acting alone and in combination[a] (Wood and McCrae 1979)

Enzyme	Cotton solubilization (%)	
	F. solani	*T. koningii*
Cellobiohydrolase (CBH)	2	1
Endoglucanases (EG)	1	1
β-Glucosidase	1	nil
CBH + EG	58	53
CBH + β-glucosidase	18	20
CBH + EG + β-glucosidase	71	72
Original culture filtrate	71	71

[a] All enzymes were present in the same proportions in which they were present in 1 ml of original cell-free culture filtrates. Incubations were for 7 days at 37 °C.

ity, the individual enzyme components have until recently been considered unable to break down the crystalline cellulose to any substantial extent. However, when the components are recombined they act in a synergistic manner to render the cellulose soluble (Coughlan 1985, Eriksson and Wood 1985, Finch and Roberts 1985). It should, however, be mentioned (cf. Sect. 2.5.3) that CBH-I alone, according to Nummi et al. (1983), can hydrolyze cotton and *Valonia* cellulose (Chanzy et al. 1983). Enari and Niku-Paavola (1987) claim that some endo-type activity is included in the CBH-I function.

The first to suggest that a form of synergism was involved in cellulose degradation seems to have been Reese et al. (1950). They observed that some microorganisms were able to attack native cellulose, while others could degrade only cellulose derivatives. Their suggestion, therefore, was that native cellulose was degraded by a complex of enzymes where one, designated the C_1-enzyme, was assumed to break the intercrystalline hydrogen bonds.

Gilligan and Reese (1954), by fractionating culture filtrates of *Trichoderma viride* and *Myrothecium verrucaria*, found that the extent of hydrolysis of cell wall cellulose by the combined fractions was somewhat higher than the calculated sum of the extent of hydrolysis caused by the individual fractions (Mandels and Reese 1964). Li et al. (1965) found that the endoglucanase in combination with one, at that time unknown, enzyme in fractions of *T. viride* culture solutions, acted synergistically in degradation of substrates such as cotton and Avicel. Since then various investigators, and probably most of all Wood and co-workers (Eriksson and Wood 1985), have shown that the individual exoglucanases, endoglucanases, and β-glucosidases degrade practically none of the crystalline cellulose, while mixtures of the three hydrolytic enzymes cause extensive hydrolysis. Particularly endo- and exoglucanases are involved in this synergistic action. A demonstration of this is given in Table 2.15 where Wood and McCrae (1979) have demonstrated the synergistic effect between cellobiohydrolases and endoglucanases and to a limited extent also between

Table 2.16. Degradation of dewaxed cotton by the endo-1,4-glucanases from *S. pulverulentum* in the presence of the exo-1,4-β-glucanase[a]

Endo-1,4-β-glucanases	Degradation (%) measured by			
	Nelson-Somogyi method		Orcinol method	
	24 h	48 h	24 h	48 h
T_1	43	18	100	140
T_{2a}	165	124	272	209
T_{2b}	40	42	51	79
T_{3a}	392	206	847	219
T_{3b}	142	130	177	173

[a] Results are measured as the production of degradation products additional to the numerical summation of endo- and exo-activities (Streamer et al. 1975).

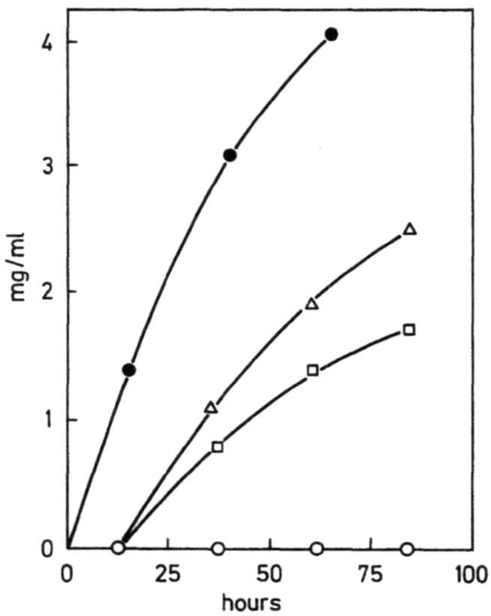

Fig. 2.32. Synergistic effects in hydrolysis of cellulose by components of *T. emersonii* filtrates. (●) Full complex added at zero time; (○) endocellulase alone added at zero time; (□) endocellulase added at zero time and exocellulase added at 12 h; (△) endocellulase added at zero time and exocellulase plus β-glucosidase added at 12 h (McHale and Coughlan 1980)

cellobiohydrolase and β-glucosidase from *T. koningii* and *Fusarium solani*. A similar synergism was demonstrated between the endo- and exoglucanases from *Sporotrichum pulverulentum* (Streamer et al. 1975; Table 2.16). The synergistic effect of the individual hydrolytic enzymes in culture filtrate from *Talaromyces emersonii* is shown in Fig. 2.32.

Until now, synergism between endo- and exoglucanases has been shown for enzymes from *T. viride* and *Penicillium funiculosum* (Selby 1969), *T. koningii*, and *Fusarium solani* (Wood 1969, 1975, 1980, Wood and McCrae 1972, 1975, 1977a, b), *T. koningii* (Halliwell and Riaz 1971), *S. pulverulentum* (Streamer et al. 1975), *Talaromyces emersonii* (McHale and Coughlan 1980), and

Table 2.17. Synergistic effects shown by combinations of cellobiohydrolase with endoglucanase +β-glucosidase from different sources in solubilizing cotton cellulose[a] (Wood 1980)

Source of cellobiohydrolase	Source of endoglucanase and β-glucosidase	Solubilization of cotton (%)
T. koningii	*T. koningii*	54
T. koningii	*F. solani*	79
P. funiculosum	*P. funiculosum*	72
P. funiculosum	*T. koningii*	51
F. solani	*F. solani*	59
F. solani	*T. koningii*	51
T. koningii	*M. verrucaria*	20
T. koningii	*Stach. atra*	21
P. funiculosum	*M. verrucaria*	11
F. solani	*M. verrucaria*	20
T. koningii	*M. echinata*	20
T. koningii	*G. roseum*	15
T. koningii	*R. albus*[b]	1

[a] All assays contained the same endoglucanase activity. Neither of the enzymes could degrade cotton significantly when acting alone.
[b] Contained no β-glucosidase.

Sclerotium rolfsii (Sadana and Patil 1985). The real synergism between endo- and exoglucanases seems to appear only on crystalline substrates, is very low with amorphous cellulose, and absent with soluble derivatives (Streamer et al. 1975, Wood and McCrae 1979).

If the hydrolysis of crystalline cellulose can be explained in terms of a sequential attack on the substrate by randomly acting endoglucanases and exo-acting cellobiohydrolases, it is reasonable to expect that an endoglucanase from one fungus should act in synergism with a cellobiohydrolase from another. Such cross-synergism has indeed been demonstrated (Wood 1975). However the extent of cooperation seems to vary considerably. Thus, endo- and exoglucanases from the fungi *T. koningii*, *F. solani*, and *P. funiculosum* can be exchanged and give rise to extensive solubilization of highly ordered cellulose. These three fungi have in common that they all produce both endo- and exoglucanases. However, when exoglucanases of these fungi are mixed with endoglucanases isolated from culture filtrates of fungi that apparently do not produce exoglucanases, namely, *Myrothecium verrucaria*, *Stachybotrys atra*, *Gliocladium roseum*, and *Memnoniella echinata* (Wood 1980), synergism is low as is evident in Table 2.17. Although the last four organisms mentioned above grow well on cotton, only endoglucanases were found in their culture filtrates (Wood 1975). Wood (1983) has demonstrated that when the cellobiohydrolases from *T. koningii*, *F. solani*, and *P. funiculosum* are acting together with the endoglucanases of the rumen anaerobic bacteria *Ruminococcus albus*, *R. flavefaciens*, and *Bacteroides succinogenes*, no synergistic effect is obtained. None of the rumen bacteria are known to produce a mixture of cellulases that contains exoglucanases. Recently, however, Gow and Wood

Table 2.18. Percentage hydrolysis of cotton by cellulases of *C. thermocellum* and *T. koningii* acting alone and in combination (Gow and Wood 1988)

Additions to *C. thermocellum* cellulase or *T. koningii* cellobiohydrolase	*C. thermocellum* cellulase (A)	*T. koningii* cellobiohydrolase (B)	A + B
None	8	0	40
10 mM DTT	0	0	35
7 mM $CaCl_2$	20	0	55
10 mM DTT + 7 mM $CaCl_2$	35	0	68

(1988) have demonstrated a synergistic action between cellulase components from the anaerobic thermophilic bacterium *Clostridium thermocellum* and *T. koningii*. This is the first demonstrated synergism between fungal and bacterial cellulases. The cellulase complex of *C. thermocellum* cannot extensively break down highly ordered cellulose unless Ca^{2+} and dithiothreitol (DTT) are present. However, when the cellobiohydrolase from *T. koningii* is combined with the *C. thermocellum* cellulase, cellulose is broken down also in the absence of Ca^{2+} and DTT. The synergistic effect of the two cellulase preparations are demonstrated in Table 2.18. Earlier results have indicated that synergism between endo- and exoglucanases from different organisms only occurs in cases where both organisms produce a full cellulase complex. It is not known whether the *C. thermocellum* cellulase complex also contains an exoglucanase.

In another study of the synergistic phenomena, Wood (1980) isolated endo- and exoglucanases from culture filtrates of *T. koningii* grown on a variety of cellulosic substrates. An endoglucanase obtained in the culture solution with one substrate was mixed with the cellobiohydrolase obtained from another substrate, and the extent to which cotton was solubilized by the enzyme mixture was measured. It was found that all the cellobiohydrolases showed the same capacity for hydrolyzing cotton cellulose when acting in cooperation with the endoglucanase induced on this substrate. However, there were considerable differences in cotton solubilization when the cellobiohydrolase induced on cotton was mixed with an endoglucanase induced on a different substrate. It thus seems that the synergism is dictated primarily by the endoglucanase component, the nature of which may determine whether or not a complex formation with the cellobiohydrolase can take place in the attack on the substrate (Wood 1975, 1980). Wood (1985a) also observed that both of the cellobiohydrolases of *P. funiculosum* can act in synergism with the endoglucanase produced by this organism to solubilize cotton, but that only one of the two cellobiohydrolases can cooperate with the endoglucanases of *F. solani* and *T. koningii*.

All of the synergistic phenomena described here so far are of the endo/exo-type. However, in addition, two unusual types of synergism have been documented. Kanda et al. (1976) reported that the combined action of *Irpex lacteus* endoglucanases caused considerable hydrolysis of crystalline cellulose.

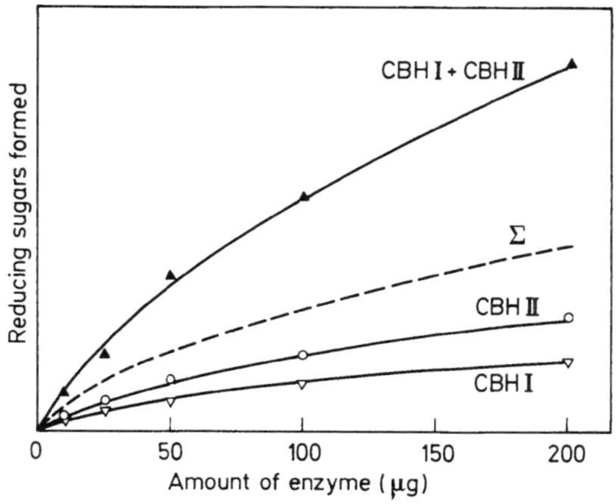

Fig. 2.33. Synergism exhibited toward crystalline cellulose (Avicel) incubated with the cellobiohydrolase *CBH-I* (△) and *CBH-II* (○) alone and in combination (▲). The non-synergistic sum of the activities of CBH-I and CBH-II is indicated by Σ (Fägerstam and Pettersson 1980)

The two endoglucanases differed in the way they attacked soluble cellulose. This synergistic effect is difficult to interpret and there is presently no explanation for this phenomenon. Further confirmation is, therefore, awaited.

Another type of unusual synergistic interaction, one of the exo/exo-kind, was first reported by Fägerstam and Pettersson (1980) who found that the hydrolysis of crystalline cellulose by the combined action of the cellobiohydrolases CBH-I and CBH-II of *T. reesei* was more then twice that calculated from the extent of hydrolysis by each of the separate enzymes (Fig. 2.33). This finding was confirmed by Brown and Gritzali (1984) who found that the exo/exo-synergism is important in the intact cellulase system as it degrades crystalline cellulose. Variation of the fraction of CBH protein represented by CBH-I and CBH-II (together with fixed levels of endoglucanase and β-glucosidase) revealed an optimum composition of the system acting on microcrystalline cellulose. However, on amorphous cellulose CBH-II only exhibited the optimum hydrolysis rate as one might expect from the five- to ten fold greater activity for CBH-II compared to CBH-I. The synergistic activity of the cellulase components of the *Trichoderma* system is thus a fundamental property.

Synergism of cellulases from *T. reesei* in degradation of crystalline cellulose has also been studied by Henrissat et al. (1985). The action of CBH-I and CBH-II and the endoglucanases, EG-I and EG-II purified from *T. reesei*, was evaluated against various substrates. The authors report that CBH-I degraded the β-D-glucan from barley in a typical endo-pattern. With cellulose as substrate, the synergism between EG-I or EG-II and CBH-I depended on the structural and ultrastructural features of the substrate. Synergistic degradation of cellulose also was obtained with mixtures of CBH-I and CBH-II. Synergism between the endoglucanases and CBH-II, on the other hand, follows the normal patterns for endo/exo-cooperation. The results from all these studies are summarized in Figs. 2.34–2.37. EG-I and CBH-I were mixed in the proportions 4 : 1, 1 : 1, and 1 : 4. The activity of these mixtures was assayed with various cellulosic substrates (Fig. 2.34). EG-I and CBH-I acted synergistically with

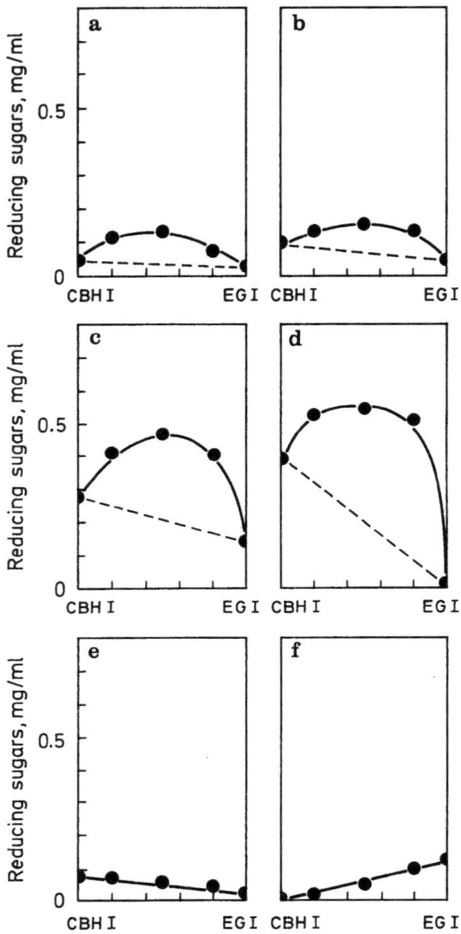

Fig. 2.34. Degradation of various celluloses by CBHI-EGI mixtures. **a** Filter paper; **b** Avicel; **c** homogenized Avicel; **d** bacterial microcystalline cellulose; **e** Valonia microcrystals; **f** CM-cellulose. The *dotted lines* represent the theoretical hydrolysis values expected for a noncooperative degradation (Henrissat et al. 1985)

some of the substrates [filter paper, Avicel, homogenized Avicel, and bacterial microcrystalline cellulose (MCC)]. As can be seen in Fig. 2.34 the synergistic action was dependent on enzyme composition. The most effective mixture was a 1 : 1 composition of the two enzymes. However, the EG-I and CBH-I do not cooperate on the highly crystalline *Valonia* cellulose or on the soluble CMC. The synergistic degradation of Avicel or bacterial MCC catalyzed by CBH-I supplemented with EG-II is presented in Fig. 2.35.

CBH-II appears to have a different behavior compared with CBH-I. CBH-II combined with EG-I or EG-II also degrades the insoluble substrates in a synergistic fashion (Fig. 2.36). However, unlike the mixtures of CBH-I with the endoglucanases, the maximum activity is obtained at a composition of 95 : 5 of CBH-II and the two endoglucanases, respectively. Finally, the synergistic effects obtained with mixtures of CBH-I and CBH-II are presented in Fig. 2.37. A distinct synergistic effect is obtained at a maximum in the proportion 1 : 4,

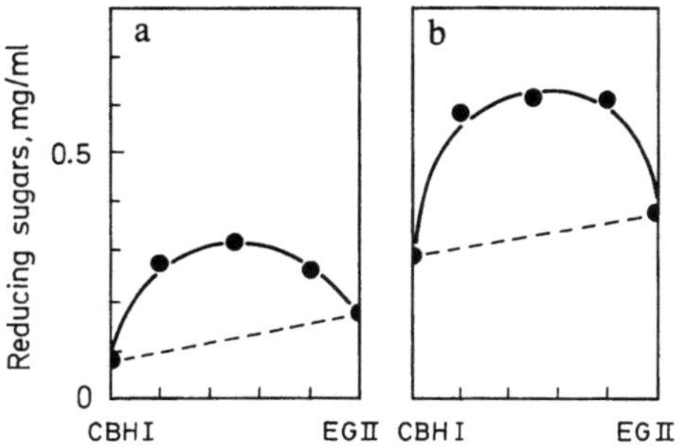

Fig. 2.35. Synergistic degradation of (**a**) Avicel and (**b**) bacterial microcrystalline cellulose by various CBH-I – EG-II mixtures. The *dotted lines* represent the theoretical hydrolysis values expected for a non-cooperative degradation (Henrissat et al. 1985)

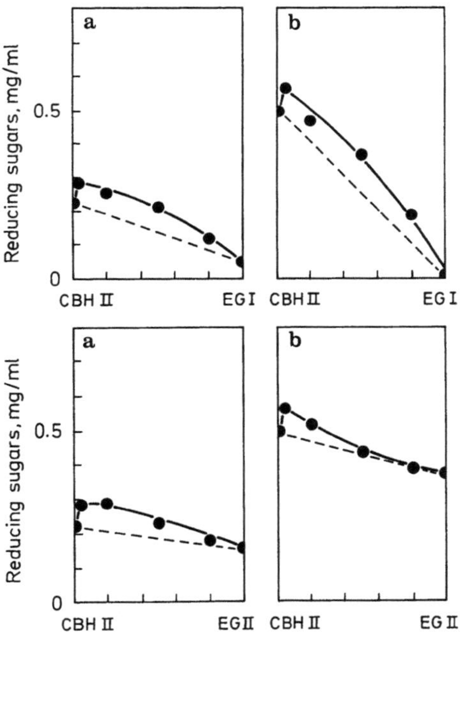

Fig. 2.36. Degradation of (**a**) Avicel and (**b**) bacterial microcrystalline cellulose by various CBH-II – EG-I and CBH-II – EG-II mixtures. The *dotted lines* represent the theoretical hydrolysis values expected for a non-cooperative degradation (Henrissat et al. 1985)

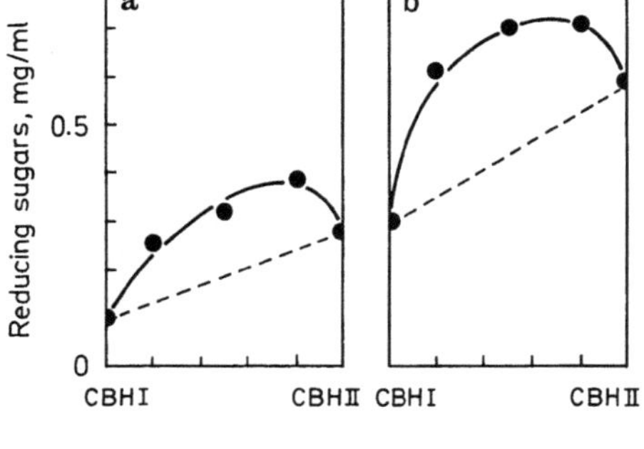

Fig. 2.37. Synergistic degradation of (**a**) Avicel and (**b**) bacterial microcrystalline cellulose by CBH-I – CBH-II mixtures. The *dotted lines* represent the theoretical hydrolysis values expected for a non-cooperative degradation (Henrissat et al. 1985)

of CBH-I : CBH-II. The strong synergistic effects obtained with CBH-I and CBH-II toward crystalline cellulose have been discussed by Henrissat et al. (1985). According to unpublished data by van Tilbeurgh et al. concerning the mechanism of action of the two exoenzymes, CBH-II behaves more typically as an exoenzyme than CBH-I. Chanzy et al. (1984) have shown that CBH-I has an endo-absorption pattern on cellulose and that it also has a significant lack of specificity for the penultimate glucosidic bond of oligocellodextrins (van Tilbeurgh et al. 1982). Also, in the paper by Henrissat et al. (1985) CBH-I was shown to catalyze the attack on barley β-D-glucan in an endo-fashion. Although the endo-character of CBH-I is now clearly established, its exact mechanism of action is incompletely understood. Despite the endo-character of CBH-I, a simple endo/exo-synergism between CBH-I and CBH-II does not fully describe the synergistic action exhibited by these two enzymes. CBH-I demonstrates an unusually strong and irreversible binding to cellulose (Reese 1982, Chanzy et al. 1983). In contrast to classical enzyme-substrate interaction, Sprey and Lambert (1983) even suggest that there is a lectin-type binding between cellulases and the substrate cellulose. It is suggested by Henrissat et al. (1985) that the synergism between CBH-I and CBH-II could be explained in terms of competitive adsorption which may enhance the turnover of the enzymes. These authors also point out the possibility of formation of a binary enzyme complex at the cellulose surface between the two enzymes. However, the most likely explanation of the synergistic action exhibited by CBH-I and CBH-II seems to be the mixture of both an endo- and an exo-type character of the function of both of these enzymes (cf. Sect. 2.5.3). In view of these results and results from their own group Enari and Niku-Paavola (1987) suggested the scheme for hydrolysis of cellulose as is presented in Fig. 2.38. According to this scheme, hydrolysis of insoluble native cellulose is accomplished by a synergistic action of the two enzymes CBH-I and CBH-II. The endoglucanases participate in the hydrolysis only at a later stage when soluble cellodextrins are hydrolyzed. This is clearly a very different picture compared to earlier and indicates that the final word on the true enzymatic mechanisms of hydrolysis of crystalline cellulose has not yet been spoken.

Synergistic action between two immunologically unrelated cellobiohydrolases (I and II) isolated from the extracellular cellulase system of the fungus *Penicillium pinophilum* was recently demonstrated by Wood and McCrae (1986a). The maximum rate of attack on the microcrystalline cellulose Avicel was exhibited when the two enzymes were mixed in the ratio 1 : 1. The authors suggest, in an attempt to explain the phenomenon of synergism between the two cellobiohydrolases, that they may be two stereo-specific enzymes concerned with the hydrolysis of the two different configurations of nonreducing groups that would exist in cellulose. So far only one type of cellobiohydrolase has been isolated from culture solutions of *F. solani* and *T. koningii*. Only CBH-II from *P. pinophilum* acts synergistically with the cellobiohydrolases of *T. koningii* and *F. solani* to solubilize Avicel. CBH-II showed no synergistic action with the endoglucanases of *T. koningii* of *F. solani* but CBH-I did.

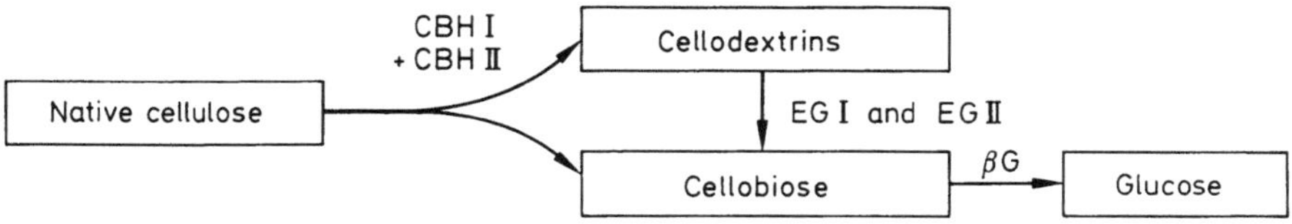

Fig. 2.38. Suggested scheme for the enzymatic hydrolysis of cellulose. Enzymes: *CBH* cellobiohydrolase; *EG* endo-β-glucanase; β G β-glucosidase (Enari and Niku-Paavola 1987)

Studies of synergism involving exoglucanases have normally been carried out with cellobiohydrolases. However, the exo-enzyme 1,4-β-glucan-glucohydrolase, found in *P. funiculosum* and *T. viride*, is unable to act synergistically with endoenzymes. Thus, Wood and McCrae (1982) found no increase in the solubilization of dewaxed cotton when the enzyme from *P. funiculosum* was added to a mixture of endoglucanase and β-glucosidase from the same organism. This is very much in contrast to the marked increase obtained in solubilization when a cellobiohydrolase was used.

In addition to the synergistic phenomena described above, other synergistic effects are also involved in cellulose degradation. Thus, cellobiose, which is one of the principal degradation products in an endo- and exoglucanase attack on cellulose, also inhibits these enzymes strongly. As low as 0.01% concentration of cellobiose is enough to cause a 79% inhibition of endo- and exoglucanases from *T. koningii* in their attack on cotton cellulose (Wood and McCrae 1978). Due to its strong inhibitory effect, cellobiose is a key product in cellulose degradation that is essential for the fungi to eliminate. Thus, the white-rot fungus *S. pulverulentum* has no less than four different ways to control the cellobiose concentration in its environment (Eriksson 1978; Fig. 2.9). The first of these is via the hydrolytic enzyme β-glucosidase, which hydrolyzes cellobiose and higher cellodextrins to glucose. Therefore, as would be expected, addition of β-glucosidase to the two other hydrolytic types of enzymes from *T. reesei* gives rise to more than an additive increase in hydrolytic activity (Gong et al. 1978).

Another way in which fungi deplete their cellobiose is via transglucosylation reactions. Thus, when the cellobiose concentration is high enough it has been demonstrated in *S. pulverulentum* that the endoglucanases, via transglucosylation, lower the cellobiose concentration (Fig. 2.9).

As described in Sect. 2.5.1, at least two oxidative enzymes are involved in cellulose degradation by *S. pulverulentum*. The first of these, cellobiose oxidase, is able to catalyze the oxidation of higher cellodextrins to their corresponding gluconic acids utilizing molecular oxygen (Ayers et al. 1978, Ayers and Eriksson 1982). This enzyme has not been isolated from culture filtrates of *T. reesei*, but oxidation products such as gluconolactone, cellobionolactone, and cellobionic acid have been identified by Vaheri (1982a, b).

The second oxidative enzyme from *S. pulverulentum* is cellobiose: quinone oxidoreductase (Westermark and Eriksson 1974a, b). The reaction catalyzed by this enzyme is shown in Fig. 2.11. This enzyme is of importance for the degradation of both lignin and cellulose. It is relatively specific for cellobiose but

Table 2.19. Degradation of cotton cellulose by enzymes from *S. pulverulentum* in oxygen and nitrogen atmosphere, respectively (Eriksson et al. 1974)

Tube number	Enzyme preparation	Cellulose degradation weight loss (%)
1	Concentrated culture solution	52.1 (Oxygen atmosphere)
2	Concentrated culture solution	21.5 (Nitrogen atmosphere)
3	Mixture of endo- and exoglucanases	20.0
4	Endo-1,4-β-glucanase	0.0
5	Exo-1,4-β-glucanase	0.0

much less specific for its quinone substrates. Thus, it can reduce both ortho- and paraquinones. The two oxidative enzymes also seem to be important in preventing enzyme inhibition by cellobiose. It was thus found by Eriksson et al. (1974) that there is considerably more degradation of cotton fibers by cell-free culture filtrates of *S. pulverulentum* in an oxygen than in a nitrogen atmosphere (Table 2.19). The cellobionic acid formed on oxidation of cellobiose by the two oxidative enzymes of *S. pulverulentum* can be hydrolyzed by β-glucosidase (Eriksson 1978). However, the product of this degradation, gluconolactone, strongly inhibits β-glucosidase (Deshpande et al. 1978). Bruchmann et al. (1987) purified β-glucosidase and the enzyme lactonase from a commercially available cellulase preparation, Novozym 188, an enzyme preparation from *Aspergillus niger* (Novo, Copenhagen, Denmark). The rate of cleavage of different substrates by β-glucosidase decreases in the order cellobiose > cellobionic acid > cellobionolactone. When the specific activity of the purified lactonase was studied with different lactones as substrates, it was shown that only glucono-1,5-lactone and cellobionolactone were hydrolyzed. It was also demonstrated that only glucono- and cellobionolactones were able to inhibit β-glucosidase. Furthermore, cellobionolactone inhibits all cellulolytic activities. Lactonase decreases the inhibitory effects of both glucono- and cellobionolactones on β-glucosidase and particularly the effects of cellobionolactone on the complete cellulase system and is therefore a component contributing to the synergistic attack of this system.

2.9 Examples of Applications of Cellulolytic Microorganisms and Their Enzymes

Approximately 0.1% of the solar energy incident upon the earth is fixed by green plants through photosynthesis (Reese et al. 1972). The annual production of this process amounts to about $15-20\times10^{10}$ tonnes of organic plant substance, half of which is cellulose. The cost of many important fermentation products depends very strongly upon the cost of the carbohydrate raw material. Lignocellulose conversion offers a potential for less expensive carbon sources in such fermentation processes. To realize this potential, two crucial objectives must be fulfilled.

1. The development of effective and economic pretreatment techniques to increase the yield of sugars from the hydrolysis of cellulose and hemicelluloses.
2. An integrated use of all the components of lignocellulosic materials must be obtained. However, such an ideal biomass refinery process has yet to be constructed.

2.9.1 Fermentation of Lignocellulosic Materials to Protein

The human population is increasing at a rate that raises doubt that the inherited systems of agriculture and animal husbandry will be sufficent to nourish all. The world population is expected to be 6 billion by the year 1997 and anticipated food production will be roughly 3900 million metric tonnes. If efficient processes allowing conversion of lignocellulosic materials into protein for feed and food could be developed, a valuable resource would have been created. However, meaningful studies of the food value of specific cellulose decomposing organisms are limited (Ghose and Pathak, 1975).

Only two processes are currently in operation for the production of microbial protein from wood sugars. These processes are the so-called Candida and Pekilo processes, based on the yeast *Candida utilis* and the fungus *Paecilomyces varioti*, respectively (Forss et al. 1974, Jones 1974). The Candida process for protein production started at the beginning of this century, while the Pekilo process is more recent. In both of these processes, utilizable substrates are mainly monomeric sugars, hexoses as well as pentoses and acetic acid present in spent liquors from the sulfite pulp cooking process. Disaccharides and higher oligosaccharides are utilized by these organisms only to a limited extent.

However, two additional processes for a direct conversion of solid lignocellulosic wastes to protein by fungi have been developed to pilot plant scale. The fungi employed in these processes are the white-rot fungus *Sporotrichum pulverulentum* (Ek and Eriksson 1980) and the mold *Chaetomium cellulolyticum* (Moo-Young et al. 1977). The two processes have been developed at the Swedish Pulp and Paper Research Institute (STFI) and the University of Waterloo, Canada, respectively. On evaluation, the STFI process was not found to be economically feasible since the protein produced could not compete on a cost basis with soybean protein, unless the substrate used in the fermentation had a negative value. At present, this probably applies to all microbial protein production processes based on lignocellulosic substrates, at least with Western economics.

One example of a negative-value substrate is dissolved substances in waste waters from mechanical pulp production. In all the different processes for production of very high yield pulps [such as mechanical (RMP), thermomechanical (TMP), and chemithermomechanical (CTMP) pulps], organic materials in the form of sugars, low molecular mass lignins, extractives, etc. are released from the wood and appear in the circulating white-water system. This is particularly true for the TMP and CTMP processes. The sugars dissolved in the white water are mixtures of monomers and oligomers, all excellent substrates for fungi producing extracellular enzymes for hydrolysis of the polymeric material to monomers. The fungus chosen for the STFI process, *S. pulverulentum*, is a white-rot fungus and is particularly well suited for the purpose of protein production from wood components since it produces enzymes for degradation of all the components in lignocellulosic materials.

The STFI approach to protein production based on lignocellulosic material was thus changed from the conversion of solid lignocellulosics into protein to a water purification process, in which dissolved substances in the white water from mechanical pulp production were used as substrates. The process has been tested at a pilot plant scale in a newsprint paper mill with a residence time of 17 h in the fermenter. No buildup of organic materials takes place in the white-water system. Considerable reduction of the residence time is expected if part of the mycelium produced in the continuous process is recirculated to the fermenter. This would also help in keeping the fermenter volume down. For a more or less completely closed white-water system of a big newsprint mill, fermenter volumes of 200–400 m^3 are necessary. The fungal mycelium, which occurs in the form of pellets of 0.2–0.4 mm in diameter, can be easily separated from the culture solution by filtration.

The water purification and protein production process has also been operated on a 25 m^3 scale. In this particular case, the waste water from a board factory was used as substrate (Ek and Eriksson 1980). The resulting protein was tested in feeding trials at the Agricultural University, Uppsala, Sweden. These feeding trials were essentially positive, particularly with ruminants (Thomke et al. 1980). However, monogastric animals could not easily digest the fungal cell walls. In spite of the positive feeding trials with ruminants, the production of fodder protein by the *S. pulverulentum* fungus is economically feasible only if the cost of alternative water purification is also taken into account.

However, there are other possibilities to use the fungal mycelium produced in addition to its use as cattle feed. It can thus be incorporated in the paper without influencing paper quality. As a rule of thumb, 1 kg sugar gives rise to 0.5 kg fungal mycelium. In a closed RMP process, where ca. 3% of the wood is converted to water-soluble substances, the fungal mycelium will correspond to approximately 1.5% of the paper weight.

A flow sheet of a completely closed white-water system of a newsprint pulp and paper mill with a fermenter inserted as a "kidney" is presented in Fig. 2.39. There are several obvious advantages of closing a paper mill system, using the described white-rot fungus-based process. These advantages are as follows:

1. Water purification. Buildup of organic compounds in the white-water system is prevented and a need for external water purification is eliminated.

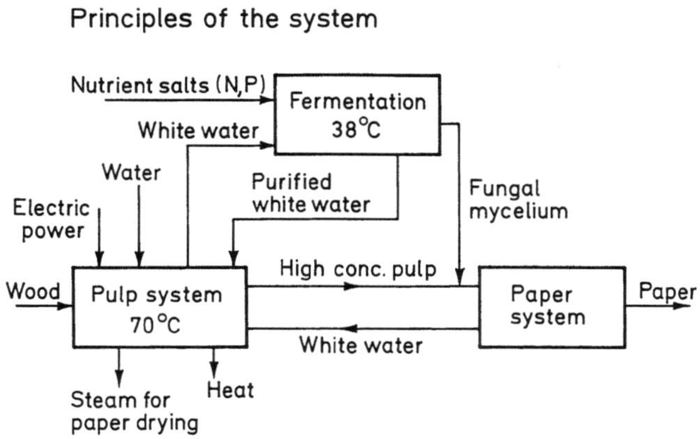

Fig. 2.39. Flow sheet of a closed white-water system of a newsprint pulp and paper-mill system. Buildup of organic substance is prevented by addition to the system of a fermenter with the white-rot fungus *Sporotrichum pulverulentum*. The fermenter acts as a "kidney" and depletes the circulating water of the dissolved material (Eriksson 1985)

2. A higher process temperature is reached in the pulping part of the mill, which can result in energy savings.
3. Increased paper production by incorporation of fungal mycelium in the paper. Alternatively, however, the fungus can be used as cattle feed.
4. Savings in water consumption. Only about 10% of the water used in an open pulp and paper mill system is necessary.

As stated, the process has been run on a large pilot scale but has not yet come into technical operation. This delay is apparently due to the normal reticence, often associated with the introduction of a new technique, and is understandable, particularly in existing mill systems where some reconstruction would be necessary. However, for new installations the technique developed should have many advantages, particularly in environmentally sensitive regions, and in areas of water shortage. New pilot runs are now planned especially to test recirculation of part of the fungal mycelium to decrease the residence time in the fermenter and thus the fermenter volume.

2.9.2 Enzymatic Saccharification of Lignocellulosic Materials for Ethanol Production

A significant amount of research has been directed toward saccharification of lignocellulosic materials and fermentation of the sugars into ethanol. The technology for such conversion has therefore advanced much faster than one would have expected a few years ago. New approaches have emerged in acid hydrolysis, showing greater promise for a process with improved economical feasibility. In enzyme technology, systems with a real potential for near-term applications have been developed.

The obstacles to enzymatic saccharification of lignocellulosics are the lignin and the crystalline cellulose (Millet et al. 1976, Lee et al. 1982). Various pretreatment techniques have been used to enhance the susceptibility of the substrate to enzymes. These include mechanical (Caulfield and Moore 1974, Tassinari et al. 1982) and chemical (Caulfield and Moore 1974, Knappert et al. 1980, Wilke et al. 1981, Saddler et al. 1982) methods as well as a steam-explosion pretreatment (Saddler et al. 1982). In two independent studies (Vallander and Eriksson 1985, Chum et al. 1985), it was found that the steam-explosion technique was the most efficient. Steam-exploded wheat straw offers an easily accessible substrate for enzymes, and up to 90% yield of total sugars has been achieved.

It is expensive to produce cellulases, and it is therefore important that this cost is kept at a minimum. This goal can be reached both by improved production processes and by increased recirculation of the enzymes to obtain as much sugar as possible per unit enzyme. It is evident from the presentation in this chapter that cellulolytic enzymes are produced by both fungi and bacteria. The productivity can be increased in many ways, such as by selection and genetic manipulation, to obtain strains of microorganisms hyperproducing with respect to one or several of the three different hydrolytic enzymes, endoglucanase, exoglucanase, and β-glucosidase, which in concert generate glucose from cellulose. Cloning of genes coding for cellulolytic enzymes has been tried for this purpose but as yet no real breakthroughs for enhanced production of cellulases have been achieved in this way (Béguin et al. 1987).

Enzymatic hydrolysis of cellulose proceeds under considerably milder conditions than acid hydrolysis. The enzymatic hydrolysis is carried out at temperatures between 40–50 °C and at pH values around 5. However, if thermophilic, anaerobic bacteria are employed for a direct conversion of lignocellulosics into ethanol, temperatures of around 70° are common (Wiegel and Ljungdahl 1986).

Residence times in the reactor for enzymatic hydrolysis of around 24 h can be necessary, which is considerably longer than for acid hydrolysis. As is shown in Fig. 2.40, the enzymatic hydrolysis of steam-exploded wheat straw is an initially fast process which gradually slows down (Deshpande and Eriksson 1984). The rate of reducing sugar formation was also low initially if the exploded straw had been previously treated with enzymes, washed, and resuspended before being subjected to further enzyme treatment. The obvious interpretation of these sugar release patterns is that the most readily accessible hemicelluloses and amorphous cellulose are hydrolyzed first at a high rate. Once these materials are removed, lignin will form a barrier, denying the enzyme access to the remaining polysaccharides. Furthermore, residual cellulose will be increasingly more crystalline than the cellulose first available. Also, the low saccharification rate of Avicel (microcrystalline cellulose) is due to the high crystallinity of this material. Therefore, the pretreatment step must remove the lignin and destroy the crystalline nature of the cellulose if a fast and a high-yield enzyme saccharification process is to be achieved.

As already stated, reuse of the enzymes is necessary to obtain an economically feasible process. It has been shown by Deshpande and Eriksson (1984)

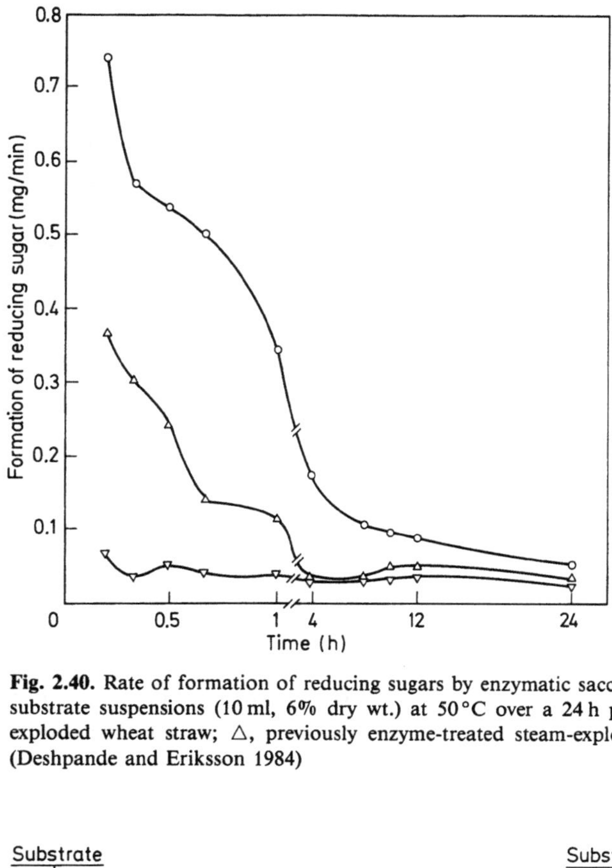

Fig. 2.40. Rate of formation of reducing sugars by enzymatic saccharification of three different substrate suspensions (10 ml, 6% dry wt.) at 50 °C over a 24 h period. Substrates: ○, steam-exploded wheat straw; △, previously enzyme-treated steam-exploded wheat straw; □, Avicel (Deshpande and Eriksson 1984)

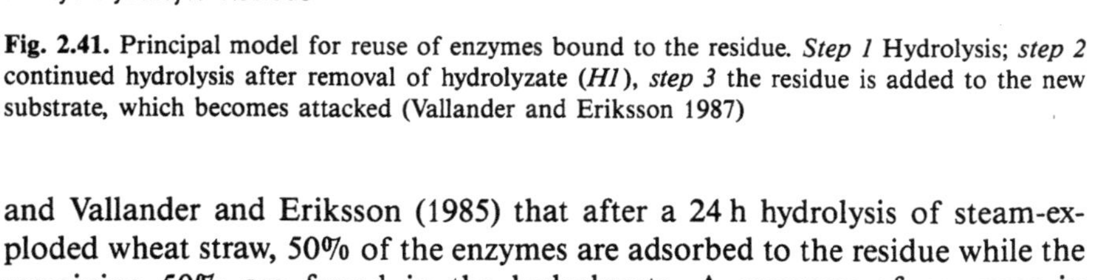

Fig. 2.41. Principal model for reuse of enzymes bound to the residue. *Step 1* Hydrolysis; *step 2* continued hydrolysis after removal of hydrolyzate (*H1*), *step 3* the residue is added to the new substrate, which becomes attacked (Vallander and Eriksson 1987)

and Vallander and Eriksson (1985) that after a 24 h hydrolysis of steam-exploded wheat straw, 50% of the enzymes are adsorbed to the residue while the remaining 50% are found in the hydrolyzate. A recovery of enzymes in hydrolyzates through contact with fresh substrate has been described by several authors (Castanon and Wilke 1980, Sinitsyn et al. 1983, Toyama et al. 1983, Deshpande and Eriksson 1984). However, to recover the enzymes adsorbed on the residue has turned out to be more difficult. Sinitsyn et al. (1983) found that enzymes were recoverable from the hydrolysis residue simply by adjusting the pH to neutrality. Deshpande and Eriksson (1984) reported that most of the adsorbed activity could be recovered by washing the residue with 0.5 M phosphate buffer, pH 7. The main disadvantages with these procedures for enzyme recovery are the extra costs incurred for chemicals and for dilution of the enzymes. Vallander and Eriksson (1987) have therefore proposed another method for recycling of enzymes from both the hydrolyzate and the solid residue. Recovery of the enzymes from the hydrolyzate is influenced by the type of contact with new substrate, contact time, and the concentration of sugars in the hydrolyzate. Enzymes adsorbed on the lignin-rich residue can be transferred to and utilized for hydrolysis of new substrate simply by bringing the solid materials together. In Fig. 2.41 the principal model for reuse of enzymes bound to the residue is presented.

The importance of reutilization of enzymes binding to the solid residue is stressed by recent findings by Vallander and Eriksson (to be published). It thus seems as if the enzymes binding to the solid residue are more efficient in hydrolyzing the solid substrate than enzymes free in the hydrolyzate. The special adsorption domain in the cellulases which bind these enzymes to cellulose and thereby increase their activity toward the substrate, is likely to be the explanation of this phenomenon. The enzymes free in the hydrolyzate are likely to have lost the binding domain (cf. Sect. 2.5.3).

An economically viable process for production of ethanol from lignocellulosic materials must also allow for the fermentation of pentoses since these normally make up 25% by weight of agriculture wastes and hardwoods. Pentose fermenting organisms that might be used in such a process include *Pachysolen tannophilus*, *Pichia stipitis* (Bruinenberg et al. 1984), and different anaerobic thermophilic bacteria (Wiegel and Ljungdahl 1986). However, there are still many problems to be solved before pentose-fermenting microorganisms can be used industrially.

Industrial production of ethanol from lignocellulosic materials implies continued research and development, particularly in those areas involving pretreatment, hydrolysis, enzyme reutilization, and pentose fermentation. Efficient distillation techniques already exist, for example, the Biostil process developed by Alfa-Laval, Sweden (Rosén 1982).

Chapter 3

Biodegradation of Hemicelluloses

3.1 The Structures of Hemicelluloses

Schulze (1891) gave the name *hemicelluloses* to those polysaccharides which could be extracted from plants by aqueous alkali. The name seemed appropriate since these polysaccharides were thought to be intermediates in cellulose biosynthesis and were found in close association with cellulose in the cell wall. We now know that the hemicelluloses are not precursors of cellulose and that they represent a distinct and separate group of plant polysaccharides which have no part in cellulose biosynthesis.

The hemicelluloses are composed of both linear and branched heteropolymers of D-xylose, L-arabinose, D-mannose, D-glucose, D-galactose and D-glucuronic acid. These individual sugars may be acetylated or methylated. Most hemicelluloses contain two to six of these sugars. Hemicelluloses from hardwoods have average degrees of polymerization (DP) of 150–200. Except for galactose-based hemicelluloses with linkages of 1,3-β-character most hemicelluloses are based on a 1,4-β-linkage of their backbone sugars. The amount of hemicellulose in dry wood is usually between 20% and 30% (Sjöström 1981). There are characteristic differences in the composition and structure of the hemicelluloses in softwood and hardwood.

3.1.1 Softwood Hemicelluloses

Galactoglucomannans

Galactoglucomannans are the principal hemicelluloses in softwoods (Table 3.1). Their backbone is a linear chain built up by 1,4-linked β-D-glucopyranose and β-D-mannopyranose units.

There are two different types of galactoglucomannans, one with a low galactose content and one with a high galactose content. In the low galactose glucomannans the ratio galactose : glucose : mannose is about 0.1 : 1 : 4, while in the high galactose glucomannans the corresponding ratio is 1 : 1 : 3. The ga-

```
→4-β-D-Glup-1→4-β-D-Manp-1→4-β-D-Manp-1→4-β-D-Manp-1→     (1)
                      6                2(3)
                      ↑                 |
                      1               Acetyl
                  α-D-Galp
```

lactose polymer is often refered to as glucomannan. The α-D-galactopyranose residue is linked to the backbone by 1,6-bonds. The mannose and glucose units in the backbones are partially substituted in C-2 and C-3 positions by acetyl groups, approximately 1 per 3–4 hexose units (**1**).

Arabinoglucuronoxylan

Another major hemicellulose in softwoods is arabinoglucuronoxylan. Its backbone is composed of 1,4-linked β-D-xylopyranose units. The chain is substituted at C-2 by 4-*O*-methyl-α-D-glucuronic acid groups with approximately two such units per ten xylose units. The xylose backbone is also substituted by α-L-arabinopyranose units, on the average 1.3 residues per ten xylose units (**2**).

→4-β-D-Xyl*p*-1→[4-β-D-Xyl*p*-1]$_2$→4-β-D-Xyl*p*-1→4-β-D-Xyl*p*-1→[4-β-D-Xyl*p*-1]$_5$→

2 ↑ 1 (from 4-*O*-Me-α-D-Glu*p*A); 3 ↑ 1 (from α-L-Ara*f*) (**2**)

4-*O*-Me-α-D-Glu*p*A α-L-Ara*f*

Arabinogalactan

In both softwoods and hardwoods arabinogalactan is only a minor component. However, in the heartwood of larch trees exceptionally large amounts of water-soluble arabinogalactan appear. The backbone of this galactan is built up by 1,3-linked β-D-galactopyranose units. Almost every galactose unit is substituted in the C-6 position. The substituents are mainly galactopyranose residues linked in the 1,6-position but there are also L-arabinose units (**3**).

→3-β-D-Gal*p*-1→3-β-D-Gal*p*-1→3-β-D-Gal*p*-1→3-β-D-Gal*p*-1→3-β-D-Gal*p*-1→

6 6 6 6 6

↑ ↑ ↑ ↑ ↑

1 1 1 1 1

β-D-Gal*p* β-D-Gal*p* (β-D-Gal*p*)$_n$ R L-Ara*f* (**3**)

6 6 6 3

↑ ↑ ↑ ↑

1 1 1 1

β-D-Gal*p* β-D-Gal*p* β-D-Gal*p* β-L-Ara*p*

R = β-D-Galactopyranose or, less frequently, L-Arabinofuranose or D-Glycopyranosyluronic acid

Other Polysaccharides

In addition to the above hemicelluloses, softwood also contains other polysaccharides, albeit in minor quantities. It deserves also to be mentioned that compression wood contains approximately 10% 1,4-β-D-galactan and 2%–5% 1,3-β-D-glucan.

3.1.2 Hardwood Hemicelluloses

Glucuronoxylan

A major component of hardwood hemicelluloses is an *O*-acetyl-4-*O*-methyl-glucurono-β-D-xylan often called glucuronoxylan. The xylose-based hemicelluloses in both softwoods and hardwoods are usually termed xylans.

The xylan content varies in different hardwood species between 15%–30%. The backbone consists of β-D-xylopyranose units linked by 1,4-bonds. Seven of ten xylose units are substituted by an acetyl group at C-2 or C-3. One 4-*O*-methyl-α-D-glucuronic acid residue per ten xylose units is 1,2-linked to the backbone (**4**).

```
┌ →β-D-Xylp-1→ ┐  4-β-D-Xylp-1→4-β-D-Xylp-1→4-β-D-Xylp-1→        (4)
│    3 (2)     │                    2
│      |       │                    ↑
└    Acetyl    ┘ 7                  1
                              4-O-Me-α-D-GlupA
```

Glucomannan

Glucomannan is another hemicellulose in hardwoods. This polysaccharide comprises 2%–5% of the wood and is composed of β-D-glucopyranose and β-D-mannopyranose units linked by 1,4-bonds. Depending on the wood species, the glucose : mannose ratio varies between 1 : 1 and 1 : 2.

Other Polysaccharides

In addition to glucuronoxylan and glucomannan, small amounts of miscellaneous polysaccharides are present in hardwoods. These polysaccharides are to some extent of the same type as those present in softwoods. They may be of importance for the living tree.

The major hemicellulose components of both softwoods and hardwoods are summarized in Table 3.1. The data concerning softwood and hardwood hemicelluloses presented in this table and above were collectively derived from a variety of sources, mainly review articles by Timell (1965 b), Richards and Whistler (1973), Aspinall (1973), Coté (1974), Preston (1974), and Sjöström (1981). The most exhaustive reviews of the hemicelluloses in wood are those by Timell (1964, 1965 a).

Table 3.1. The major hemicellulose components (Sjöström 1981)

Hemicellulose type	Occurrence	Amount (% of wood)	Composition: Units	Molar ratios	Linkage	Solubility[a]	$\overline{DP_n}$
Galactoglucomannan	Softwood	5 – 8	β-D-Manp	3	1.4	Alkali, water	~200
			β-D-Glcp	1	1.4		
			α-D-Galp	1	1.6		
			Acetyl	1			
Galactoglucomannan	Softwood	10 – 15	β-D-Manp	4	1.4	Alkaline borate	~200
			β-D-Glcp	1	1.4		
			α-D-Galp	0.1	1.6		
			Acetyl	1			
Arabinoglucuronoxylan	Softwood	7 – 10	β-D-Xylp	10	1.4	Alkali	~200
			4-*O*-Me-α-D-GlcpA	2	1.2	dimethylsulfoxide*	
			α-L-Araf	1.3	1.3	water*	
Arabinogalactan	Larchwood	5 – 35	β-D-Galp	6	1.3	Water	350
			α-L-Araf	2/3	1.6		
			β-L-Arap	1/3	1.3		
			β-D-GlcpA	Little	1.6		
Glucuronoxylan	Hardwood	15 – 30	β-D-Xylp	10	1.4	Alkali	200
			4-*O*-Me-α-D-GlcpA	1	1.2	dimethylsulfoxide*	
			Acetyl	7			
Glucomannan	Hardwood	2 – 5	β-D-Manp	1 – 2	1.4	Alkaline borate	200
			β-D-Glcp	1	1.4		

[a] The asterisk (*) represents a partial solubility.

3.2 Biosynthesis of Hemicelluloses

From all recent reviews it is clear that very little is known about biosynthesis of hemicelluloses (Delmer 1974, Robinson 1977, Darvill et al. 1979, Delmer 1982b). This is in contrast to the knowledge of cellulose synthesis where much more is now known (cf. Sect. 2.1.2).

As is evident from the above structures, almost all of the polymers characterized as hemicelluloses contain more than one type of monosaccharide residue. It is obvious that the biosynthesis of such polymers requires an ingenious coordination of the variety of enzymes and substrates acting in concert. It is conceivable that a loss of even one of these enzyme activities may render the entire synthesis inoperable.

While cellulose is synthesized from UDP-glucose, both UDP-glucose and GDP-glucose are involved in the synthesis of hemicelluloses. For the formation of mannose-containing hemicelluloses (galactoglucomannans and glucomannans) GDP-glucose is the principal nucleotide. In the synthesis of hemicelluloses varying enzymatic reactions are involved such as epimerization, dehydrogenation, and decarboxylation (Fig. 3.1).

In terms of contribution to the biomass of the world, the hemicellulosic components of wood are the most abundant next to cellulose and lignin. In spite of this, studies of in vitro biosynthesis of hemicelluloses in wood cells are almost nonexistent. Concerning the site of synthesis of these matrix components, it is believed that, in contrast to cellulose, synthesis of the hemicelluloses occurs intracellularly, namely, within the Golgi apparatus. Via Golgi-derived vesicles the hemicelluloses are secreted to the cell walls through fusion of the vesicles with the plasma membrane. These conclusions are reached mainly from autoradiographic and biochemical studies. The literature where these conclusions are described has been reviewed by Robinson (1977) and Darvill et al. (1979).

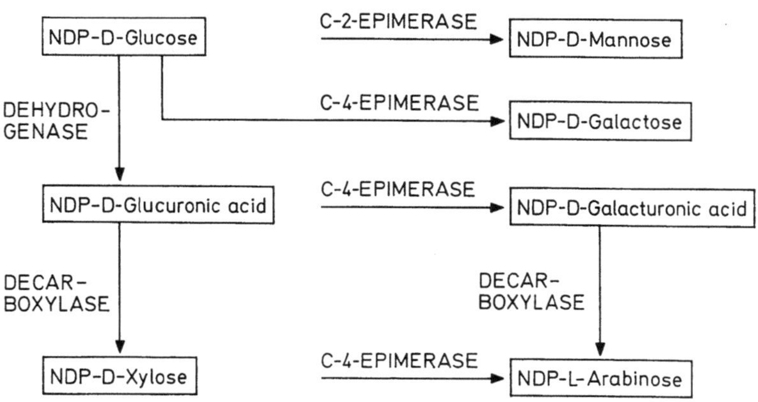

Fig. 3.1. Simplified representation of the formation of hemicellulose precursors from UDP-D-glucose or GDP-D-glucose. Note that NDP (nucleotide diphosphate) means either UDP or GDP (Sjöström 1981)

3.3 Xylanolytic Enzymes

For enzymatic hydrolysis of cellulose only enzymes hydrolyzing 1,4-β-glucosidic linkages are required. Since the structure of xylans is variable, involving both linear 1,4-β-linked chains of xylose but also branched heteropolysaccharides (cf. Sect. 3.1), a more complex assembly of enzymes is required than for cellulose hydrolysis.

A complete degradation of a branched acetyl xylan requires a concerted action of several different hydrolytic enzymes, i.e., endo-1,4-β-xylanase (EC 3.2.1.8), β-xylosidase (EC 3.2.1.37), α-glucuronidase (EC 3.2.1), α-L-arabinofuranosidase (EC 3.2.1.55), and acetylesterase (EC 3.1.1.6) or acetylxylanesterase (Fig. 3.2). The enzymatic hydrolysis of xylans has recently been reviewed by Biely (1985), Dekker (1985), and Poutanen (1988a).

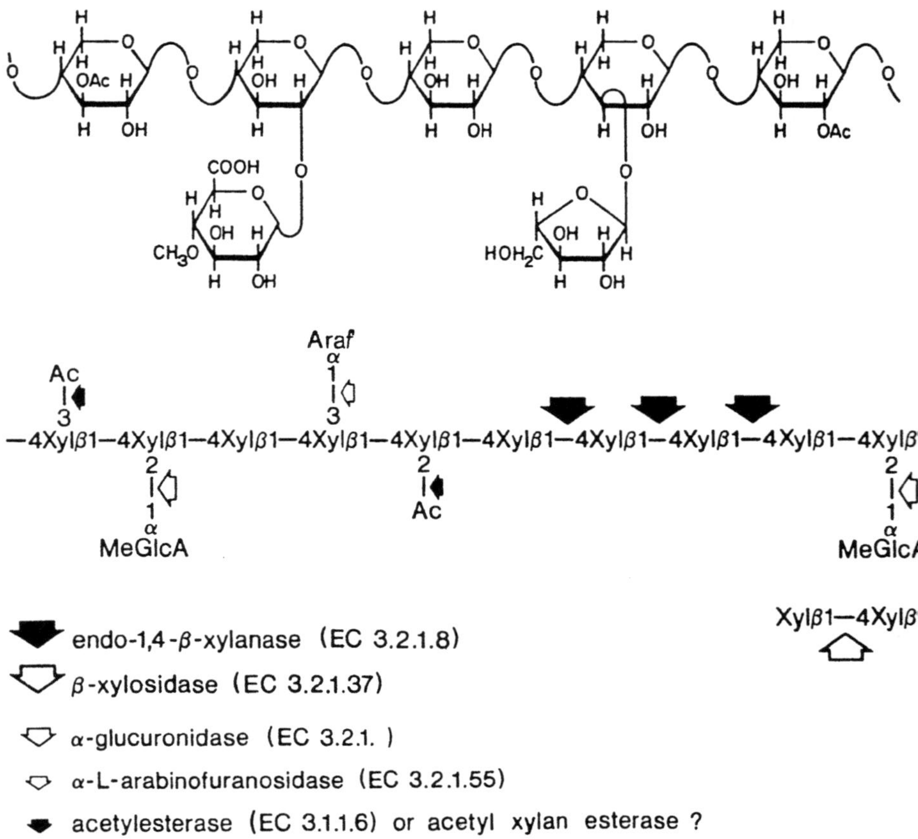

Fig. 3.2. A hypothetical plant xylan and the sites of its attack by microbial xylanolytic enzymes. The fragment comprising five D-xylose units is presented in the *upper part* of the figure. *Ac,* Acetyl Group; *Araf,* L-arabinofuranose; *MeGlcA,* 4-*O*-methyl-D-glucuronic acid; *Xyl,* D-xylose (Biely 1985)

The endoxylanases are the best characterized and most widely studied of the xylanolytic enzymes (Woodward 1984, Frederick et al. 1985, Fournier et al. 1985, Shei et al. 1985, Lappalainen 1986). These enzymes launch an endwise attack on the backbone of xylans to produce both substituted and non-substituted shorter oligomers, xylobiose, and xylose. To convert the water-soluble oligomers, dimers, etc. to xylose, β-xylosidases are employed (Rodionova et al. 1983, Matsuo and Yasui 1984a, Poutanen and Puls 1988). The enzymes releasing the substituents on the xylan backbone, namely, α-L-arabinosidase, α-D-glucuronidase, and acetylxylanesterase, act in synergism with the endoxylanases and xylosidases. Their participation is necessary to achieve a total hydrolysis of xylans to monosaccharides (Biely 1985, Poutanen 1988a).

3.3.1 Endo-1,4-β-D-Xylanases

Endo-1,4-β-xylanases (EC 3.2.1.8) are produced by both fungi and bacteria. However, xylanases of the former have been more extensively studied. The most studied and best characterized of the fungal endoxylanases have been those of *Aspergillus niger* (Table 3.2). Production and characterization of endoxylanases from several *Trichoderma* strains, i.e., *Trichoderma viride, T. reesei, T. harzianum, T. koningii*, and *T. lignorum* have also been reported (Table 3.3).

In addition to the studies of *A. niger* and *Trichoderma* endoxylanases, studies of this enzyme have been carried out with many other fungi, such as in the brown-rot fungi *Coniophora cerebella* (King and Fuller 1968), *Tyromyces palustris* (Ishihara et al. 1975), and *Poria placenta* (Micales et al. 1987). The endoxylanases from *Irpex lacteus* (*Polyporus tulipiferae*) have been studied by Hoebler and Brillouet (1984), Kanda et al. (1985), and Brillouet (1987). Other fungal strains from which endoxylanases have been studied are *Sporotrichum dimorphosphorum* (Comtat and Joseleau 1981, Comtat 1983) and *Gliocladium virens* (Takahashi and Kutsumi 1979), *Ceratocystis paradoxa* (Dekker and Richards 1974, 1975a, b, c), *Neurospora crassa* (Mishra et al. 1984), *Termitomyces clypeatus* (Mukherjee and Sengupta 1985), *Talaromyces byssoclamydoides* YH-50 (Yoshioka et al. 1981), and *Chaetomium cellulolyticum* (Dubeau et al. 1987).

Endoxylanases from bacteria have also been produced, isolated, and characterized. The most often studied bacterial xylanases are from the genera *Bacillus* and *Streptomyces* (Table 3.4). In addition to the xylanases listed in Table 3.4, the xylanases from *Bacillus subtilis* have been studied by Lyr (1972) and those from a thermophilic, acidophilic *Bacillus* sp. by Uchino and Nakane (1981). Molecular cloning of a *B. subtilis* xylanase gene in *Escherichia coli* has been carried out by Bernier et al. (1983). The xylanase gene from *B. subtilis* has been isolated and its complete nucleotide sequence determined by Paice et al. (1986). Studies of other bacterial xylanases involve those of *Thermonospora* sp. (Ristproph and Humphrey 1985) and *Cellulomonas uda* (Rappi and Wagner 1986).

Table 3.2. Characteristics of *Aspergillus niger* endoxylanases (Poutanen 1988a)

MW (kDa)	pI	pH-optimum	Substrate	K_m (mg ml)	Main end products	Remarks	Reference
20.8	6.7	4–6	Soluble xylan, xylo-oligomers		X_2, X	No activity toward insoluble xylan, arabinose not released	Frederick et al. 1981
13	8.6	6.0	Soluble xylan			Requires branch points in	
13	9.0	5.5	Insoluble branched xylan		X_3, X_5	substrate, arabinose not released	Frederick et al. 1985
14	4.5	4.9	$>X_5$, soluble xylan		X_2–X_6	Arabinose not released	Shei et al. 1985
28	3.65	5.0	$>X_5$, soluble xylan		X_3, X_5		Fournier et al. 1985
13	4.0		Xylo-oligosaccharides			Transferase activity No cellulase activity	Vrsanska et al. 1982
31	nd	4.0	Larchwood xylan preferred to		X, X_2, Ar	Also cellulase activity	John et al. 1979
50	nd	5.5–6.0	oligosaccharides			Also cellulase activity	
50	nd	4.0–4.5					
33	4.2	4.0	Carboxymethyl xylan, arabino- and glucuronoxylans	3	$X_2>X_3>X_1$, oligomers	No cellulase activity Traces of arabinose liberated No transferase End product inhibition	Gorbacheva and Rodionova 1977a, b
nd	nd	nd	X_5		X_2, X_3	High transferase activity	Conrad and Nöthen 1984

Table 3.3. Characteristics of *Trichoderma* endoxylanases (Poutanen 1988a)

Microbial origin	MW (kDa)	pI	pH-optimum	Substrate	K_m (mg ml⁻)	Main end products	Remarks	Reference
Trichoderma viride	nd	nd	3.5	Soluble xylan, glycol xylan			The enzyme was crystallized No activity toward insoluble xylan	Hashimoto et al. 1971
Trichoderma viride	17.8	9.2	4.8	Hardwood and spruce xylans		X_2, X, 4-*O*-MeGlcA-substituted oligomers		Sinner and Dietrichs 1975, 1976
Trichoderma viride	nd	8.5, 6.0, 5.3	nd	Activity detection in gel with dyed xylan		nd		Biely et al. 1985a
Trichoderma viride	20	8.5	4.5 – 5	Rye flour arabinoxylan	3.3	Arabinosyl-substituted oligomers	Also hydrolyzed CMC	Gibson and McCleary 1987
Trichoderma reesei	nd	7.6, 5.4, 4.7, 4.3, 4.0	nd	Wood xylan		nd	Three most acidic also hydrolyzed CMC	Kolarova and Farkas 1983
Trichoderma reesei	8.5 11.7 14.4 40.7	10.3 8.6 6.6 8.5	nd	Not reported			Arabinose-liberating enzymes	Dekker 1985

Table 3.3 (continued)

Microbial origin	MW (kDa)	pI	pH-optimum	Substrate	K_m (mg ml^{-})	Main end products	Remarks	Reference
Trichoderma reesei	nd	8.5, 4.8		Detection in gel				Hrmova et al. 1986
Trichoderma reesei	22	6.5	4–5	Insoluble birch xylan		X_1, X_2, oligosaccharides	Higher xylanase activity	Lappalainen 1986
	32	4.3	4–5				High laminarinase activity	
Trichoderma harzianum	20	9.4	5	$>X_4$		$X-X_4$	Transferase activity	Tan et al. 1985a, b
	29	9.5	5	$>X_3$			Substrate inhibition	
Trichoderma harzianum	22	8.5	4.5–5	$>X_3$			Arabinose not released	Wong et al. 1986
Trichoderma koningii	29	7.24	4.9–5.8	Oat-straw arabinoxylan	1.4	X_2, X_3, X, Ar	Arabinose released, but no α-arabinofuranosidase activity	Wood and McCrae 1986b
	18	7.30	4.9–5.5		4.2	X_2, X_3, ArX_4		
Trichoderma pseudokoningii	15	9.6	5.0	Arabinoxylan of corn cell walls		Xylo-oligomers	Solubilizing activity	Baker et al. 1977
Trichoderma lignorum	21		3.5			X_1, X_2, substituted oligomers	Arabinose not liberated	John et al. 1981
	20		6.5	Arabinoxylan				

Table 3.4. Characteristics of extracellular bacterial endoxylanases (Poutanen 1988a)

Microbial origin	MW (kDa)	pI	pH-optimum	Substrate	K_m (mg ml^{-1})	Main end products	Remarks	Reference
Bacillus circulans	85	4.5	5.5 – 7	Larchwood xylan		$X - X_4$		Esteban et al. 1982
	15	9.1	5.5 – 7			$X_2 - X_4$		
Bacillus coagulans	22	10	6.0		10			Esteban et al. 1983
Bacillus pumilus	24	nd	6.5	Larchwood xylan, long chains preferred		$X_2 - X_5$	Trans-xylosidase activity	Panbangred et al. 1983
Bacillus spp.	nd	nd	6 – 7			X, X_2		Okazaki et al. 1984
Streptomyces lividans	43	5.2	6.0	Soluble oat spelts	0.78	Mainly X_2, some X_n, little X		Morosoli et al. 1986a
Streptomyces sp. E-86	40.5	7.3	5.5 – 6.2	Corn-cob arabinoxylan		X and X_2, arabinosubst. oligosacch.	Arabinose not liberated	Kusakabe et al. 1977a, b
Streptomyces sp. No 3137	50	7.1	5.5 – 6.5					
	25	10.14	5.0 – 6.0	Xylan $\gg X_4 \gg X_3$		X and X_2		Marui et al. 1985
	25	10.26	5.0 – 6.0					
Streptomyces sp. KT-23	43	6.9	5.5			Mainly X_2, little X		Nakajima et al. 1984
Clostridium acetobutylicum	65	4.45	5.0	Larchwood xylan	6.0	$X_5 - X_2$	High activity toward CMC	Lee et al. 1987
Costridium acetobutylicum	29	8.50	5.5 – 6.0	Larchwood xylan	6.7	X_3 and X_2	No activity toward CMC	

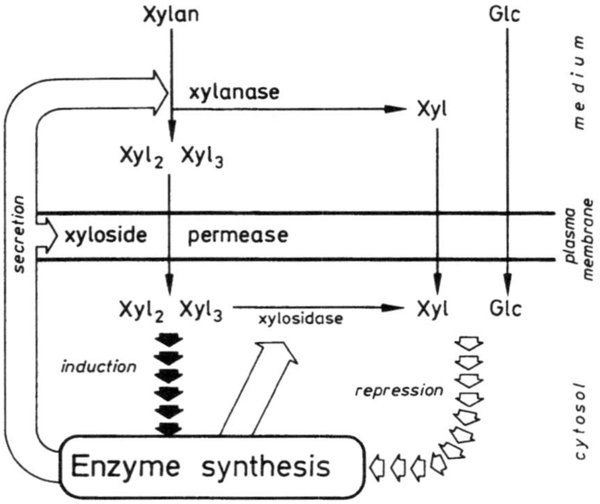

Fig. 3.3. The xylanolytic system of *Cryptococcus albidus*: cellular localization of enzyme components and regulation of their synthesis. *Glc*, D-glucose; *Xyl*, D-xylose; Xyl_2, xylobiose, Xyl_3, xylotriose (Biely 1985)

Anaerobic bacteria have also been reported to produce xylanases. Anaerobic fermentation of xylans has thus been carried out using *Clostridium acetobutylicum* (Lee et al. 1985, 1987, Lemmel et al. 1986), *Thermoanaerobacter ethanolicus* (Wiegel et al. 1983), and *Clostridium thermocellum* (Wiegel et al. 1985). Xylanases produced by the rumen microflora have been characterized by Morrison (1975), Pettipher and Latham (1979), and Brice and Morrison (1982). Formation of esterase, xylanase, and xylosidase by cellulolytic anaerobes, namely, *Acetivibrio cellulolyticus*, *A. cellulosolvens*, and *Bacterioides cellulosolvens* as well as the thermophilic anaerobe *C. thermocellum*, were studied by Khan et al. (1987).

Different genera of yeasts have also been reported to produce extracellular xylanases (Fig. 3.3). Thus *Cryptococcus* xylanases have been studied by Notario et al. (1979), Biely et al. (1980), and Yasui et al. (1984). Xylanases from *Aureobasidium* were studied by Leathers et al. (1986) and from *Trichosporon* by Stüttgen and Sahm (1982).

The xylanases, the most studied and of most practical importance, as is also the case with the cellulases, are those produced by mycelial fungi. Practically all cellulase-producing microorganisms also produce xylanases and vice versa (cf. Tables 2.1, 2.9, and 2.13). The information in the literature whether or not xylanolytic and cellulolytic systems are under separate or common regulatory control varies. It was thus proposed by Eriksson and Goodell (1974) that in the white-rot fungus *Polyporus adustus* cellulases, mannanases, and xylanases are under the control of a single common, regulatory gene. These authors isolated from *P. adustus* a group of pleiotropic mutants lacking cellulase, mannanase, and xylanase activities, suggesting that these enzymes have some common requirements necessary for their synthesis. Revertants

Table 3.5. Effect of cellulose, glucomannan, xylan, and glucose on enzyme synthesis in wild type of the white-rot fungus *Polyporus adustus* (Eriksson and Goodell 1974)

Substrate	Enzyme					
	Cellulase	Mannanase	Xylanase	Pectinase	Laminarinase	Aryl β-glucosidase
Glucose	1.5	0.7	10.7	5.5	28.9	1.27
Cellulose	40	13.5	18.2	2.0	18.1	0.23
Glucomannan	139	32.6	36.0	6.6	33.9	1.11
Xylan	2.0	0.6	5.5	1.6	–	0.53

Note: Enzyme activities given are the maxima measured during the 19-day cultivation. Three flasks of each medium were used. Enzyme activities are in enzyme units per liter of culture solution except for pectinase where the units are relative.

could be isolated, which indicated that a mutation was not a deletion of the cellulase, xylanase, and mannanase genes. In the wild-type *P. adustus*, the synthesis of these enzymes appears to be induced by cellulose and glucomannan (Table 3.5). The explanation favored by Eriksson and Goodell (1974) was that a single regulatory protein governs the induction of cellulase, mannanase, and xylanase, and that this protein is defective in the mutants.

In contrast to these findings, Hrmovà et al. (1986) suggested that in *T. reesei* the syntheses of cellulases and xylanases are under separate control. Cellulose- and xylan-degrading enzymes of *T. reesei* QM 9414 were induced by sophorose, xylobiose, cellulose, and xylan and were analyzed by isoelectric focusing. It was found that the sophorose-induced enzyme system contained two types of endo-1,4-β-glucanases (EC 3.2.1.4), one specific for cellulose and the other nonspecific, hydrolyzing both cellulose and xylan and also exo-1,4-β-glucanases (EC 3.2.1.91), that is, all types of glucanases produced during growth on cellulose. Specific endoxylanases (EC 3.2.1.8) present in the cellulose-containing medium were less abundant in the sophorose-induced enzyme system. Xylobiose and xylan induced only specific endoxylanases. On the basis of these results, Hrmovà et al. (1986) concluded that the syntheses of cellulases and xylanases in *T. reesei* QM 9414 are under separate control and that nonspecific endoglucanases are constituents of the cellulose-degrading enzyme system.

Tangnu et al. (1981) studied xylanase and cellulase production of *T. reesei* Rut C30, a noncatabolite repressed mutant of *T. reesei*. These authors found that both of these enzymes were constitutive and adaptive. In this mutant almost equal amounts of xylanase were produced if Solka Floc (cellulose) or glucose were used as substrates (Table 3.6). The production of xylanase by *Streptomyces xylophagus* and *Chaetomium trilaterale* are also included for comparison.

Although there is still some confusion in the literature as to whether xylanases are produced constitutively or inductively, some recent reports demonstrate that these enzymes are inducible at least in some yeasts and fungi. Sol-

Table 3.6. Comparison of xylanase-producing strains (Tangnu et al. 1981)

Run No.	Substrate	Strain	Xylanase activity (IU ml^{-1})	Remarks
1	Solka Floc (10 g l^{-1})	Rut C-30	114	pH≥5.0, temperature = 25 °C, proteose peptone = 1 g l^{-1} Produced 90 IU ml^{-1} xylanase activity when glucose instead of cellulase was used as substrate
2	Larchwood xylan (10 g l^{-1})	*Streptomyces xylophagus*	38.8	pH 7.4, temperature = 30 °C, bactopeptone = 3 g l^{-1}
3	Larchwood xylan (10 g l^{-1})	*Chaetomium trilaterale*	6.6	pH 7.0, temperature = 28 °C, yeast extract = 1 g l^{-1}

Table 3.7. Effect of carbon source on the production of D-xylan-degrading enzymes[a] by *Trichoderma lignorum* (Defaye et al. 1985)

Carbon source	Cell dry-weight (g l^{-1})	Extracellular protein (mg ml^{-1})	D-Xylanase activity (U ml^{-1})	β-D-Xylosidase activity (mU ml^{-1})	α-L-Arabinofuranosidase activity (mU ml^{-1})
L-Arabino-D-xylan	12.0	0.34	104	215	367
D-Xylan	10.7	0.10	65	80	48
Cellulose	13.7	0.11	14	50	68
Soluble starch	10.5	0.19	12	16	77
L-Arabino-D-galactan	2.4	0.11	5	12	173
Cellobiose	8.4	0.14	4	13	68
Sucrose	1.7	0.08	4	12	36
D-Xylose	6.0	0.19	4	23	107
Xylitol	8.9	0.12	3	13	186
D-Glucose	8.2	0.16	5	12	60
No addition	1.6	0.07	0	0	26

[a] Enzyme activities, cell dry weight, and extracellular protein were measured after 72 h.

uble fragments of D-xylan, such as xylobiose and higher homologous oligosaccharides, were effective as inducers of xylanase in the yeast *Cryptococcus albidus* (Biely et al. 1980). 4-Thiocellobiose, as a nonmetabolizable substrate analog was found to be an excellent inducer of both cellulases and xylanases in the basidiomycete *Schizophyllum commune* (Rho et al. 1982). In *Trichoderma lignorum*, xylan-degrading enzymes can be induced by several oligo- and polysaccharides (Defaye et al. 1985). Optimal induction is obtained when 1-arabino-D-xylan is used as a carbon source. Enzymes are also produced by cells grown on glycerol in the presence of 4-thioxylobiose and methyl-

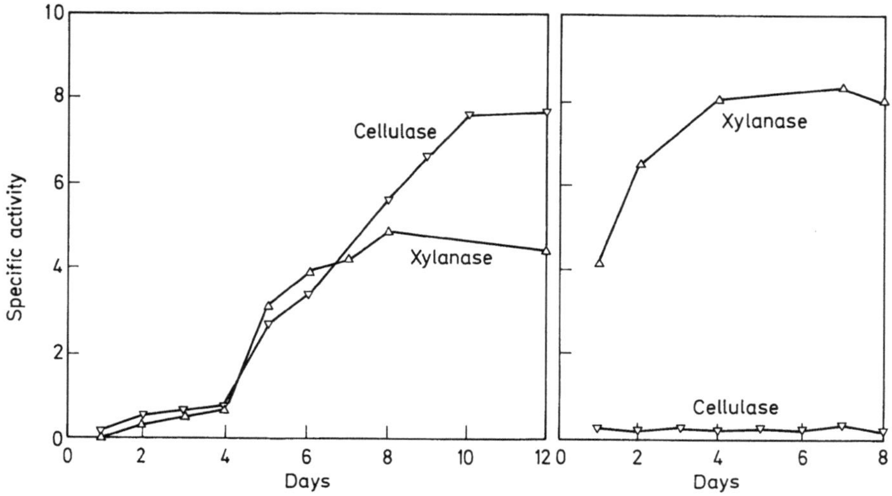

Fig. 3.4. Production of cellulase and xylanase by a *Monilia* sp. (*left*) in the presence of cellulose (Avicel), and (*right*) in the presence of xylan (Gong et al. 1981)

β-D-xylopyranoside. The effect of carbon source on the production of xylan-degrading enzymes by *T. lignorum* is presented in Table 3.7.

In the fungus *Aureobasidium pullulans*, the natural inducers of xylanase appear to be specifically derived from arabinoxylan (Leathers et al. 1986). Starch, cellulose, and sugars often associated with hemicelluloses such as glucose, galactose, mannose, and glucuronic acid, fail to serve as inducers. Xylose, xylobiose, and arabinose, on the other hand, were identified as natural and direct inducers of xylanase from this color variant strain of *A. pullulans*. The induction of xylanase was subject to glucose repression in this strain.

In the yeast *Cryptococcus flavus* xylanase was produced inductively when the organism was grown in a medium containing xylose, xylooligosaccharides, or xylan. To obtain a more effective inducer, a variety of derivatives of xylose, xylobiose, and xylan were tested for their inducing ability. It was found that β-methylxyloside was a nonmetabolizable inducer that possessed a higher inducing ability than other compounds. In their nongrowing phase, cells grown on a glucose medium produced significant amounts of xylanase simultaneously with the uptake of β-methylxyloside added.

Studies of the regulation of xylanases and cellulases in fungi are often complicated by a simultaneous production of these enzymes and sometimes also by substrate cross-specificity of cellulases and xylanases. However, when grown on xylan only, several fungal species produce specific xylanases with little or no cellulases, while, when grown on cellulose, cellulases are produced together with xylanases (Fig. 3.4; Gong et al. 1981, Stewart et al. 1983, Hrmovà et al. 1986).

The activity of endoxylanases is often determined by measuring reducing sugars released from solid xylan substrates (Poutanen 1988a). However, other

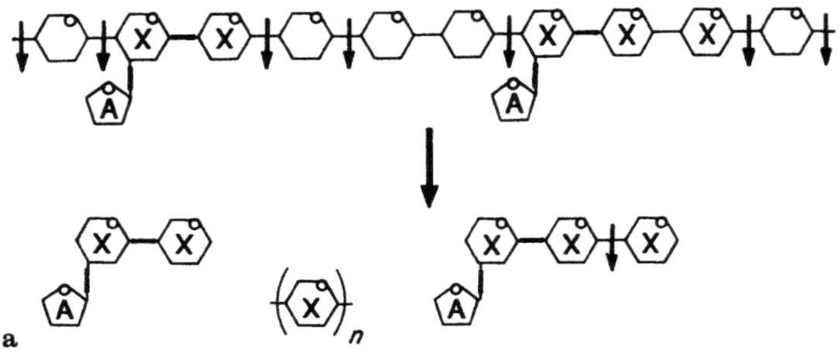

a

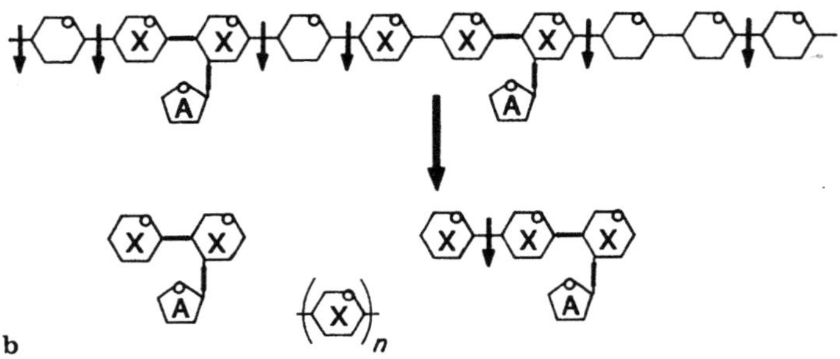

b

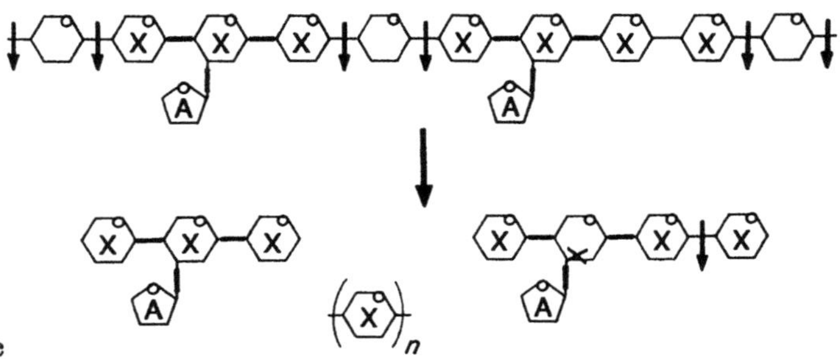

c

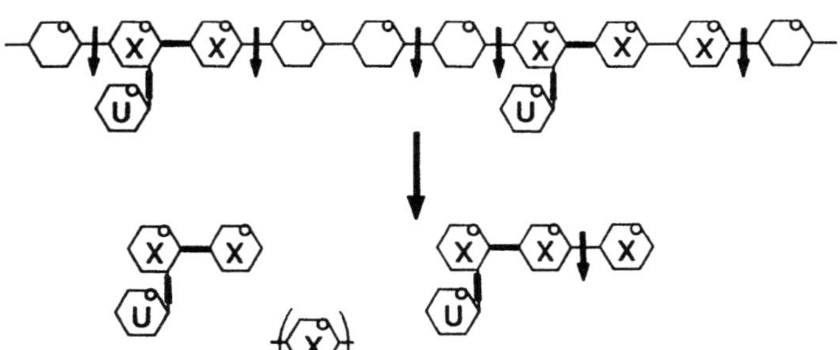

d

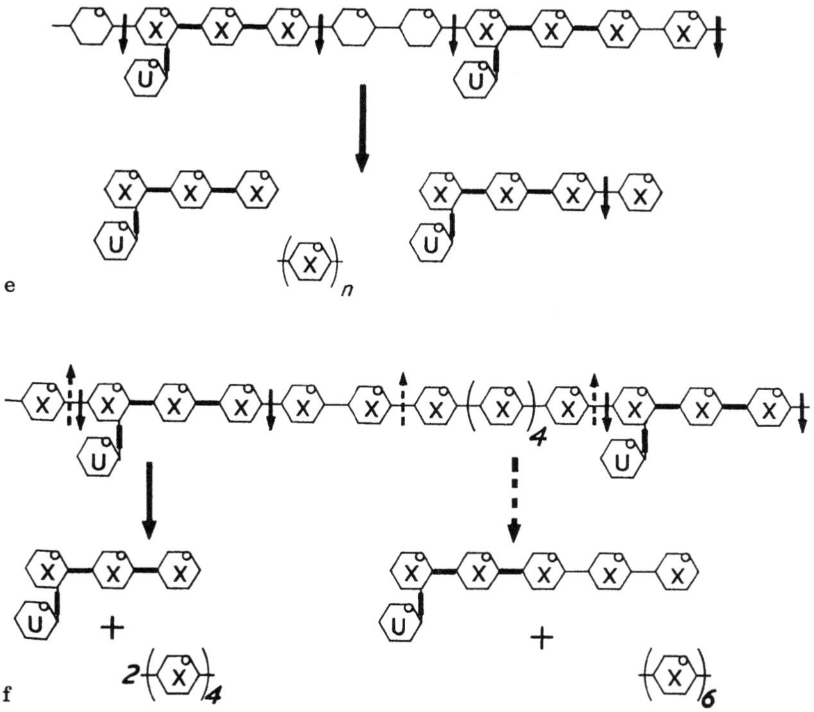

Fig. 3.5. Enzymatic degradation of heteroxylans describing the mode of action of endo-1,4-β-D-xylanases from **a** *Ceratocystis paradoxa* (Dekker and Richards 1975 a, b, c), *Sporotrichum dimorphosphorum* (Comtat and Joseleau 1981), and *Streptomyces* sp. (Kusakabe et al. 1977 b); **b** *Aspergillus niger* (Takenishi and Tsujisaka 1973); **c** *Cephalosporium sacchari* (Dekker et al. 1975); **d** *Trichoderma viride* (Sinner and Dietrichs 1976); **e** *Aspergillus niger* and *Oxiporus* sp. (Sinner and Dietrichs 1976), and *Sporotrichum dimorphosphorum* (Comtat and Joseleau 1981); and **f** *Trametes hirsuta* (Kubačková et al. 1979). *X* 1,4-linked β-D-xylosyl residue; *A* 1,3-linked α-L-arabinofuranosyl residue; *U* 1,2-linked 4-*O*-methyl-α-D-glucopyranosyl uronic acid residue; ↓, site of cleavage; heavy bonds, enzyme-resistant linkages (Dekker 1985)

methods have also been used, such as viscometry (Kundu et al. 1980) or the so-called nephelometric assay (Nummi et al. 1985). However, there are many factors causing variation in the determination of xylanase and xylosidase activity, as has been shown by Khan et al. (1986). The results obtained show that the reported activities in international units (U) varied by a factor of 3 to 107, depending on the dilution of enzymes prior to assay. Determination of xylanase was also affected by the type of substrate used, varying from 1.5 to 104 U ml^{-1} among three different xylan preparations from larch wood. The units of enzymatic activity obtained appeared to vary with the availability of easily degradable xylo-oligosaccharides in the substrate. It was therefore clear that defining of the composition and the use of standardized methods for preparation of the xylan substrate would help in minimizing variations due to substrate. Also, enzyme dilution needs to be highly standardized to minimize variations.

Differences obtained in the measurement of xylanase activities are clearly also dependent upon differences in the mode of action of different xylanases from different sources and also upon the considerable heterogeneities in xylans from different sources. Dekker (1985) classifies the endoxylanases as debranching and nondebranching enzymes according to their ability to release L-arabinose from arabinoxylans. Both types are also capable of attacking glucuronoxylans and unsubstituted 1,4-β-D-xylans. A list of microorganisms producing debranching endoxylanases is presented by Dekker (1985).

According to Dekker (1985), the nondebranching group of xylanases are by far the most common. These enzymes degrade heteroxylans randomly. Such enzymes have been studied particularly by Reilly and co-workers, who isolated several different xylanases from a commercial *A. niger* enzyme preparation that did not debranch arabinoglucuronoxylans (Fournier et al. 1985, Frederick et al. 1985, Shei et al. 1985). The mode of action on heteroxylans of several different endoxylanases from various microorganisms is presented in Fig. 3.5.

3.3.2 1,4-β-D-Xylosidases

β-D-Xylosidase (1,4-β-D-xylan xylohydrolase, EC 3.2.1.37) hydrolyzes xylo-oligosaccharides to xylose and are essential for the complete degradation of xylans. β-Xylosidases have been purified and characterized from many fungi and also from some bacteria (Table 3.8).

Reese et al. (1973) tested 250 microorganisms for their ability to produce β-xylosidases. The microorganisms represented a wide variety of genera (bacteria, phycomycetes, ascomycetes, basidiomycetes, and fungi imperfecti). It was found that none of the bacteria (21 strains of 13 genera) secreted β-xylosidase when grown on xylan plus methyl β-D-xylopyranoside. However, other enzymes such as β-xylanase, β-1,3-glucanase, and amylase were secreted by some of these bacteria. Most fungi produced no xylosidase detectable by the assay procedure used but the black *Aspergilli* produced appreciable amounts of enzyme, much of which remained intracellular. The same authors found that the β-xylosidase is located in the fungal mycelium during the early stages of growth and is secreted into the medium only later in the incubation cycle, so that little enzyme remains bound after 6 days of cultivation. The results obtained with *Botryodiplodia* (Fig. 3.6) are typical of most fungi-producing β-xylosidases. In the black *Aspergilli*, however, more of the enzyme remains associated with the mycelium (Fig. 3.6).

β-Xylosidase activity is normally assayed using the substrate p-nitrophenyl β-D-xylopyranoside (PNPX) at pH 4.5 for fungal enzymes and around 6 for bacterial enzymes (Poutanen 1988a). Methyl β-D-xylanopyranoside has also been used as substrate in assaying for β-xylosidase. Xylose is then determined by the nitrosalisylic acid method or by the Somogyi-Nelson method (Nelson 1944, Somogyi 1952). The natural substrate, xylobiose, has also invariably been used as substrate (Rodionova et al. 1983).

Table 3.8. β-Xylosidases (Poutanen 1988a)

Microbial origin	MW (kDa)	pI	pH-optimum	K_m[c] (mM)	Remarks	Reference
Aspergillus niger	253[a] 122[b]	4.9	3.8 – 4.0	nd	Also α-arabinosidase, β-glucosidase, β-galactosidase activities, transferase activity	Rodionova et al. 1983
Aspergillus niger	78[a]	nd	6.7 – 7	nd	Hydrolyzed xylotriose better than xylobiose	John et al. 1979
Aspergillus fumigatus	360[a] 90[b]	5.4	4.5	2.0		Kitpreechavanich et al. 1986
Penicillium wortmanni	100[a] 102[b]	5.0	3.3 – 4.0	0.12	Also α-arabinosidase activity	Deleyn et al. 1978, 1982
Penicillium wortmanni	110[a], 115[b] 180[a], 100[b] 195[a], 105[b] 210[a], 101[b]	3.7 4.8 4.3 4.6	4 – 4.5 3 – 4 3 – 4 3 – 4			Matsuo et al. 1987
Trichoderma reesei	90[b]	–	–			Lappalainen 1986
Trichoderma viride	101[a] 102[b]	4.45	3.5	5.8	Transferase activity	Matsuo and Yasui 1984a
Chaetomium trilaterale	240[a]	4.86	4.2	2.8[d]	Also high β-glucosidase activity	Uziie et al. 1985a, b
Emericella nidulans	118[b]	3.25	4.5 – 5.0	6.6		Matsuo and Yasui 1984b
Malbranchea pulchella var. *sulferrea* No. 48	26[a]	–	6.2 – 6.8	3.1	No β-glucosidase or transferase activity	Matsuo et al. 1977a, b
Sclerotium rolfsii	170[b]	6.8	4.5	0.04		Lachke et al. 1985
Bacillus pumilus	110[a], 60[b]	4.4	7.0 – 7.3	1.5	Intracellular	Kersters-Hilderson et al. 1969, 1982

[a] Determined by gel filtration.
[b] Determined by SDS-PAGE.
[c] For p-nitrophenyl β-D-xyloside.
[d] For phenyl β-D-xyloside.

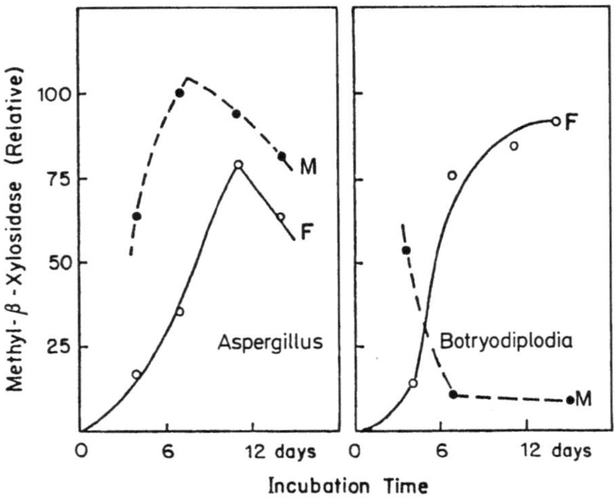

Fig. 3.6. Location of β-D-xylopyranosidase during growth of *Aspergillus niger* QM877; *Botryodiplodia* sp. QM7092. β-D-Xylopyranosidase activity (vs methyl β-D-xylopyranoside) of mycelium (*M*, ●) and extracellular culture fluid (*F*, ○) (Reese et al. 1973)

Reese et al. (1973) found that β-xylosidase is an inducible enzyme in the many fungi they tested. When the organisms were grown on starch, no β-xylosidase was produced. However, addition of methyl β-D-xylopyranoside to the starch culture led to enzyme formation and secretion. Xylanolytic enzymes are probably induced by the oligosaccharides resulting from xylan hydrolysis. Xylose is also an inducer in some organisms. However, both xylan and xylose are less effective inducers than methyl β-D-xylopyranoside. Xylosidase production is also catabolite repressed by both glucose and xylose.

The most comprehensive biochemical research has dealt with the intracellular β-xylosidase of *Bacillus pumilus* (Panbangred et al. 1984). These authors isolated two β-xylosidase genes from *B. pumilus* and compared their gene products. The β-xylosidase I and II were purified to homogeneity and compared. Both were dimer enzymes consisting of 65 000–70 000 dalton subunits. Specific activity and the K_m value of β-xylosidase I to p-nitrophenyl β-D-xyloside as substrate were respectively 100 and 1/40 times those of β-xylosidase II. The mobilities of the two β-xylosidases on polyacrylamide electrophoresis were also different. β-Xylosidase I converts oligosaccharides to xylose, while β-xylosidase II had little activity on xylobiose. The results obtained suggest that β-xylosidase I is the main enzyme for xylan hydrolysis in *B. pumilus*.

The β-xylosidase of *B. pumilus* does not transfer the xylose residue to nucleophiles other than water and operates with inversion of product configuration (Kersters-Hilderson et al. 1976), suggesting a single displacement mechanism without formation of an intermediate enzyme-xylosyl complex (Kersters-Hilderson et al. 1978). The action patterns of the enzyme are characteristic of an exohydrolase that hydrolyzes the 1,4-β-D-linkage of xylo-oligosaccharides starting at the nonreducing end. Thin-layer chromatography showed that xylotetraose was hydrolyzed to triose and then to xylobiose, i.e., xylobiose was not released directly from xylotetraose (Van Doorslaer et al. 1985).

The active site of the *B. pumilus* β-xylosidase has also been studied and found to be composed of an extended region such that at least two xylose residues of a xylo-oligosaccharide can bind at each subsite (Kersters-Hilderson et al. 1978, 1980, Van Doorslaer et al. 1980).

The fungal β-xylosidases exhibit several different traits compared to the *B. pumilus* enzyme. The molecular weights are generally higher (cf. Table 3.8) and they generally also have transferase activities. The formation of β-1,3- or β-1,4-xylobiose in transfer reactions of fungal β-xylosidase purified from a commercial hemicellulase preparation with phenyl β-D-xyloside substrate was first reported by Claeyssens et al. (1966). Takenishi et al. (1973) observed a strong transfer activity in a β-xylosidase purified from *Aspergillus niger*. Claeyssens et al. (1980) also analyzed the transfer reaction products formed from aryl β-D-xylopyranoside using a purified β-xylosidase from *Penicillium wortmanni*. The transfer products were xylobioses with β-1,3- and β-1,4-linkages. Matsuo et al. (1987) separated four different types of β-xylosidases from *P. wortmanni*. Each of these enzymes were purified to homogeneity and their physicochemical, enzymatic properties as well as immunological relationships were studied (Win et al. 1987). A strong xylosyl-transfer activity was noted with various xylo-oligosaccharides as substrates for each type of the *P. wortmanni* xylosidases.

Most of the reported β-xylosidases show highest activity toward xylobiose and no activity toward xylan. The activity toward xylo-oligosaccharides is often reported to decrease rapidly with increasing chain lengths (Rodionova et al. 1983, Matsuo and Yasui 1984a, Van Doorslaer et al. 1985). Matsuo and Yasui (1984a) reported that *T. viride* β-xylosidase hydrolyzed xylo-oligosaccharides at a velocity rate ranking in the order $X_2 > X_3 > X_4 > X_5$. On the other hand, the same authors (Matsuo and Yasui 1984b) found that with the *Emericella nidulans* enzyme this order was $X_3 > X_2 > X_4 > X_5$.

Poutanen and Puls (1988) purified and partly characterized a β-xylosidase from *T. reesei*. The enzyme was found to be a glycoprotein with a molecular weight of 100 000 as determined by SDS-gel electrophoresis. The pH optimum was 4 and the isoelectric point was at pH 4.7. The enzyme was competitively inhibited by xylose with K_i 2.3 mM. The purified enzyme also showed α-arabinosidase activity.

In an incisive study of the β-xylosidase from *Aspergillus niger* van Tieghem, Takenishi et al. (1973) found that the purified enzyme was most active at pH values between 3 and 4 and was stable over a range of pH values between 4–7 at 30 °C for 24 h. The enzyme showed glycosyltransferase activity and synthesized xylo-oligosaccharides from xylobiose and xylotriose. A purified β-xylosidase split only β-xylosidic linkages and showed no activity with substances having other glycosidic linkages. The activity toward β-xylo-oligosaccharides decreased with increasing polymerization of the substrates.

The specificities of β-xylosidases seem to vary a great deal with the source of the enzyme. Thus, Fukuda et al. (1969) found that a purified β-xylosidase from *Charonia lampas*, a marine gastropod, hydrolyzed both β-xylosidic and β-glucosidic linkages. From kinetic studies it was concluded that the hydrolysis of the different linkages might be due to the activity of one single enzyme.

The major synthetic products from xylo-oligosaccharides by the *A. niger* van Tieghhem β-xylosidase were substances containing β-1,4-linkages, xylotriose from xylobiose and xylotetraose and xylopentaose from xylotriose. This suggests that the enzyme is capable of transferring a xylosyl unit preferentially to the C_4 of appropriate xylo-oligosaccharides. This seems to be characteristic of the *A. niger* β-xylosidase, since it is well known that a sugar residue is generally transferred to various positions of sugars in the transglycosylation by other glycosidases. The question of specificity of various β-glycosidases seems to be open, as has been discussed in great detail by Reese et al. (1973).

Mesophilic and thermophilic cellulolytic anaerobic bacteria have the ability to degrade both cellulose and hemicelluloses. Although extensive work has been carried out on cellulose-degrading enzymes produced by these anaerobes (Sect. 2.6.2), relatively little is known about the enzyme systems involved in the hydrolysis of hemicelluloses. However, particularly thermophilic anaerobes must be regarded as very promising for biotechnical conversion processes. Wiegel et al. (1985) studied differences in xylan degradation by various noncellulolytic thermophilic anaerobes and *Clostridium thermocellum*. It was found that *C. thermocellum* (60 °C) first utilized the high-molecular weight fraction (DP between 15 and 40 xylose units). Xylose and xylo-oligomers (DP = 2 – 5) accumulated while the bacterium was not growing. Whereas xylan was hydrolyzed and the small oligoxylans were degraded after more than 100 h of incubation, xylose was not significantly utilized. In contrast, *C. thermohydrosulfuricum* (70 °C) and *Thermoanaerobicum brockii* (70 °C) utilized xylose first and then the xylo-oligomers of DP = 2 – 5. However, xylo-oligomers of DP > 6 were utilized only slowly. The strains *Thermoanaerobacter ethanolicus* (70 °C), *Thermobacteriodes acetoethylicus* (70 °C), and *C. thermosaccharolyticum* (60 °C) utilized xylose preferentially. Xylo-oligomers of DP = 2 – 6 and higher were apparently concomitantly utilized without significant differences. In contrast to *C. thermocellum*, the noncellulolytic organisms studied grow during xylan hydrolysis, producing ethanol, lactate, acetate, CO_2, and H_2.

In a study by Khan et al. (1987), three mesophilic cellulolytic anaerobes, namely, *Acetivibrio cellulolyticus*, *A. cellulosolvens*, and *Bacteroides cellulosolvens* and the thermophilic anaerobe *C. thermocellum* were examined for their production of xylanase, xylosidase, and esterase activities when grown on media containing cellulose. Xylanase and xylosidase activities produced by these anaerobes were stimulated by the presence of ascorbic acid, cysteine, or dithiothreitol. All the strains used in this study were unable to utilize glucose or xylose, and their growth on xylan was limited. Esterase, xylanase, and xylosidase appeared to be produced by the anaerobes as a part of their cellulolytic enzyme system. Xylan seems to be degraded during their growth on cellulose. The production of xylanase, xylosidase, and esterase by three of

Table 3.9. Xylanase, xylosidase, and esterase formation by three cellulolytic anaerobes in media containing different carbon sources (Khan et al. 1987)

Anaerobe	Substrate[a]	Xylanase	Xylosidase ($IU\ ml^{-1}$)[b]	Esterase
A. cellulolyticus	Cellobiose	0.36	0.21	0.9
	Cellulose, cotton	0.48	0.29	2.5
	Cellulose, wood	0.46	0.38	2.8
	Xylan (larchwood)[c]	0.21	0.10	1.6
B. cellulosolvens	Cellobiose	0.24	1.40	2.2
	Cellulose, cotton	0.38	0.86	2.6
	Cellulose, wood	0.48	1.18	2.9
	Xylan (larchwood)[c]	0.20	0.60	0.5
C. thermocellum	Cellobiose	0.29	6.65	1.1
	Cellulose, cotton	0.78	3.87	1.8
	Cellulose, wood	0.95	4.12	1.8
	Xylan (larchwood)	0.12	1.40	0.4

[a] Initial substrate concentration, $2\ g\ l^{-1}$ (w/v).
[b] Activities are expressed in international units, and calculated as μmoles of xylose released per minute in case of xylanase and μmoles of p-nitrophenol released per minute in case of xylosidase and esterase.
[c] From P-L Biochemicals Inc., Milwaukee, Wisconsin.

the studied cellulolytic anaerobes in media containing different carbon sources is given in Table 3.9. The formation of these three enzymes appears to be constitutive since they were produced in media containing cellobiose or cellulose and since the presence of xylan had little or no effect on their formation. As these bacteria do not actively utilize xylose, the usefulness of xylanase enzymes is not clear. The stimulation of xylanase and xylosidase activities in the anaerobes by the presence of an antioxidant agent such as ascorbic acid, or a reducing agent such as cysteine or DTT, indicates the instability of these enzymes under oxidative conditions.

3.3.3 α-Arabinosidase

L-Arabinose residues are widely distributed in heteropolysaccharides and glycoconjugates, although the quantities of L-arabinose and L-arabinan in living tissues are relatively small. The α-L-arabinofuranosyl side chains of xylans are usually α-1,3-linked to the β-1,4-xylopyranosyl backbone (Aspinall 1980). Although the frequency of L-arabinose side chains is generally relatively low, they restrict the enzymatic hydrolysis of hemicelluloses by xylanases (Brice and Morrison 1982).

The definition of α-L-arabinofuranosidase (EC 3.2.1.55) is that it hydrolyzes (terminal) nonreducing α-L-arabinofuranosyl groups of α-L-arabinofuranosides, arabinans, arabinoxylans, and arabinogalactans (IUB 1984). Recent reviews of α-L-arabinofuranosidases have been presented by Kaji (1984) and Poutanen (1988a).

Table 3.10. *α*-L-Arabinofuranosidases (Poutanen 1988a)

Microbial origin	MW (kDa)	pI	pH	K_m (mM)	Substrate specificity	Reference
Aspergillus niger	53[a]	3.6	3.8–4.0	4.9[c]		Kaji et al. 1969, Kaji and Tagava 1970
Corticium rolfsii	–	–	2.5	2.9[c]	Only arabinose produced from beet arabinan and arabinoxylan	Kaji and Yoshihara 1971
Streptomyces sp.	92[b]	4.4	6.0	3.6[d]	Produced arabinose from beet arabinan and arabinoxylan	Kaji et al. 1981
Streptomyces purpurascens	495[a] 62[b]	3.9	6.5	0.08[d]	Hydrolyzed only arabinooligomers, not arabinan or arabinoxylan	Komae et al. 1982
Bacillus subtilis	65[b]	5.3	6.5		Produced arabinose from beet arabinan	Weinstein and Albersheim 1979
Ruminococcus albus	305[a] 75[b]	–	6.8	1.3[d]	Produced arabinose from beet arabinan	Greve et al. 1984

[a] Determined by gel filtration.
[b] Determined by SDS-PAGE.
[c] For phenyl *α*-arabinofuranoside.
[d] For 4-nitrophenyl *α*-L-arabinofuranoside.

A number of organisms, including fungi, bacteria, actinomycetes, protozoa, and plants, release L-arabinose from L-arabino-containing polysaccharides or from simple substrates (Dekker and Richards 1976). However, α-arabinosidases have been purified and characterized from only a few of these organisms (Table 3.10). Fungal α-arabinosidases have often been reported to be produced by fungal pathogens or in pectolytic systems (Fielding and Byrde 1969, Kaji and Yoshihara 1970, Kaji 1984).

In assays for α-arabinofuranosidase activity phenyl α-L-arabinofuranoside, p-nitrophenyl α-L-arabinofuranoside, and beet arabinan have been used as substrates (Kaji 1984). Reducing sugars released have been determined by the Nelson-Somogyi method (Nelson 1944, Somogyi 1952). It is, however, difficult to evaluate how specific these assay techniques might be, since β-xylosidases of *A. niger* (Rodionova et al. 1983) and *Penicillium wortmanni* (Deleyn et al. 1978) were also reported to show activity toward p-nitrophenyl α-L-arabinofuranoside. Arabinose has also been reported to be liberated by purified xylanase of *A. niger* (John et al. 1979) and *Trichoderma koningii* (Wood and McCrae 1986b). These xylanases did not, however, show activity toward p-nitrophenyl α-L-arabinofuranoside or arabinosylotriose or tetraose. Moreover, most xylanases produce a mixture of arabinose-substituted xylo-oligomers from arabinoxylans.

In most studies, α-arabinosidases have been induced by beet arabinan (Kaji and Yoshihara 1970, Weinstein and Albersheim 1979, Kaji et al. 1981). In contrast, in her studies of *T. reesei* α-arabinosidase, Poutanen (1988b) used a medium containing glucose and distiller's spent grain as the main sources of carbon and nitrogen, respectively.

Some physicochemical properties of α-L-arabinofuranosidases are given in Table 3.10. It should be particularly observed that the molecular weights of the studied enzymes vary considerably. Although several α-arabinofuranosidases have been purified and partly characterized, relatively little has been reported about their role in the hydrolysis of xylan. All α-arabinosidases studied hydrolyze beet arabinan to arabinose with the exception of the enzyme of *Streptomyces purpurascens* (Komae et al. 1982). Some enzymes have also been re-

Table 3.11. Hydrolysis of wheat straw arabinoxylan by *T. reesei* enzymes (Poutanen 1988b)

Enzyme dosage (nkat g^{-1} substrate)			Hydrolysis products (% dry wt.)		
α-Arabinosidase	Xylanase	β-Xylosidase	Xylose	Xylobiose	Arabinose
15000[a]	37000[a]	500[a]	69	0	8
500	0	0	0	0	1
500	10700[b]	0	18	24	3
5000	0	0	1	<1	4
0	10700[b]	0	11	20	0

[a] Activities of crude culture filtrate.
[b] Partly purified by gel chromatography according to Lappalainen (1986). Substrate concentration $10\,g\,l^{-1}$, pH 4.0, temperature 45 °C, hydrolysis time 48 h.

ported to produce arabinose from arabinoxylan (Kaji and Yoshihara 1971, Komae et al. 1982). In her studies of *T. reesei* α-arabinosidase, Poutanen (1988b) found that the purified enzyme, MW 53 kDa, was able to release arabinose from long-chain arabinoxylan, although arabinose-substituted xylo-oligosaccharides produced by endoxylanases were the preferred substrates. An increased production of xylose, xylobiose, and arabinose reflected the synergistic action between α-arabinosidase and endoxylanases when they were used simultaneously to hydrolyze wheat straw arabinoxylan (Table 3.11). This demonstrates clearly that α-arabinosidase is an essential part of the xylanolytic system of *T. reesei*. The *T. reesei* enzyme also hydrolyzes beet arabinan to give arabinose as an end product.

3.3.4 α-D-Glucuronidase

In birch wood xylan the main substituents on the β-D-xylopyranosidic backbone are 4-*O*-methylglucurono and *O*-acetyl residues (Timell 1967). It was suggested by Timell (1962) that the 4-*O*-methylglucuronic acids side chains in xylan are a steric hindrance for the hydrolysis of the xylan backbone by endoxylanolytic enzymes in the vicinity of the branching points. The 4-*O*-methyl-D-glucuronic acid residues are linked α-glycosidically to the *O*-2 position. The degree of branching varies depending on the source of the xylan hemicellulose in question.

The importance of α-1,2-glucuronidase for the hydrolysis of acidic xylo-oligomers was pointed out by Puls et al. (1978). The presence of such an enzyme activity in *Trichoderma reesei* QM 9414 culture solution grown on hemicellulose was already reported by Dekker (1983). The enzyme liberated glucuronic acid from birch wood xylan.

α-Glucuronidase is the least studied of the enzymes participating in degradation of xylans. One of the reasons for this is the lack of a suitable activity assay method. However, Puls et al. (1987) used the trisaccharide 2′-*O*-(4-*O*-methyl-α-D-glucupyranoxyluronic acid)-xylobiose as a standard substrate for α-glucuronidase assay. The compound was obtained from alkali-extracted (4-*O*-methyl-D-glucurono)-D-xylan from birch wood. A partial hydrolysis with aqueous 45% formic acid of this preparation yielded a mixture of neutral and acidic xylo-oligomers. The detailed preparation of the α-glucuronidase substrate from alkali extracted birch wood glucuronoxylans is described by Puls et al. (1987). The same assay technique for α-glucuronidase was used by Poutanen et al. (1987).

Schmidt et al. (1979), when screening for fungal production of xylanolytic enzymes, found that there was strong evidence for α-glucuronidase production by the white-rot fungi *Agaricus bisporus* and *Pleurotus ostreatus*. Both fungi grow on beech wood holocellulose and on the water extract of steam-pretreated birch wood. The same authors found that both organisms proved to be only weak producers of xylanase and β-xylosidase activity compared with other fungi. Growth and enzyme activities of the two fungi in these media were studied by Puls et al. (1987). These authors also reported that they were the first to isolate an α-glucuronidase from a microorganism and that the enzyme had been neither purified nor characterized until then. *A. bisporus* and *P. ostreatus* produce α-glucuronidase in both of the media mentioned (Table 3.12). As can be seen in Table 3.12 *A. bisporus* was the more active fungus and alkali-extracted xylan was the best of the two substrates for enzyme production. Since maximum activity of α-glucuronidase is obtained with *A. bisporus* on beech wood xylan medium this enzyme was purified from such *A. bisporus* culture solutions. The enzyme was found to act in synergism with xylanases and to liberate 4-*O*-methylglucuronic acid from 4-*O*-methylglucuronic acid substituted xylo-oligomers. In this work the enzyme test mentioned above for α-glucuronidase activity was developed. The substrate used, 2-*O*-(4-*O*-methyl-α-D-glucupyranosyluronic acid)-D-xylobiose, is not degraded by xylanases or β-xylosidases.

The uronic acid substituent is bound by an α-1,2-linkage to a nonreducing xylose residue. Anion-exchange chromatography in acetate medium was used as a tool for detection of degradation products. When hardwood xylan was incubated with a purified α-glucuronidase as a first step of hydrolysis, no effect was found. This indicates that this enzyme is not able to remove the uronic acid side groups from intact xylans.

The conclusions from the studies of α-glucuronidase in the two microbial xylanolytic systems were the following:

1. Most fungi seem to excrete too little α-1,2-glucuronidase for a complete cleavage of the uronic acid side chains that protect the neighboring β-1,4-linkages of the xylan backbone from cleavage by xylanases and/or xylosidases.
2. *A. bisporus* and *P. ostreatus* are weak producers of xylanase and β-xylosidase enzymes. However, they produce a surplus of α-glucuronidase.
3. When hardwood xylan is incubated with xylanolytic enzymes, xylose yield increases when α-glucuronidase activity is present.
4. The (4-*O*-methylglucurono)-D-xylan is an artificial substrate. A natural substrate would be (*O*-acetyl-4-*O*-methylglucurono)-D-xylan, suggesting that esterase activity must be considered as an additional enzyme for the total degradation of xylans.

A search for organisms that produce all the necessary enzymes for a total hydrolysis of xylan was the main goal in the thesis work by Poutanen (1988a). In a study of the xylanolytic enzymes produced by *T. reesei* QM 9414, *Aspergillus awamori* VTT-D-75028, *Fusarium oxysporum* VTT-D-80134, *Bacillus subtilis* ATCC 12711, and *Streptomyces olivochromogenes* ATCC 21713, it was found that the culture solutions produced differed with respect to β-xylosidase and side-group cleaving activities. *T. reesei* gave the highest xylanase activity, and all the fungi produce β-xylosidase, whereas in the bacterial culture filtrates this activity was negligible (Table 3.13). Only *T. reesei* seemed to produce all the side-group cleaving activities assayed (α-arabinosidase, α-glucuronidase, and acetyl esterase). *F. oxysporum* and *S. olivochromogenes* produce esterase, while all side-group cleaving activities were low in the culture

Table 3.12. Enzyme activities and growth of *Agaricus bisporus* and *Pleurotus ostreatus* in media with beechwood xylan or extract of steamed birchwood (Puls et al. 1987)

Fungus	Substrate	Time of incubation (days)	Enzyme activities[a]			Growth (mg mycelial dry matter l^{-1} culture)	pH
			α-Glucuronidase (% degradation of (4-*O*-Me-α-D-GpA)-xylobiose	Xylanase (nkat ml^{-1})	β-Xylosidase (nkat ml^{-1})		
Agaricus bisporus	Beechwood xylan	0	n.t.	n.t.	n.t.	50	4.5
		7	n.t.	3.3	n.t.	90	4.8
		14	98	3.6	0.02	240	4.6
		27	99	2.7	0.10	590	3.5
		41	97	5.2	0.19	390	3.5
	Extract of steamed birchwood	0	n.t.	n.t.	n.t.	50	4.2
		7	n.t.	4.9	n.t.	490	3.3
		14	34	6.7	0.05	1150	2.6
		28	59	7.5	0.14	750	2.6
		42	90	6.7	0.19	490	2.9
Pleurotus ostreatus	Beechwood xylan	7	n.t.	4.0	n.t.	50	4.9
		41	24	0.7	0.02	660	5.8
	Extract of steamed birchwood	7	n.t.	4.8	n.t.	130	4.0
		42	15	0.6	0	880	3.3

[a] n.t. = Not tested.
1 nkat activity ≅ production of the reducing equivalent of 1 nmol xylobiose per second.

Table 3.13. Comparison of activities of enzymes in the xylanolytic system in different culture filtrates (Poutanen et al. 1987)

Organism	Xylanase ($nkat\ ml^{-1}$)	β-Xylosidase ($nkat\ ml^{-1}$)	Acetyl xylan esterase ($nkat\ ml^{-1}$)	Glucuronidase ($nkat\ ml^{-1}$)	Arabinosidase ($nkat\ ml^{-1}$)
Trichoderma reesei	2170	190.0	366	12	272
Aspergillus awamori	200	18.0	3	0.1	7
Fusarium oxysporum	62	0.9	53	0	6
Bacillus subtilis	311	0.4	1	0	7
Streptomyces olivochromogenes	90	0.4	28	0	2

Table 3.14. The composition of the enzymatic hydrolyzates of steamed birchwood hemicellulose[a] (Poutanen et al. 1987)

Source of enzyme	Initial pH	Amount of xylanase used ($nkat\ g^{-1}$ of substrate dry wt.)	(% of substrate dry wt.)			
			Xylose	Xylobiose	Acetic acid	4-*O*-Methyl glucuronic acid
T. reesei	5.0	600	45.2	2.8	10.0	1.8
		2500	51.6	1.2	10.8	5.4
A. awamori	5.0	600	19.6	6.8	3.2	0
		2500	38.6	3.2	8.8	0.1
F. oxysporum	5.0	600	2.8	29.0	10.8	0
		1000	3.0	31.0	10.8	0
B. subtilis	6.5	600	1.2	4.4	1.0	0
		2500	1.8	5.0	1.4	0
S. olivochromogenes	6.5	600	7.6	32.0	10.0	–
		2500	13.6	40.0	10.8	2.7

[a] Temperature 45 °C, hydrolysis time 24 h.

filtrates of *A. awamori* and *B. subtilis* (Table 3.13). That the enzyme production is important for the degradation of hardwood xylan is evident from Table 3.14, where the composition of the enzymatic hydrolyzates of steam birch wood hemicellulose is presented when similar amounts of xylanase produced by the different organisms studied were used. It can be seen in this table that *T. reesei*, producing the most complete enzyme mixture, also gives rise to the most complete hydrolysis of the substrate used.

The purification of the α-glucuronidase of *T. reesei* gave an enzyme with an apparent molecular weight of 70 kDa. The α-glucuronidase purified from *A. bisporus* (Puls et al. 1987) seems to be of much higher molecular weight, i.e., 450 kDa. The partially purified enzyme from *T. reesei* hydrolyzed (4-*O*-methylglucuronosyl)-xylobiose to yield 4-*O*-methylglucuronic acid and xylobiose.

3.3.5 Acetyl (Xylan) Esterase

Acetyl groups have long been known to be present as substituents in various plant polysaccharides, notably hemicelluloses and pectins (Hägglund et al. 1956, Kertesz 1963, Timell 1964, 1965a, b). In glucuronoxylans 60%–70% of the xylose residues are esterified. Bouveng (1961) developed techniques by which it could be shown that the acetyl groups are assigned to C-2 and C-3 of the xylose residues. The galactoglucomannans of softwoods are also acetylated but to a lesser extent (Lindberg et al. 1973a, b).

When wood is processed under alkaline conditions, the acetyl groups of the hemicelluloses are removed, while under acid conditions the rates of deacetylation and xylose production from birch wood xylan were essentially the same (Maloney et al. 1985). However, in steaming processes most of the acetyl groups remained bound to the solubilized and partially hydrolyzed xylan and were not removed until enzymatic hydrolysis was undertaken by *T. reesei* enzymes (Poutanen et al. 1986).

Frohwein et al. (1963) demonstrated the existence of esterases in unpurified enzyme preparations. These enzymes were shown to hydrolyze acetylated carbohydrates, i.e., acetyl-mannose, acetyl-glucose, acetyl-maltose, and acetyl-cellobiose. Reese et al. (1969) used cellobiose octa-acetate as an inducer for the production of cellulases. The acetyl group was split off by esterases, and the slow supply of cellobiose induced cellulases without causing catabolite repression which often occurs when cellobiose is supplied in excess (cf. Sect. 2.7).

Williams and Withers (1981) isolated more than 100 bacterial cultures from bovine rumen contents by enrichment techniques in polysaccharide-containing media based on their ability to degrade plant cell-wall structural polysaccharides. Thirty of the more active xylan- and arabinan-degrading isolates were studied for both the production and activity of the principal polysaccharidase and glycosidase enzymes participating in the hydrolysis of the plant cell wall polysaccharides. Polysaccharidase and glycosidase activities of 20 of these bacterial isolates grown in xylan-containing media are presented in Table 3.15. Esterase activity could be traced in all but one of these 20 isolates.

For the assay of xylan esterase activity, Williams and Withers (1981) used a spectrophotometric assay in which the rate of p-nitrophenol released from p-nitrophenyl acetate was determined. Biely et al. (1985) used a solution of nondialyzable fraction of birch hemicellulose. One unit of xylan esterase was defined as the amount of enzyme liberating one µmol acetic acid from 10 mg of the substrate dissolved under standardized conditions. Poutanen and Sund-

Table 3.15. Polysaccharidase and glycosidase activities in isolates following growth in xylan-containing media (Williams and Withers 1981)

Isolate	Polysaccharidase: Cell bound[a] Xylanase	Cell bound[a] Arabinanase	Pectinase	CMCase	Extracellular[b] Xylanase	Extracellular[b] Arabinanase	Glycosidase activity[c]: Esterase	α-L-arabino-furanosidase	α-L-fucosidase	β-L-fucosidase	β-D-fucosidase	α-D-galactosidase	β-D-galactosidase	α-D-glucosidase	β-D-glucosidase	β-D-glucuronidase	α-D-mannosidase	β-D-mannosidase	Alkaline phosphatase	α-D-xylosidase	β-D-xylosidase
X1	68.2	–		+	26.5	35	4.2	37.7	–	–	10.8	10.8	52.1	–	12.6	23.4	–	32.3	77.3	–	–
X2	66.8	22.3		+	12.5	10	7.5	118.2	12.3	6.4	23	44.9	160.5	23.5	57.8	21.9	15.5	8.6	7032	25.1	272.7
AGA3 (i)	44.1	81.8	+	+§	–	15	2.6	13.3	–	–	–	128.3	8.3	15	–	16.7	8.3	5	20	–	–
FB19d	18	–	+	+§	2.5	–	9.9	40.2	11.8	18.9	21.3	35.5	–	73.3	26	4.7	4.7	23.6	196.3	11.8	23.6
P2	189.8	–			27.5	10	3.4	18.8	–	–	–	–	221.5	99.3	–	20.9	8.4	–	2931	–	428
P5	25	–			5	–	1.3	29.7	–	–	13.5	17.1	290.2	121.3	14.4	53.9	27	30.6	701.8	18	–
RL3	<5	12.5			15	5	1.5	–	–	–	10.3	–	187.2	23.4	–	41.2	11.2	–	464	–	4.7
RL16B	140	38			15	10	5.9	1167	10.2	4.5	33.7	76.4	1043	74.1	35.9	38.2	112.3	10.1	–	–	161.7
RL16C	15.6	10.4		+	–	–	4.7	19	–	17.7	–	4.1	95.3	–	–	36.8	–	–	296.8	–	8.2
HRL	3.9	–	+	+§	223	10	6.7	4163	40.8	6.3	1505	–	1078	581	37.6	10.4	3.1	3.1	30.3	2.1	7.3
HRL2A	42.6	17	+	+§	12.5	5	1.3	3629	–	–	76.4	13.5	2126	152.8	11.2	61.8	169.6	10.1	22.5	5.6	330
F1	24.4	–	+		7.5	–	–	37.3	8.3	11.7	15.2	15.2	14.5	26.9	3.5	17.9	8.3	8.3	38	11.7	4.1
F6 (i)	<5	34.8		+	20	20	1.4	44.9	–	–	17.3	–	521.9	–	79.5	15	–	–	230.4	–	3.5
F15B	14.4	–	+		2.5	20	1.1	23.6	–	–	52	4.7	2033	14.2	7.1	54.3	22.4	4.7	941.4	–	–
FBHay A	13.9	7	+	+	15	5	2.1	42.8	–	–	17.4	26.4	378.6	160.4	–	38.5	–	5.5	588	–	4.6
FBHay B	<5	–	+		–	25	2.9	829.6	–	–	50.1	456.3	801.9	110.6	43.2	13.8	67.4	10.4	44.9	–	917.7
Ferm 6	<5	7.1	+		–	25	14.3	13.1	2.8	–	–	–	–	438	8.4	1.8	15	–	19.7	–	53.4
4	35.4	35.4		+	45	45	1.3	94	7.5	1.9	7.5	7.5	–	48.7	1.9	7.5	15	–	202	–	–
8A	38.7	7.7	+	+	42.5	55	2.3	12.3	–	–	–	–	1094	12.3	–	10.2	4.1	–	476	6.1	–
H2b	9	35.9			350	50	2.5	23132	31.9	4.7	883.4	26	3077	5750	102.9	49.7	37.8	3.5	147.8	–	14.2

[a] Units (μg) reducing sugar formed mg^{-1} protein h^{-1}. [b] Units (μ) reducing sugar formed ml^{-1} supernatant h^{-1}. [c] Units (nmol) p-nitrophenol released mg^{-1} protein h^{-1}. [d] +, Activity >1 cp reduction in viscosity mg^{-1} protein; §, ml supernatant h^{-1}.

berg (1988) also used birch wood hemicellulose as substrate for acetyl xylan esterase activity. Acetyl esterase activity was determined by these authors using α-naphtyl acetate. The release of free α-naphtol was measured spectrophotometrically.

The existence of acetyl xylan esterases in fungal cultures was first reported by Biely et al. (1985 b, 1986). The acetyl esterases of the fungi *Trichoderma reesei*, *Aspergillus niger*, *Schizophyllum commune*, and *Aureobasidium pullulans* were found to have a high capacity to de-acetylate steamed xylan (Biely et al. 1985 b). Partially purified esterases and xylanases from *T. reesei* and *S. commune* exerted a synergistic action in the hydrolysis of acetylated xylan and de-acetylated xylo-oligomers, xylose, and acetic acid were produced (Biely et al. 1986). The de-acetylation caused by the esterases creates new sites on the xylan backbone for attack by the xylanases. These in turn were then able to produce shorter xylo-oligomers, which are the preferred substrates for esterases.

Poutanen and Sundberg (1988) purified an acetyl esterase from *T. reesei*. The production of esterase activity by *T. reesei* became evident when a rapid accumulation of acetic acid was observed when acetylated xylo-oligomers were hydrolyzed with culture solution from this fungus (Fig. 3.7 a, Poutanen et al. 1986). Esterases are also produced by *A. awamori*, albeit to a lesser extent, as is evident from Fig. 3.7 b. The acetyl esterase from *T. reesei* has a molecular weight of 45 000 as determined by SDS gel electrophoresis or 67 000 as determined by gel filtration. The esterase activity was separated into two isoenzymes by chromatofocusing with isoelectric points at pH 6.8 and 6.0, respectively.

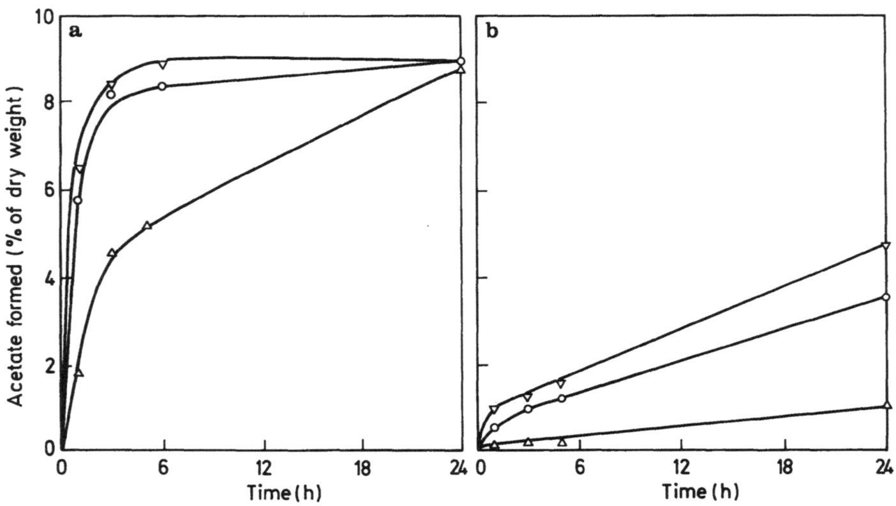

Fig. 3.7. The liberation of acetic acid in the hydrolysis of steamed birchwood hemicellulose fraction (acetylated xylo-oligomers). Temperature 45 °C, initial pH 5. Xylanase (nkat g^{-1} substrate): (△——△) 1000, (○——○) 5000, (▽——▽) 10000. **a** *T. reesei* enzyme; **b** *A. awamori* enzyme (Poutanen et al. 1986)

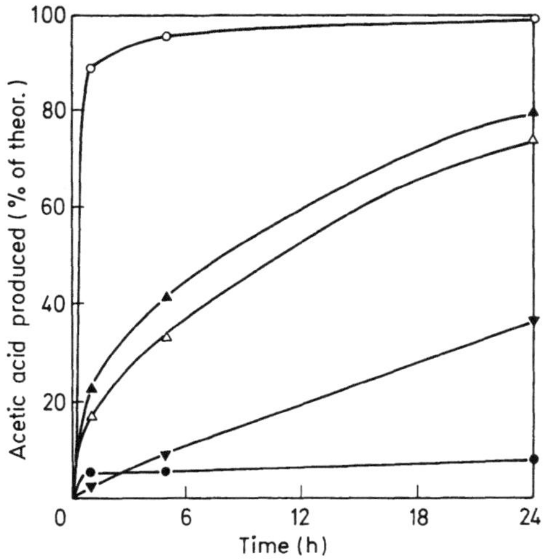

Fig. 3.8. Formation of acetic acid in the hydrolysis of steamed acetylated xylo-oligomers by enzymes of *T. reesei*. Substrate concentration 10 mg ml^{-1}, initial pH 5, temperature 45 °C. (●——●) Purified naphthylacetyl esterase (5 nkat ml^{-1}); (▼——▼) partly purified endoxylase preparation (150 nkat ml^{-1} xylanase), (△——△) esterase and xylanase (5 and 150 nkat ml^{-1}, respectively); (▲——▲) esterase, xylanase, and β-xylosidase (5, 150, and 5 nkat ml^{-1}, respectively), and (○——○) culture filtrate (0.5 mg ml^{-1} protein, 5 nkat ml^{-1} esterase, 5 nkat ml^{-1} β-xylosidase, 150 nkat ml^{-1} xylanase (Poutanen and Sundberg 1988)

The enzyme showed activity toward naphtyl acetate, triacetin, and glucose and xylose acetates. However, acetic acid was liberated from acetylated xylo-oligomers only to a small extent. A liberation of acetic acid from oligomeric substrates was enhanced by the addition of other enzymes in the xylanolytic system (Fig. 3.8).

The results presented indicate that acetyl esterase from *T. reesei* has a high specificity toward short xylo-oligomers compared to long xylo-oligomers. A synergistic action with depolymerizing enzyme(s) is needed to achieve a high degree of de-acetylation and for the production of xylose.

3.4 Mannan-Degrading Enzymes

Galactoglucomannans and glucomannans in softwoods and hardwoods are both branched heteropolysaccharides, and several enzymes are therefore involved in their complete degradation.

An enzyme preparation suitable for complete hydrolysis of wood mannans requires a concerted action of the following hydrolytic enzymes: endo-1,4-β-mannanase (EC 3.2.1.78), β-mannosidase (EC 3.2.1.25), β-glucosidase (EC 3.2.1.21), and α-galactosidase (EC 3.2.1.22). The enzymatic hydrolysis of mannans has been reviewed by Dekker and Richards (1976) and Dekker (1985).

The endomannanases are the best characterized and most widely studied of the mannan-degrading enzymes. These enzymes attack in an endwise manner the backbone of the mannans to produce shorter substituted and nonsubstituted oligomers, mannobiose, and mannose. To convert the water-soluble oligomers and dimers to mannose, β-mannosidases are employed (Reese and Shibata 1965, Wan et al. 1976, Rättö and Poutanen 1988). The enzymes releasing the glucose and galactose substituents on this mannan backbone, i.e., β-glucosidase and α-galactosidase, act in synergism with the endomannanases and β-mannosidases.

3.4.1 Endo-1,4-β-D-Mannanases

Endo-1,4-β-D-mannanases (EC 3.2.1.78) are produced by various fungi, bacteria, marine alga, germinating seeds of terrestrial plants, and by various invertebrates. An extensive description of organisms producing endomannanases has been given by Dekker and Richards (1976). Fungal mannanases from *Tyromyces palustris* have been studied by Ishihara and Shimizu (1980), from *Aspergillus tamari* by Civas et al. (1984b), and from *Penicillium purpurogenum* by Kusakabe et al. (1988). Endomannanases have also been purified from commercial *Aspergillus* enzyme preparations (Eriksson and Winell 1968, Yamazaki et al. 1976). Reese and Shibata (1965) screened a large number of fungi, particularly *Aspergillii* and *Penicillii*, for their production of mannanases.

Pavlova and Tinyanova (1980) studied 112 soil microorganisms for their capability of growing on mineral medium-containing mannan as a sole carbon source. Twenty-three of these strains were found to produce extracellular mannanase. The 10 most active strains were identified as belonging to *Pseudomonas*, *Actinomycetes*, *Bacillus*, *Flavobacterium*, and *Arthrobacter* species. Araki and Kitamikado (1981) assayed the culture fluids of 117 bacterial strains isolated from natural habitats for β-mannanase activity. The highest β-mannanase productivity was detected in the strain F-25 of *Aeromonas* sp. isolated from the intestinal contents of the fresh water fish *Salmo gairdnerii*. Endomannanases produced by *Bacillus subtilis* (Emi et al. 1972), *Aeromonas hydrophila* (Araki 1983), and *Streptomyces* sp. (Takahashi et al. 1984) have been studied in more detail.

Endomannanase activity has been routinely assayed using Remazol brilliant blue (RBB)-dyed carob galactomannan as substrate (McCleary 1988a). However, in most cases the Somogyi-Nelson reducing sugar assay has been the widely employed assay technique (Dekker and Richards 1976). Some physicochemical properties of endomannanases from different microbial sources are presented in Table 3.16.

The physicochemical properties of endomannanases from some plant sources have also been investigated. These properties are very similar to those from microbial origin. Thus, an endomannanase from konjac tubers, *Amorphophallus konjac*, has a pH optimum at pH 4.7 and is stable between pH 4–8 (Sugiyama et al. 1973). Shimahara et al. (1975) isolated two endomannanases

Table 3.16. Physicochemical properties of endomannanases from some different microbial sources

Microbial origin	MW (kDa)	pI	pH-opt.	pH-stability	References
Aspergillus niger	42	4.1	3 – 3.8	2.0 – 9.2	Eriksson and Winell (1968)
A. niger van Tieghem		3.4	3.5	4.0 – 7.0	Tsujisaka et al. (1972)
A. wentii	42 – 47				Keilich et al. (1970)
A. niger		4.1	3.6	5.0 – 9.0	Yamazaki et al. (1976)
Tyromyces palustris	61	3.5	3.5		Ishihara and Shimizu (1980)
Irpex lacteus	53	5.0, 5.5	3.0		McCleary (1988a)
Bacillus subtilis	22	5.2	6.0	5.0 – 9.5	Emi et al. (1972)
Streptomyces sp.	42.9	3.7	6.8		Takahashi et al. (1984)

from konjac tubers with molecular weights of 32 000 and 29 000, respectively. An endomannanase purified from alfalfa seeds by Villarroya and Petek (1976) was found to have a pH optimum at 5.2 and a molecular weight of 40 000.

Endomannanases of microbial origin have been reported to be both inductive and constitutive enzymes. Reese and Shibata (1965) tested over 290 cultures for their ability to produce endomannanase in shake flasks on a basal medium to which mannan had been added. Most of the strains investigated produced at least trace amounts of the desired enzyme. All of the active microorganisms produce small amounts of endomannanase even when grown on simple sugars or on starch. However, when mannan or mannan-containing carob gum were used as substrates, the organisms were greatly stimulated in their enzyme production, the increase being 10- to more than 100-fold (Table 3.17).

Lyr (1963) observed that some basidiomycetes produce larger amounts of endomannanase when grown on cellulose, compared to when grown on mannan. This behavior was in some instances also observed by Reese and Shibata (1965). This effect has to some extent been discussed by Eriksson and Goodell (1974), who suggested that cellulases, mannanases, and xylanases in *Polyporus adustus* are under the control of a single, common, regulatory gene (cf. Sects. 2.7 and 3.3.1).

Rättö and Poutanen (1988) studied mannanase production of two bacterial and two fungal strains, i.e., *Bacillus subtilis*, *Streptomyces olivochromogenes* and *Trichoderma reesei*, *Aspergillus awamori*. The highest activity was obtained with *B. subtilis*; the activity increased tenfold when galactomannan was added to the medium.

Endomannanases are usually secreted into the medium in which the microorganism is cultivated. However, some exceptions have been noted, namely, for the bacteria *Sporocytofaga myxococcoides* (Clermont et al. 1970), *Aerobacter mannanolyticus* (Inoue and Inoue 1957), and in *Xanthomonas campestris*

Table 3.17. Effect of growth substrate on production of β-1,4-mannanase (Reese and Shibata 1965)

Organisms	β-1,4-Mannanase, U ml^{-1} (max.)					
	QM No.	Mannose	Glucose	Starch	Mannan	Carob
Aspergillus fumigatus	45h	0.6	1.0	0.8	18	32[a]
A. giganteus	7974	0.0	0.0	0.0	12	3[a]
A. luchuensis	873	1.7	0.9	–	18	30
A. nidulans v. *echinulatus*	1909	0.0	0.0	–	10	3
Bacillus subtilis	B1564	1.5	1.5	–	125	60
Cladosarum olivaceum	2000	1.8	0.8	0.6	9	7[a]
Gelatinosporium sp.	684	1.5	1.3	–	18	19
Fusicoccum sp.	2045	1.2	0.8	1.0	5	72
Paecilomyces varioti	3408	3.0	2.8	2.8	60	265[a]
Penicillium funiculosum	474	0.4	0.4	0.4	49	8[a]
P. ochro-chloron	477	0.6	0.0	–	72	18
P. verrruculosum	7999	0.3	0.0	0.2	25	142[a]
P. wortmanni	1859	2.1	0.0	–	530	400

[a] In these, the carob gum was partially hydrolyzed (enzymes) before addition to the medium.

(Dekker and Candy 1979), where the enzymes were either intracellular or membrane-bound.

Hydrolysis by endomannanases of the mannans occurring in wood is effected by the degree and pattern of substitution of these polysaccharides. Products arising from degradation of several D-mannans indicate that they are mostly degraded by fungal endomannanases to manno-oligosaccharides of DP$\geq$2 and mixed D-mannose-oligosaccharides containing D-galactose (from galactomannan) or D-glucose (from glucomannans). The purified endomannanase from *Aspergillus niger* (Eriksson and Winell 1968, Yamazaki et al. 1976) degraded different mannan compounds into mannobiose and mannotriose. No activity toward mannobiose could be demonstrated for the enzyme. Kusakabe et al. (1988) isolated seven different oligosaccharides from the hydrolyzate of konjac glucomannan by a purified endomannanase from *Penicillium purpurogenum* No. 618. These oligosaccharides were identified as M–M, G–M, G–M–M, M–G–M, G–G–M, M–G–M–M, and M–G–G–M; where G- and M-represent β-1,4-D-glucopyranose and β-1,4-D-mannopyranose, respectively. The mode of attack of the endomannanases from *P. purpurogenum* is presented in Fig. 3.9.

An *Aspergillus niger* endomannanase has been purified by McCleary (1979). The enzyme has been demonstrated to depolymerize effectively even highly substituted galactomannans. Its action on galactomannans, glucomannans, and manno-oligosaccharides has been studied in detail, and a computer model has been developed to simulate the subsite-binding requirements and catalytic behavior (McCleary 1979, McCleary et al. 1983, McCleary and Matheson 1983). McCleary (1988a) found that the endomannanase from *Irpex lacteus*, in common with the enzyme from *A. niger*, has a greater ability to hy-

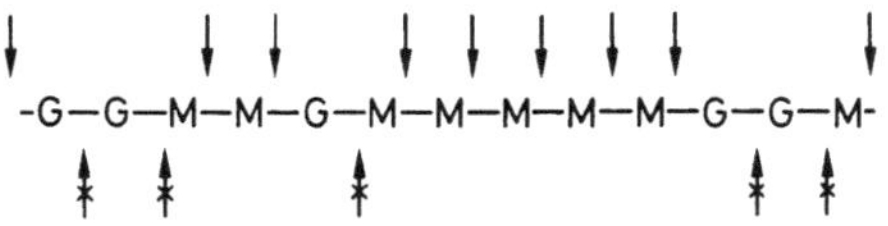

Fig. 3.9. Hydrolysis of konjac glucomannan by a purified endomannanase from *P. purpurogenum*. *G*-, β-1,4-D-glucopyranosidic linkage; *M*-, β-1,4-D-mannopyranosidic linkage. ↓ and ↑ indicate the mannanase hydrolyzable and nonhydrolyzable linkages, respectively (Kusakabe et al. 1988)

drolyze highly substituted galactomannans than do the enzymes from legume seeds or from *Bacillus subtilis.*

Coulombel et al. (1981) studied two endomannanases, one from *Streptomyces* and one from fenugreek seeds (*Trigonella foenum-graecum*). By using [^{3}H]-mannobiose as a labeled acceptor, they could demonstrate transfer reactions catalyzed by two endomannanases, using mannotetraose and mannopentaose as substrates. Both enzymes hydrolyze at random the β-1,4-mannan from ivory nut (*Phytelephas macrocarpa*) as well as oligomannosides with a degree of polymerization higher than 5. However, the hydrolysis of tetramer and pentamer substrates are not of this type. None of the two enzymes hydrolyze mannotriose or mannobiose and the rate of hydrolysis of substrates with a higher degree of polymerization increases with the chain lengths of these substrates. The endomannanase from *Streptomyces* degrades mannotetraose to mannobiose and mannotriose without any liberation of mannose. To explain the accumulation of mannotriose without any release of mannose in the hydrolyzates of mannotetraose by the *Streptomyces* endomannanase it was postulated by Coulombel et al. (1981) that one mannose unit was transferred from the tetramer to mannobiose, the major product of hydrolysis. The hypothesis was confirmed by using a radioactively labeled mannobiose as acceptor. Thus, in early stages of incubation of mannotetraose with the enzyme, radioactive mannotriose appears in the presence of labeled mannobiose. The whole reaction scheme is summarized in Fig. 3.10. Similarly, incubation of mannopentaose with the mannanase leads to labeled mannotriose in the presence of radioactive mannobiose. The *Trigonella* endomannanase also behaves as a true endopolysaccharidase. It does not hydrolyze mannotriose or mannobiose but only higher oligomannosides. Also, this enzyme catalyzes transglycosylation reactions but in a way different from that of the endomannanase of *Streptomyces*, while transfer reactions of the *Streptomyces* enzyme involve only a single mannose unit. The endomannanase from fenugreek seedlings is able to transfer oligomannose residues, a mannobiose group from mannotetraose, and a mannotriose group from mannopentaose.

Although several endomannanases have been reported to be produced by fungi, bacteria, and higher plants, no exotype enzyme has been reported except in one case. Thus Araki and Kitamikado (1982) describe the purification and characterization of a novel exo-β-mannanase from *Aeromonas* sp. F-25. The evidence they present is derived from the following experiments. The action of the enzyme was studied with reduced manno-oligosaccharides as substrates.

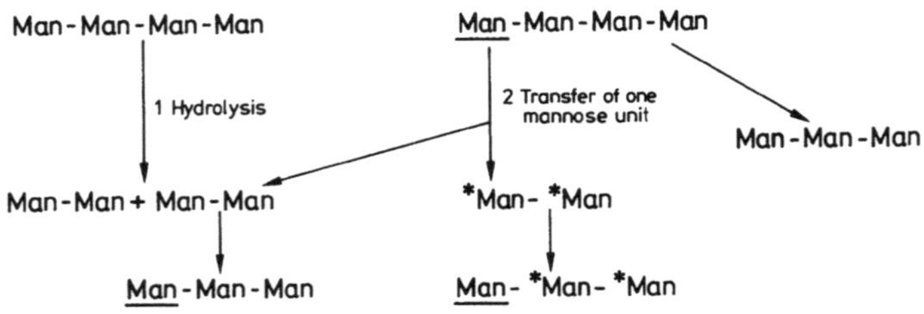

Fig. 3.10. Reaction pattern of *Streptomyces* β-mannanase on mannotetraose *Man*- mannose residue involved in transglycosylation: **Man*,[^{3}H]mannose residue (Coulombel et al. 1981)

The enzyme attacked M_3H and M_5H to give M_2 and M_1H and formed $M-2$ and M_2H from M_4H. On the basis of these results, the enzyme was characterized as one splitting off M_2-residues from the nonreducing end of reduced manno-oligosaccharides. The enzyme was also demonstrated to have the ability to cause transglycosylation. Based on these results, the enzyme was characterized as a novel exo-β-mannanase, successively removing M_2 residues from the nonreducing end of the β-mannan molecule. No experiment was carried out to ascertain that the enzyme removed their products with conversion of configuration. This is a well-known and often used characteristic of an exo-polysaccharidase. It appears that this enzyme will need further characterization until it can be classified as a true exo-β-mannanase.

3.4.2 1,4-β-D-Mannosidase

β-Mannosidases (β-D-mannoside mannohydrolase, EC 3.2.1.25) split off 1,4-linked β-D-mannosyl groups from the nonreducing end of their substrates, manno-oligosaccharides and mannose-containing glycopeptides (Dey 1978). Mannosidases have been demonstrated to be produced by microorganisms and plants and also to be present in animal tissues (Dey 1978).

The most convenient assay for β-mannosidase activity is to use p-nitrophenyl β-D-mannopyranoside as substrate and to measure the rate of cleavage of glycosidic bonds by the rate of release of p-nitrophenol (McCleary 1988a). Another assay technique would be to use reduced 1,4-β-D-mannooligosaccharides as substrate and to measure the increase in D-mannose reducing sugar equivalents by the Somogyi/Nelson method.

β-Mannosidase is one of the three enzymes (endomannanase, β-mannosidase, and α-galactosidase) involved in the hydrolysis of galactomannans. Preparations produced from microorganisms normally vary in the relative amounts of each of these components. Since β-mannosidase is the enzyme responsible for the hydrolysis of mannobiose and mannotriose (produced by endomannanase) to mannose, its absence in mannanase preparations results in accumulation of oligosaccharides during mannan hydrolysis (Reese and Shibata 1965; Fig. 3.11). Reduced mannobiose was found to be highly resistant to hydrolysis

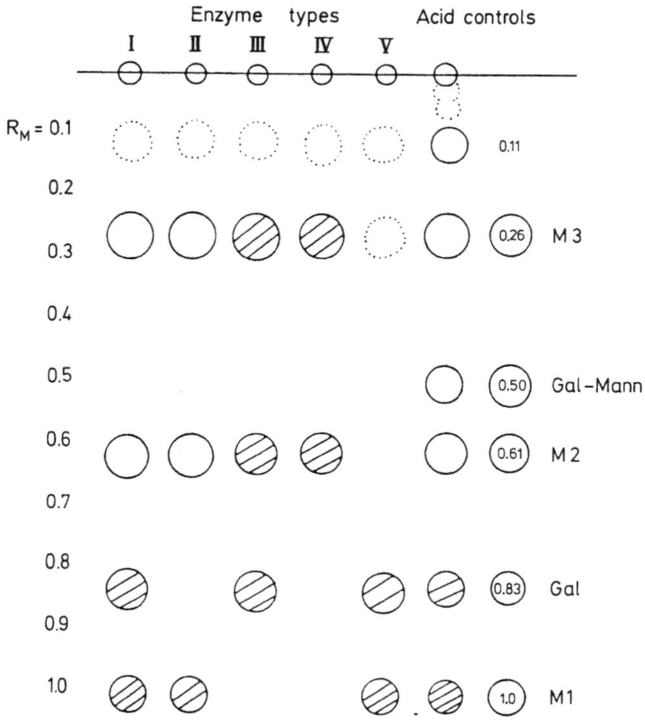

Fig. 3.11. Hydrolysis products of carob gum (diagrammatic). Enzyme type: *I* mannanase and β-mannosidase + α-galactosidase: *P. wortmanni*; *II* mannanase + β-mannosidase (no galactosidase): *Paecilomyces varioti*; *III* mannanase (no mannosidase) + galactosidase: *A. giganteus*; *IV* mannanase (no mannosidase, no galactosidase): *Bacillus subtilis*; *V* mannanase weak relative to both glycosidases: snail (*Helix*). Authentic markers are *M2* = β-1,4-dimannose; *M3* = β-1,4-trimannose; *Gal-Man* = 6-*O*-α-galactopyranosyl-D-mannose. *Cross-hatching* is an indication of the intensity of the spots. *Dotted circles* represent weak components, visible only under ultraviolet light. In all tests, only a limited hydrolysis has occurred (Reese and Shibata 1965)

with mannosidase. Reese and Shibata (1965) reported that this resistance is common to all reduced dimers with which they have worked (cellobiitol, lactitol, maltitol). Reduced trimers, on the other hand, were as readily hydrolyzed as the trimers themselves. The action of β-mannosidase on reduced mannotriose was not inhibited by γ-mannolactone, but δ-lactone was an inhibitor, very active and specific for the β-mannosidases.

Specific glycosidases are important tools for the study of the anomeric configuration and sequential arrangement of sugar units in heteropolysaccharides. For these glycosidases to be really important tools, it is essential that they are free of contamination by other glycosidases. Wan et al. (1976) have isolated a highly purified form of β-mannosidase from the fruiting bodies of the fungus *Polyporus sulphureus*. The authors report that this fungus is unique among reported sources of β-mannosidases in that it has the advantage of being almost free of α-mannosidase activity. Also, a single fruiting body often supplies several kilograms of starting material and in addition this mushroom is both easily found and identified. The purified enzyme liberates β-D-mannopyranosyl units from various natural substrates. The molecular weight of the enzyme is estimated on a Sephadex G-100 column to be in the order of 64 000. The isoelectric point was found to be at pH 3.9. With p-nitrophenyl β-D-mannopyranoside as substrate, the apparent K_m was found to be 1.6 mM. The enzyme is stable over a pH range of 2.8 – 8.5 and the reported pH optimum is at pH 2.8.

A β-mannosidase from *Aspergillus niger* was purified to apparent homogeneity and partly characterized by Elbein et al. (1977). The enzyme is a glycoprotein and has an estimated (gel filtration on Sephadex G-200) molecular weight of 120 000. The purified β-mannosidase had a high activity toward p-nitrophenyl β-D-mannoside but did hydrolyze the corresponding α-D-mannoside or any other tested p-nitrophenyl glycoside. The enzyme activity against 1,4-β-mannotriose was higher than against the corresponding mannobiose.

The activity toward these compounds was greatly lowered if they were reduced with $NaBH_4$. The apparent K_m for p-nitrophenyl β-D-mannoside was 2.0 mM. Preliminary studies of the β-mannosidase suggested that the glycopart of the enzyme was composed of the sugars mannose and acetylglucosamine. The effect of various carbon sources on the production of β-mannosidase by *A. niger* was also studied. In general, locust bean gum and guar flour were the best substrates for enzyme formation, although they did not necessarily give the maximum amount of cell biomass. Mannose or starch, as carbon sources, did not give rise to β-mannosidase production.

A β-mannosidase with similar properties as that from *A. niger* was isolated from the edible mushroom *Tremella fuciformis* by Sone and Misaki (1978). The molecular weight of the enzyme was estimated by Sephadex G-100 chromatography and found to be 140 000. With p-nitrophenyl β-D-mannoside as substrate, the K_m was 2.1 mM. The rate of hydrolysis of 1,4-β-linked-D-manno-oligosaccharides appears to be markedly reduced with an increase in the degree of polymerization. Thus, β-mannotriose is hydrolyzed at the rate of 30%, and β-mannotetraose at a rate of 13% relative to mannobiose. The enzyme activity is not inhibited by manno-δ-lactone, which is similar to the result found by Reese and Shibata (1965) but different from the β-mannosidase from hen oviduct (Sukeno et al. 1972). The pH optimum of the *T. fuciformis* β-mannosidase was at pH 5.

β-Mannosidases from other sources have also been purified and partly characterized. Thus, Muramatsu and Egami (1967) purified the enzyme from the liver of *Turbo cortunus*, a common marine gastropod in Japan. Sukeno et al. (1972) purified and partly characterized both an α- and a β-mannosidase from hen oviduct. The molecular weight of the β-mannosidase was found to be about 100 000. Houston et al. (1974) purified and partially characterized a β-mannosidase from malted barley. This enzyme had a molecular weight of about 88 000 as determined by acrylamide gel electrophoresis.

3.4.3 α-Galactosidase

α-Galactosidases (α-D-galactoside galactohydrolases EC 3.2.1.22) have been reported to occur widely in microorganisms, plants, and animals (Dey and Pridham 1972). The α-galactosidases in plant seeds have been the most widely studied (McCleary 1988c, McCleary et al. 1976).

Not many studies have been carried out to evaluate whether α-galactosidases are constitutive or induced in fungi and bacteria. Reese and Shibata (1965) made an attempt to obtain a preparation of α-galactosidase free of mannanase. More than 40 organisms were grown on galactosides (melibiose, melibiose octaacetate, and raffinose) and on sucrose monopalmitate and tested for activity against methyl α-D-galactoside. *Aspergillus fumigatus* grown on sucrose monopalmitate, produced the most α-galactosidase, and the enzyme was secreted into the medium.

Hogness and Battley (1957) studied the mechanism by which α-galactosides induce the synthesis of both α- and β-galactosidase in *E. coli* and *Aerobacter aerogenes*. The latter bacterium was mainly used since all attempts to extract α-galactosidase from *E. coli* failed, although the enzyme activity could be measured in whole cells. Both α- and β-galactosidases could be easily separated and were physicochemically distinctly different proteins. The α-galactosidase catalyzes the hydrolysis of melibiose, methyl, ethyl, phenyl, and o-nitro-phenyl α-D-galactosides. Addition of any one of the above α-galactosides to an exponentially growing culture of *A. aerogenes* causes as much as 100 fold increase in the rate of α-galactosidase production. D-galactose induces both of the galactosidases, while the β-galactosides tested (lactose, methyl β-D-galactoside, etc.) induce only β-galactosidase activity.

Several different substrates have been used for the assay of α-galactosidase activity. Thus melibiose and raffinose, presumed to be natural substrates for plant α-galactosidases, are commonly used to determine the enzyme activity (Dey and Pridham 1972). p-Nitrophenyl α-D-galactopyranoside and galactomannans have also been used as substrates (McCleary 1988b).

Bahl and Agrawal (1969) purified and characterized an α- and a β-galactosidase from *Aspergillus niger*. Both enzymes were highly specific for the anomeric configurations of the glycosidic linkages. The α-galactosidase hydrolyzed methyl α-galactopyranoside, melibiose, raffinose, and stachyose and removed 37%–40% of the total terminal 1,6-α-linked galactose residues from galactomannans from guar and locust bean.

Suzuki et al. (1970) purified and crystallized an α-galactosidase enzyme from mycelium of the fungus *Mortierella vinacea*. The crystalline α-galactosidase was free of protease and also of other glycosidase activities. A single protein band was obtained on examination on polyacrylamide gel electrophoresis. The crystalline preparation contains 10.8% neutral sugars and 2.7% D-glucoseamine, and is thus a glycoprotein. It hydrolyzes the usual α-D-galactopyranosides but does not liberate D-galactose from the galactomannan of guar gum. The enzyme also catalyzed transglycosylation reactions, and when melibiose was used as substrate two new oligosaccharides with mobilities corresponding to mannotriose and verbascotetraose were detected. With raffinose as substrate two new oligosaccharides with mobilities corresponding to stachyose and verbascose were obtained.

Lee and Wacek (1970) purified both an α- and a β-galactopyranosidase from a commercial, crude enzyme preparation (Rhozyme HP-150, *A. niger*). The α-galactosidase was obtained in a state free of other undesirable glycosidases. However, it was evident that there were several components of α-galactosidases which seem to differ in their action on different substrates.

Civas et al. (1984a, b) studied both mycelial α-galactosidase and secreted α-galactosidase activities from the fungus *Aspergillus tamarii*. Two α-galactosidases were purified from the mycelial extract to the extent that each of the enzymes showed a single protein band of α-galactosidase activity when examined by polyacrylamide gel electrophoresis. They catalyzed the hydrolysis of O-nitrophenyl α-D-galactoside, melibiose, raffinose, and stachyose but did not attack the galactomannans. The molecular weights were rather similar or 265000 ± 5000 and 254000 ± 5000. Polyacrylamide gel electrophoresis in the presence of sodium dodecyl sulfate (SDS) in each case showed a single protein band with molecular weights corresponding to 88000 and 77500, respectively. The purified enzymes were glycoproteins where the glyco-part consisted of N-acetylglucosamine, mannose, glucose, and galactose in the estimated molar proportions of 1:9:5:8 in α-galactosidase I.

In a second paper, Civas et al. (1984b) purified both an α-D-galactosidase and a β-D-mannanase secreted into the growth medium of *A. tamarii* cultivated on galactomannan. Both enzymes, which were purified to the extent that they showed a single protein band on polyacrylamide gel electrophoresis, were glycoproteins containing N-acetylglucoseamine, mannose, and galactose in the molar proportions of 1:6:1.5 for the α-galactosidase enzyme. This enzyme has a molecular weight of 56000 as determined by polyacrylamide gel electrophoresis in the presence of SDS. This α-galactosidase differed markedly from the mycelial forms I and II studied in the earlier paper (Civas et al. 1984a), with regard to both its kinetic and its structural properties. The pH optimum for the enzyme was 4.8.

3.5 Examples of Applications of Hemicellulolytic Organisms and Their Enzymes

The biotechnical utilization of lignocellulosic materials, i.e., using microorganisms or their enzymes, for conversion and upgrading of these resources, implies some kind of pretreatment to make the biomass more easily available. In several independent studies the conclusion has been reached that steam explosion is the most efficient of these techniques (Chum et al. 1985, Vallander and Eriksson 1985, Brownell and Saddler 1987).

Tanahashi et al. (1982) reported that lignin and hemicelluloses are hydrolyzed to low molecular-weight products during steam explosion. Brownell and

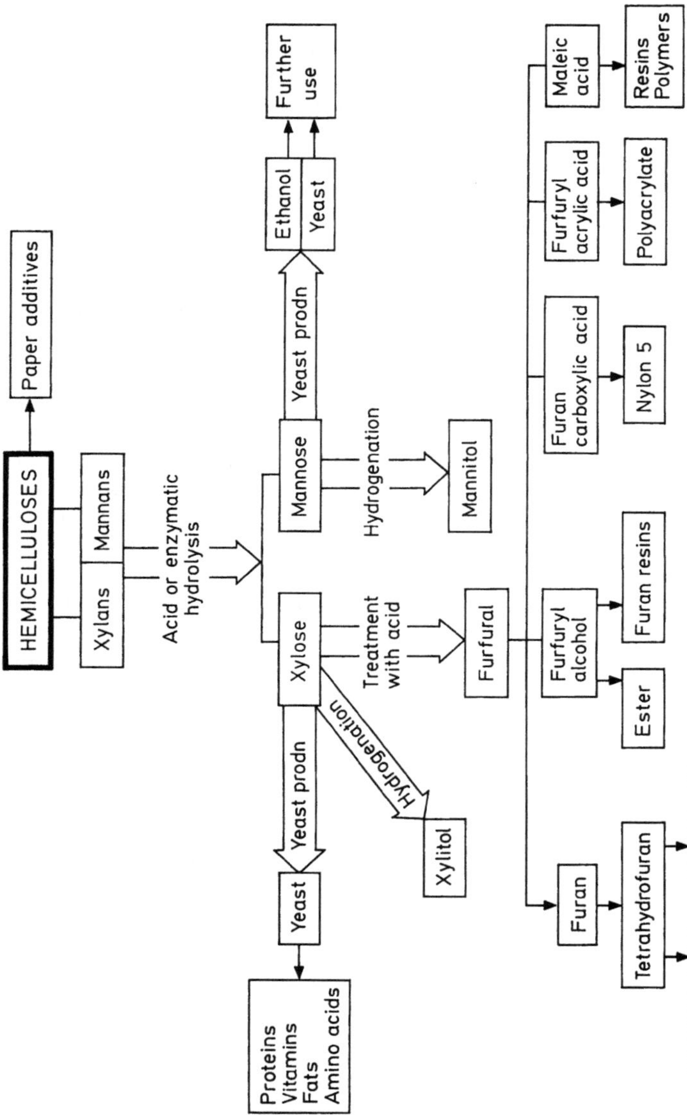

Fig. 3.12. Principles of utilization of hemicelluloses (Wegener 1982)

Saddler (1987) demonstrated that most of the pentosan content in the solid residue after steam explosion can be washed out with water. One use for enzymes participating in hydrolysis of xylans (Sect. 3.3) would be for conversion of xylans into monomeric sugars. Some attempts have been made to utilize immobilized xylanolytic enzymes (for conversion of water-soluble xylans) for this purpose (Puls et al. 1978, US Patent 1981, Allenza et al. 1986, Shimizu and Ishihara 1987).

A scheme for the principles of hemicellulose utilization has been set up by Wegener (1982, Fig. 3.12). The main use of hemicellulolytic enzymes would be to convert polymeric and oligomeric substances from the hemicelluloses into monomeric sugars. The continued upgrading of the resulting materials would be a mixture of biotechnical and chemical treatments. Utilization of the hexoses for conversion into alcohol would involve no problems and could be done by existing techniques (cf. Sect. 2.9.2). However, fermentation of pentoses into ethanol is still unsolved on a technical scale, although much effort has been devoted to such developments (Dekker 1986, du Preez et al. 1986). Pentoses, however, can be used as substrates for microorganisms for the conversion into rich protein fodder, and various processes have been developed for this purpose (Eriksson 1985).

One potential use for xylanolytic enzymes in particular could be for treatment of cellulosic pulps for removal or partial hydrolysis of residual xylans. Several studies have thus been undertaken where xylans have been removed by xylanolytic enzymes for upgrading of dissolving pulps (Paice and Jurasek 1984a). Viikari et al. (1986, 1987) and Chauvet et al. (1987) have used xylanolytic enzymes for pretreatments of pulps prior to bleaching. The reason for this is that xylans and lignin are at least closely associated if not bound to each other in wood.

A similar idea was used by Eriksson et al. (1983) in their approach to microbial delignification of wood chips using cellulase mutants of white-rot fungi. The first mutants used were also xylanase-less and caused only a very limited delignification. However, "construction" of Cel^- strains with increased xylanase production also enhanced the delignification caused by these organisms compared to the xylanase-less (Eriksson 1985). It is, however, doubtful whether treatment of pulp for improving dissolving pulp quality or for improving bleaching would ever be economically feasible since enzymes are expensive and in these cases could not be reutilized.

The use of xylanolytic enzymes has also been suggested for processing of cereals since pentosans make up the bulk of dietary fibers of wheat flour and bran (Neukom et al. 1977). McCleary et al. (1986) used xylanases for treatment of dough. They found that this treatment decreases water absorption and also changes the consistency and loaf texture and structure.

In her thesis, Poutanen (1988a) points to the importance of β-xylosidase and in particular side chains removing glycosidases for a complete saccharification of xylans. So far very little has been reported on the production and use of these enzymes. However, this may improve since it has been an industrial interest to produce xylose from xylans for further conversion to xylitol (Nevalainen 1987). Research on potential applications of individual enzymes involved in xylan and mannan conversion has only started. It seems likely that the interest in these enzymes will increase considerably when more industrial applications can be visualized.

Chapter 4

Biodegradation of Lignin

During photosynthesis an estimated $10^{11}-10^{12}$ tons of terrestrial biomass is produced every year (Bassham 1975, Bisaria and Ghose 1981) and 15%–36% of this lignocellulosic material in gymnosperms and angiosperms is lignin (Higuchi 1981a, 1985a). The lignin content of angiosperms (hardwoods) is generally lower than that of gymnosperms (softwoods) (Table 1.1).

Lignin is a phenylpropanoid structural polymer of vascular plants which gives the plants rigidity and binds the fiber cell walls together. Lignin also decreases water permeation across cell walls of xylem tissue and makes the wood more resistant to attack by microorganisms (Kirk 1971, Sarkanen and Ludwig 1971, Vance et al. 1980, Higuchi 1981a). Although quite resistant to microbial attack, lignin is ultimately degraded to humus, water, and carbon dioxide following the death of the plant tissues (Hurst and Burges 1967, Christman and Oglesby 1971, Zeikus 1981). The humus found in the upper layer of soils is important for the structure of the soil and increases aeration and moisture-holding capacity. The humus also functions as a basic ion exchanger and is able to store and release nutrients to the surroundings, that is, to growing plants and trees (Christman and Oglesby 1971). The degradation of humus in temperate climates is slow; its lifetime is between 360 and 2860 years depending on temperature, humidity, and type of soil and vegetation (Hurst and Burges 1967).

Research into lignin degradation is important for several reasons. Next to cellulose, lignin is the most important renewable material but it also constitutes a barrier that must be breached before most cellulose is accessible to enzymes. Lignin also is a hindrance for efficiently utilizing the cellulose in the production of ethanol, sugars, and protein (Eriksson 1985, Eriksson and Kirk 1985). For the biotechnical production of useful products from lignin and for the improvement of existing pulping and bleaching processes in the pulp and paper industry, a detailed knowledge of the basic mechanisms of biological lignin degradation is necessary. Similarly, in the upgrading and delignification of straw and sugarcane bagasse for cattle feed or delignification of wood for paper production a thorough knowledge of ligninolytic organisms is needed.

In the late 1960s the chemical structure of lignin became clear (Freudenberg 1965, Freudenberg and Neish 1968, Nimz 1974, Adler 1977), but the complexity of this structure made progress in understanding its biodegradation rather slow (Higuchi 1971, Kirk 1971). After development of more specific assays for lignin degradation using ^{14}C-lignins (Haider and Grabbe 1967, Haider and Trojanowski 1975, Kirk et al. 1975), the unusual and unexpected physiological requirements for lignin degradation by white-rot fungi were reported by Kirk et al. (1978a). *Phanerochaete chrysosporium* (asexual stage

= *Sporotrichum pulverulentum*) has been the major fungus used in these investigations, but other white-rot fungi have also been studied. Detailed chemical analyses were also made on lignin isolated from wood degraded by white-rot and brown-rot fungi. These analyses were valuable for a better understanding of the metabolic pathways and enzyme mechanisms involved in the degradation of lignin. During 1970–1978 numerous studies on phenoloxidases (laccase and peroxidase) and oxidoreductases from white-rot fungi were reported, but their true importance in lignin degradation could not be elucidated at that time. During these years, research on the use of naturally occurring white-rot fungi and cellulase-less mutants of white-rot fungi were reported for the production of biomechanical pulp and for purification of bleach liquors from pulping (Ander and Eriksson 1978, Crawford 1981 b, Kirk and Chang 1981). Book chapters on advances in the understanding of the chemistry and physiology of fungal lignin degradation were published when the study of lignin degradation was primarily conducted in relatively few laboratories (Kirk et al. 1977, Kirk et al. 1978 b, Kirk 1981).

After an international seminar on Lignin Biodegradation at Madison, Wisconsin, in 1978 (Kirk et al. 1980), the interest in this field grew rapidly, and many reviews and books have been published: Ander and Eriksson (1978), Crawford and Crawford (1980, 1984), Crawford (1981 b), Higuchi (1981 b), Zeikus (1981), Higuchi (1982), Kirk and Fenn (1982), Jahnshekar and Fiechter (1983), Palmer and Evans (1983 a), Kirk (1984), Paterson et al. (1984), Harvey et al. (1985 b), Leisola and Fiechter (1985 a), Buswell and Odier (1987), Evans (1987), Harvey et al. (1987 b), Kirk and Farrell (1987), Tien (1987), Kirk (1988 a, b), Umezawa (1988).

A few books on general aspects of wood biodegradation have also been published with chapters that deal with lignin degradation (Kirk 1983 a, Brown and Gritzali 1984, Kirk and Cowling 1984, Higuchi 1985 b, Wood 1985 a).

4.1 Biosynthesis and Structure of Lignin

Biosynthesis of lignin starts with the fixation of carbon dioxide in plants to form carbohydrates. These are further metabolized via the shikimic acid pathway to tyrosine and phenylalanine (Higuchi et al. 1977, Higuchi 1980, 1981 a, 1985 a, Grisebach 1981; Fig. 4.1). Ammonia is eliminated from these amino acids by tyrosine ammonia-lyase (TAL) and phenylalanine ammonia-lyase (PAL), respectively, to produce p-hydroxycinnamic acid and cinnamic acid (Gross 1980, Grisebach 1981, Higuchi 1985 a). The further enzymatic reactions of these acids and the enzymes responsible are shown in Fig. 4.2. The conversion of caffeic acid to ferulic acid is catalyzed by *O*-methyl transferases (OMT) which are common in higher plants. Sinapic acid is formed from ferulic acid by hydroxylation and methylation. In angiosperms a hypothetical ferulate-5-hydroxylase enzyme initiates the biosynthesis of sinapic acid after which the formation of syringyl lignin follows. In gymnosperms, ferulate is used directly for further biosynthesis of guaiacyl lignin.

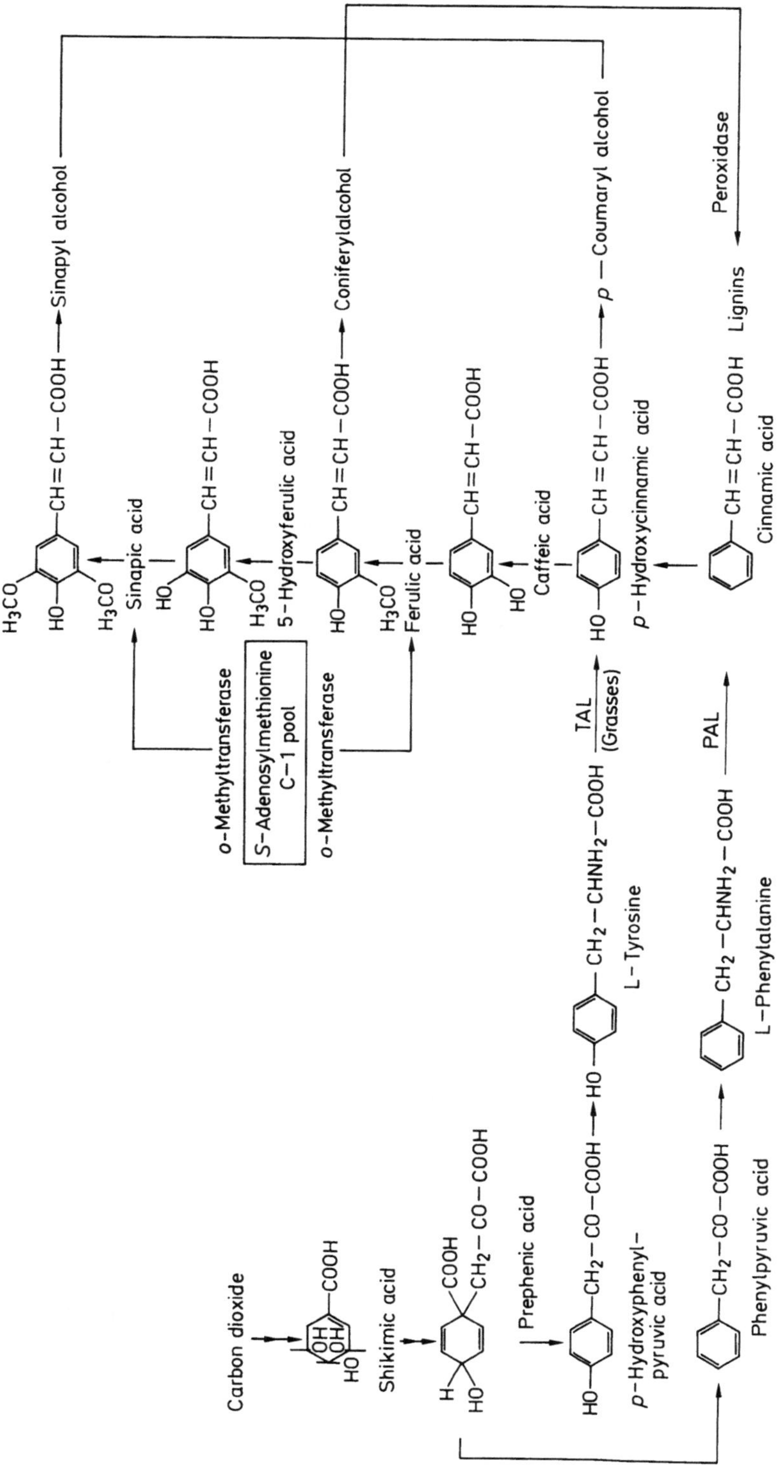

Fig. 4.1. Biosynthesis of lignin starting from carbon dioxide (Higuchi et al. 1977)

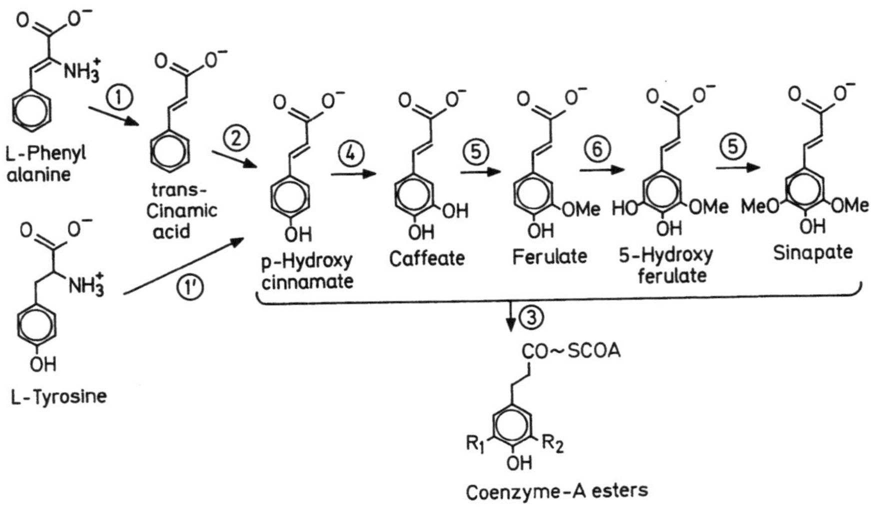

Fig. 4.2. The phenylalanine-cinnamic acid pathway to p-hydroxycinnamic, ferulic, and sinapic acids. Enzymes: *1* Phenylalanine ammonia-lyase (PAL); *1'* Tyrosine ammonia-lyase (TAL), only in grasses. *2* Cinnamate 4-hydroxylase. *3* 4-Coumarate: CoA ligase or 4-coumaryl-CoA synthase. *4* 4-Coumarate 3-monooxygenase. *5* S-adenosylmethionine (SAM): caffeate 3-*O*-methyltransferase (OMT). *6* "Ferulate-5-hydroxylase" (hypothetical) (Grisebach 1981)

The sinapic, ferulic, and p-hydroxycinnamic acids are now reduced to the lignin precursors sinapyl, coniferyl, and p-coumaryl alcohols (Figs. 4.1 and 4.3). The reduction of ferulic and sinapic acids occurs through the successive mediation of the three enzymes hydroxycinnamate: CoA ligase, cinnamoyl-CoA reductase, and cinnamyl alcohol dehydrogenase. p-Hydroxy cinnamic acid is probably reduced in a similar way (Higuchi 1980, 1985a).

The last enzymatic step in lignin biosynthesis is a peroxidase-mediated dehydrogenation of the above-mentioned cinnamyl alcohols to produce phenoxy radicals of the type shown in Fig. 4.3 for coniferyl alcohol. Formation of hydrogen peroxide probably occurs via oxidation of malate in a complex reaction (Fig. 4.4) involving NAD, O_2, NADH, and Mn^{2+} (Gross et al. 1977, Gross 1980). The superoxide radical $O_2^{\bullet -}$ seems to be formed from O_2 as an intermediate; nonenzymatic disproportionation of this radical then gives H_2O_2. The strongest H_2O_2 production is obtained if ROH (Fig. 4.4) is coniferyl alcohol, which is the most important lignin monomer. Reaction (2) in Fig. 4.4 is performed by a peroxidase-Mn^{2+} complex. This enzyme complex may have similarities with the Mn-peroxidase isolated from white-rot fungi (Sect. 4.5.2), but comparative studies have not been done.

Although laccase has been implicated in the formation of phenoxy radicals, it is now believed that peroxidase is the actual enzyme (Higuchi 1985a). For example, Harkin and Obst (1973) investigated lignifying cells using syringaldazine with and without the addition of H_2O_2 to detect laccase or peroxidase activity. Addition of a syringaldazine solution to freshly cut stems pro-

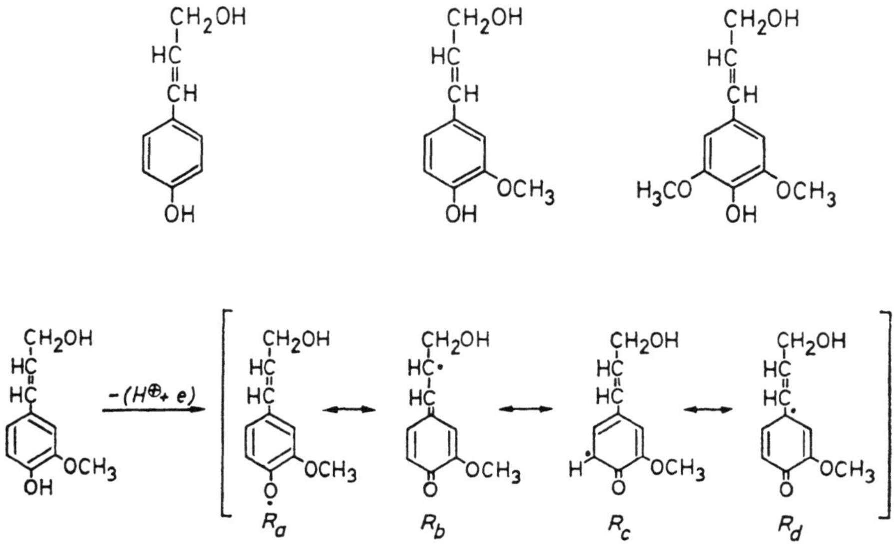

Fig. 4.3. The lignin precursors p-coumaryl alcohol, coniferyl alcohol, and sinapyl alcohol. Resonance structures of the coniferyl alcohol radical

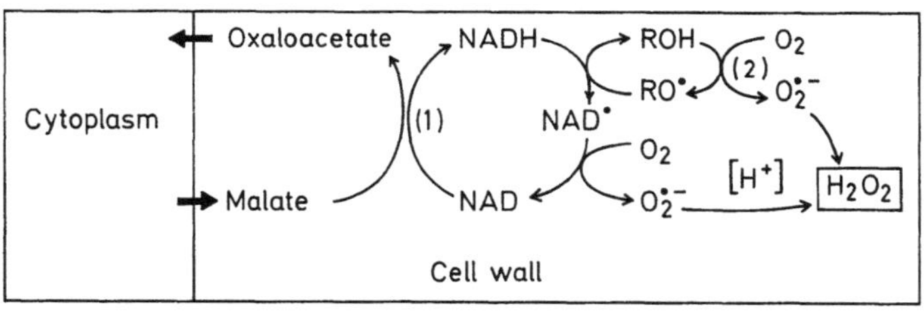

$$2\ \text{Malate} + 2\ O_2 \longrightarrow 2\ \text{Oxaloacetate} + 2\ H_2O_2$$

Fig. 4.4. Proposed scheme for the generation of hydrogen peroxide in the plant cell walls. *1* NAD: malate oxidoreductase; *2* peroxidase-Mn^{2+} complex (Gross et al. 1977, Higuchi 1985a)

duced a quinone coloration only if H_2O_2 was also added. The color was then formed immediately in the xylem tissue adjacent to the cambium, indicating that peroxidase rather than laccase is responsible for the radical formation necessary for the lignification process.

Westermark (1982) found that superoxide radicals could be stabilized by the high concentrations of Ca^{2+} that are present in nonlignified cells. Thus in the presence of 50 mM of Ca^{2+}, xanthine oxidase-generated superoxide radicals mediated a one-electron oxidation of coniferyl alcohol to produce dehydrodiconiferyl alcohol and pinoresinol. Since lignin-rich lamellae in the middle secondary wall are approximately 35 Å thick, Westermark (1982) did not believe that a peroxidase of about 55–60 Å diameter could be located within the cell wall and get access to its substrate. In the comprehensive review of Higuchi (1985a), however, it was pointed out that plants have many isoenzymes, some of which may be rather small. These isoenzymes are probably bound to the cell walls. According to Higuchi (1985a), all results taken together indicate "that lignin precursors are synthesized in Golgi bodies and/or endoplasmic reticulum, transported to the wall via vesicles, and dehydrogenatively polymerized to lignin by peroxidase bound to cell walls". When the different phenoxy radicals have been formed, they couple nonenzymatically and in a random way to form dilignols, oligolignols, and finally the lignin polymer. Gaspar et al. (1985) concluded that lignification is the result of increased intracellular, relative calcium levels and secretion of basic and acidic peroxidases in a two-step fashion.

A typical softwood (gymnosperm) lignin contains mainly coniferyl alcohol, some p-coumaryl alcohol, but no sinapyl alcohol (Sakakibara 1980). In hardwood (angiosperm) lignins there are equal amounts of coniferyl and sinapyl alcohol (46%) and a minor amount of p-coumaryl alcohol (8%). Grass and bamboo lignins are composed of all three alcohols (Paterson et al. 1984). However, in these lignins 5%–10% p-coumaric acid is esterified with the C_γ-hydroxyl group of mainly p-coumaryl alcohol side chains (Shimada 1972, Higuchi 1980).

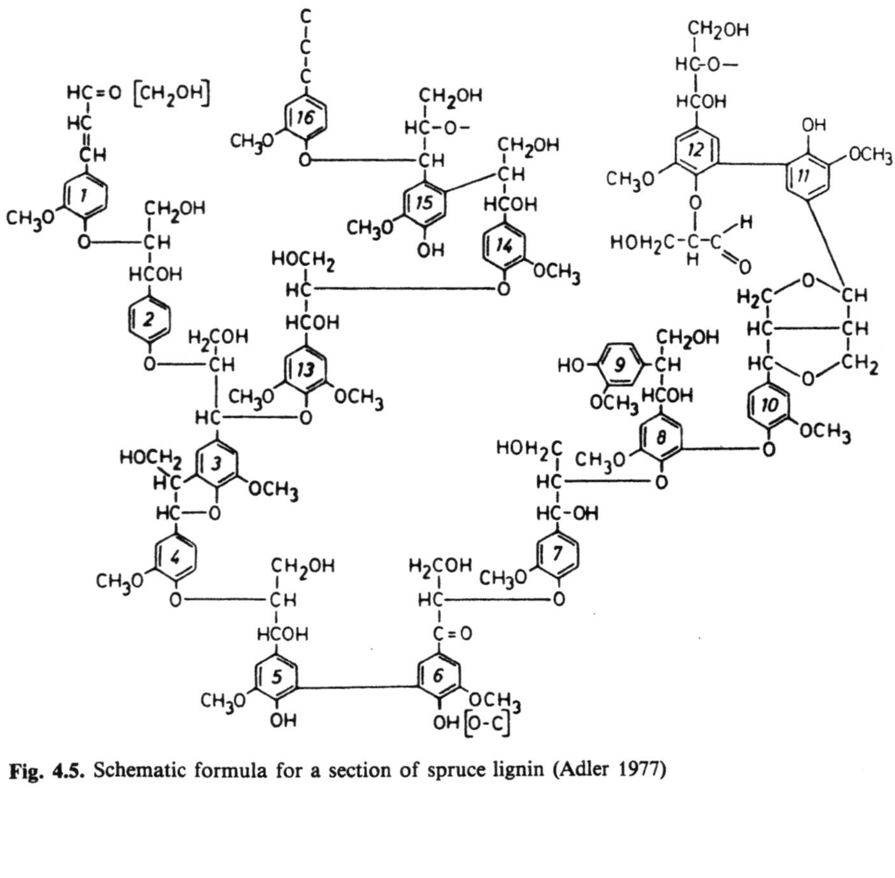

Fig. 4.5. Schematic formula for a section of spruce lignin (Adler 1977)

In Fig. 4.5, a schematic formula for a section of spruce lignin is shown (Adler 1977). Commonly occurring structures in spruce lignin have also been presented by Freudenberg (1965), Hall et al. (1980), and Sakakibara (1983), whereas Nimz (1974) studied the structure of hardwood lignin. Hardwood lignins contain about 20%–21% methoxyl groups per C_9-unit compared with 15% for softwood lignin (Sarkanen and Ludwig 1971, Shimada 1972).

There are three major intermonomer linkages in lignin: (1) The arylglycerol-β-aryl ether structure which connects about half of the phenylpropane units in spruce lignin and about 60% of these units in birch lignin (Adler 1977); (2) the phenylcoumaran structure, which comprises 9%–12% of the phenylpropane units in spruce lignin and about 6% of those in birch lignin; and (3) biphenyl structures, which involve 9.5%–11% of the units in spruce lignin and 4.5% in birch lignin. It may be pointed out that the lignin formula in Fig. 4.5, like any other lignin formula, only shows an arbitrary sequence of phenylpropane units from a very limited part of the lignin macromolecule, i.e., it is schematic-qualitative and not meant to be quantitative.

4.2 Lignin Preparations and Methods for Studying Lignin Biodegradation

Many different lignin preparations were used in earlier years to study lignin degradation: Klason lignin, kraft lignin, lignosulfonates, enzymatically liberated lignin (from brown-rotted wood), acidolysis lignin, Brauns' native lignin, cellulase lignin, dehydrogenation polymerizate (DHP), and milled wood lignin (MWL). Advantages and disadvantages of all these different lignin preparations have been discussed by Crawford (1981b), Janshekar and Fiechter (1983), and Paterson et al. (1984). It was concluded that MWL is the best lignin preparation, and that DHP is good for general use.

MWL or "Björkman lignin" is extracted from ball-milled wood using different solvents. Depending on the source, the average molecular weights of such lignins are 11000–16000 with a carbohydrate content of about 4%–9% (Björkman 1956, Chang et al. 1975, Adler 1977). MWL is considered to be relatively unmodified lignin, and is obtained in yields of up to 50% of the lignin in wood (Crawford 1981b, Janshekar and Fiechter 1983). The necessary milling time may be decreased if carbohydrates are removed by polysaccharide-degrading enzymes after milling. This gives a somewhat more representative lignin preparation than MWL (Chang et al. 1975).

By feeding plants L[U-^{14}C]-phenylalanine or the ^{14}C-labeled lignin precursors trans-cinnamic acid, p-coumaric acid, or ferulic acid, it is possible to prepare plant material labeled only in the lignin component. Cinnamic, p-coumaric, and ferulic acids give the most specific ^{14}C-incorporation into the lignin, i.e., very little ^{14}C is incorporated in the plant protein (Crawford and Crawford 1978, Crawford 1981b). One advantage in using labeled lignocellulosics such as wood or straw is the natural presence of polysaccharides which are in intimate contact with the lignin. ^{14}C-labeled MWL can also be prepared by the method of Björkman (1956). Methods for preparation of ^{14}C-labeled lignocellulosics were thoroughly reviewed by Crawford (1981 b). Due to possible protein labeling, the minimum recovery of $^{14}CO_2$ from ^{14}C-labeled lignocellulosics should be greater than 2% after growth of any microorganism.

Synthetic lignins can also be synthesized without association with polysaccharides. Such DHP lignins are synthesized by oxidative polymerization of the p-hydroxycinnamyl alcohols seen in Fig. 4.3 with peroxidase plus H_2O_2 (Haider and Grabbe 1967, Haider and Trojanowski 1975). Most often the lignin monomer used is coniferyl alcohol. DHPs are insoluble in water and have an average molecular weight of 1490–1600 (Kirk et al. 1975, Crawford 1981 b, Paterson et al. 1984). DHPs with greater molecular weights (4000–10000) have been synthesized by Odier and Heckman (unpublished results; Fig. 4.6) and Ander and Stoytschev (unpublished results) using methods described by Faix et al. (1985). In this method separate solutions of coniferyl alcohol and H_2O_2 were slowly added to horseradish peroxidase. The addition time used by Ander and Stoytschev was 5 h for a 250 ml lignin suspension, while Faix et al. (1985) working on a larger scale (final volume 5.9 l) used an addition time of 5–6 days.

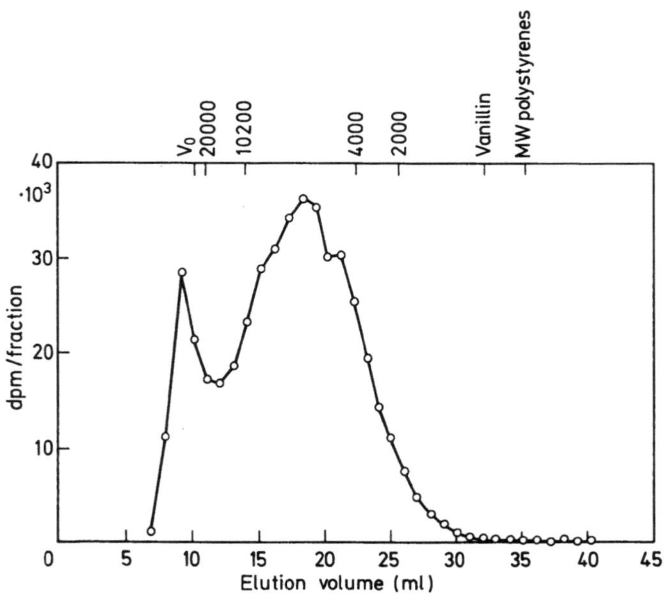

Fig. 4.6. Gel permeation chromatography of ring-labeled DHP on Sephadex LH-60 in dimethylformamide plus 0.1 M lithium chloride (courtesy of Odier and Heckman, INRA, Paris unpublished results 1987). This DHP was produced and distributed to several researchers through Ander and Eriksson, STFI, Stockholm and Odier, INRA, Paris

DHP can be prepared in three different ways: the "Zulaufverfahren", the "Zutropfverfahren", and the dialysis tube method. In the Zulaufverfahren method (Freudenberg and Neish 1968), the coniferyl alcohol solution is added all at once to the peroxidase/H_2O_2 solution. This gives a molecular weight of 1000–1200 (Higuchi 1981a, 1985a, Paterson et al. 1984). If the coniferyl alcohol is added dropwise (Zutropf method) over a long time one obtains more guaiacylglycerol-β-coniferyl ether dimers and a higher molecular weight. Faix et al. (1981) used all three precursors, p-coumaryl, coniferyl, and sinapyl alcohols, in different proportions with the Zutropfverfahren and obtained molecular weights ranging from 1270 to 2940, while Tanahashi and Higuchi (1981) had peroxidase in a cellulosic dialysis tube immersed in a solution of H_2O_2 and coniferyl alcohol. When the latter two components diffused into the dialysis tube, polymerization occurred, giving a Mw of about 6000. This preparation, however, is much less soluble in organic solvents and "is of disputed value" according to Paterson et al. (1984).

Methods for synthesis of coniferyl alcohol DHP were also studied by Brunow and Wallin (1981). They found that during polymerization the coniferyl alcohol concentration should not be above 10^{-4}M in order to ensure more than 80% of high molecular weight DHP. This DHP is similar to those made by the "Zutropf" method.

If the coniferyl alcohol used is ^{14}C-labeled in different positions, it is possible to prepare DHPs labeled in the aromatic ring, in the methoxyl carbon, or in any of the side-chain carbons (Haider and Trojanowski 1975, Kirk et al. 1975). Methoxyl-labeled coniferyl alcohol, ferulic acid, or vanillic acid may be prepared by the use of S-[methyl-^{14}C]adenosyl-L-methionine as the methyl donor plus catechol-*O*-methyltransferase (Frazer et al. 1986, Frazer and Young 1986). When biodegraded, minimum recovery of $^{14}CO_2$ from labeled DHP should be greater than 1% before it can be stated that lignin degradation has been obtained (Kirk et al. 1975). Labeled DHPs have been invaluable in the studies of the physiological demands of fungi for lignin degradation (Sect. 4.3).

Gravimetric determination of sulfuric acid-insoluble (Klason) lignin has been widely used to measure degradation of lignin in wood (Effland 1977, Yoshihara et al. 1984). Extractives and lignin degradation products, however, interfere and give too high values. Lignin in wood can also be determined by dissolution of ground wood meal (about 5 mg) in acetyl bromide plus acetic acid and measuring absorbance at 280 nm (Johnson et al. 1961). To obtain a more rapid dissolution of the wood, Iiyama and Wallis (1988) added perchloric acid to the digestion medium.

A colorimetric method to measure lignin content was developed by Garcia and Latge (1987), who used phloroglucinol-HCl plus acetylbromide to give a red-colored solution with an absorption maximum at 545 nm. The method could be used for 0.5 to 2 mg of lignin in spruce or poplar wood, and fungal mycelium did not interfere. In wheat straw, however, phenolic compounds interfere. It could be expected that with *Phanerochaete chrysosporium*, veratryl alcohol produced during secondary metabolism may also interfere.

In conclusion, it can be stated that measurement of $^{14}CO_2$ evolution from labeled lignin preparations is the most sensitive and accurate method to measure lignin degradation provided that the actual lignin preparation has a molecular weight larger than at least 1000. However, in this technique only the final product carbon dioxide is measured and nothing is known about degradation products and chemical changes in the lignin molecule. If, after fungal degradation, it is taken into consideration that mycelium and/or lignin degradation products are produced, the more simple acetyl bromide methods may also be valuable when only small samples are available.

4.3 Lignin Degradation by White-Rot Fungi

Since white-rot fungi, most of which belong to the Basidiomycetes, are the best lignin degraders among all known microorganisms, the main emphasis will be on them in this review. Some xylariaceous fungi in the *Ascomycotina* also produce a white-rot in hardwoods (Rogers 1979, Kirk 1983a). White-rot fungi degrade wood either by a simultaneous attack on lignin and polysaccharides or by preferentially degrading lignin. Different aspects of selective lignin degradation are discussed in detail in Chap. 1 and in Sect. 4.9.

In most studies of lignin biodegradation the white-rot fungus *Phanerochaete chrysosporium* has been used (Kirk 1984). In Sweden, the same fungus was isolated from chip piles and identified as *Chrysosporium lignorum* P 127-1 (Bergman and Nilsson 1966). It was later renamed *Sporotrichum pulverulentum* Novobranova, since it was also isolated in Russia and was given there a Latin description by Novobranova (1972). Since the fungus has recently been named *P. chrysosporium* and since *S. pulverulentum* was found to be the anamorph of *P. chrysosporium* (Burdsall and Eslyn 1974, Burdsall 1981), the name *P. chrysosporium* has now been generally adopted (Johnsrud and Eriksson 1985). The formation of fruit bodies by *P. chrysosporium* made classical genetics possible, and several crossings of homokaryotic basidiospore cultures have been obtained (Johnsrud and Eriksson 1985). Cellulase-less mutants of *Polyporus adustus* (Eriksson and Goodell 1974) and *S. pulverulentum* (Ander and Eriksson 1975, 1976) have been obtained by irradiating conidia with UV-light (Sect. 4.9). Techniques for isolation of mutants have also been developed by Gold and co-workers (Gold and Cheng 1978, Gold et al. 1982a, b). Formation, fusion and regeneration of protoplasts (Gold et al. 1983a), genetic recombination, and basidiospore formation (Alic and Gold 1985, Alic et al. 1987) have also been studied in *P. chrysosporium*.

4.3.1 Physiological Demands, Secondary Metabolism, and Veratryl Alcohol Production in White-Rot Fungi

As suggested by Keyser et al. (1978) the ligninolytic enzyme system of *Phanerochaete chrysosporium* is synthesized as part of several physiological

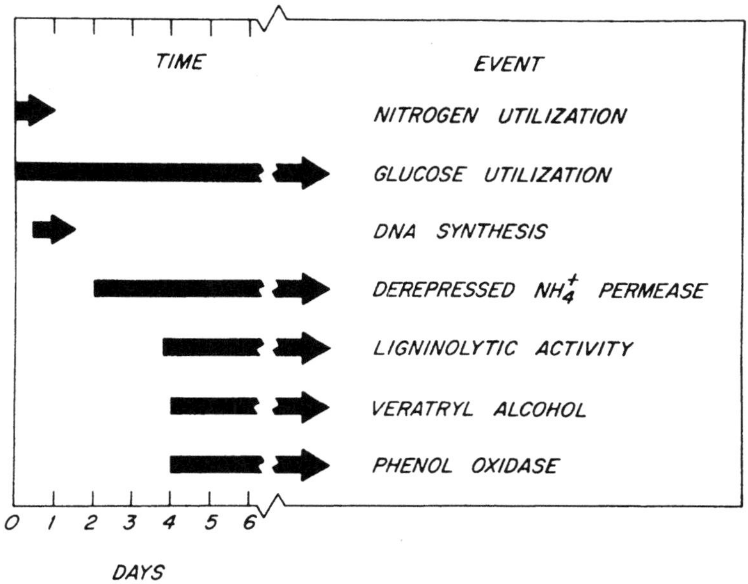

Fig. 4.7. Time relationship for the appearance of physiological events in cultures of *Phanerochaete chrysosporium* grown under lignin degradation conditions (Kirk 1980)

("secondary metabolic") events triggered mainly by nitrogen starvation. The timing of these events is shown in Fig. 4.7. Thus ligninolytic activity, veratryl alcohol production and peroxidase activity occur simultaneously on day 4 (Kirk 1980) and lignin degradation does not occur during the primary growth phase. Secondary metabolism can also be triggered by carbohydrate or sulfur starvation (Jeffries et al. 1981, Buswell et al. 1984). These findings are true for *P. chrysosporium* and many other white-rot fungi. However, some white-rotters are not so strongly regulated by nitrogen and these fungi degrade lignin under conditions of nutrient sufficiency (Freer and Detroy 1982, Leatham and Kirk 1983).

In 1978 it was reported that *P. chrysosporium* produced 0.2 mM veratryl alcohol (3,4-dimethoxybenzyl alcohol) de novo as a secondary metabolite from glucose in low nitrogen cultures (Lundquist and Kirk 1978). The significance of this metabolite has been an intriguing question for many years, and it is only recently that we are beginning to learn that the fungus uses veratryl alcohol during lignin degradation. Many years ago Russell et al. (1961) found this compound in *Polyporus versicolor* cultures but considered it to be a degradation product from lignin. Veratryl alcohol production has been reported in *Sporotrichum pulverulentum*, *Pycnoporus cinnabarinus*, *Phlebia radiata*, *Coriolus versicolor*, and *Phlebia* (*Merulius*) *tremellosus* (Hatakka 1986, Kawai et al. 1986a, Biswas-Hawkes et al. 1987). The latter authors also found veratryl alcohol oxidase (ligninase) activity in *Phlebia tremellosus* cultures. Three mutants of *P. chrysosporium* (Cel 44, Phe 3, and 104-2) with reduced or no capacity

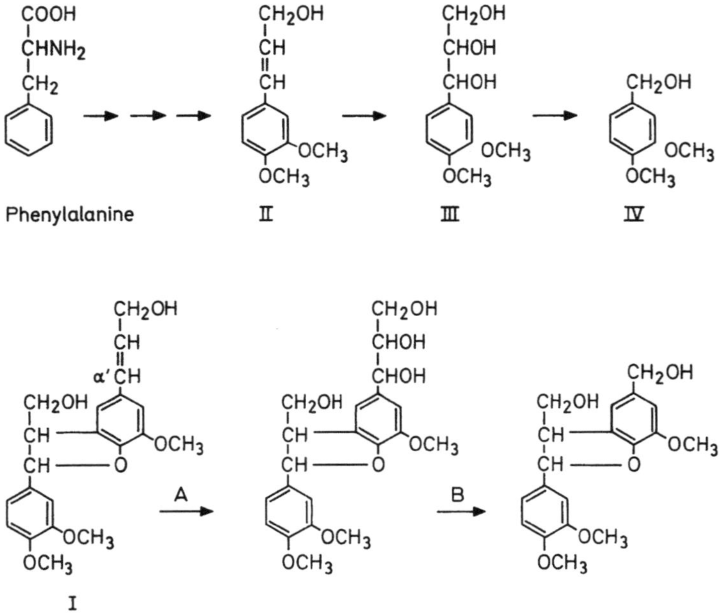

Fig. 4.8. Synthesis of veratryl alcohol (*IV*) from phenylalanine via 3,4-dimethoxycinnamyl alcohol (*II*) and veratryl glycerol (*III*) by *Phanerochaete chrysosporium*. Degradative pathway of methyl dehydrodiconiferyl alcohol (*I*) showing similarities with veratryl alcohol biosynthesis (Shimada et al. 1981)

to degrade lignin, produced little or no veratryl alcohol (Ander et al. 1980b, Gold et al. 1982b), while Liwicki et al. (1985) concluded that mutants with reduced ligninolytic activity were not necessarily defective in veratryl alcohol synthesis.

Veratryl alcohol is synthesized from phenylalanine via 3,4-dimethoxycinnamyl alcohol and veratrylglycerol by *P. chrysosporium* as shown in Fig. 4.8 (Shimada et al. 1981). In this figure some steps in the degradation of methyl dehydrodiconiferyl alcohol are also shown. Steps A and B (hydroxylation and C_α-C_β-cleavage) are similar in the degradative pathway and in the biosynthetic pathway. Further similarities in veratryl alcohol synthesis and lignin degradation are that both events are stimulated by O_2 and repressed by culture agitation and by glutamate addition (Shimada et al. 1981, Kirk and Shimada 1985).

In *P. chrysosporium* different peroxidases (Sect. 4.5) are produced only during secondary metabolism and in the early experiments lignin degradation and lignin peroxidase production were not found to be induced or stimulated by addition of veratryl alcohol, lignin degradation products, or lignin. However, Ulmer et al. (1984) found that the addition of $1-2\,g\,l^{-1}$ of straw dioxane lignin to *P. chrysosporium* cultures at inoculation, increased decolorization of Remazol Brilliant Blue R six fold. Decolorization of certain polymeric dyes is caused by enzymes of the ligninolytic system in *P. chrysosporium*, according

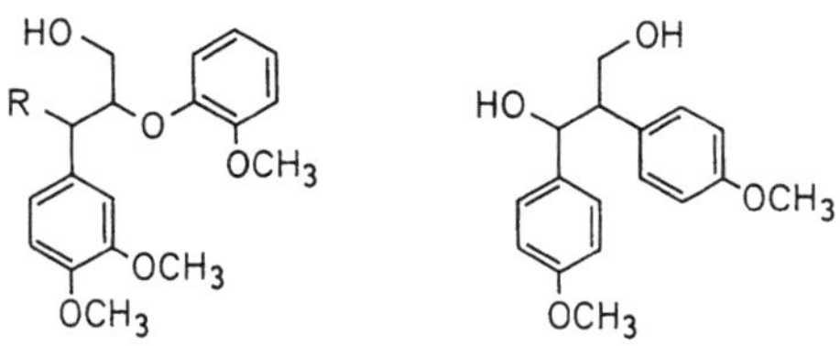

I R = H, OH
II R = O
III

Fig. 4.9. Two dimers (*I, II*) which increase production of ligninase in *Phanerochaete chrysosporium*. Dimer *III* decreased ligninase production (Faison and Kirk 1985)

to Glenn and Gold (1983). Somewhat later Leisola et al. (1984, 1985a) investigated several aromatic compounds, including veratryl alcohol, for their influence on lignin degradation rates in cultures of *P. chrysosporium*. They found that addition of such compounds gave rise to increased lignin degradation. Veratryl alcohol addition (0.25 mM) also increased the glucose oxidation rate and thus H_2O_2 production. Increased production of lignin peroxidase has also been reported after addition of both lignin and different lignin models as well as of veratryl alcohol (Faison and Kirk 1985). Thus birch and spruce MWL, DHP, and the two dimers I, II in Fig. 4.9 gave two- to fivefold increases in ligninase activity when added to 5.5-day-old *P. chrysosporium* cultures, whereas dimer III decreased ligninase activity. Activity increase with veratryl alcohol and veratraldehyde (0.2 mM final concentration) was 4.5 and 3.7 times, respectively. Recently, Faison et al. (1986) showed that veratryl alcohol probably functions via an induction type of mechanism affecting only certain ligninase isoenzymes. So far veratryl alcohol seems to be the best inducer and it may be formed from lignin during its degradation (cf. Fig. 4.8). However, cultures in the primary growth stage are not induced by veratryl alcohol to degrade lignin (Faison et al. 1986). In *Phlebia radiata*, lignin peroxidase is induced by veratryl alcohol, guaiacol, and three lignin dimers (Hatakka et al. 1987). According to Haemmerli et al. (1986a) and Tonon and Odier (1988), veratryl alcohol protects lignin peroxidase from being inactivated by hydrogen peroxide. The involvement of veratryl alcohol in lignin degradation via its cation radical is discussed in Sect. 4.5.3.

4.3.2 Influence of Nitrogen on Secondary Metabolism and Lignin Degradation

The repression of lignin degradation (^{14}C-lignin → $^{14}CO_2$) by nitrogen compounds such as NH_4NO_3 was first discovered by Kirk and co-workers (Kirk et al. 1978a, Keyser et al. 1978) with *Phanerochaete chrysosporium* and *Coriolus versicolor*. Later, Fenn et al. (1981) and Fenn and Kirk (1981) showed that many amino acids and especially glutamate, glutamine, and histidine strongly suppressed lignin degradation. When the nitrogen concentration was increased from 2.6 to 5.6 mM by addition of glutamate, the ligninolytic activity was repressed by 83%. Oxidation of ^{14}C-ring-labeled acetovanillone to $^{14}CO_2$ was

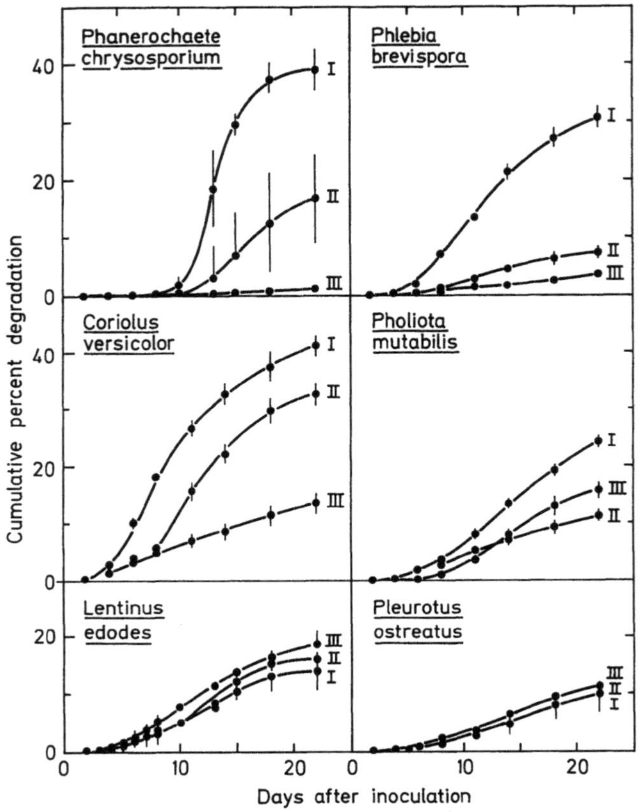

Fig. 4.10. Effect of different nitrogen concentrations on release of $^{14}CO_2$ from ring-labeled DHP by several white-rot fungi. Total nitrogen: *I*, 2.6 mM; *II*, 7.8 mM; *III*, 34.0 mM nitrogen (Leatham and Kirk 1983)

also strongly repressed by glutamate (Fenn and Kirk 1981). In those cultures, the concentration of intracellular glutamate correlates with lignin degradation; a low glutamate concentration is related to extensive lignin degradation. Addition of glutamate has been shown to decrease the level of cyclic adenosine monophosphate (cAMP) in the cells of *P. chrysosporium* (MacDonald et al. 1984). Buswell et al. (1982a) confirmed the suppressive effect of nitrogen on ligninolysis by using another strain of *P. chrysosporium*. Some ammonia-assimilating enzymes such as glutamate dehydrogenase, glutamine synthetase, and glutamate synthase were also studied by Buswell et al. (1982a). Lignosulfonate degradation by *P. chrysosporium* is greater in low nitrogen medium compared to high nitrogen medium, as reported by Kern (1983b).

Leatham and Kirk (1983) demonstrated that degradation of DHP by *Phlebia brevispora*, *Coriolus versicolor*, and *Pholiota mutabilis* is repressed by 34 and 7.8 mM nitrogen but not by 2.6 mM (Fig. 4.10). No nitrogen repression

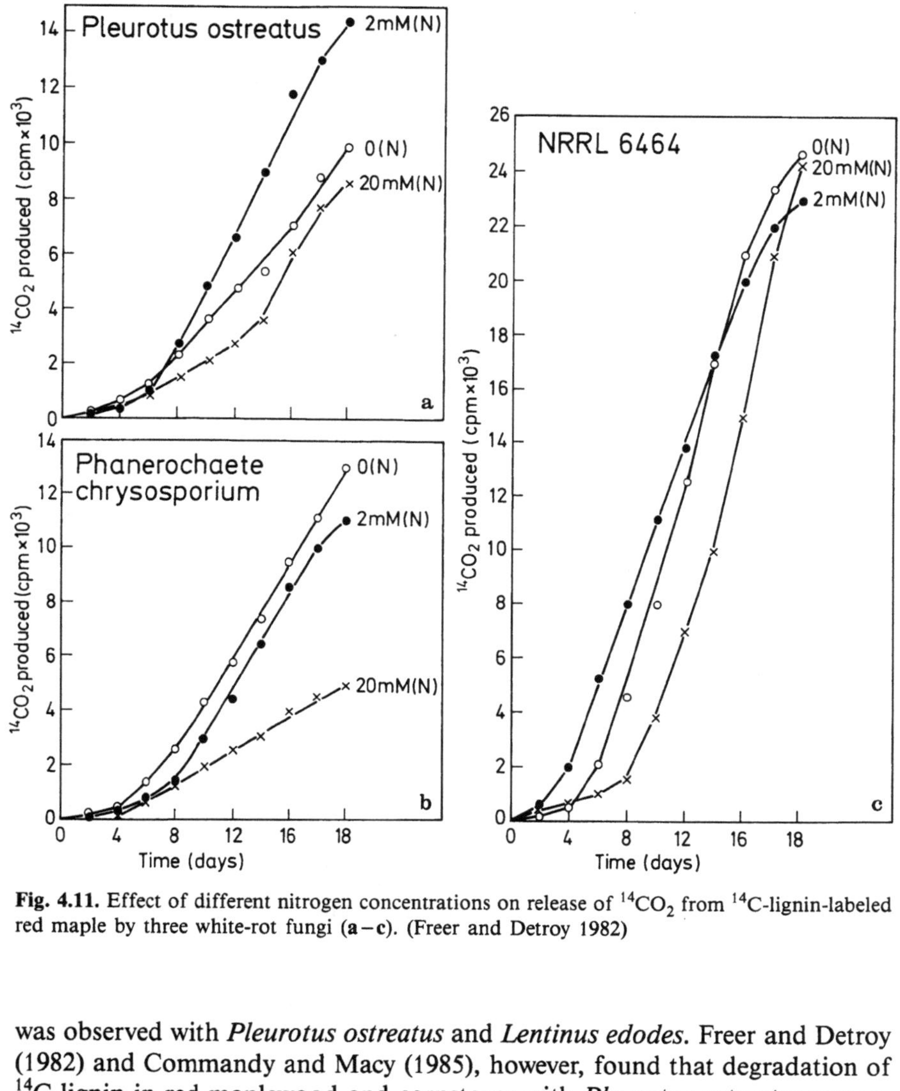

Fig. 4.11. Effect of different nitrogen concentrations on release of $^{14}CO_2$ from ^{14}C-lignin-labeled red maple by three white-rot fungi (**a**–**c**). (Freer and Detroy 1982)

was observed with *Pleurotus ostreatus* and *Lentinus edodes.* Freer and Detroy (1982) and Commandy and Macy (1985), however, found that degradation of ^{14}C-lignin in red maplewood and cornstover with *Pleurotus ostreatus* was repressed by large amounts of nitrogen as shown in Fig. 4.11 for red maple lignin. Rates of $^{14}CO_2$ evolution by an unidentified basidiomycete (NRRL 6464), isolated from cattle dung, were similar but 20 mM nitrogen delayed the onset of $^{14}CO_2$ evolution by 4 days, as shown in Fig. 4.11 (Freer and Detroy 1982). If in Figs. 4.10 and 4.11 the amount of mycelium is taken into consideration, the strong lignin degradation in the presence of little or no nitrogen is even more striking. It was also pointed out by Freer and Detroy (1982) that nitrogen affects the water solubilization of lignin by white-rot fungi. Thus, solubilization of ^{14}C-lignin by *Pleurotus ostreatus* became much stronger with no nitrogen, compared to the solubilization obtained at 2 or 20 mM nitrogen.

Using poplar wood meal and *Coriolus versicolor*, Yoshihara et al. (1987) found that good degradation of lignin occurred in the presence of 100–160 mM urea. However, greater degradation took place with a 4 mM nitrogen concentration using $NH_4H_2PO_4$. Yoshihara et al. (1987) used 1 g of poplar wood in 2 ml nitrogen-containing medium. This is very different from the conditions used by Leatham and Kirk (1983). In their study 50 µg DHP and 10 ml medium plus nitrogen were used, and strong repression of ligninolytic activity by *C. versicolor* was observed with 34 mM glutamic acid (Fig. 4.10).

Of 40 strains of white-rot fungi screened for degradation of DHP in the presence of 8 mM nitrogen as asparagine, *Polyporus platensis* showed the strongest activity (Trojanowski and Hütterman 1987). Trojanowski et al. (1985) found that protoplasts, prepared by hydrolysis of mycelium from the white-rot fungus *Heterobasidion annosum*, could release $^{14}CO_2$ from labeled synthetic lignin equally well or even better than the intact mycelium, although less than 2% $^{14}CO_2$ was obtained from ring-labeled DHP. Release of $^{14}CO_2$ per mg protein by protoplasts was considerably greater than that obtained by the untreated mycelium. The protoplasts also evolved $^{14}CO_2$ from labeled vanillic, ferulic, and veratric acids (Trojanowski et al. 1985).

Reid (1979, 1983 a, b) investigated the influence of various concentrations of nitrogen and carbon sources on the degradation of aspen lignin and DHP. His comprehensive studies show that nitrogen sources such as asparagine, casein hydrolyzate, and urea strongly inhibit aspen lignin degradation by *Phanerochaete chrysosporium*. However, the carbon : nitrogen ratio allows a better prediction of the extent of lignin degradation than do the absolute carbohydrate or nitrogen levels. Peptone and yeast extract stimulated aspen lignin degradation but repressed DHP degradation according to Reid (1983 a, b). Earlier, Ander and Eriksson (1977) found that in several white-rot fungi degradation of lignin in pine wood blocks (measured as Klason lignin) was generally stimulated by 1% malt extract while asparagine and ammonium nitrate varied in their effect on degradation of lignin.

Several other authors have investigated the degradation of ^{14}C-labeled lignins in spruce and poplar woods. Ander et al. (1983 b) found that degradation of these two lignins to $^{14}CO_2$ by *Phanerochaete chrysosporium* was suppressed by yeast extract. The degradation of the dimer [4-methoxy-^{14}C] veratrylglycerol-β-guaiacyl ether was also repressed. It was further shown that conversion of labeled spruce and poplar lignins to $^{14}CO_2$ was highest in the presence of 1.0% glucose and 2.6 mM nitrogen.

When *Heterobasidion annosum* was tested for $^{14}CO_2$ evolution from ^{14}C-labeled lignin in spruce and poplar, only 3% of total ^{14}C was released as $^{14}CO_2$ in 24 days (Bono et al. 1984). Some repression with 15 mM nitrogen compared to 1.5 mM nitrogen was observed.

Fungal degradation of lignin in wheat straw was affected by the amount of nitrogen (NH_4NO_3), and it appears that nitrogen represses lignin degradation by most fungi in this substrate (Zadrazil and Brunnert 1980). A similar conclusion was made by McCarthy et al. (1984) using two *Phanerochaete chrysosporium* strains and *Coriolus versicolor.* With the best *P. chrysosporium* strain, 50% $^{14}CO_2$ was released from ^{14}C-labeled lignin in wheat straw in 15 days in a low nitrogen medium.

All these results show that nitrogen repression of lignin degradation in white-rot fungi is common but may not always be the rule, and suggest that addition of nitrogen to certain fungal cultures may increase the efficiency of these fungi in different biotechnical applications utilizing lignin or lignin-related compounds. Such fungi are probably found in environments of high nitrogen concentrations such as cattle dung, whereas most white-rot fungi that grow in wood usually encounter low nitrogen concentrations.

The very low nitrogen content in wood, with a C:N ratio of usually about 350–500:1 (Cowling and Merril 1966), may explain the strong influence nitrogen has on lignin degradation. Primary growth is possible only for a short time during wood decay. When nitrogen becomes limiting, secondary metabolism, including lignin degradation, begins. Lignin removal exposes cellulose and hemicelluloses which are consumed for further growth of the fungus (Kirk and Fenn 1982). The small supply of nitrogen in wood is probably utilized in a process of autolysis and reuse of the nitrogenous constituents of the fungal mycelium, as suggested by Cowling and Merrill (1966) and Levi et al. (1968).

A new technique for selecting and enhancing the lignin-degrading capacity of fungi and other microorganisms was reported by Tien et al. (1987). They coupled glycine to a lignin model and obtained the glycine adduct I shown in Fig. 4.12. *Phanerochaete chrysosporium* utilized compound I as a nitrogen source, but only when the fungus entered secondary metabolism and lignin peroxidase had cleaved the glycine adduct to release glycine. Formaldehyde and veratraldehyde were also produced in the process. The similarity of this reaction to cleavage of a *β-O-4* model dimer is also shown in Fig. 4.12.

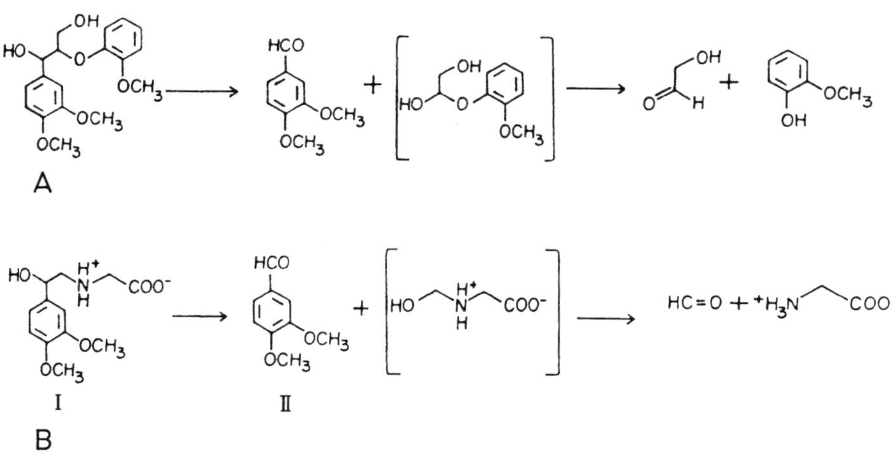

Fig. 4.12. Products formed from ligninase-catalyzed oxidation of a *β-O-4* dimer (**A**) and the glycine adduct (**B**) (Tien et al. 1987)

Using well-known methods, Kuwahara et al. (1987) irradiated conidia of *Phanerochaete chrysosporium* (isolate ME-446) with UV light and placed them on agar plates containing 30 mM diammonium-tartrate as a nitrogen source (60 mM N). Decolorization of Remazol Brilliant Blue R by the hyphae from germinating conidia was used as an indicator of production of lignin peroxidase and Mn-peroxidase. The mutant MA 4 produced the strongest activity of these peroxidases in cultures containing 60 mM nitrogen compared to 2.4 mM nitrogen while the wild-type strain produced peroxidases only in a low nitrogen medium.

4.3.3 Influence of Carbon Source on Lignin Degradation

More than 10 years ago it was found that degradation of kraft lignin, MWL, and lignosulfonates by white-rot fungi on agar plates containing 17 mM nitrogen was strongly stimulated by the presence of cellulose. This was also true for ^{14}C-labeled DHP in liquid cultures in the presence of 26 mM nitrogen (Ander and Eriksson 1975, Hiroi and Eriksson 1976, Hiroi et al. 1976, Kirk et al. 1976). The fungi used were *Phanerochaete chrysosporium*, *Coriolus versicolor*, and *Pleurotus ostreatus*. Glucose did not stimulate lignin degradation as much as cellulose, and no degradation was obtained without an extra addition of some type of carbon source. Degradation of kraft lignin, which seemingly occurred without addition of an easily available carbon source, probably reflected the weakness of the chlorine consumption method used to determine residual kraft lignin (Ander and Eriksson 1975, Hiroi and Eriksson 1976).

Later, Kirk et al. (1978a) and Keyser et al. (1978) found that in liquid cultures containing only 2.6 mM nitrogen (as asparagine plus NH_4NO_3) the percentage of $^{14}CO_2$ evolution from labeled DHP was the same for both glucose (38% ± 6%) and cellulose (41% ± 4%). The discrepancy with earlier work may be due to the rather slow growth of the fungus on cellulose with a high degree of polymerization and crystallinity. Secondary growth conditions were induced although the nitrogen concentration was as high as 17 – 26 mM. In low nitrogen media, the ligninolytic activity was mainly governed by the nitrogen concentration, and the type of carbon source was not as critical.

Kadam and Drew (1986) recently came to the same conclusion using *Coriolus versicolor*, ^{14}C-labeled kraft lignin, and 17 mM nitrogen. Cellobiose and celluloses of different crystallinity were used as co-substrates, and it was found that the more recalcitrant the carbohydrate the higher its ability to support ligninolytic activity, although release of $^{14}CO_2$ was only 5.5% or lower in 18 days.

With *Coriolus hirsutus*, the presence of glucose stimulated kraft lignin degradation on agar plates, and Saxena et al. (1985) found that the best kraft lignin degradation was obtained with 1% glucose plus 1% kraft lignin. Higher or lower glucose concentrations resulted in less lignin degradation. Kirk et al. (1978a) showed that growth on glycerol, which is a rather good carbon source for *Phanerochaete chrysosporium* ME-446, resulted in 29% ± 3% $^{14}CO_2$

from DHP in 13 days, when a nitrogen-limited medium was used. In contrast, the *Phanerochaete chrysosporium* strain INA 12 (Buswell et al. 1984) grew relatively little on glycerol and, therefore, may be able to degrade DHP also in a glycerol high nitrogen medium (26 mM). In these cultures strong lignin peroxidase activity and high concentrations of veratryl alcohol were observed, indicating conditions of secondary metabolism. Jeffries et al. (1981) found that carbohydrate (cellobiose) limitation also triggered ligninolytic activity in the presence of 7.7 mM nitrogen. Thus it may be that carbohydrate limitation induces secondary metabolism and lignin degradation follows. However, total lignin degradation (DHP → $^{14}CO_2$) was greater in the presence of high glucose or cellobiose levels plus a limiting amount of nitrogen (Jeffries et al. 1981).

Using *Polyporus* sp. A 336, Bone and Levonen-Munoz (1984) found that lignin degradation in solid state fermentation of oat straw was best in the presence of 10% glucose (w/w straw basis). Xylose and mannose were slightly less effective in supporting lignin degradation.

In *Lentinus edodes*, in which lignin degradation is not repressed by nitrogen (Leatham and Kirk 1983), addition of glucose, fructose, or xylose stimulated ligninolytic activity (DHP → $^{14}CO_2$) when the fungus was grown on an oak wood meal medium (Leatham 1986).

In general, degradation of isolated lignins in the presence of low nitrogen seems to be stimulated by addition of sugars or cellulose. However, the picture is not that clear for the effects of various nitrogen and carbon sources on lignin in wood. Hatakka and Uusi-Rauva (1983) examined eight white-rot fungi for their degradation of ^{14}C-labeled poplar lignin. Addition of glucose or cellulose to the poplar wood medium stimulated lignin degradation by *Phlebia radiata* 79, but suppressed degradation by *Coriolus versicolor* and *Pleurotus ostreatus*. For *Phanerochaete sordida* 37 and *Phanerochaete chrysosporium* P 127-1 cellulose addition had no effect, while glucose slightly suppressed lignin degradation. Selective lignin degraders did not appear to need an extra energy source in addition to the polysaccharides in the wood meal.

Reid (1979) found that ^{14}C-lignin in aspen wood was converted to $^{14}CO_2$ by *Phanerochaete chrysosporium* PRL 2750 faster when cellulose concentration increased from 0.2% to 1% or 5%. This is different from the results obtained by Hatakka and Uusi-Rauva (1983).

The variability in results is further shown by the experiments of Roch et al. (1983), who used lignin-labeled poplar wood to test the influence of 1 or 10 g l^{-1} of glucose addition to a low nitrogen medium. As shown in Fig. 4.13, with the two strains ME-446 and P 127-1 of *Phanerochaete chrysosporium*, glucose concentration had little effect on $^{14}CO_2$ release. With *Dichomitus squalens* and *Phlebia radiata*, however, $^{14}CO_2$ evolution decreased on addition of 10 g l^{-1} glucose. These results with *Phlebia* are also different from those reported by Hattakka and Uusi-Rauva (1983).

Regulation of lignin degradation in *Phanerochaete chrysosporium* is probably also affected by the formation of an extracellular polysaccharide that is formed from glucose during nitrogen starvation (Leisola et al. 1982, Eriksson

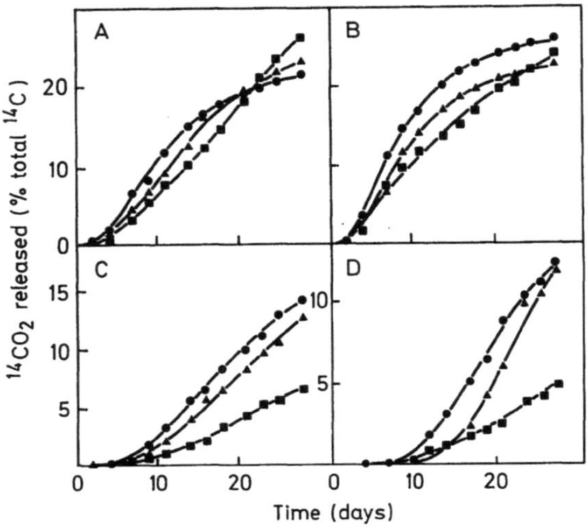

Fig. 4.13. Release of $^{14}CO_2$ from lignin-labeled poplar by **(A)** *Phanerochaete chrysosporium* ME-446; **(B)** *Phanerochaete chrysosporium* P 127-1; **(C)** *Dichomitus squalens*; **(D)** *Phlebia radiata*. (▲) No glucose added; (●) glucose 1 g l^{-1}; (■) glucose 10 g l^{-1} (Roch et al. 1983)

et al. 1986, Bes et al. 1987). The significance of this polysaccharide is discussed in Sect. 4.5.5.

In order to simulate natural conditions, Ander et al. (1981) cultivated *Phanerochaete chrysosporium* in glass Petri dishes on lignin-impregnated glass fiber disks, which were placed either on cellulose or Sephadex beads containing a low nitrogen glucose medium. After extraction, the degradation of lignin was demonstrated by gel permeation chromatography. Evans and Palmer (1983) used this technique with *Coriolus versicolor* and Whatman cellulose powder as a support for the lignin disks. In Fig. 4.14A the percentage recovery of MWL from the disks is shown. *C. versicolor* degraded the MWL much better in the presence of a low N medium as compared to a high nitrogen medium. If the Petri plates did not contain cellulose, the MWL was not degraded. Figure 4.14B shows that in the absence of cellulose, MWL was polymerized. This is probably due to strong laccase induction and no production of cellobiose: quinone oxidoreductase (Sect. 4.5.6).

4.3.4 Influence of Agitation and Other Cultivation Conditions on Lignin Degradation and Ligninase Production

In an important paper by Kirk et al. (1978a), it was shown that culture agitation of *Phanerochaete chrysosporium*, resulting in the formation of small mycelial pellets, strongly suppressed lignin degradation (^{14}C-DHP → $^{14}CO_2$). If the cultures were allowed to stand for 9 days and then shaken, 27% $^{14}CO_2$ was formed in 33 days. During this time, standing cultures released 47%

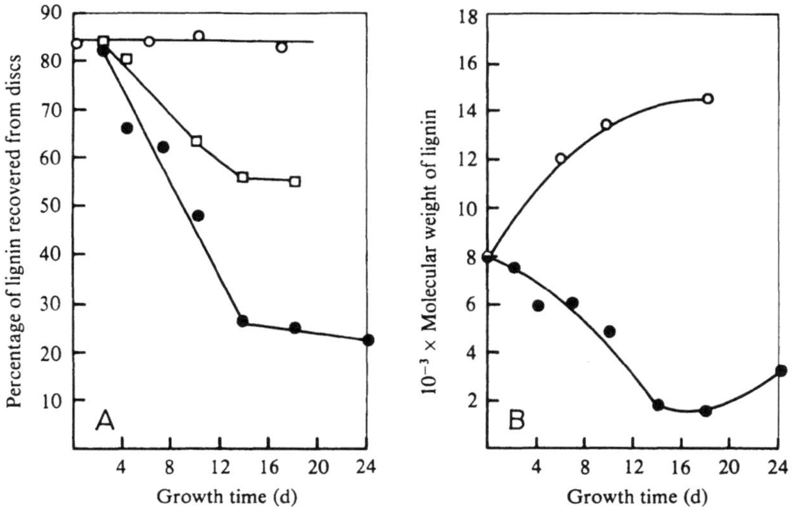

Fig. 4.14 A. Percentage recovery of MWL from lignin disks after growth of *Coriolus versicolor*. (○) Low N without cellulose; (●) low N plus cellulose; (□) high N plus cellulose. **B** Changes in the molecular weight of MWL caused by *C. versicolor* in a low N medium with (●) and without (○) the presence of cellulose (Evans and Palmer 1983)

$^{14}CO_2$. Several years later Reid et al. (1985) found that the same strain of *P. chrysosporium* (ME-446), as used by Kirk and co-workers, could degrade both DHP and lignin in aspen wood to CO_2 in agitated cultures containing single, large pellets. According to Gold et al. (1984), their strain of *P. chrysosporium*, which was derived from ME-446, could also degrade lignin in agitated cultures.

Since it is a drawback to use non-agitated cultures in biotechnological applications, research efforts have been made to develop techniques and strains that can degrade lignin and produce ligninases in agitated, submerged cultures. This would allow the use of stirred tank fermenters for large-scale production of ligninases. Leisola and Fiechter (1985 b) showed that *Phanerochaete chrysosporium* produced lignin peroxidase and degraded DHP under mild agitation conditions (150 ml medium/500-ml flasks or 50 ml medium/200-ml flasks). Increased agitation led to a complete inhibition of ligninolytic activity. Lignin peroxidase production was enhanced by addition of 1 mM veratryl alcohol or aldehyde. Using N- and C-limited agitated cultures, Leisola et al. (1985 a) found that the level of lignin peroxidase was dependent on the concentration of veratryl alcohol, on the size and concentration of pellets, and on the oxygen partial pressure. Leisola and Fiechter (1985 b) suggested that several events, such as reduction of veratraldehyde to the alcohol, the rate of glucose oxidation, and production of extracellular lignin peroxidase, could be used to estimate sensitivity of the cells to agitation when considering their lignin-degrading capacity. Waldner et al. (1988) found that *Bjerkandera adusta* released more than 40% $^{14}CO_2$ from labeled lignin of wheat straw in agitated cultures in 6 days. Before adding the straw lignin the fungus was grown for 3 days and

Table 4.1. Effect of different concentrations of Tween 80 on ligninase activity in *Phanerochaete chrysosporium* BKM-F-1767 in agitated submerged cultures[a] (Jäger et al. 1985)

Tween 80 concentration (%)	Ligninase activity ($U\,l^{-1}$)	Day of maximum activity[b]
0	0	
0.01	87±23	5
0.05	134±45	5–6
0.10	125±34	6
0.20	123±42	7

[a] Culture volume 30 ml in 125-ml flasks.
[b] Days after inoculation.

then the pellets were concentrated by sedimentation and decantation of the supernatant liquid.

Jäger et al. (1985) used various detergents to obtain lignin peroxidase activity and lignin degradation in agitated cultures. Conditions used were: 30 ml medium/125 ml, 300-ml medium/1000-ml, and 600 ml medium/2-l flasks. The detergents used were: methylbenzethonium chloride, Tween 20, Tween 80, 3-[(3-colamidopropyl)-dimethylammonio]1-propanesulfonate (CHAPS) and sodium dodecylsulfate. The best concentration for Tween 80 was 0.05% – 0.1% (Table 4.1). The volume of the culture or flask size had no influence on lignin peroxidase activity. It may be mentioned here that Reid et al. (1985) used 0.1% Tween 80 in preparing their spore inoculum. This may well be the reason for lignin degradation in their agitated cultures. Tran and Chambers (1987) recently calculated a probable final concentration of 0.01% Tween 80 in Reid's cultures. This should give enhanced lignin peroxidase activity according to Table 4.1 if it is assumed that the different strains reacted similarly (Jäger et al. 1985).

The mutant SC26 of *Phanerochaete chrysosporium* BKM-F-1767 can also produce lignin peroxidase in shaking flasks and in a disk fermenter (rotating biological contactor) after addition of Tween 80 and/or veratryl alcohol or benzyl alcohol according to Kirk et al. (1986a, b, c). These authors obtained up to 800 – 1500 $U\,l^{-1}$ lignin peroxidase. This corresponds to 13.3 – 25.0 nkat ml^{-1} (1 International Unit = 16.67 nanokatals; 1 katal is the amount of enzyme that converts 1 mol s^{-1} of substrate to product). With *Phlebia radiata* (Hatakka et al. 1987) the addition of Tween 80 was essential for production of lignin peroxidase in shaking cultures. Asther et al. (1987, 1988) studied lignin peroxidase production with *Phanerochaete chrysosporium* INA-12 in shaking cultures and found that different combinations of Tween 80, oleic acid, olive oil, and tall oil would increase lignin peroxidase activity up to maximally 30 nkat ml^{-1} (1800 $U\,l^{-1}$).

Paszczynski et al. (1986) cultivated BKM-F-1767 in an ordinary low nitrogen medium (Kirk et al. 1978a) and found that lignin peroxidase and Mn-peroxidase could be produced in a 20-l carboy containing 1 l medium. Linko et al. (1986) immobilized spores of *Phanerochaete chrysosporium* both in-agarose and agar gel beads and produced lignin peroxidase for 20 days in a horizontal column reactor. Later, Linko (1988a, b) obtained 730 U l^{-1} lignin peroxidase in a 10-l bioreactor with nylon-web-immobilized *P. chrysosporium*. During the first 5.5 days a total of about 6200–10000 Units of lignin peroxidase was produced. Kirkpatrick and Palmer (1987) immobilized *P. chrysosporium* in cubes of polyurethane foam and harvested lignin peroxidase every 24–48 h, while Willershausen et al. (1987) used a bioreactor originally designed for tissue cultures. After 6–7 days, 1 l medium containing 200–230 U l^{-1} of lignin peroxidase was obtained by the latter authors. Janshekar and Fiechter (1988) found among several parameters that polyethylene glycol increased the yield of lignin peroxidase in a 42-l stirred tank reactor.

4.3.5 Influence of Oxygen and Carbon Dioxide on Lignin Degradation. Anaerobic Degradation of Lignin and Related Compounds

The influence of oxygen on lignin biodegradation has been investigated with *Phanerochaete chrysosporium* and many other white-rot fungi. Originally, it was shown by Kirk et al. (1978a) that *P. chrysosporium* degraded DHP to $^{14}CO_2$ much better in the presence of 100% O_2 than in air (21% O_2). No $^{14}CO_2$ was produced from DHP in the presence of 5% O_2 plus 95% N_2. In these experiments 10-ml cultures in 125-ml flasks were used. Bar-Lev and Kirk (1981) found that oxygen both induces ligninolytic activity and increases this activity when already present. However, the optimal oxygen partial pressure was found to be about 0.5 atm (50% oxygen) for lignin degradation by *P. chrysosporium*. Also, oxidation of [^{14}C]-labeled glucose to $^{14}CO_2$ was highest at 50% O_2. The effect of increasing O_2 concentration, however, was not as strong for glucose oxidation as it was for lignin oxidation. According to Shimada et al. (1981), the concentration of oxygen strongly influences the amount of veratryl alcohol produced by *P. chrysosporium*.

Later, Leisola et al. (1983a, b) found that reducing the culture depth in stationary cultures from the 4–10-mm, depths commonly used to less than 1.5-mm, markedly increased glucose consumption and lignin degradation. The reason for this is probably oxygen limitation in the larger volume cultures. In the very shallow cultures (3 ml medium/ 200 ml Erlenmeyer flasks), the fungus only poorly degrades ^{14}C-labeled straw and poplar lignin in 100% O_2 atmosphere (Leisola et al. 1984), probably reflecting O_2 toxicity.

The influence of different oxygen concentrations on lignin degradation in whole wood has also been investigated for several fungi. Reid and Seifert (1980) showed that degradation of aspen wood lignin by *Phanerochaete chrysosporium* was equally good in 1 or 2 atm O_2. However, 3 atm O_2 inhibited fungal growth. Several other fungi were later investigated by Reid and Seifert (1982) and Hatakka and Uusi-Rauva (1983). As shown in Table 4.2, all fungi, except *Pleurotus ostreatus* and *Grifola frondosa*, evolved much more $^{14}CO_2$ from labeled aspen lignin in an oxygen atmosphere than in air (Reid and Seifert 1982). The same general pattern was also obtained for degradation

Table 4.2. Degradation of ^{14}C-labeled aspen wood lignin by white-rot fungi in the presence of air or oxygen measured as percent $^{14}CO_2$ evolution and percent loss of Klason lignin (Reid and Seifert 1982)

	$^{14}CO_2$ (%)		Klason lignin loss (%)	
	Air	O_2	Air	O_2
Phanerochaete chrysosporium	10.8	35.2	13	40
Coriolus versicolor	14.6	35.5	24	46
Gloeoporus dichrous	9.7	18.1	22	24
Polyporus brumalis	16.6	33.0	19	33
Merulius tremellosus	14.0	22.3	30	40
Pycnoporus cinnabarinus	13.6	22.6	18	37
Lentinus edodes	9.7	18.0	18	41
Bondarzewia berkeleyi	9.0	13.8	25	27
Pleurotus ostreatus	11.7	11.6	17	17
Grifola frondosa	9.2	10.6	8	15

of Klason lignin by the fungi investigated by Hatakka and Uusi-Rauva (1983). Lignosulfonate degradation by *P. chrysosporium* was better in 100% oxygen than in air if a 30-ml medium in 250-ml flasks was used (Kern 1983b).

In substrates such as wood and straw with reduced oxygen concentrations, a higher lignin degradation was obtained in an atmosphere of 100% oxygen. Thus, Yu and Eriksson (1985) showed that the weight loss in birch and spruce wood caused by three white-rot fungi was greater in O_2 than in air. These authors also found that thinner wood blocks (wafers) were easier to degrade than large ones; ^{14}C-labeled straw gave the most $^{14}CO_2$ in an O_2 atmosphere.

In solid state fermenters not only oxygen concentration but also CO_2 concentration is important. Thus, Kamra and Zadrazil (1985) investigated lignin degradation in wheat straw by *Stropharia rugosoannulata*, while Levonen-Munoz and Bone (1985) cultivated *Phanerochaete chrysosporium*, *Polyporus tulipiferae*, and *Polyporus* sp. A 336 to determine lignin degradation with different concentrations of oxygen, nitrogen, and CO_2. Changes in the atmosphere (O_2 and CO_2) during wood decay were comprehensively investigated also by Kozlik and Schánĕl (1974a, b). Generally, it seems that optimal conditions must be found for every fungus, substrate, or cultivation technique. The importance of oxygen is discussed further in connection with the activity of lignin-degrading enzymes (Sect. 4.5).

From the above and Sect. 4.1 it is clear that lignin biosynthesis and degradation are dependent on oxygen for different enzymes to be active. H_2O_2 used by peroxidases in the xylem tissues of trees, and by the white-rot fungi, is produced from O_2 by oxidases. As shown below, lignin biodegradation is oxidative and several important oxidoreductases are dependent on oxygen or H_2O_2 for their activity. Thus, it is to be expected that lignin is not degraded anaerobically.

Hackett et al. (1977) and Zeikus et al. (1982) studied degradation of labeled synthetic lignins in anaerobic lake sediments during 41–110 days. Degradation to $^{14}CO_2$ and/or $^{14}CH_4$ was limited to the low molecular weight material. Anaerobic degradation of labeled lignocelluloses and synthetic lignins, kraft lignin, and beech wood have also been studied with the result that after several weeks or months very little high molecular weight lignin was degraded anaerobically (Holt and Jones 1983, Odier and Monties 1983, Benner et al. 1984, Benner and Hodson 1985, Colberg and Young 1985 a).

Anaerobic biodegradation of lignin-related monomers and oligomers, under methanogenic conditions, however, is possible according to comprehensive research by Colberg, Young, and co-workers (Healy and Young 1979, Healy et al. 1980, Colberg and Young 1982, 1985 b, Grbic-Galic and Young 1985). It was shown that not only lignin-related aromatic acids (monomers) but also dimeric, trimeric, and possibly tetrameric compounds were anaerobically degradable to methane and carbon dioxide (Colberg and Young 1982, 1985 b). During these fermentations, a great number of organic compounds were produced, which indicates a possible future for the manufacture of chemicals from lignin wastes and for disposal of toxic pollutants (Hanselman 1982, Kaiser and Hanselman 1982, Taylor 1983, Colberg and Young 1985 b).

4.4 Chemistry of Lignin Degraded by White-Rot Fungi

Much of the knowledge about the chemistry of lignin biodegradation comes from studies of the structure of fungal-degraded polymeric lignin isolated from decayed wood and from identification of the low molecular weight lignin degradation products isolated from this decayed wood (Reviews: Kirk 1984, Chen and Chang 1985). As an equally important part, the degradation of lignin-related dimers has also been studied (Enoki et al. 1980, 1981 a, b, Goldsby et al. 1980, Enoki and Gold 1982, Kirk 1983 b, 1984, Higuchi 1985 b). Most of the results come from experiments with *Phanerochaete chrysosporium*, although other fungi have also been used.

When lignin in wood is degraded, only the surface of the lignin is affected and small amounts of degradation products such as vanillin and vanillic acid are formed. However, the degraded part of the lignin is more soluble and can be extracted from the unchanged lignin with neutral organic solvents (Chen and Chang 1985). Thus different woods decayed by several white-rot fungi have been extracted successively with petroleum ether, chloroform, acetone, methanol, and aqueous dioxane. The isolated, partly degraded polymeric lignin and the low molecular weight degradation products in the methanol extract were further separated on the basis of molecular weight. Degraded lignins from spruce (*Picea glauca*) and birch (*Betula papyrifera*) have been isolated and characterized (Kirk and Chang 1974, 1975, Chen et al. 1982, Chua et al. 1982, Tai et al. 1983 a, b, Terazawa et al. 1983, 1987). The fungi used were *Coriolus versicolor*, *Dichomitus squalens*, and *Phanerochaete chrysosporium*.

Taken together this comprehensive research shows that lignin biodegradation is mainly oxidative since the oxygen content of the isolated polymeric

Table 4.3. Elemental analyses and methoxyl content of decayed lignins and milled wood lignins from birch and spruce (Chen and Chang 1985)

Source	C (%)	H (%)	O (%)	OCH_3 (%)
Spruce				
MWL	62.85	6.08	31.07	15.11
MWL	62.81	5.88	31.31	15.24
Lignin decayed by				
Polyporus anceps	58.71	4.99	36.30	11.21
Coriolus versicolor	57.97	4.70	37.23	11.33
Poria subacida	58.54	5.23	36.22	11.75
Birch				
MWL	57.47	6.22	36.31	20.52
Lignin decayed by				
Phanerochaete chrysosporium				
M-HA-1 [a]	55.29	5.87	38.84	12.50
M-HA-2	49.09	4.81	46.10	7.52
M-HA-3	63.74	6.67	29.59	11.31

[a] M-HA- indicates the high molecular weight acidic fraction of the methanol extract of decayed wood.

lignin is generally higher than the corresponding lignin from sound wood (Table 4.3). There is also a decrease in the methoxyl content.

Analyses of spruce lignin degraded by *Dichomitus squalens* and *Coriolus versicolor* (Kirk and Chang 1974, 1975) showed a decrease in the methoxyl content, but new aromatic hydroxyl groups replacing the methoxyls were not found. This differs from the results with the brown-rot fungus *Gloeophyllum trabeum* (*Lenzites trabea*) which gives rise to the formation of new phenolic hydroxylic groups by both demethylation and aromatic hydroxylation (Kirk 1975). After decay by both brown- and white-rot fungi there is an increase in the content of α-carbonyl and carboxyl groups. For lignin rotted by white-rot fungi it was suggested that in addition to side-chain oxidation there was also demethylation followed by cleavage of aromatic nuclei in the polymer (Kirk and Chang 1975). This idea of ring cleavage in the polymer was partly based on the finding that the degraded lignins contained a greater number of total (aromatic plus aliphatic) carboxyl groups compared with the lower number of aromatic carboxyl groups. Aliphatic carboxyl groups would be formed during aromatic ring cleavage, and a decreased number of aromatic rings were also found. With the discovery of lignin peroxidases (ligninases) (Glenn et al. 1983, Tien and Kirk 1983), which cleave bonds between the α- and β-carbons in the side chains, it is now believed that side-chain cleavage is the main depolymerizing feature of the ligninolytic system (Chen and Chang 1985, Kirk 1987, Robert and Chen 1987, Terazawa et al. 1987).

Chen et al. (1982, 1983a) further divided the methanol extract from decayed wood into acidic and phenolic fractions. In the low molecular weight acid fraction the chemical structures of 28 compounds were determined, ten of which were aromatic acids. The most common products were vanillic acid,

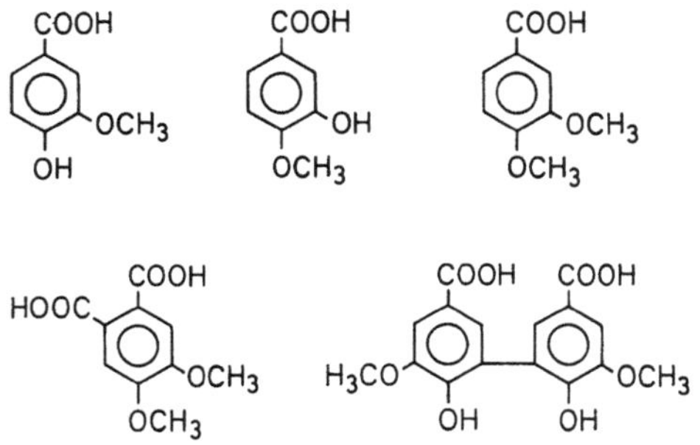

Fig. 4.15. The most common products (vanillic acid, isovanillic acid, veratric acid, m-hemipinic acid, and dehydrodivanillic acid) isolated from the low molecular weight acid fraction of the methanol extract after degradation of spruce wood by *Phanerochaete chrysosporium* (Chen et al. 1982)

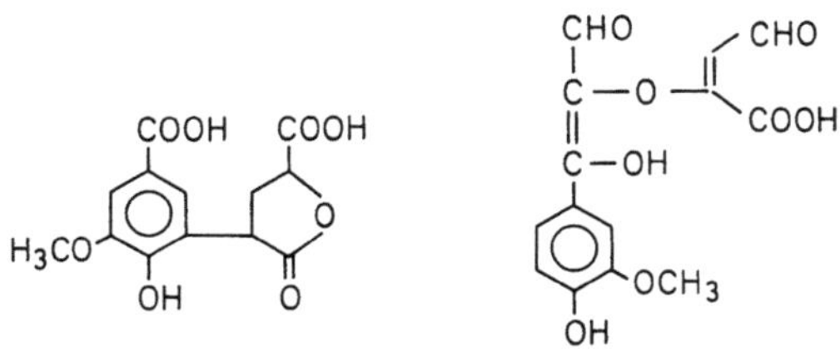

(1) (2)

Fig. 4.16. Two ring-cleavage products identified in extracts of spruce wood degraded by *Phanerochaete chrysosporium* (Kirk 1984)

isovanillic acid, veratric acid, m-hemipinic acid, and dehydrodivanillic acid (Fig. 4.15). These compounds are probably formed by C_α-C_β cleavage and C_α oxidation. Veratric acid may be formed by de novo synthesis from glucose (Lundquist and Kirk 1978) or by 4-*O*-methylation of vanillic acid (Chen et al. 1982) in a similar way as for syringic acid (Eriksson et al. 1984). Among identified ring-cleavage products those presented in Fig. 4.16 have been obtained in small amounts, and may come from 5-5 (biphenyl) and β-*O*-4 (β-aryl ether) structures. However, it cannot be deduced from these structures whether ring cleavage occurs in the polymer or not. A possible pathway for formation of compound (2) is shown in Fig. 4.17 (Chen et al. 1983a, Kirk 1984). Some ring cleavage in the polymer of lignosulfonate by protocatechuate 3,4-dioxygenase was recently reported (Wojtas-Wasilewska and Luterek 1987, Wojtas-Wasilewska et al. 1988).

Hiroi and Tamai (1983) studied degradation of beech wood meal by the white-rot fungus *Grifola frondosa* for 1 – 6 months and found syringic acid, vanillic acid, and a small amount of vanillin in the water-soluble fraction of the decayed wood meal. Dioxane lignin prepared from the decayed wood was lower in methoxyl content and higher in oxygen content than the corresponding lignin from sound wood. Using rice straw and cotton stalks, Nagieb et al. (1988) obtained similar results with *Phanerochaete chrysosporium* and *Coriolus versicolor.* The content of phenolic hydroxyl groups per C_9 unit of the degraded lignins was much higher than in the control lignins.

Oxidation of the α-carbon appears to be common in fungal lignin degradation (Crawford 1981b, Kirk 1984). However, it has been found that metabolism

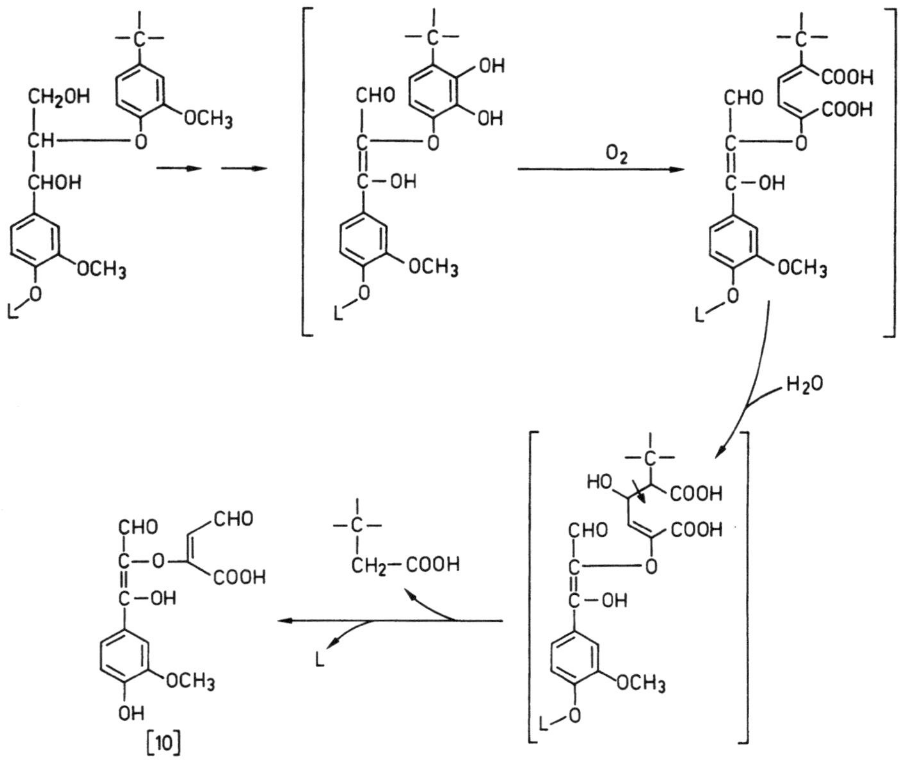

Fig. 4.17. Proposed scheme for degradation of lignin by hydroxylation/demethoxylation and ring cleavage to form one of the ring-cleavage structures in Fig. 4.16 (Kirk 1984). More recent results suggest that demethoxylation is not a prerequisite for C_α-C_β-cleavage or ring-cleavage (Umezawa and Higuchi 1989, Robert and Chen 1989)

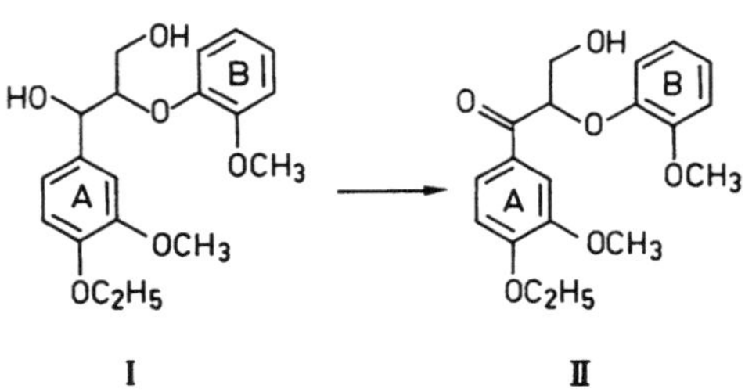

Fig. 4.18. Oxidation of the *β*-*O*-4 dimer *I* to the corresponding dimer *II* containing an *α*-carbonyl group (Fenn and Kirk 1984)

of the model compound 1-(3-methoxy-4-ethoxyphenyl)-2-(o-methoxyphenoxy)propane-1,3-diol (Fig. 4.18) was suppressed after oxidation of the *α*-carbon to a carbonyl group (Fenn and Kirk 1984). Similar results have also been obtained by Enoki and Gold (1982) and Kirk and Nakatsubo (1983) using dimers of the *β*-1 type. Surprisingly, chemical C_α oxidation of spruce lignin enhanced its depolymerization by ***Phanerochaete chrysosporium***, and it was proposed that C_α oxidation may direct the degradation against the aromatic rings, and that C_α oxidation in the lignin polymer increases the accessibility for depolymerizing enzymes. The reason for this may be that C_α oxidation removes one center of asymmetry at C_α (Fenn and Kirk 1984).

4.5 Biochemistry of Lignin Degradation

4.5.1 Oxygen Radicals and Lignin Degradation

When it was found that lignin degraded by white-rot fungi had a higher oxygen content than sound lignin, and that oxygen stimulated lignin degradation, there seemed to be a possibility for monooxygenases and dioxygenases to be involved in lignin degradation. However, the only dioxygenase studied to date has been in one white-rot fungus, namely, *Pleurotus ostreatus* (Wojtas-Wasilewska et al. 1983, 1988). As an alternative it was proposed that reduced, highly reactive oxygen species were responsible for the changes detected in degraded lignin (Hall 1980). The reduction of oxygen is catalyzed by many flavoprotein oxidases; the one-, two-, or four-electron reduction of O_2 gives superoxide ($O_2^{\bar{\cdot}}$), hydrogen peroxide (H_2O_2), or water, respectively (Halliwell 1981, Fee 1982, Walling 1982). The hydroxyl radical, $^{\bullet}OH$, can be formed according to the Fenton reaction:

$$Fe^{2+} + H_2O_2 \rightarrow Fe^{3+} + OH^- + {}^{\bullet}OH \ .$$

Several investigations have been undertaken to determine whether reduced oxygen species, or excited oxygen like singlet oxygen ($^1O_2$), could give rise to the reactions observed during lignin degradation by *Phanerochaete chrysosporium*. It was first suggested that $^1O_2$ was involved (Nakatsubo et al. 1981b), but this was later rebutted by Kirk et al. (1983a) and Kutsuki et al. (1983). The reasons were that photosensitized riboflavin (used in the studies) probably produces not only singlet oxygen but also other oxygen radicals, and that a chemical system generating $^1O_2$ specifically ($HOCl + H_2O_2$) did not give the same cleavage products from a model dimer as the fungal cultures (Kirk et al. 1983a). In another paper, Gold et al. (1983c) investigated oxidation of ^{14}C-labeled lignins by UV/H_2O_2, Fenton's reagent, photosensitized riboflavin, and UV- and γ-irradiation. Especially ^{14}C-methoxy labeled lignin but also ring- and side-chain labeled lignins were extensively degraded by both photosensitized riboflavin and UV/H_2O_2. With Fenton's reagent, rapid degradation was obtained. These results suggested that the hydroxyl radical $^{\bullet}OH$ is at least partly involved in lignin degradation, but the studies did not show if or how the fungus produces $^{\bullet}OH$. Further evidence for the involvement of $^{\bullet}OH$ in lignin degradation was reported by Forney et al. (1982b), Kutsuki and Gold (1982), and Bes et al. (1983). In the first two papers, ethylene production from the 2-keto-4-methiolbutyric acid (KTBA) and methional were used to show $^{\bullet}OH$ production in ligninolytic cultures of *P. chrysosporium*. Although KTBA has been used as a sensitive measure of ligninolytic activity (Kelley and Reddy

1982), methional and KTBA are not considered to be specific for •OH detection (Pryor and Tang 1978, Faison and Kirk 1983). It was shown in the paper by Kelley and Reddy (1982) that evolution of $^{14}CO_2$ from ^{14}C-labeled synthetic lignin by *P. chrysosporium* started at day 3, which is 3–4 days earlier than the appearance of ethylene from KTBA. This indicates that ethylene formation is not related to lignin degradation. Also, different scavengers used may not be specific for the hydroxyl radical. Oxidation of [7-^{14}C]-benzoate to $^{14}CO_2$, which was used by Bes et al. (1983), is perhaps specific for •OH. Winston and Cederbaum (1982) stated that the oxidant releasing $^{14}CO_2$ from labeled benzoic acid appears to resemble •OH in its oxidizing properties. However, unequivocal evidence that •OH is formed in a particular process is extremely difficult to obtain. The •OH can only be detected by the reactions it undergoes, yet none of these are known to be absolutely specific, as noted by Winterbourn (1987).

Regarding H_2O_2, evidence indicates that this two-electron reduced oxygen species is necessary in fungal lignin degradation (see also Sect. 4.5.5). Faison and Kirk (1983) studied production of H_2O_2, $O_2^{\bar{\bullet}}$, and •OH in connection with fungal oxidation of synthetic lignin to CO_2. They added o-dianisidine and peroxidase to obtain a color reaction due to H_2O_2 production by the fungus *Phanerochaete chrysosporium*. The H_2O_2 appeared at day 3, just before the [^{14}C]-lignin began conversion to $^{14}CO_2$. Addition of catalase to destroy H_2O_2 in the cultures resulted in strong inhibition of lignin degradation. Also, addition of superoxide dismutase decreased lignin degradation. Thus, H_2O_2 and perhaps also $O_2^{\bar{\bullet}}$ are important for lignin degradation. Production of •OH, which was measured by benzoate decarboxylation, suggested that •OH was formed during the nonligninolytic primary growth phase and peaked at day 4. The •OH synthesis was not stimulated by O_2. This indicates that •OH is probably not involved in lignin degradation (Faison and Kirk 1983).

In another paper, Kirk et al. (1985) compared degradation of two nonphenolic lignin model dimers in cultures of *Phanerochaete chrysosporium* by purified ligninase and Fenton's reagent, respectively. The ligninase and the fungal culture formed similar degradation products that were different from those obtained with Fenton's reagent (•OH). With ESR spectroscopy, •OH was detected in the Fenton system but not in the ligninase system. This indicates that Fenton chemistry is of little importance in the fungal degradation of β-1 and β-*O*-4 substructure models and of similar structures in lignin. In ESR experiments with ligninase and 5,5-dimethyl-1-pyrroline *N*-oxide, no •OH formation was detected (Kirk et al. 1985).

Some evidence for enzymatic production of •OH or $O_2^{\bar{\bullet}}$ was obtained by Palmer and Evans (1983a, b) and Evans et al. (1984). Using similar purification methods that Fåhraeus and Reinhammar (1967) had used for laccase, they purified a heme protein from culture broths of *Coriolus versicolor*. Besides laccase A and B with a molecular weight of 62000, a red-brown fraction (not studied by Fåhraeus and Reinhammar) was isolated. It had a Soret band at 442 nm and α and β bands at 588 and 560 nm, respectively, indicating a heme group. The MW of this protein was 53700. Since H_2O_2 had no effect on the absorption spectrum, the enzyme obviously did not bind H_2O_2, and was therefore not considered to be an ordinary catalase or peroxidase. Oxygen radical release, considered most likely to be $^{\bullet}OH$, was measured as ethylene formation from methional in a protein concentration-dependent fashion. However, it was not possible to determine whether there was a direct release of $^{\bullet}OH$ from the protein or if the iron moiety of the protein reacted with H_2O_2 in a Fenton-like reaction (Evans et al. 1984). An extracellular lignin peroxidase was later isolated from *C. versicolor* by the same research group (Dodson et al. 1987). The enzyme, which was purified, does not appear to be the same as the described enzyme above (see Table 4.8).

Two cellobiose oxidizing enzymes from *Phanerochaete chrysosporium*, namely, cellobiose: quinone oxidoreductase and cellobiose oxidase, have been reported to produce both $O_2^{\bar{\bullet}}$ and H_2O_2 (Morpeth 1985, Morpeth and Jones 1986). Indications for $O_2^{\bar{\bullet}}$ production by cellobiose oxidase were also obtained by Eriksson (1981 b) and Ayers and Eriksson (1982). For further discussion of these enzymes see Sect. 4.5.5.

4.5.2 The Phenoloxidases Laccase and Peroxidase Including Manganese-Dependent Peroxidase (Mn-Peroxidase)

The importance of phenoloxidases in lignin degradation has been discussed for a long time (Kirk and Kelman 1965, Kirk 1971, Ander and Eriksson 1976, 1978). About all white-rot fungi produce extracellular phenoloxidases as found by Bavendamm (1928) and Davidson et al. (1938) using color formation around mycelia on gallic and tannic acid media. These enzymes catalyze the removal of one electron from phenolic hydroxyl groups and from amino groups of a wide range of phenols and aromatic amines (Mayer and Harel 1979, Sjoblad and Bollag 1981). Recently, new types of peroxidases (ligninases and Mn-dependent peroxidases) have been detected in white-rot fungi (Glenn et al. 1983, Kirk and Tien 1983, Kuwahara et al. 1984). The ligninases, or lignin peroxidases as they are best called, use both phenolic and nonphenolic compounds as substrates and produce aromatic cation radicals when a nonphenolic substrate is oxidized (Kersten et al. 1985). These enzymes are discussed in Sect. 4.5.3. All the above types of enzymes (laccase, Mn-peroxidase, and lignin peroxidase) may be defined as phenoloxidases, although Mn-peroxidase and lignin peroxidase were not discussed before 1983.

Laccase has a broad substrate specificity and is classified as EC 1.10.3.2 benzenediol: O_2 oxidoreductase, but it also has some properties similar to EC 1.10.3.1 (1,2-benzenediol: O_2 oxidoreductase) or EC 1.14.18.1 (monophenol, L-dopa: O_2 oxidoreductase) according to Enzyme Nomenclature Int. Union of Biochemistry 1984 (Webb 1984). The EC number of peroxidase (e.g., horseradish peroxidase) is EC 1.11.1.7 donor: hydrogen peroxide oxidoreductase. Mn-peroxidase (Glenn et al. 1986, Paszczynski et al. 1986) and lignin peroxidase (Glenn et al. 1983, Tien and Kirk 1983) belong to this same class of enzymes. (They do not as yet have an EC number.) However, in some early

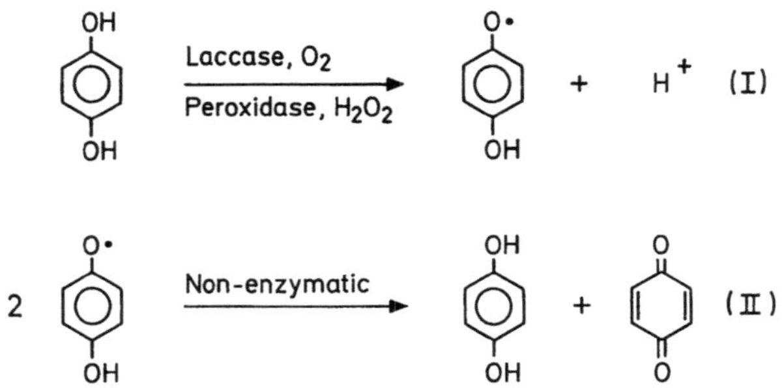

Fig. 4.19. Oxidation of hydroquinone by laccase or peroxidase to give a phenoxy radical which disproportionates into hydroquinone and p-benzoquinone

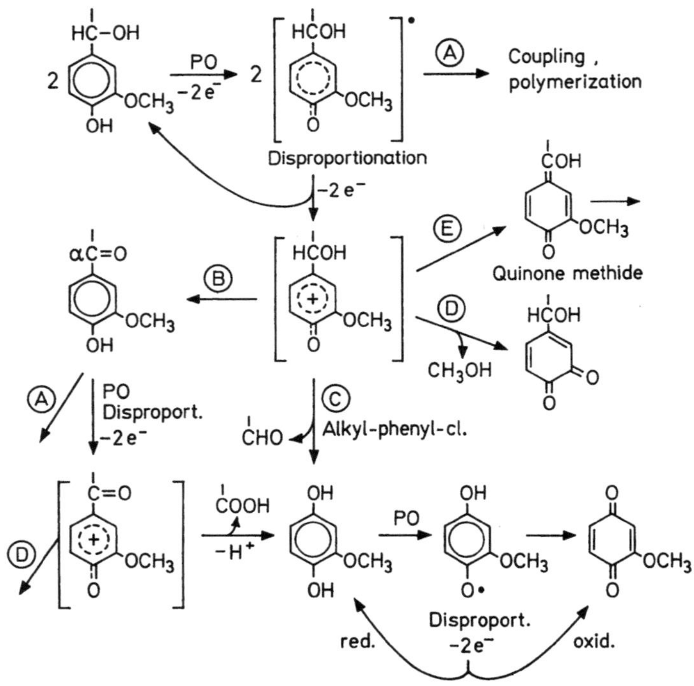

Fig. 4.20. Different reactions (A–E) after single-electron oxidation of lignin-related phenols, catalyzed by laccase and peroxidase (After Kirk and Shimada 1985)

papers, lignin peroxidase was classified as an oxygenase, since oxygen seemingly was incorporated into the substrate by the enzyme (Gold et al. 1984, Tien and Kirk 1984, Andersson et al. 1985, Renganathan et al. 1985). This is discussed in Sect. 4.5.3.

In Fig. 4.19, a simple reaction (I) catalyzed by laccase or peroxidase is shown. Reaction II is a disproportionation which is nonenzymatic. Following the single electron oxidation, many different reactions can occur. Most of these reactions are summarized in Fig. 4.20, which is a simplified reaction scheme adopted from Kirk and Shimada (1985). The reactions include: (A) oxidative coupling and polymerization, (B) oxidation of α-carbinol groups to α-carbonyl structures, (C) alkyl-phenyl cleavage, (D) demethoxylation to produce methanol, (E) other reactions initiated by quinone methide formation.

Earlier reasons for believing that laccase or peroxidase are involved in lignin degradation were that (1) lignin contains phenolic hydroxyl groups which constitute a substrate, (2) white-rot fungi which degrade lignin also produce laccase and/or peroxidase extracellularly, (3) lignin degradation is oxidative. Brown-rot fungi, which have little effect on lignin (Sect. 4.7.1), do not produce phenoloxidases extracellularly. For these reasons laccase and peroxidase from many white-rot fungi have been purified and characterized. Some of the earliest purifications were done with laccase from *Coriolus versicolor* by Mosbach (1963) and by Fåhraeus and Reinhammar (1967) who reported the separation of two blue proteins, laccase A and B. Laccase A contains four Cu-atoms per molecule and 14% carbohydrate and has a molecular weight of 60000 – 64400. Laccase B has a lower specific activity, is more acidic, but has otherwise similar properties as laccase A (Fåhraeus and Reinhammar 1967). Laccase has also been isolated and characterized from other white-rot fungi (cf. the comprehensive review by Mayer and Harel 1979). Some properties of laccases are shown in Table 4.4.

The laccases from *Agaricus bisporus* (Wood 1980) and *Lentinus edodes* (Leatham and Stahmann 1981) have a molecular weight of 100000; the *Agaricus* laccase contains two Cu-atoms per molecule (Table 4.4). Laccase from *Pleurotus ostreatus* was reported to have a molecular weight of 59000 (Sannia et al. 1986).

The influence of laccase on degradation of lignin and lignin model compounds has been investigated especially in Japan. The main conclusions drawn from these early experiments are that lignin is both polymerized and depolymerized by laccase (Ishihara and Miyazaki 1972, 1974, Konishi et al. 1972, 1974). Recently, Geiger et al. (1986a, b) isolated laccase from both *Rigidosporus lignosus* and *Phellinus noxius*. These white-rot fungi are pathogens of the rubber tree, *Hevea brasiliensis*. The molecular weights for the laccases were 52000 – 55000 (*Rigidosporus*) and 70000 (*Phellinus*). With guaiacol as a substrate, pH optima for these enzymes were 5.6 – 6.0 and 4.6, respectively. The *Rigidosporus* laccase was used for experiments with a lignin isolated from *Hevea brasiliensis* with thioglycolic acid. At pH 6 and after 9-day incubation, the amount of low molecular weight lignin material increased by 6.8% compared to a control. A low molecular weight lignin fraction incubated in the same way was polymerized.

Arora and Sandhu (1985, 1986, 1987) cultivated a great number of soil and wood-rotting fungi on different lignin media and measured laccase production. Generally, a positive correlation between laccase production and lignin degradation was obtained. Using thioglycolic acid as an inhibitor of laccase activity, Haars and Hütterman (1980) found that lignosulfonate (LS) polymer-

Table 4.4. Properties of laccases isolated from different white-rot fungi. These enzymes are dependent on O_2. C_α-C_β cleavage in phenolic dimers mainly by laccase III

Enzyme	Molecular weight	Central atom	Carbohydrate (%)	Important reactions	References (fungal origin)
Laccase I	120000	0.58 Cu		Laccase I – III: C_α-oxidation, alkyl-aryl cleavage, C_α-C_β cleavage in phenolic dimers	Kawai et al. (1987a, b, 1988), Morohoshi and Haraguchi (1987a, b), Morohoshi et al. (1987a, b, c), Wariishi et al. (1985), (*C. versicolor*)
II	210000	0.01 Cu?			
III	130000	2.36 Cu			
III	66000	2 Cu	10% – 15%		
Oxidase (laccase)	61000 (No heme protein)	Cu/Fe	Glycoprotein		Niku-Paavola (1987) (*P. radiata*)
Laccase A, B	62000	4 Cu	14%	Induced with 2,5-xylidine	Fåhraeus and Reinhammar (1967), Evans (1985) (*C. versicolor*)
Laccase	100000	2 Cu	15%		Wood (1980) (*A. bisporus*)
Laccase	100000				Leatham and Stahmann (1981) (*L. edodes*)

ization by *Heterobasidion annosum* was prevented, resulting in depolymerization. The degradation products had a molecular weight lower than 30000 compared to the original molecular weight 2×10^6. Thioglycolic acid does not affect growth or morphology of *Heterobasidium annosum* (Haars and Hütterman 1980). With *Phanerochaete chrysosporium*, however, no polymerization of LS was detected either in the presence or absence of this inhibitor, but release of $^{14}CO_2$ from ^{14}C-LS decreased in the presence of thioglycolic acid (Kern 1983b). Since *P. chrysosporium* produces little laccase in addition to the different peroxidases, the effect of thioglycolic acid as a Cu-chelating agent on these latter enzymes should also be investigated.

The polymerization/depolymerization question also has been investigated by Kaplan (1979). He used both laccase (from *Coriolus versicolor*) and horseradish peroxidase, different lignin model compounds, and lignin preparations. No degradation products were detected under the conditions employed. Leonowicz et al. (1985) found that laccase from *C. versicolor* both polymerized and depolymerized lignosulfonate, although the former reaction was predominant. Using *Pleurotus ostreatus*, similar results were obtained by Wojtas-Wasilewska et al. (1980).

Mutants of white-rot fungi have also been used to study the importance of phenoloxidases in lignin degradation. Using *Phanerochaete chrysosporium* P 127-1 spores and UV-light, Ander and Eriksson (1976) produced a phenoloxidase-less mutant, Phe 3, which did not degrade lignin or other major wood components in birch, pine, or spruce wood blocks (Table 4.5). However, a

Table 4.5. Weight losses[a] and losses of wood components[b] caused by *Phanerochaete chrysosporium* (P127-1) wild-type WT, "phenoloxidase-less" mutant Phe 3 and the revertant Rev 9 after 46 days (Ander and Eriksson 1976)

		Birch	Pine	Spruce
Weight loss (%)	WT	25.9	7.1	6.9
	Phe 3	0.6	0.9	1.2
	Rev 9	11.0	10.0	6.8
Lignin loss (%)	WT	25	10	10
	Phe 3	−6	−3	4
	Rev 9	33	17	7
Glucan loss[c] (%)	WT	10	5	5
	Phe 3	1	3	−2
	Rev 9	3	13	10
Xylan loss (%)	WT	25	NT	NT
	Phe 3	−1	NT	NT
	Rev 9	29	NT	NT
Mannan loss (%)	WT	NT	23	17
	Phe 3	NT	2	4
	Rev 9	NT	17	29

NT = not tested.
[a] Based on original dry weight.
[b] Based on original amount of component.
[c] Represents cellulose loss.

Table 4.6. Endo-1,4-β-glucanase production by *Phanerochaete chrysosporium* wild-type WT, Phe 3, and Rev 9 in cellulose shake flasks with or without 0.25% kraft lignin and 10^{-3} M phenols[a]

Addition	Endo-1,4-β-glucanase production (U ml^{-1})		
	WT	Phe 3	Rev 9
Ethanol and/or boiled laccase	0.47	0.92	0.56
Vanillic acid	0.22	0.06	0.64
Vanillic acid + 2 μg laccase	NT	0.14	NT
Vanillic acid + 8 μg laccase	NT	0.44	NT
p-Hydroxybenzoic acid	0.18	0.05	0.52
Ferulic acid	0.03	0.02	0.08
Kraft lignin	1.08	0.02	1.30
Kraft lignin + 2 μg laccase	NT	0.03	NT
Kraft lignin + 8 μg laccase	NT	0.12	NT
Kraft lignin + 16 μg laccase	NT	0.20	NT

NT = not tested.
[a] The phenols were added 1 day after inoculation, whereas kraft lignin was present during sterilization. Laccase was added 6 h after addition of vanillic acid, the addition of laccase to the kraft lignin flasks being carried out at the same time (Ander and Eriksson 1976).

phenoloxidase-positive revertant, Rev 9, had recovered the ability to degrade all wood components. Growth in shake flasks on glucose or cellulose by Phe 3 was normal. Cellulase production by the mutant was inhibited by vanillic acid and kraft lignin (Table 4.6), but addition of pure laccase to the cultures with such media resulted in cellulase production in relation to the amount of laccase added (Ander and Eriksson 1976). The results indicate that phenoloxidases have a regulatory role on the formation of at least the polysaccharide-degrading enzymes. Other results seem to indicate that treatment of kraft lignin with laccase decreases the chlorine number (Kyrklund and Strandell 1967; cf. Hiroi and Eriksson 1976) of the kraft lignin. The polymerizing effect of laccase on kraft lignin may partly explain why some degradation of kraft lignin was obtained with Phe 3 after addition of laccase (Ander and Eriksson 1976).

With the development of low nitrogen media for lignin degradation (Kirk et al. 1978a) and a sensitive technique to use ^{14}C-labeled DHP, lignin degradation by the phenoloxidase-less mutant Phe 3 was again tested. Cultivation was done on a 10 mM dimethylsuccinate medium containing 1% glucose and low nitrogen according to Fenn and Kirk (1979) with O_2-flushing. The unexpected result was that Phe 3 released 53% and the wild-type 63% $^{14}CO_2$ from ^{14}C-ring-labeled DHP in 21 days (Ander, unpublished results). Also on a cellulose medium, strong release of $^{14}CO_2$ by Phe 3 was obtained (57% vs 46% for the wild type). Tests for phenoloxidase activity using syringaldazine addition to the intact cultures, showed that not only wild-type *Phanerochaete chrysosporium* P 127-1, but also the mutant Phe 3 had phenoloxidase activity in the low nitrogen medium. Obviously the old-type high nitrogen media, which did not allow secondary metabolism, strongly suppressed phenoloxidase production in Phe 3. All syringaldazine tests on wood blocks, however, were negative for Phe 3, since poor growth and no lignin degradation occurred in the wood blocks.

An interesting question arises about the amount of laccase, Mn-peroxidase, and/or lignin peroxidase production in Phe 3. To date, this has not been investigated in detail. In wild-type *Phanerochaete chrysosporium* P 127-1, mainly manganese peroxidase and lignin peroxidase have been detected (Ander and Eriksson 1976, 1987). Ander et al. (1980b) and Eriksson et al. (1983) also found laccase activity, although the latter authors used 15 mM H_2O_2 in their peroxidase assays which probably inhibits both Mn-peroxidase and lignin peroxidase and overestimates the relative laccase activity. The lignin peroxidase is now assayed with 0.54 or 0.15 mM H_2O_2 (Tien and Kirk 1984, Faison and Kirk 1985), while 0.03 (Ander and Eriksson 1987) or maximally 0.75 mM H_2O_2 (Glenn and Gold 1985) is optimal for Mn-peroxidase, at least that from *P. chrysosporium*.

Kelley et al. (1986) found that glucose oxidase negative mutants of *Phanerochaete chrysosporium* did not degrade DHP, nor did they produce lignin peroxidase or Mn-peroxidase. It is not known whether Phe 3 is glucose oxidase negative.

In 1982 Gold and co-workers reported experiments with a pleiotropic mutant of *Phanerochaete chrysosporium* lacking phenoloxidase activity (Gold et al. 1982b). The results indicated that a variety of secondary metabolic functions (Sect. 4.3.1) were affected in this mutant. These functions included phenoloxidase production, fruit body formation, lignin degradation, and veratryl alcohol production. The mutant could not evolve $^{14}CO_2$ from ring-, methoxy- or side-chain labeled DHPs. (Lignin peroxidase and Mn-peroxidase had not been discovered or characterized at the time this work was done.) Also in the mutant Phe 3, production of veratryl alcohol as well as sporulation were strongly reduced (Ander et al. 1980b).

Liwicki et al. (1985), in the search for phenoloxidase-negative mutants, used γ irradiation of asexual spores of *Phanerochaete chrysosporium* and spread the spores on o-anisidine agar plates. Twenty-nine mutants were isolated and tested for release of $^{14}CO_2$ from ^{14}C-labeled lignin in wheat straw. Although all of the mutants lacked phenoloxidase according to the o-anisidine plate test, some of them could degrade lignin, whereas others could not. All mutants degrading lignin also produced veratryl alcohol. The results are somewhat obscured by the authors' inability to detect phenoloxidase activity in *P. chrysosporium* with the p-phenylenediamine test (as described by Wood and Goudenough 1977). It would have been useful to look also for possible phenoloxidase activity in intact liquid cultures of the mutants. Syringaldazine added to whole cultures may be used for qualitative measurement of peroxidase activity in *P. chrysosporium* and may in a better way reveal peroxidase activity in seemingly phenoloxidase-less mutants. Faison and Kirk (1983), however, added o-dianisidine to ligninolytic cultures of *P. chrysosporium* and detected no endogenous peroxidase activity in their cultures. The reason for these results may be instability caused by reduction of the quinone formed from

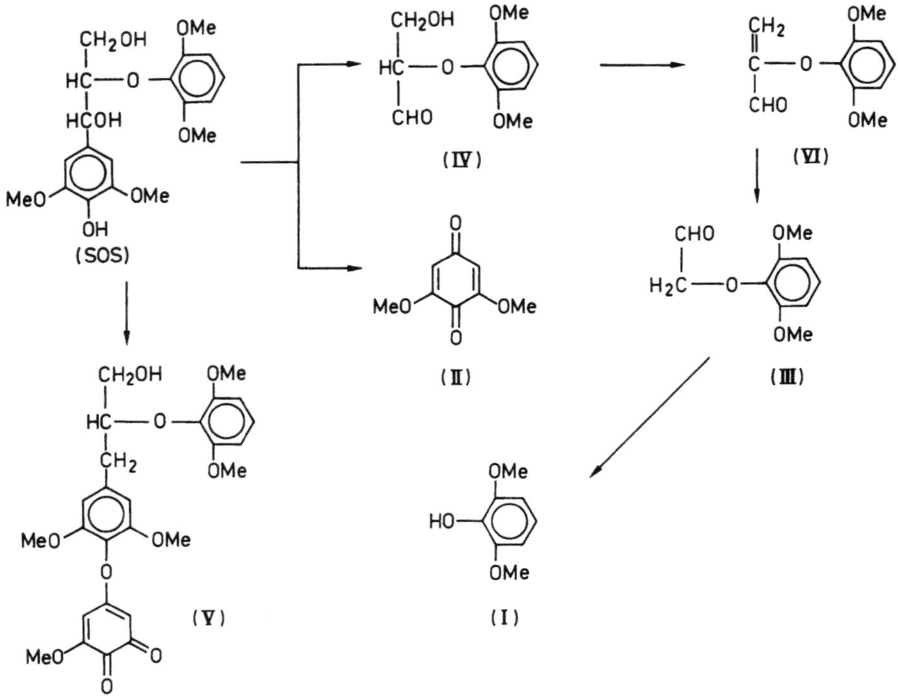

Fig. 4.21. Products obtained at incubation of syringylglycerol-β-syringyl ether (SOS) with laccase IIIc purified from *Coriolus versicolor* (Morohoshi and Haraguchi 1987b). See text for further details

o-dianisidine. Ander and Eriksson (1987) found that in crude samples of *P. chrysosporium* ME-446 and SC26, the purple color from syringaldazine oxidation could vary due to the presence of quinone reducing enzymes. Peroxidase and/or laccase activity may even go undetected. Therefore, a method which measures $^{14}CO_2$ evolved from the ^{14}C-labeled COOH group of vanillic acid was developed (Ander and Eriksson 1987). In this method, released $^{14}CO_2$ is irreversibly trapped in NaOH, and there is no dependence of a stable quinone color. By choosing an acid pH (4–5) and excluding NAD(P)H, interference from vanillate hydroxylase (Buswell et al. 1981) can be avoided.

Laccase has been separated into several isoenzymes, and certain fractions of laccase have been found that can cleave C_α-C_β bonds in the side chains of syringylglycerol-β-guaiacyl ether (Wariishi et al. 1985) and β-1 lignin substructures (Kawai et al. 1986b). Morohoshi et al. (1983) isolated laccase I, II, and III from culture solutions of *Coriolus versicolor* using chromatography on Sephadex G50 and DEAE Sephadex A-25. Laccase I and II both polymerized and depolymerized different isolated lignins such as Brauns native lignin, MWL, and a lignin carbohydrate complex (Morohoshi and Haraguchi 1987a, Morohoshi et al. 1987a, c). Laccase III depolymerized MWL and a water-solu-

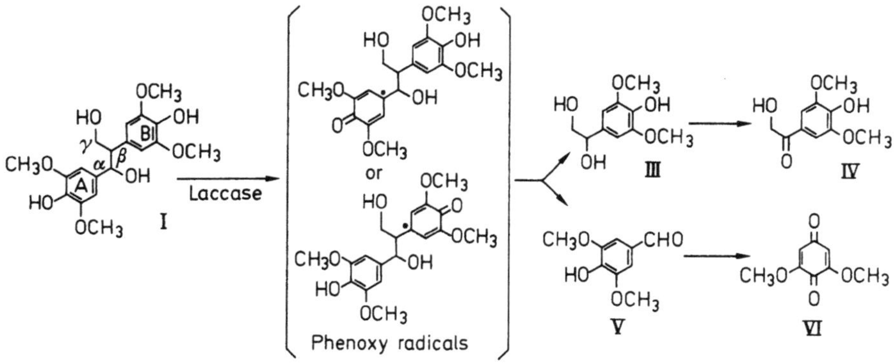

Fig. 4.22. Proposed pathway for degradation of a phenolic β-1 syringyl lignin substructure (I) by laccase III or IIIc from *Coriolus versicolor* (Kawai et al. 1987b). See text for further details

ble lignin fraction from residual wood and was considered to be one of the ligninolytic enzymes by Morohoshi et al. (1987a). Laccase I mostly polymerized lignin while laccase II was intermediary in its activity.

Laccase III was further purified by isoelectric focusing, giving the laccase fraction IIIc (Morohoshi and Haraguchi 1987b), which was used for incubation with syringylglycerol-β-syringyl ether (SOS) at pH 4. After 30 s, alkyl-phenyl cleavage occurred, and 2,6-dimethoxy-p-benzoquinone(1,4) (II) and 2-(2,6-dimethoxyphenoxy)-3-hydroxy propanal (IV) were formed as shown in Fig. 4.21. After 8 h, compounds (VI), (III), and 1,3-dimethylpyrogallol (I) were identified, and compound (V) was detected as a coupling product between the phenoxy radical of SOS and 1,3-dimethylpyrogallol (I).

Using laccase fractions III or IIIc, Kawai et al. (1987b) studied degradation of the phenolic β-1 lignin substructure (I), as shown in Fig. 4.22. Via the phenoxy radicals, compound I was cleaved between the C_α- and the C_β-carbons in the side chain, and syringylglycol (III) plus syringaldehyde (V) was formed. Further laccase-catalyzed degradation gave α-hydroxyacetosyringone (IV) and 2,6-dimethoxy-p-benzoquinone (VI). A corresponding nonphenolic β-1 model (II) with an ethoxy group was not degraded by laccase (Kawai et al. 1987b). In whole cultures, however, nonphenolic dimers are degraded by *Coriolus versicolor* (Kamaya and Higuchi 1984c, Kawai et al. 1985a, b). C_α-C_β cleavage of compound (I) in Fig. 4.22 was probably similarly catalyzed by horseradish peroxidase (Kamaya and Higuchi 1984b).

Syringylglycerol-β-guaiacylether (Fig. 4.23, I) was cleaved by laccase III to give 48% C_α-C_β cleavage and 9% alkyl-phenyl cleavage (Wariishi et al. 1987). Final monomer products were guaiacol (VIII), syringaldehyde (VII), a paraquinone (IX), and an orthoquinone (XIV). Kawai et al. (1988) recently investigated the reactions of laccase with a β-1 lignin model compound with a similar structure as the one in Fig. 4.22 but containing only one phenolic hydroxyl group. It was found that laccase catalyzed C_α-C_β cleavage, alkyl-aryl cleavage, and C_α oxidation.

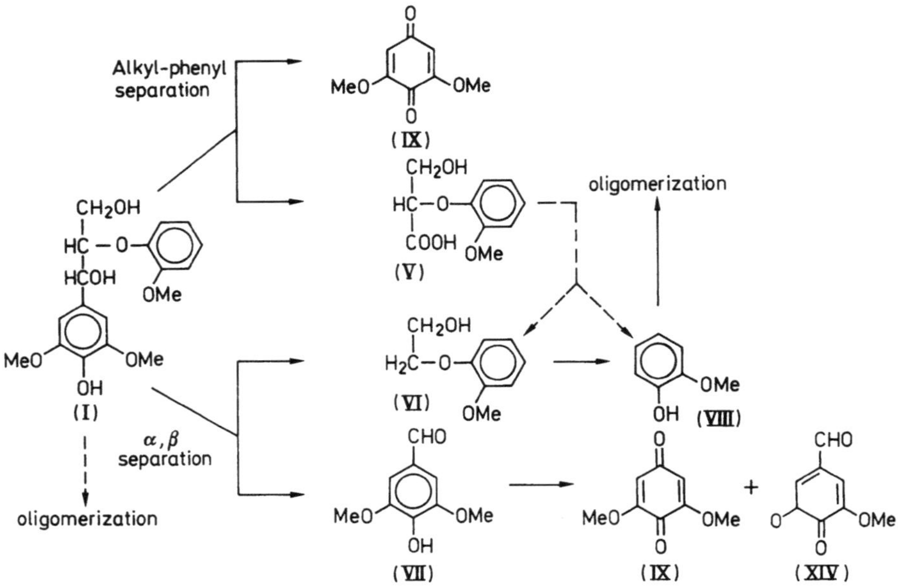

Fig. 4.23. Alkyl-phenyl cleavage and C_α-C_β cleavage of syringylglycerol-β-guaiacylether (I) by laccase III (Wariishi et al. 1987). See text for further details

It can be concluded from all the above results that laccase is of importance in degradation of terminal phenolic units in lignin. Nonphenolic dimers or similar structures in lignin, however, are not degraded by laccase III but may be degraded in cooperation with lignin peroxidase so that new phenolic hydroxyl groups are formed.

In order to clone the laccase III gene, laccase III was purified on a 30-l scale by Morohoshi et al. (1987b). The molecular weight of the enzyme was found to be 66000; it contained two Cu atoms per molecule (Table 4.4). It was mentioned that laccase I and III corresponded to laccase B and A, respectively, purified by Fåhraeus and Reinhammar (1967). From the mycelium of *Coriolus versicolor*, mRNA was prepared and translated into five proteins, one of which had a molecular weight of 60000, probably corresponding to laccase III (MW 66000) containing 10%–15% carbohydrate (Morohoshi et al. 1987b).

Haraguchi et al. (1983) have studied degradation of MWL with *Xylobolus frustulatum*, a white-rot fungus with low phenol oxidase activity. It was found that an intracellular enzyme preparation had a polymerizing activity similar to that of laccase against syringic and vanillic acids at pH 5. However, at pH 7 the vanillic acid was strongly degraded as evidenced by the UV spectrum. No degradation of MWL occurred at pH 4, but at pH 6, in the presence of NADH, an appreciable depolymerization to low molecular weight products was obtained with the crude enzyme, as evidenced by gel filtration. Also at pH 7 more degradation products were formed with than without NADH (Haraguchi et al. 1983). This seems to be a new type of enzyme activity that

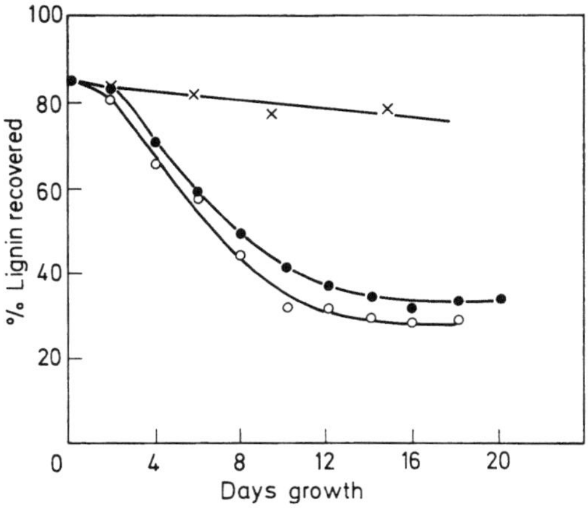

Fig. 4.24. Degradation of lignin in lignin-impregnated glass fiber disks by *Coriolus versicolor* with or without laccase antibodies. (×) culture plus catalase; (●) control culture; (○) culture with laccase antibodies (Evans 1985)

is clearly different from the activities of laccase- and peroxidase-type enzymes generally discussed in connection with lignin degradation.

A useful technique to study the importance of laccase in lignin degradation was reported by Evans (1985). She used rabbit antibodies raised against laccase A, which was obtained as a homogeneous protein. The antibodies reacted with both laccase A and B. Using lignin-impregnated glass fiber disks (Sect. 4.3.3, Ander et al. 1981) it was found that *Coriolus versicolor* could degrade MWL even when the laccase was inhibited by antibodies as shown in Fig. 4.24. After 9 days, control plates with active laccase developed a red-brown ring which did not appear in cultures with added antibodies. In vitro, the antibodies reduced the activity of laccase A and B to 5%–8% of the original activity. Addition of catalase to the lignin glass fiber disks resulted in a strongly reduced lignin degradation rate (Fig. 4.24). Now two heme proteins that contain protoporphyrin IX and are dependent upon hydrogen peroxide for activity (Table 4.7) have been isolated from *C. versicolor* (Evans et al. 1984, Dodson et al. 1987). The existence of these enzymes may explain the influence of catalase; the results suggest that peroxidase(s) are more important than laccase for lignin degradation by *C. versicolor.*

Evans (1985) also investigated the degradation of MWL in 50% dioxane-water with the same laccase as used in the antibody experiment with or without H_2O_2. A degradation to smaller M_r-fragments was found in the presence of laccase plus H_2O_2, whereas without H_2O_2 or with antibodies present the lignin appeared unchanged. It was suggested that under conditions of more hydrogen peroxide than oxygen, laccase may catalyze the production of hydroperoxide radicals, which may effect cleavage of C_α-C_β bonds. Under high oxygen conditions, hydroxyl radicals may be produced, which cause the oxidation of lignin to quinones and/or phenoxy radicals that polymerize (Evans 1985). It thus appears possible to regulate polymerization or depolymerization of lignin both in vitro and in vivo by manipulating the oxygen and H_2O_2 concentrations. Although the experimental conditions for the MWL experiment were

Table 4.7. Properties of peroxidases isolated from different white-rot fungi (the names ligninase and lignin peroxidase are used interchangeably in the literature)

Enzymes	Molecular weight	Carbohydrate (%)	Cofactors	Important substrates	Important reactions	References (fungal origin)
Lignin peroxidase (Ligninase)	42000	~15	H_2O_2 heme	Veratryl alcohol	C_α-C_β cleavage in non-phenolic substrates	Tien and Kirk (1984) (*P. chrysosporium*)
Lignin peroxidase Diarylpropane oxygenase	41000		H_2O_2 heme	Veratryl alcohol	C_α-C_β cleavage in non-phenolic substrate	Gold et al. (1984) (*P. chrysosporium*)
Lignin peroxidase (Ligninase I)	42–43000	21	H_2O_2 heme	Veratryl alcohol	C_α-C_β cleavage in non-phenolic substrates	Paszczynski et al. (1986) (*P. chrysosporium*)
Lignin peroxidase (Ligninase)	39–42000		H_2O_2 heme	Veratryl alcohol	C_α-C_β cleavage in non-phenolic substrates	Leisola et al. (1985a, c) (*P. chrysosporium*)
Lignin peroxidase	43–45000		H_2O_2 heme	Veratryl alcohol		Jönsson et al. (1987) (*C. versicolor*)
Mn-peroxidase	46000		Mn^{2+}, H_2O_2 heme	Phenols[a]	$Mn^{2+} \rightarrow Mn^{3+}$	Glenn and Gold (1985) (*P. chrysosporium*)
Peroxidase-M2	45–47000	17	Mn^{2+}, H_2O_2 heme	Phenols[a]	$Mn^{2+} \rightarrow Mn^{3+}$	Paszczynski et al. (1986) (*P. chrysosporium*)
Peroxidase	46000		Mn^{2+}, H_2O_2 heme	Phenols[a]		Leisola et al. (1985c) (*P. chrysosporium*)
Mn-peroxidase	49000		Mn^{2+}, H_2O_2 heme	Phenol red		Johansson and Nyman (1987) (*C. versicolor*)
Extracellular peroxidase	50000	Glycoprotein	H_2O_2 heme	Veratryl alcohol	C_α-C_β cleavage in a nonphenolic substrate	Dodson et al. (1987) (*C. versicolor*)
Extracellular heme-protein (Peroxidase-type)	53700	Glycoprotein	H_2O_2? heme		Methional→ethylene	Evans et al. (1984) (*C. versicolor*)
Lignin peroxidase (Ligninase, LI-LIII)	40–45000	Glycoprotein	H_2O_2 heme	Veratryl alcohol		Niku-Paavola (1987) (*P. radiata*)

[a] For substrates see Table 4.8.

not described by Evans, the conversion between laccase and peroxidase activity deserves further attention.

Before the purification and characterization of the manganese-dependent peroxidase (Mn-peroxidase) from *Phanerochaete chrysosporium* (Kuwahara et al. 1984), little work had been published on fungal peroxidases and their influence on lignin degradation. However, some work had been done in Poland with lignosulfonates as substrates. Thus, Lobarzewski et al. (1982) and Lobarzewski (1984, 1987) studied the molecular weight changes of lignosulfonate after treatment with immobilized peroxidase from *Coriolus versicolor*, or with immobilized fungal mycelium. Intracellular peroxidase was separated from laccase using a technique based on the affinity of the peroxidase for vanillin attached to silanized glass beads (Lobarzewski 1981). In *C. versicolor* media containing low concentrations of carbon, nitrogen, or sulfate, Paszczynski and Lobarzewski (1984) found induction of peroxidase activity.

Manganese-dependent peroxidase was first purified and characterized by Gold and co-workers (Kuwahara et al. 1984, Glenn and Gold 1985). Together with lignin peroxidase, this enzyme was found in the culture medium of *P. chrysosporium* and was separated from lignin peroxidase by chromatography on blue agarose (Kuwahara et al. 1984). The activity of Mn-peroxidase can be measured with phenol red in the presence of lactate, 0.1 mM $MnSO_4$, and 0.05 mM H_2O_2 at pH 4.5. The dependence on H_2O_2 and Mn^{2+} for enzyme activity is shown in Fig. 4.25 (Glenn and Gold 1985). The Mn-peroxidase could oxidize NADH also in the absence of H_2O_2, which instead was produced as a result of the enzyme reaction (Glenn and Gold 1985, Paszczynski et al. 1985). This may be one of the sources of H_2O_2 required for lignin degradation by the fungus (see Sect. 4.5.5). A Mn-peroxidase recently purified from *Coriolus versicolor* (Johansson and Nyman 1987) has maximal activity in the presence of 0.020 mM H_2O_2 and 0.30 mM $MnSO_4$ and has a MW of 49000.

Mn-peroxidase and lignin peroxidase from *Phanerochaete chrysosporium* BKM-1767 also have been separated, purified, and characterized by Paszczynski et al. (1985, 1986). Different properties of these enzymes obtained from different sources are shown in Table 4.7.

In Table 4.8, oxidation of different substrates by Mn-peroxidase, ligninase, and horseradish peroxidase is compared (Paszczynski et al. 1986). It should be observed that ligninase does not oxidize 2,6-dimethoxyphenol, vanillylacetone, or syringaldazine in contrast to Mn-peroxidase. It was also found that Mn-peroxidase plus Mn^{2+}, without hydrogen peroxide but in the presence of NADH oxidizes NADPH, GSH, dithiothreitol, and dihydroxymaleic acid, thus forming hydrogen peroxide at the expense of oxygen. Both Glenn et al. (1986) and Paszczynski et al. (1986) suggested that Mn-peroxidase oxidizes Mn^{2+} to Mn^{3+}, although Glenn et al. (1986) found that an Mn^{3+}-lactate complex was responsible for all oxidations of the substrates. This complex can diffuse through a semipermeable membrane and oxidize, for example, the polymeric dye Poly B (and possibly lignin in wood). Since Paszczynski et al. (1986) used a tartrate buffer, the Mn-peroxidase in their system may form an Mn^{3+}-tartrate complex, which acts similarly to the lactate complex.

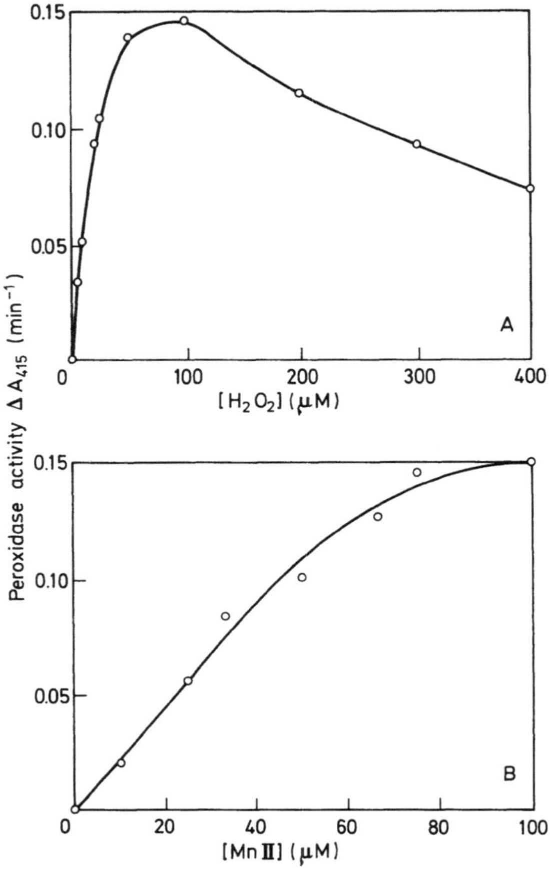

Fig. 4.25. Dependence of Mn-peroxidase on H_2O_2 (**A**) and Mn^{2+} (**B**) for enzyme activity (Glenn and Gold 1985)

Huynh et al. (1986b) and Bono et al. (1987) have shown that both Mn^{3+}-acetate as well as Mn-peroxidase plus H_2O_2 and $MnSO_4$ oxidize many different phenols, such as guaiacol, vanillyl alcohol, and vanillyl acetone. Later, Forrester et al. (1988) and Hammel et al. (1989) showed that suitably chelated Mn^{3+} could also oxidize nonphenolic substrates like veratryl alcohol.

In Fig. 4.26, the catalytic cycle and the five oxidation states of the iron-heme enzyme Mn-peroxidase are shown (Gold et al. 1987, Wariishi et al. 1988). The catalytic cycle involves oxidation of the native Fe^{3+} enzyme by H_2O_2 (ROOH) to form Compound I of the enzyme. This is reduced back to the native enzyme in two separate, single electron steps via Compound II. Both Mn^{2+} and phenols can reduce Compound I to Compound II, but only Mn^{2+} readily reduces Compound II to the native enzyme. Thus, Mn^{2+} is essential for the catalytic activity of the Mn-peroxidase (Wariishi et al. 1988). Further spectral characterization of Mn-peroxidase and comparison with other peroxidases were done by Mino et al. (1988). The results indicate that the heme envi-

Table 4.8. Different substrates oxidized by Mn-peroxidase, ligninase, and horseradish peroxidase (HRP)[a]

Substrate	Mn-peroxidase	Ligninase	HRP
1,2-Dihydroxybenzene (catechol)	+	+ + +	+ + +
1-Methoxyphenol (guaiacol)	+ + +	+	+ + +
2,6-Dimethoxyphenol	+ + +	−	+ + +
3-tert-Butyl-4-methoxyphenol	+ +	+ +	nd
3,5-Di-tert-butyl-4-methoxyphenol	+ +	+	±
4-Hydroxy-3-methoxybenzaldehyde (vanillin)	+	+	nd
4-Hydroxy-3,5-dimethoxybenzaldehyde (syringyl aldehyde)	+ + +	±	+ + +
4-Hydroxy-3-methoxybenzyl alcohol (vanillyl alcohol)	±	±	nd
4-Hydroxy-3,5-dimethoxybenzyl alcohol (syringyl alcohol)	+ + +	±	+
4-Hydroxy-3-methoxybenzoic acid (vanillic acid)	+	+	nd
4-Hydroxy-3,5-dimethoxybenzoic acid (syringic acid)	+ + +	+ +	nd
1-[4-Hydroxy-3-methoxyphenyl]ethanone (acetovanillone)	+	nd	nd
1-[4-Hydroxy-3,5-dimethoxyphenyl]ethanone (acetosyringone)	+ +	nd	nd
1-[4-Hydroxy-3-methoxyphenyl] ethanol (apocynol)	±	−	nd
1-[4-Hydroxy-3,5-dimethoxyphenyl] ethanol	+	nd	nd
4-Hydroxy-3-methoxycinnamic alcohol (coniferyl alcohol)	+ + +	+ +	nd
2-Methoxy-4-propenylphenol (isoeugenol)	+ +	nd	nd
4-[4-Hydroxy-3-methoxyphenol]-3-buten-2-one (vanillylacetone)	+ + +	−	+ + + +
4-[4-Hydroxy-3,5-dimethoxyphenyl]-3-buten-2-one (syringylacetone)	+ + +	−	+ + +
3-(4-Hydroxy-3-methoxyphenyl)-2-propenoic acid (ferulic acid)	+ +	nd	nd
3-(4-Hydroxy-3,5-dimethoxyphenyl)-2-propenoic acid (sinapic acid)	+ + +	+	+ +
4,4′-Dihydroxybiphenyl	+ + +	nd	+ + +
Vanillinazine	+ +	−	+ +
Syringaldazine	+ + +	−	+ + + +
Glyoxal bis(2-hydroxyanil)	+ +	nd	nd
Guaiacylglycerol-β-guaiacyl ether	+	nd	nd
Quercetin	+ + +	+ +	nd
Rutin	+	nd	nd
Chlorogenic acid	+ + +	nd	nd
Disyringylmethane	+ + +	−	+ + + +

[a] Symbols for rate of reactions: + + + + (very fast), + + + (fast), + + (moderate), + (slow), ± (very slow), − (no change), nd = not determined (Paszczynski et al. 1986).

ronment of Mn-peroxidase resembles those of horseradish peroxidase, cytochrome c peroxidase, and lignin peroxidase.

The role of Mn-peroxidase in lignin degradation is not clear, although one role may be to form H_2O_2. The Mn-peroxidase shows no affinity for nonphenolic compounds, which are readily attacked by ligninase. The Mn-peroxidase oxidizes syringyl compounds better than guaiacyl compounds (Paszczynski et al. 1986). The finding of Blanchette (1984b, Chap. 1.1) that manganese often accumulates as manganese dioxide in white-rotted wood suggests that Mn-peroxidase is of great importance in lignin degradation. Thus Kern (1989) found that MnO_2 stabilizes lignin peroxidase.

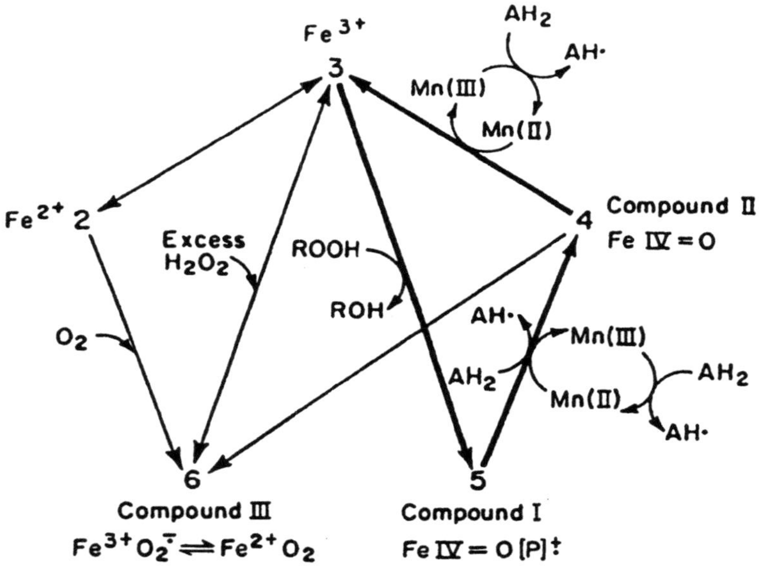

Fig. 4.26. The catalytic cycle and the five oxidation states of Mn-peroxidase. $ROOH = H_2O_2$; AH_2 = phenolic substrate (Gold et al. 1987)

An extracellular peroxidase from the white-rot fungus *Phellinus weirii* (*Inonotus weirii*) has recently been investigated by Mustranta (1987). She found that this fungus grew poorly on a low nitrogen medium (Kirk et al. 1978a) and did not produce the extracellular peroxidase on such a medium. Instead, the enzyme was produced during active growth on glucose or xylose in a high nitrogen medium. The molecular weight was 42000 and the optimal pH was 3. However, the enzyme was not stimulated by Mn^{2+}, it did not oxidize veratryl alcohol, and lignin or veratryl alcohol did not induce enzyme production (Mustranta 1987). Regarding these properties it may be concluded that the *Phellinus weirii* peroxidase is different from both Mn-peroxidase and ligninase.

4.5.3 Ligninase/Lignin Peroxidase

A lignin-degrading enzyme, named ligninase or lignin peroxidase, was discovered in *Phanerochaete chrysosporium* and the first publications on this new enzyme were later published at the same date, namely, August 12, 1983, by two independent research groups (Glenn et al. 1983, Tien and Kirk 1983). Information about the enzyme also appeared at the Lignin Biodegradation Meeting in Japan, May 31–June 2, 1983 (Gold et al. 1983b, Kirk and Tien 1983). Ligninase may be used as a trivial name, although the name lignin peroxidase reflects that the enzyme is a peroxidase.

The extracellular lignin peroxidase has been purified and characterized (Gold et al. 1984, Tien and Kirk 1984). As seen in Table 4.7, it has a molecular weight of 41000−42000 and an isoelectric point of 3.5. The enzyme is a glycoprotein that contains about 15% carbohydrate (or 11% according to Paszczynski et al. 1986) and a heme prosthetic group, which is iron protoporphyrin IX, giving the native enzyme a Soret band at 407 nm. A lignin peroxidase from *P. chrysosporium* has now been crystallized (Troller et al. 1988). The color of the crystals is light yellow to dark brown due to the heme group. A useful assay for the enzyme is the oxidation of 0.4 mM veratryl alcohol to veratraldehyde (measured at 310 nm) in sodium tartrate buffer at pH 3, in the presence of 0.54 mM H_2O_2 (Tien and Kirk 1984) or 0.15 mM H_2O_2 (Faison and Kirk 1985). Later, 2 mM veratryl alcohol and 0.4 mM H_2O_2 were used (Kirk et al. 1986a).

The ligninase can partly depolymerize lignin, as shown by Glenn et al. (1983) and Tien and Kirk (1983). The latter authors methylated free phenolic hydroxyl groups of spruce and birch lignins with $^{14}CH_3I$ and used the largest molecules (>1500) as substrates for ligninase. From the ^{14}C-labeled spruce and birch lignins small amounts of labeled veratraldehyde were obtained. The birch lignin also gave syringaldehyde methyl ether. The low molecular weight compounds obtained from the ^{14}C-methylated lignins after ligninase treatment were separated from the high molecular weight material using gel filtration, as shown for birch lignin in Fig. 4.27 (Kirk and Tien 1983, Tien and Kirk 1983, Kirk et al. 1984). The depolymerization products accounted for about

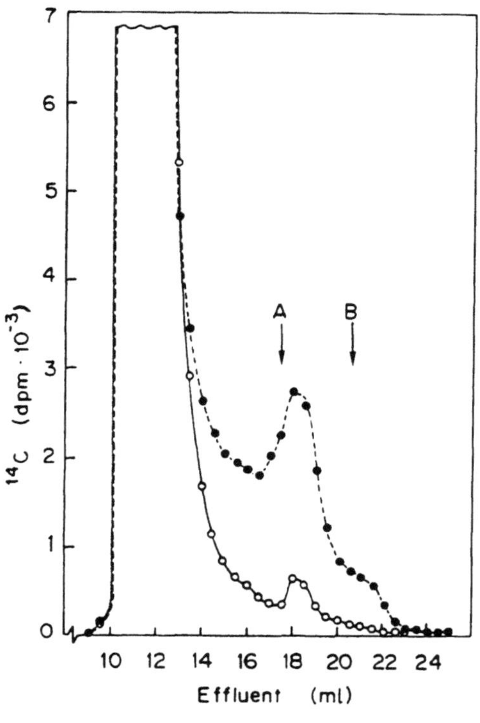

Fig. 4.27. Gel filtration of ^{14}C-methylated birch MWL after treatment with crude extracellular ligninase + H_2O_2 for 1 h at pH 3. Conditions: Sephadex LH-20 in N,N-dimethylformamide. Calibrants *A*, MW=348; *B*, MW=166. (●) Active enzyme; (○) boiled enzyme (Kirk et al. 1984)

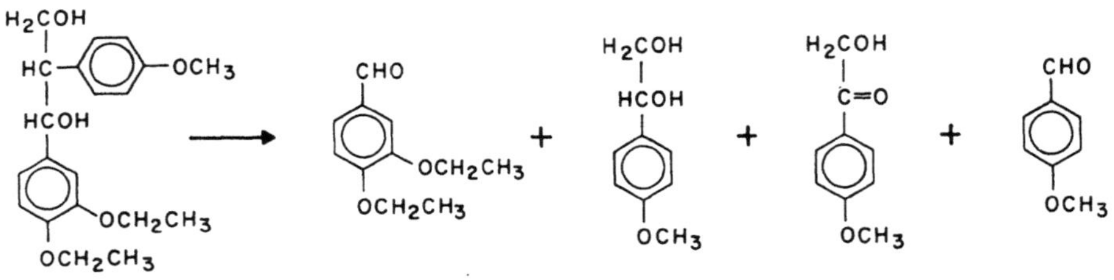

I (42.1%) → VI (32.7%) + VIII (16.7%) + IX (1.6%) + XI (5.1%)

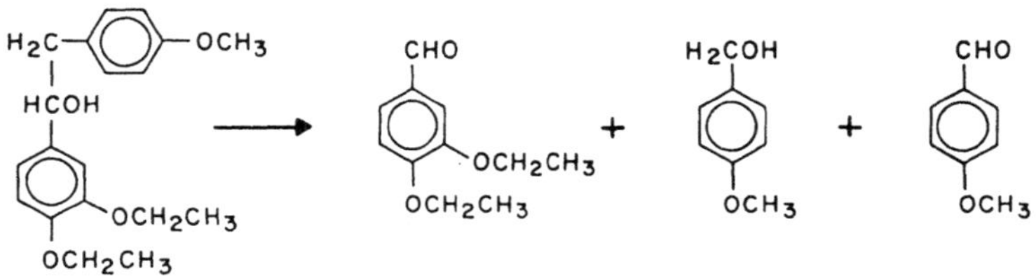

II (50.5%) → VI (21.3%) + XII (7.3%) + XI (2.8%)

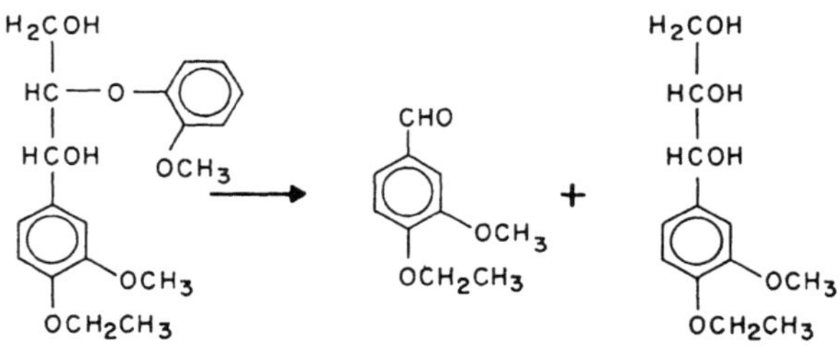

V (71.0%) → VII (15.0%) + X (6.1%)

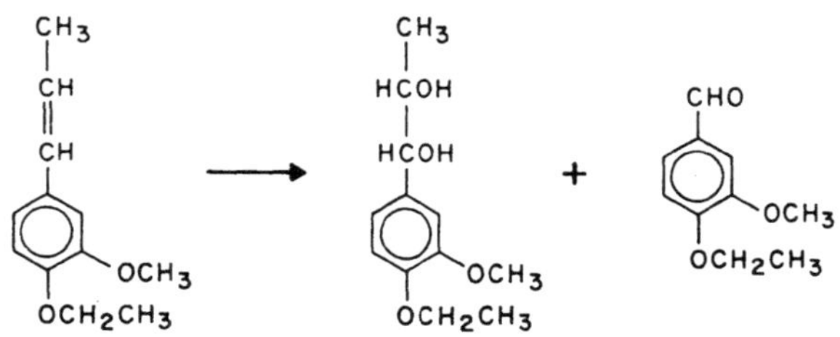

III (43.5%) → IV (6.8%) + VII (6.3%)

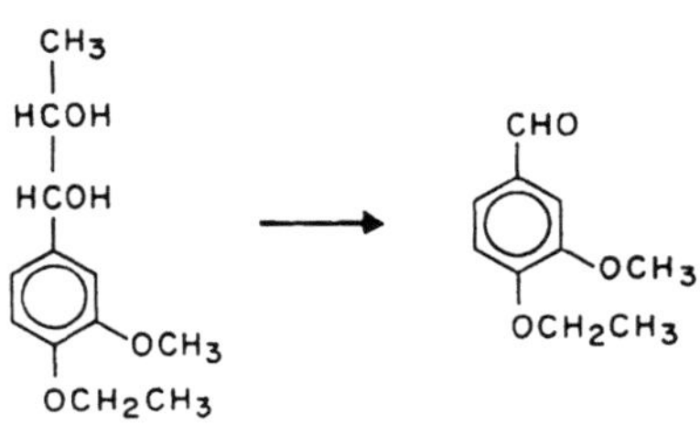

IV (63.5%) → VII (16.8%)

22% and 6% of the original ^{14}C in the spruce and birch lignins, respectively. One effect of using methylated lignin may be that the lower content of free hydroxylic groups compared to the original lignins may decrease the polymerizing effect of ligninase that is normally apparent (Haemmerli et al. 1986a, Kirk 1986). Possibilities to prevent polymerization by ligninases are discussed in Section 4.5.6. The intrinsic phenoloxidase activity of ligninase catalyzed a rapid oxidation of guaiacol to colored products, suggesting oxidative coupling and polymerization (Tien and Kirk 1984). This probably helps to explain why phenoloxidases have been considered necessary for lignin degradation for so many years (Ander and Eriksson 1978).

In common with other peroxidases, the lignin peroxidase is quite nonspecific. It releases ethylene from 2-keto-4-thiomethyl butyric acid (KTBA) and also oxidizes a variety of lignin model compounds, including those shown in Fig. 4.28 (Glenn et al. 1983, Gold et al. 1984). The yield of products obtained from the diarylpropane I, the diarylethane II, and the diol IV indicate that C_α-C_β cleavage is the dominant reaction. The production of 6% phenylglycerol X from the β-ether dimer V shows that β-aryl ether cleavage also occurs (Gold et al. 1984).

After the discovery of the ligninase, many laboratories have directed their research on lignin toward the nature and function of this enzyme. Thus, ligninases have been found in various fungi, including *Coriolus* (*Trametes*) *versicolor* (Dodson et al. 1987, Jönsson et al. 1987, Waldner 1987, Waldner et al. 1988), *Phlebia radiata* (Hatakka et al. 1987, Niku-Paavola 1987, Kantelinen et al. 1988), *Phlebia* (*Merulius*) *tremellosus* (Biswas-Hawkes et al. 1987), *Pleurotus ostreatus* (Waldner 1987), *Panus tigrinus* (Golovleva et al. 1986), and possibly *Pleurotus sajor-caju* (Fukuzumi 1987). The properties of some of these enzymes are given in Table 4.7.

The research groups already mentioned and others have continued with more detailed studies of lignin peroxidase, and improved methods for separation of lignin peroxidase isoenzymes have been reported. Renganathan et al. (1985) compared the yields of lignin peroxidase from *Phanerochaete chrysosporium* cultivated in media buffered with acetate, succinate, dimethylsuccinate, tartrate, or citrate and found that the highest yield of lignin peroxidase was obtained in the acetate cultures. After separation on DEAE-Sepharose and Sephadex G-100, these workers found three molecular forms of lignin peroxidase. Using a Pharmacia FPLC Mono Q anion exchange column, Kirk et al. (1986a) separated the lignin peroxidases in a 20-fold concentrated extracellular fluid of *P. chrysosporium* BKM-F-1767 into ten different peaks with 409 nm absorbance; six had veratryl alcohol-oxidizing activity. The predominant peak was called H8 (Hemeprotein 8); this enzyme fraction has been used in much of the further research (Farrell 1987a, b, Kirk 1987).

Using analytical isoelectric focusing, Leisola et al. (1987) separated 21 extracellular proteins of *P. chrysosporium*. Fifteen of these oxidized veratryl al-

Fig. 4.28. Products identified, and proposed pathways for the oxidation of lignin model compounds by purified lignin peroxidase from *Phanerochaete chrysosporium* (Gold et al. 1984)

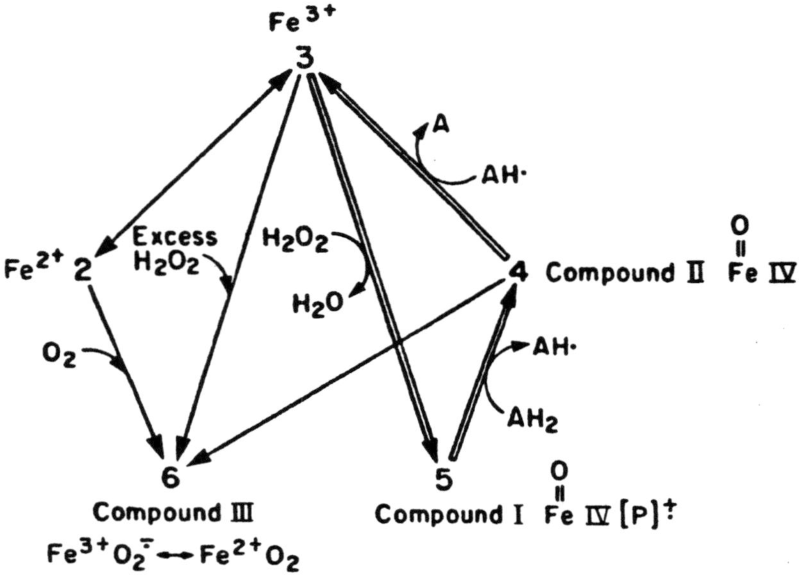

Fig. 4.29. The five redox states of lignin peroxidase. Reaction paths *3→5→4→3* indicate one catalytic cycle of the enzyme, AH_2 = substrate, which may be phenolic or nonphenolic (Renganathan and Gold 1986)

cohol, while six were Mn-dependent peroxidases. A great degree of homology concerning immunological properties and amino acid composition was present in the two types of peroxidases. However, antiligninase antibodies did not react with Mn-peroxidases, and differences in the amount of glycine, leucine, and isoleucine in the two types of peroxidases were obtained.

Characterization of *P. chrysosporium* lignin peroxidase by (electronic) absorption spectra, electronic paramagnetic resonance (EPR), and resonance Raman spectra (Andersson et al. 1985, 1987, Kuila et al. 1985) has shown that the ferric iron of lignin peroxidase is predominantly high-spin pentacoordinate at 25 °C, whereas at temperatures <2 °C the high-spin hexacoordinate state dominates. This temperature-dependent change of coordinate state of lignin peroxidase is similar to that of cytochrome c peroxidase (Andersson et al. 1987) but was not observed with Mn-peroxidase (Mino et al. 1988).

The five redox states of lignin peroxidase have been characterized by Renganathan and Gold (1986), as shown in Fig. 4.29. The native Fe^{3+} enzyme is oxidized by one molar equivalent of H_2O_2 to Compound I (2 e^- oxidized). Veratryl alcohol, which is a two-electron substrate, in 0.5 molar equivalent reduces Compound I to Compound II, while further addition of 0.5 molar equivalent of veratryl alcohol gives the native lignin peroxidase. Thus, dehydrogenation of veratryl alcohol was suggested to proceed via two single electron oxidation steps with the formation of a cation radical intermediate. This is in contrast to results by Tien et al. (1986) who were unable to detect free radi-

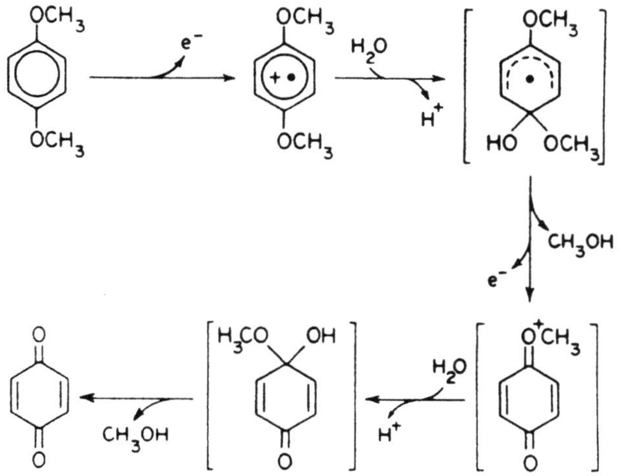

Fig. 4.30. Hypothetical scheme showing two sequential one-electron oxidations of 1,4-dimethoxy-benzene and addition of water with loss of methanol. The intermediate cation radical and the products p-benzoquinone and methanol have been identified (Kersten et al. 1985)

cals such as cation radicals by ESR (EPR) measurements during the action of ligninase on veratryl alcohol; only veratraldehyde and no free radicals were detected. As discussed below, Haemmerli et al. (1987) and Leisola et al. (1985b, 1987), in addition to veratraldehyde, also reported the formation of ring-cleavage products during ligninase oxidation of veratryl alcohol. Tien et al. (1986) observed the following features for ligninase oxidation of veratryl alcohol: (1) there is a 1 : 1 stoichiometry in the oxidation of veratryl alcohol by H_2O_2; (2) the enzyme operates through a ping-pong mechanism during steady-state catalysis; (3) peroxidase-like Compounds I and II (cf. Fig. 4.29) are observed in transient state kinetic experiments; (4) free radicals are formed with some substrates but apparently not with veratryl alcohol.

Some characteristics of the mechanisms for lignin peroxidase action were also reported somewhat earlier. Thus, Kersten et al. (1985) with ESR spectroscopy for the first time showed that 1,4-dimethoxybenzene was converted to p-benzoquinone and methanol via an aryl cation radical (Fig. 4.30) and proposed that C_α-C_β cleavage and all other reactions catalyzed by lignin peroxidase proceed via aryl cation radicals. Later, Kersten et al. (1987) found that also horseradish peroxidase plus H_2O_2 converted 1,2,4,5-tetramethoxybenzene to 2,5-dimethoxy-p-benzoquinone and methanol. The intermediate cation radical reacts with H_2O which results in demethoxylation of the aromatic substrate. It was shown by Kersten et al. (1987) that lignin peroxidase can oxidize methoxybenzenes with a higher redox potential than those oxidized by horseradish peroxidase. In order to further elucidate lignin peroxidase mechanisms, Hammel et al. (1985) synthesized the model dimer 1-(3,4-dimethoxyphenyl)-2-phenylethanediol (dimethoxyhydrobenzoin, DHMB), which in air is quantitatively cleaved by lignin peroxidase to give only benzaldehyde and

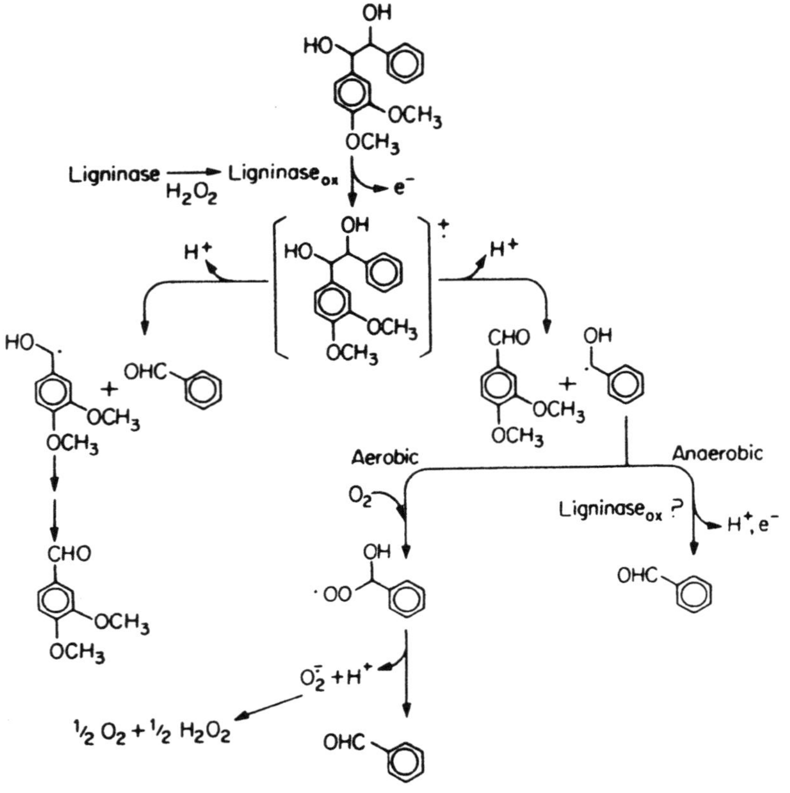

Fig. 4.31. Proposed scheme for aerobic and anaerobic cleavage of dimethoxyhydrobenzoin by lignin peroxidase (Hammel et al. 1985)

veratraldehyde unlike other dimers used (Fig. 4.31). It was proposed that lignin peroxidase and H_2O_2 attack the methoxylated ring of DMHB to yield a cation radical, which is then split into either benzaldehyde and an α-hydroxy(dimethoxylbenzyl) radical (after addition of O_2) or into veratraldehyde plus an α-hydroxybenzyl radical (Fig. 4.31). In aerobic reactions O_2 adds to the two carbon-centered radicals α-hydroxy(dimethoxybenzyl) radical and the α-hydroxybenzyl radical, producing veratryl alcohol and benzaldehyde, respectively (Hammel et al. 1985). This suggests that incorporation of oxygen is a secondary reaction not performed directly by the lignin peroxidase. It was proposed that similar mechanisms apply to the enzymatic C_α-C_β cleavage of the side chains in polymeric lignin.

Initially, lignin peroxidase was classified as an oxygenase since under aerobic conditions one atom of molecular oxygen was incorporated into some of the cleavage products (Gold et al. 1984, Kuwahara et al. 1984, Tien and Kirk 1984). However, the role of molecular oxygen in lignin peroxidase reactions has now been investigated using m-chloroperoxybenzoic acid (mCPBA), which can

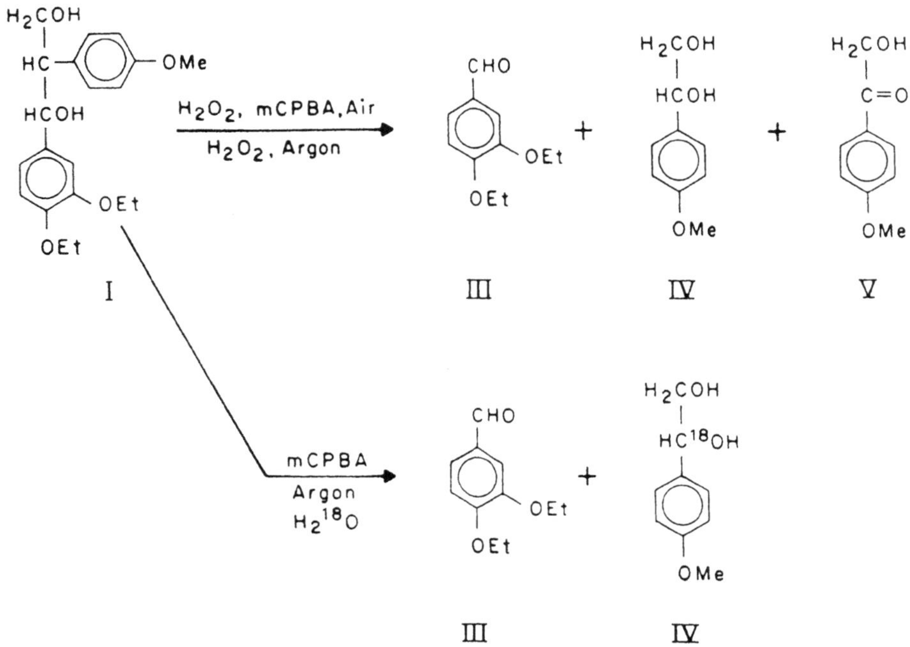

Fig. 4.32. Oxidation of the diarylpropane I by lignin peroxidase in the presence of m-chloroperoxybenzoic acid or H_2O_2 under different conditions (Renganathan et al. 1986)

give truly anaerobic conditions not possible to obtain with H_2O_2 (Renganathan et al. 1986). In the presence of mCPBA or H_2O_2 the dimeric diarylpropane I (Fig. 4.32) was oxidized to 3,4-diethoxybenzaldehyde III, the phenylglycol IV, and the phenylketol V. Under anaerobic conditions, but in the presence of mCPBA and $H_2{}^{18}O$, ^{18}O was incorporated into the α-position (originally the β-position) of the phenylglycol IV. This strongly indicates that molecular oxygen is not required for the reaction and that the lignin peroxidase should indeed be classified as a peroxidase (Renganathan et al. 1986).

The formation of carbon-centered free radicals and the addition of O_2 was also investigated by Hammel et al. (1986a). Using three β-1 linked dimeric lignin models, they found that lignin peroxidase catalysis gives rise to carbon-centered and peroxyl radicals as the major products of C_α-C_β cleavage. According to Hammel et al. (1985, 1986a) and Renganathan et al. (1986) it is now clear that it is the β-carbon of a dimer which via a hydroperoxyl radical is hydroxylated by O_2 under aerobic conditions. This hydroxylation occurs after the ligninase-induced formation of a cation radical and is not enzyme-dependent.

In addition to Kersten et al. (1985), Palmer and co-workers were among the first to suggest the importance of radical cation intermediates in lignin degradation and also that the ligninase is to be characterized as a peroxidase (Harvey et al. 1985a, 1986, 1987a, b, Schoemaker et al. 1985, 1988, Palmer et

Table 4.9. Stimulation of the oxidation of 4-methoxymandelic acid and anisyl alcohol by ligninase in the presence of the mediators 1,4-dimethoxybenzene or veratryl alcohol (Harvey et al. 1986)

	Substrate	Yield of anisaldehyde (nmol)
No mediator	4-Methoxymandelic acid	36
	anisyl alcohol	20
1,4-Dimethoxybenzene as mediator	4-Methoxymandelic acid	244
	anisyl alcohol	142
Veratryl alcohol as mediator	4-Methoxymandelic acid	548
	anisyl alcohol	82

al. 1987). Thus, Harvey et al. (1985a) showed that the single electron transfer reagent tris(phenanthroline)iron(III) performs the same reactions as those catalyzed by lignin peroxidase but without the involvement of H_2O_2 or oxygen. It was suggested that the oxidation reactions are mediated via a single electron transfer between the aromatic ring and Compound II of the lignin peroxidase. In a discussion letter Schoemaker et al. (1985) made the following summary of their opinions on the mechanisms of enzymatic lignin degradation: After a one-electron oxidation of lignin or lignin degradation products initiated by ligninase, cation radicals are produced. These radicals may act as electron transfer agents to induce the formation of cation radicals even in more remote lignin structures, thus causing degradation in those parts of the lignin which are not located near the hyphal surface.

Harvey et al. (1986) found that oxidation of 4-methoxymandelic acid and anisyl alcohol by lignin peroxidase was stimulated by the presence of catalytic amounts of 1,4-dimethoxybenzene (DMB) or veratryl alcohol (Table 4.9). It was proposed that these two latter dimethoxylated compounds act as one-electron oxidants for the monomethoxylated compounds 4-methoxymandelic acid and anisyl alcohol, which are slowly oxidized by lignin peroxidase without the presence of these mediators. It appears as if methoxyl groups activate the substrates, whereas C_α-carbonyl groups (electron-withdrawing) inactivate them (Harvey et al. 1986, Kirk et al. 1986d). In Fig. 4.33 the suggested role for the veratryl alcohol and its cation radical is shown. However, the actual formation of this cation radical has not yet been demonstrated. It may be that it is too short-lived (Tien et al. 1986, Harvey et al. 1987a, Tien 1987).

Kirk and Farrell (1987) suggest that the veratryl alcohol protects the enzyme from inactivation by 4-methoxymandelic acid or anisyl alcohol. Using *Phanerochaete chrysosporium* ME-446, Tonon and Odier (1988) found that 1 mM veratryl alcohol protected ligninase from being inactivated by 0.024 mM H_2O_2 in glucose cultures. Only the isoenzyme of the highest specific activity (M_r 41 800) was found to be protected by veratryl alcohol. Glycerol cultures, which produced less H_2O_2, needed less veratryl alcohol for protection of the activity of the enzyme.

Results obtained by Higuchi and co-workers (Higuchi 1986, Umezawa et al. 1986a) have shown that the degradation patterns of the arylglycerol-β-aryl

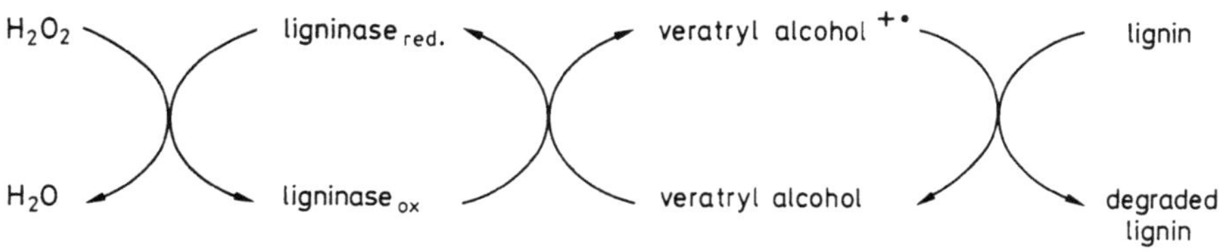

Fig. 4.33. Hypothetical scheme for degradation of lignin by lignin peroxidase with veratryl alcohol as mediator (Harvey et al. 1986)

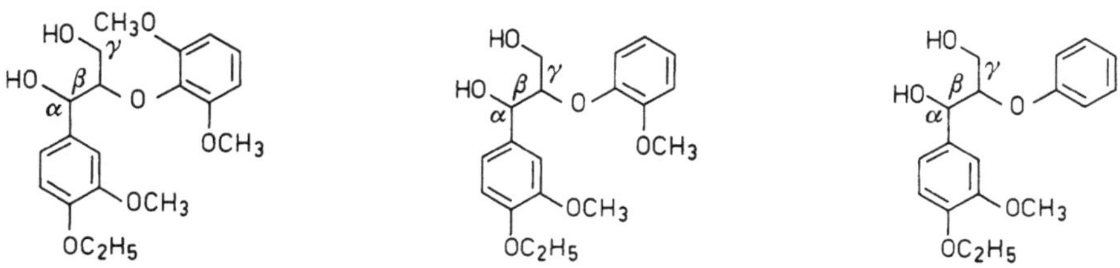

[28] [29] [30]

Fig. 4.34. Arylglycerol-β-aryl ethers [28–30] with two, one, or no methoxyl group on the B-ring, and used for cleavage studies with lignin peroxidase (Higuchi 1986)

ethers [28–30] (Fig. 4.34) by lignin peroxidase are strongly influenced by the methoxyl groups in the aromatic ring. A β-syringyl ether [28] was mainly degraded via ring cleavage while a β-guaiacyl ether [29] and a β-phenyl ether [30] without methoxyl groups were degraded via C_α-C_β cleavage by lignin peroxidase (Fig. 4.34). Aromatic rings with two methoxyl groups are attacked by lignin peroxidase easier than aromatic rings with only one methoxyl group (Yokota et al. 1988).

The mechanism for lignin peroxidase action on six nonphenolic dimeric model substrates was investigated by Kirk et al. (1986d). The substrates used (I–VI) and some of the degradation products are shown in Fig. 4.35. The dimethoxylated ring is the A-ring, whereas the β-ether-linked ring is called the B-ring. Compound (I) was rapidly oxidized almost completely to veratraldehyde (VII) plus less than 2% of veratrylglycerol (VIII) and 20% of compound (IX). Traces of p-benzoquinone, methanol, and compound (II) were also detected. Model compound (II), containing a C_α-carbonyl group, was also readily oxidized by ligninase. Three major products were identified: compound (IX) (50% conversion), p-benzoquinone, and methanol. This means that in the presence of a C_α-carbonyl group, the A-ring oxidation is strongly retarded, and instead the B-ring is attacked, giving a rather strong demethoxylation to methanol (50 mol% at the end of the reaction) of the B-ring methoxyl. Together with β-ether cleavage, this reaction is favored in C_α-carbonyl compounds (see also Higuchi 1986).

Model compound (III), which is more lignin-related than compound (I), was also rapidly oxidized by the ligninase, and veratraldehyde was the major

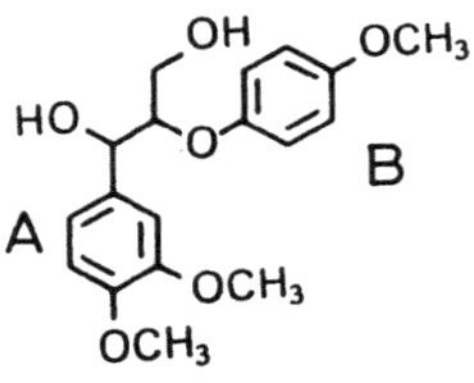
(I)

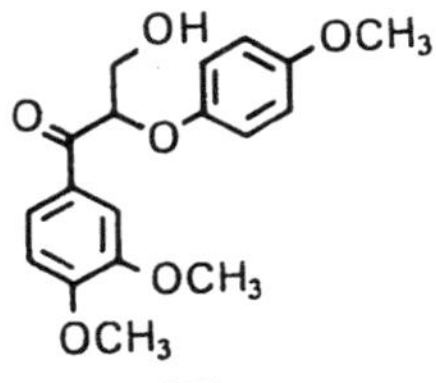
(II)

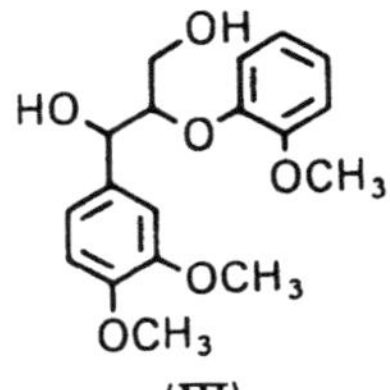
(III)

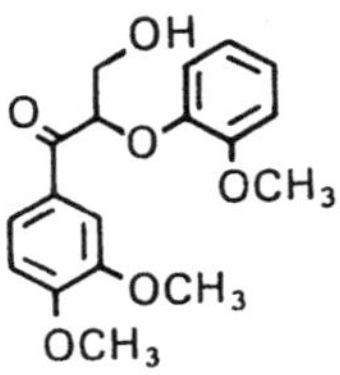
(IV)

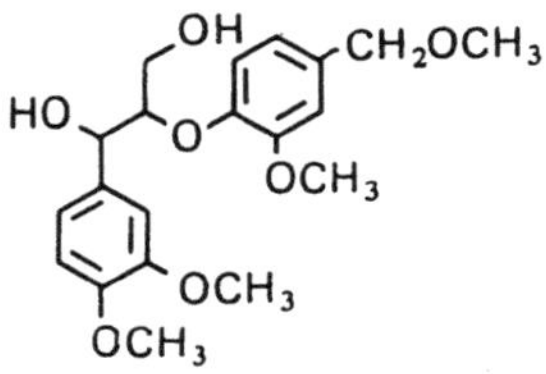
(V)

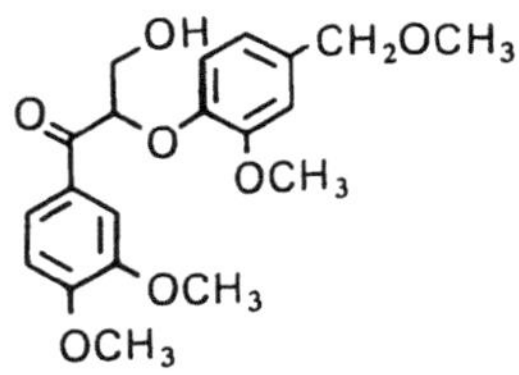
(VI)

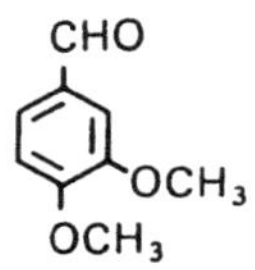
(VII)

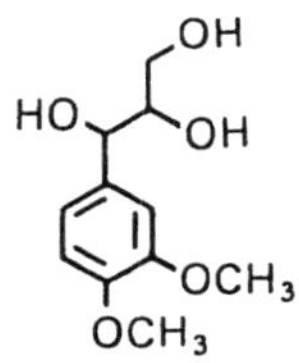
(VIII)

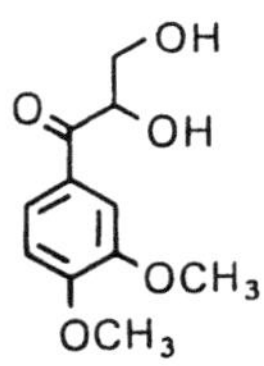
(IX)

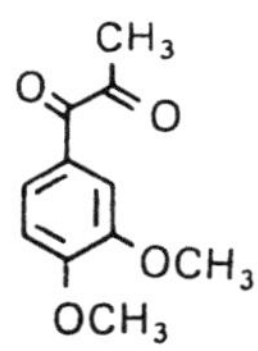
(X)

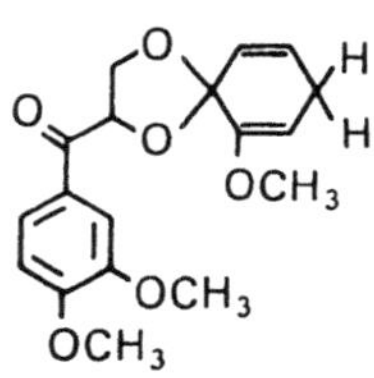
(XI)

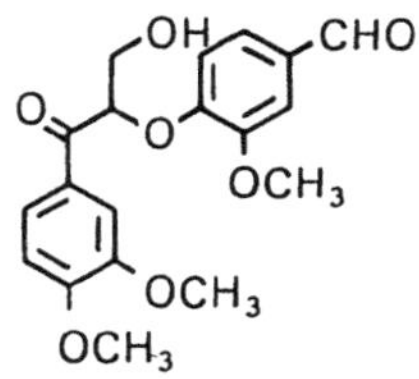
(XII)

Fig. 4.35. Arylglycerol-β-aryl ethers (*I–VI*) used for studies on ligninase-catalyzed cleavage. The *A*-ring is dimethoxylated, whereas the *B*-ring has different substituents (Kirk et al. 1986d). See text for further details

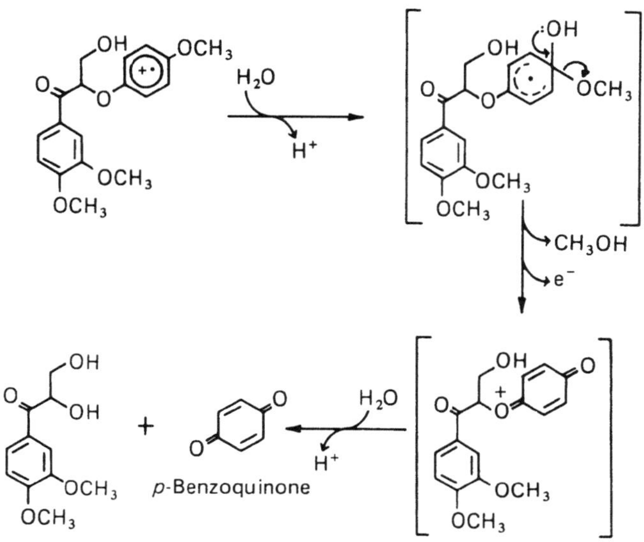

Fig. 4.36. Proposed mechanism for cleavage of compound (II) in Fig. 4.35 to give p-benzoquinone, methanol, and compound (*IX*) (Kirk et al. 1986d)

product together with the C_α-carbonyl compound (IV). Small amounts of veratrylglycerol (VIII) and compound (IX) as well as a minor amount of methanol were also detected. No products were found from ring B.

Compound (IV) in Fig. 4.35 was a poor substrate for ligninase although it was slowly converted to a new, relatively unstable compound suggested to be compound (XI). This was further degraded to give guaiacol and the diketone (X). Model compounds (V) and (VI) which have a methoxymethyl group on the B-ring are most similar to lignin, and compound (V) was readily oxidized by the ligninase, giving 50 mol% veratryldehyde and 23 mol% of product (VI), while compound (VI) was oxidized at about the same rate as compound (IV); the only identified product was (XII).

In Fig. 4.36, a possible mechanism for formation of product (IX), p-benzoquinone, and methanol from the carbonyl compound (II) by ligninase is shown (Kirk et al. 1986d). The reaction is initiated by the formation of a cation radical in the B-ring (detected by ESR at room temperature without spin-trapping agents). Since small amounts of methanol were formed from compounds (III) and (IV) it may be that the ligninase is responsible for some of the methanol produced during lignin degradation. Lignin in wood (Kirk and Chang 1974) as well as DHP and MWL (Ander and Eriksson 1985) are all strongly demethoxylated by white-rot fungi. Mn-peroxidase and laccase have also been reported to have some demethoxylating ability, at least with phenolic lignin degradation products as substrates (Ander and Eriksson 1985, Ander et al. 1985).

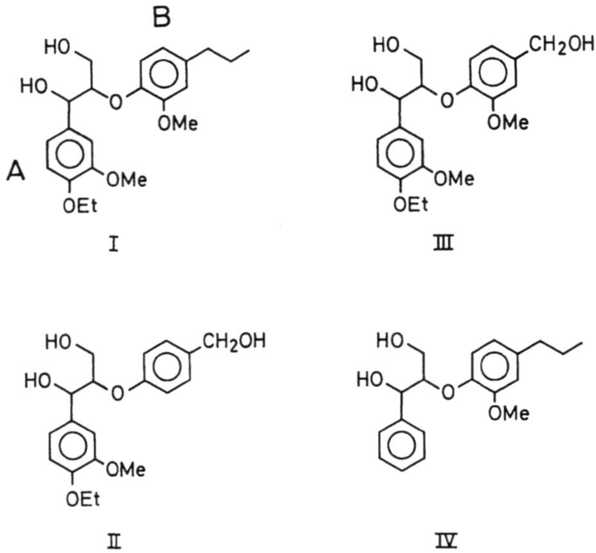

Fig. 4.37. Arylglycerol-β-aryl ethers *I–IV* used for studies of lignin peroxidase oxidation mechanisms (Miki et al. 1986)

Huynh et al. (1986a) studied the transformations of some *O*-arylpropane lignin model compounds by lignin peroxidase. The compounds used were not of the usual β-*O*-4 type studied by Kirk et al. (1986d) and Miki et al. (1986). With the compound 3-(4′-methoxyphenoxy)-1,2-propanediol, formation of p-benzoquinone, glycerol, and two cyclic ketals was found. The reaction mechanism for formation of p-benzoquinone plus glycerol (Huynh et al. 1986a) is probably the same as for conversion of 1,4-dimethoxybenzene to p-benzoquinone and methanol by lignin peroxidase (Kersten et al. 1985).

Miki et al. (1986) investigated the mechanisms of lignin peroxidase oxidation of the four different β-aryl ether dimers shown in Fig. 4.37. The isolated products indicate that both dimers I and III were cleaved at their α-β bonds and at the β-ether bonds, while in compound II only C_α-C_β cleavage occurred. From compound IV the resulting products indicated only β-ether bond cleavage. It appears as if the minimum requirement for enzymatic formation of an aryl cation radical is two alkoxy ring substituents. Thus dimers I and III have two substituents in both A and B rings and are oxidized to the corresponding cation radicals. In dimer II only ring A forms a cation radical, since only that ring has two alkoxy groups. The results of Miki et al. (1986), summarized in Fig. 4.38, show that if an aryl cation radical is formed on the A ring then C_α-C_β cleavage yields benzaldehyde plus glycolaldehyde and a phenol (pathway a). Formation of a cation radical on ring B results in an intramolecular attack by the 1-hydroxyl, releasing methanol and possibly a benzodioxane radical intermediate. This may give rise to incorporation of ^{18}O from $H_2{}^{18}O$ into the 1-position of the phenylglycerol, resulting in a catechol (pathway b in

Fig. 4.38). In pathway c, the 3-hydroxyl attacks the aryl cation which is followed by rearrangement and side-chain cleavage. As indicated in pathway b, it is not necessary to open the aromatic ring to form a phenylglycerol as was suggested by Umezawa and Higuchi (1985a, b). Using *Phanerochaete chrysosporium* they found that the dimer 4-ethoxy-3-methoxyphenylglycerol-β-guaiacyl ether was ring-cleaved to give a cyclic carbonate of 4-ethoxy-3-methoxy-phenylglycerol, which was converted to 4-ethoxy-3-methoxyphenylglycerol. A ring-cleavage product from the B ring of a dimer was also isolated and identified by Umezawa and Higuchi (1985b) for the first time.

Higuchi (1986), Umezawa et al. (1986a, b), and Umezawa and Higuchi (1986) further investigated ring cleavage in β-*O*-4 lignin model dimers by lignin peroxidase (Fig. 4.39). It was reported from these studies that the following esters of arylglycerol were formed, namely, the cyclic carbonates (V) and (V′), formate (VI), and methyl oxalate (VII). Although Leisola et al. (1985b) reported ring cleavage of veratryl alcohol by crude lignin peroxidase, this was the first demonstration of lignin peroxidase-catalyzed ring cleavage in lignin-related dimers. The β-guaiacyl ether (I) in Fig. 4.39 gave veratraldehyde as the main product, whereas β-(2,6-dimethoxyphenyl)ether (II) was attacked in the B ring, giving essentially no veratraldehyde but mainly ring-cleavage products. Indeed, aromatic ring cleavage was found to be one of the main reactions catalyzed by lignin peroxidase (Umezawa et al. 1986a, b). This means that conventional dioxygenases may not be involved in ring cleavage of lignin-related dimers or lignin, although Wojtas-Wasilewska and Luterek (1987) reported some decrease in absorbance at 280 nm after incubation at pH 8 of Na-lignosulfonate with an intracellular protocatechuate 3,4-dioxygenase from *Pleurotus ostreatus*. Since pH 8 was used for the incubation, interference with laccase and/or lignin peroxidase may be negligible. The authors proposed that the dioxygenase can cleave aromatic rings in the intact lignosulfonate polymer. However, this does not necessarily mean that lignin in wood or water-insoluble lignins are cleaved in the aromatic rings by this intracellular dioxygenase.

Umezawa and Higuchi (1986) proved that demethylation or demethoxylation is not a prerequisite for ring cleavage by lignin peroxidase. Metabolism of methoxylated phenols by bacteria usually proceeds via demethylation to yield orthophenolic groups before attack by dioxygenases to cleave the ring (Cain 1980).

Based on further ^{18}O incorporation studies with $H_2{}^{18}O$ and $^{18}O_2$, Umezawa and Higuchi (1987a, b) obtained additional evidence for the lignin peroxidase mechanisms of ring cleavage of the dimers presented in Fig. 4.39. The mechanism involves one-electron oxidation of the aromatic ring to a cation radical, attack of a nucleophile (H_2O or the hydroxyl groups at the C_α or C_γ position of the propyl side chain), and coupling with O_2. Using another β-*O*-4 ether dimer, Miki et al. (1987) also identified aromatic ring-cleavage products, some of them analogous to the compounds V–VII in Fig. 4.39.

In summary, it can be concluded that many reactions follow after lignin peroxidase-catalyzed oxidation of aromatic rings to cation radicals. Most im-

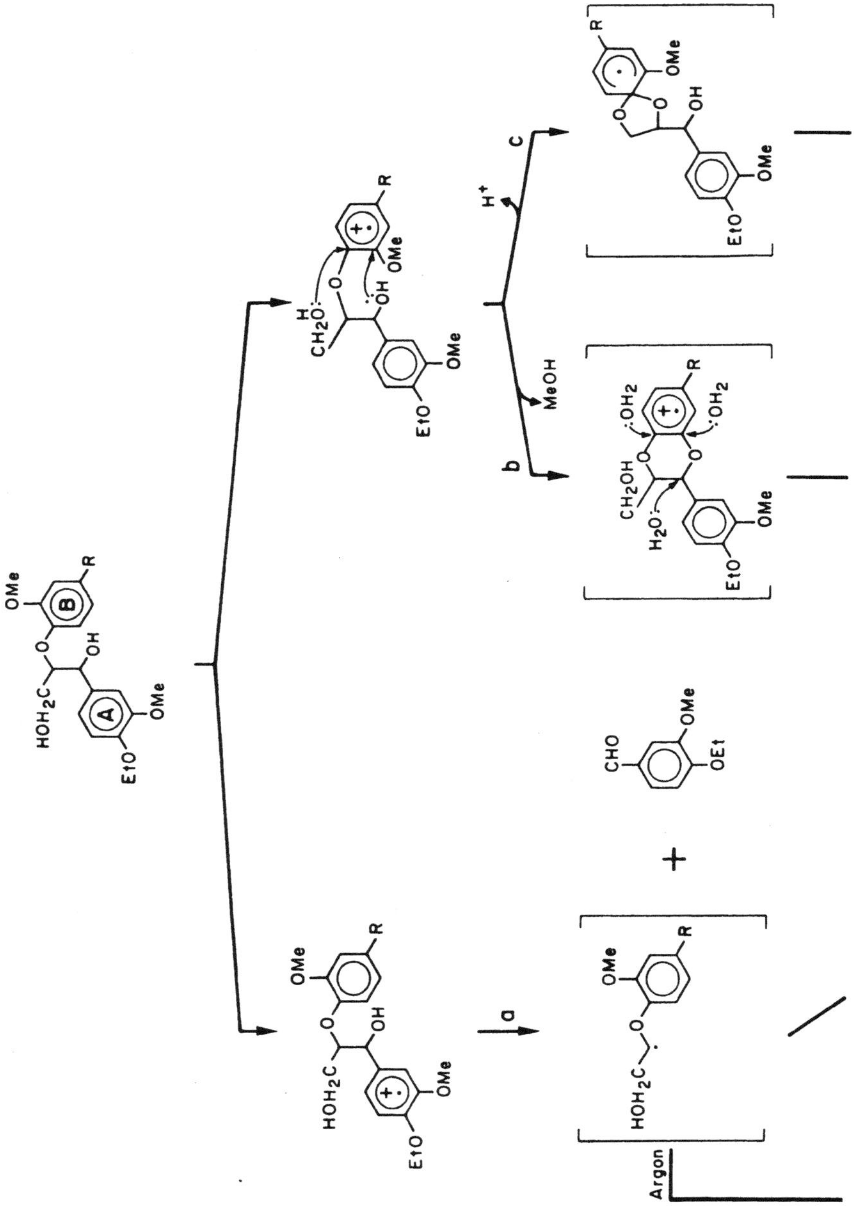

Fig. 4.38

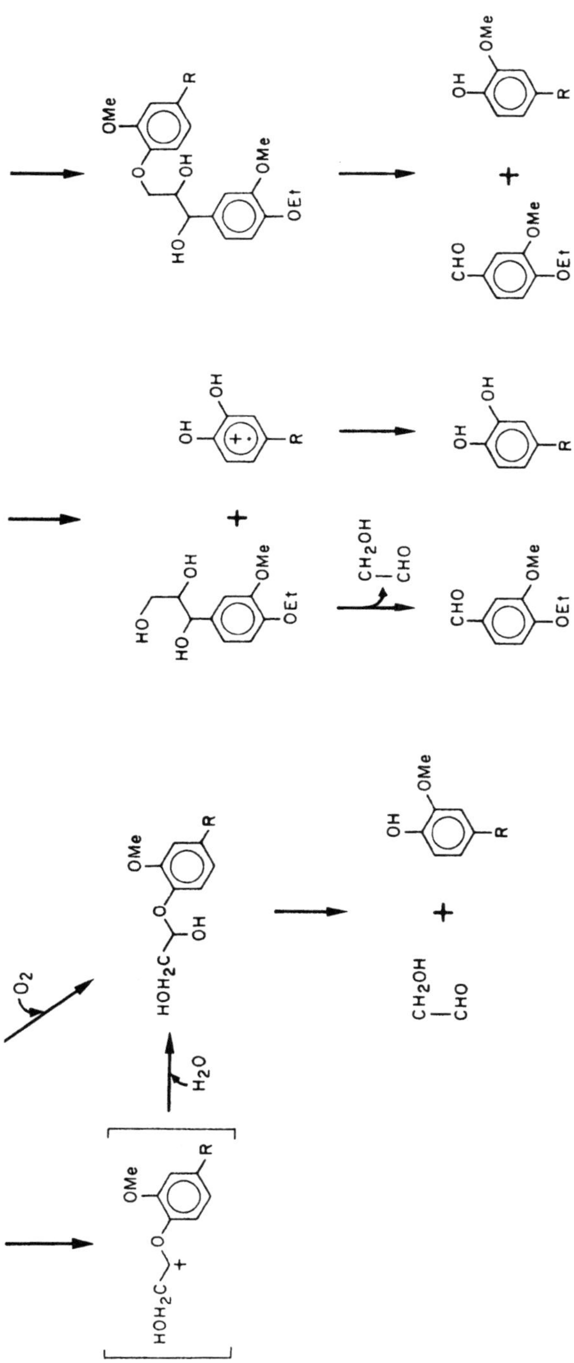

Fig. 4.38. Proposed pathway for the oxidative cleavage of β-aryl ether dimers under argon by lignin peroxidase. $R = -CH_2OH$, $-CH_2CH_2CH_3$ (Miki et al. 1986)

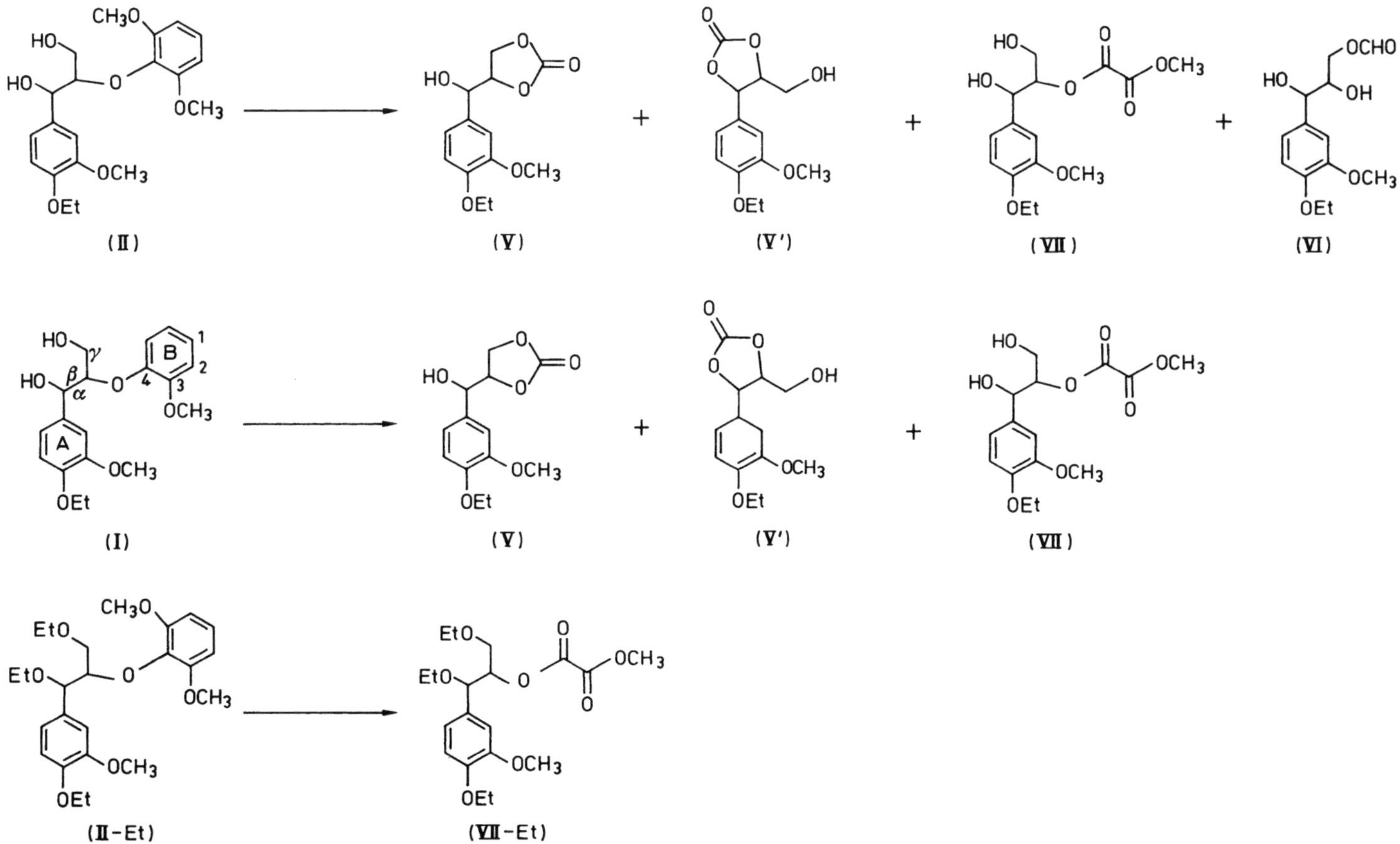

Fig. 4.39. Cleavage of *β-O-4* lignin dimers by lignin peroxidase to yield the esters: cyclic carbonates (*V*) and (*V'*), formate (*VI*), and methyl oxalate (*VII*) (Umezawa et al. 1986b)

portant of these reactions are cleavage of C_α-C_β bonds, β-*O*-4 bonds, and aromatic rings. However, ring cleavage does not occur without the presence of both H_2O_2 and oxygen, which adds to carbon-centered radicals (Miki et al. 1987, Umezawa and Higuchi 1987a, b, Kirk 1988a, b). In the absence of oxygen, radical-radical coupling will occur, leading to polymerization.

Whether an aromatic ring is oxidized by lignin peroxidase depends on its oxidation potential. C_α-carbonyl groups (electron-withdrawing) inactivate aromatic nuclei to lignin peroxidase whereas alkoxyl groups activate them. This partly explains why syringyl lignin is easier to degrade than guaiacyl lignin.

Some of the lignin peroxidase catalyzed reactions liberate phenolic units. Such phenols, as well as the approximately 10% phenolic units in lignin, have a lower oxidation potential and will be preferentially oxidized not only by lignin peroxidase but also by laccase and Mn-peroxidase (Higuchi 1986, Kirk 1988b), perhaps leading to coupling/polymerization reactions. In vitro this is a major reaction of laccase and all peroxidases.

Ring-Cleavage and Other Reactions of Veratryl Alcohol

Ever since the discovery in *Phanerochaete chrysosporium* of veratryl alcohol as a secondary metabolite (Lundquist and Kirk 1978), much interest has been focused on the biosynthesis and metabolism of this substance. The metabolism of a similar compound, veratric acid, by the white-rot fungus *Pycnoporus cinnabarinus* was investigated by Hatakka (1985). Under shaking conditions, substantial reduction to veratraldehyde and veratryl alcohol was found, but the further metabolism or ring-cleavage studies of the latter compound were not pursued. Based on gas chromatographic studies it was found that the main pathway of veratric acid metabolism by *P. cinnabarinus* in stationary cultures with cellulose as carbon source was demethoxylation to vanillic acid, which was further metabolized through methoxyhydroquinone.

Leisola et al. (1985b) found that both *P. chrysosporium* and lignin peroxidase could cleave the veratryl alcohol ring to produce the two isomeric lactones IIIa and IIIb (Fig. 4.40). Further studies (Haemmerli et al. 1987, Leisola et al. 1988) of the oxidation of veratryl alcohol by lignin peroxidase showed formation of five products: veratraldehyde, two quinones, and the same lactones as found previously (Fig. 4.41). The quinones and the lactones were not formed if an argon atmosphere was used instead of oxygen. In similar experiments, a third lactone (Fig. 4.42) was detected by Shimada et al. (1987b). Although Tien et al. (1986) suggested that veratryl alcohol is oxidized to the aldehyde via a direct two-electron step or via two rapid and consecutive one-electron steps without formation of other products, Schoemaker et al. (1985) and Palmer et al. (1987) proposed the distinct intermediacy of both the veratryl alcohol cation radical (Va) and a corresponding veratryl alcohol radical (Vb) (Fig. 4.41). Based on these latter suggestions and on their own experiments, the scheme for oxidation of veratryl alcohol by lignin peroxidase shown in Fig. 4.41 was proposed by Haemmerli et al. (1987). The radical (Vb) is attacked by

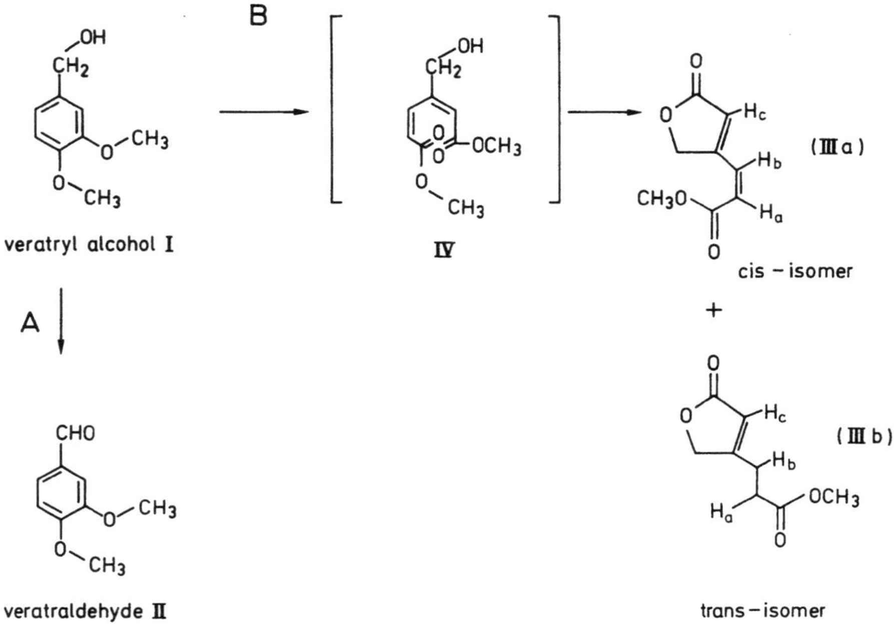

Fig. 4.40. Proposed mechanism (pathway B) for cleavage of veratryl alcohol to yield the two cis- and trans-isomers (*IIIa, IIIb*) Pathway A gives veratraldehyde (II) (Leisola et al. 1985b)

O_2 to release superoxide anion $O_2^{\bullet-}$ simultaneously with production of veratraldehyde (VI). Under anaerobic conditions only veratraldehyde is formed. The perhydroxy radical and H_2O can further react with the cation radical (Va), which will lead to ring opening and formation of two lactones (IIIa and IIIb in Fig. 4.40). If the cation radical (Va) reacts with the nucleophilic solvent water, at least the two phenoxy radicals (VII) and (VIII) may be formed. According to Haemmerli et al. (1987), these may then react further to the quinones I and II in Fig. 4.41. Methoxyl cleavage will yield methanol. The quinone I and the isomeric lactones IIIa and IIIb are formed not only in *Phanerochaete chrysosporium* cultures but also in *Phlebia radiata* cultures (Kantelinen et al. 1988).

Shimada et al. (1987b) showed that ^{18}O from $H_2{}^{18}O$ was incorporated at either the C_3 or C_4 carbon of the veratryl alcohol cation radical (Fig. 4.42). ^{18}O from $^{18}O_2$ was incorporated at the carbon atom which did not receive the ^{18}O from $H_2{}^{18}O$. These results indicate that the superoxide anion radical is not participating in the formation of a dioxetane bridge as in route B (Fig. 4.42). In addition to the lactones IIa and IIb also detected by Haemmerli et al. (1987), lactone IIc (Fig. 4.42) was found by Shimada et al. (1987b).

In summary, Leisola et al. (1987) suggested that in the presence of O_2 veratryl alcohol is degraded to CO_2 via two quinones and two lactones. Under secondary metabolic conditions, different radical species like ver-alc$^{+\bullet}$, ver-

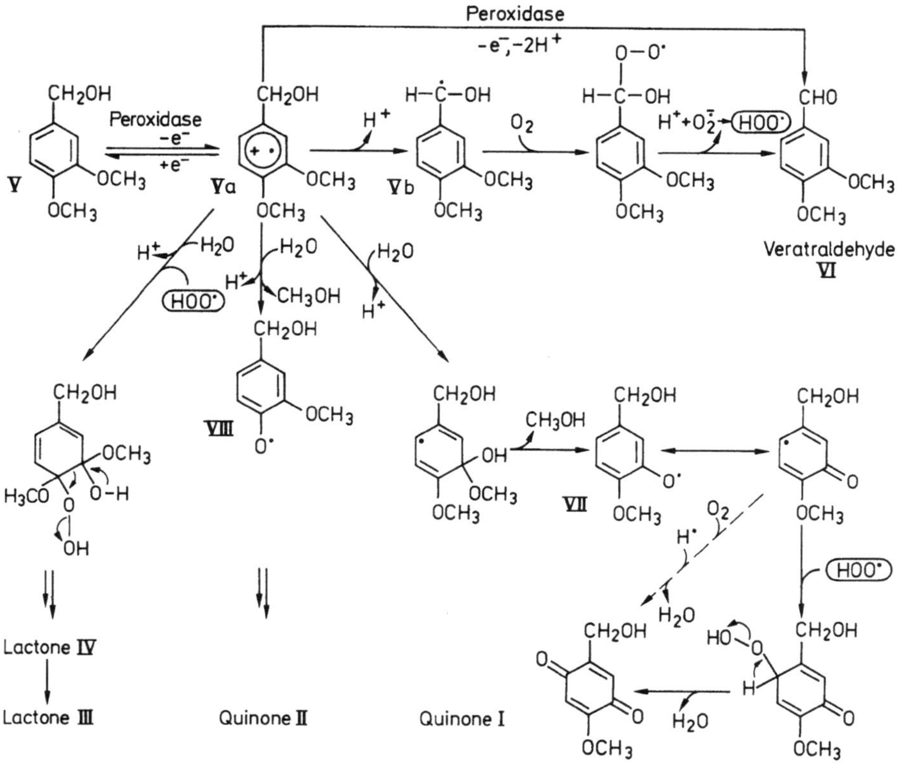

Fig. 4.41. Possible reactions of the veratryl alcohol cation radical and a hypothetical mechanism for the formation of quinone I (Haemmerli et al. 1987)

alc•, and •OOH are present all the time and are ready to attack lignin. This would explain how lignin located at some distance from the mycelium can be oxidized (Leisola et al. 1987, Palmer et al. 1987). Acids, aldehydes, or ketones will probably be reduced to the corresponding alcohols, since it has been reported (Fenn and Kirk 1984, Kirk et al. 1986d) that C_α-oxidation retards degradation of lignin dimers and that aldehydes and ketones are not substrates for lignin peroxidase. The alcohols formed are good substrates for oxidative attack and will be metabolized further (Waldner 1987, Leisola et al. 1988, Schoemaker et al. 1989). Vanillic acid is readily reduced to the corresponding aldehyde and alcohol in noncarbon-limited cultures of *P. chrysosporium* P 127-1. In C-limited cultures, however, the main pathway is decarboxylation of vanillic acid to methoxyhydroquinone (Ander et al. 1980b, 1983a, b).

Final Comments on Degradation of Dimeric Lignin-Related Substances

Major contributions to the understanding of lignin biodegradation have been made by Higuchi and co-workers (Higuchi 1981b, 1985b, 1986). As discussed

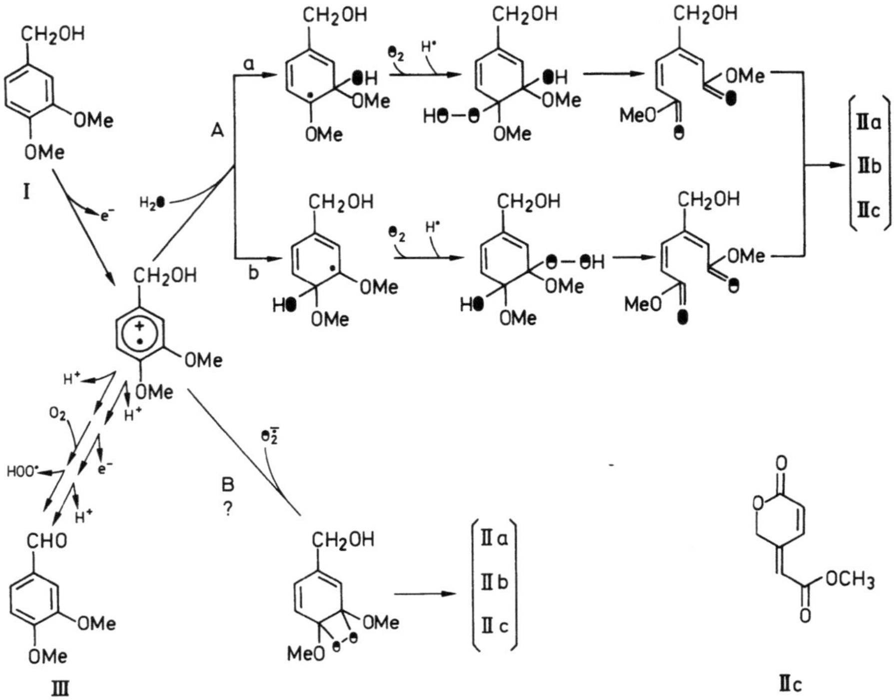

Fig. 4.42. Formation of the veratryl alcohol cation radical by lignin peroxidase and oxygenation by $^{18}O_2$ or $H_2^{18}O$ followed by ring cleavage. Compounds *IIa* and *IIb* are the lactones *IIIa* and *IIIb* in Fig. 4.40. (●) $H_2^{18}O$; (○) $^{18}O_2$ (Shimada et al. 1987b)

before, much of their recent work has dealt with lignin peroxidase and aromatic ring cleavage. In addition, whole cultures of *Phanerochaete chrysosporium* (Nakatsubo et al. 1981a, Umezawa et al. 1982, Kamaya and Higuchi 1983, Umezawa et al. 1983a, b, Kamaya and Higuchi 1984a, b, Umezawa and Higuchi 1984, Habe et al. 1985c, Umezawa and Higuchi 1985a, b, c, d, Umezawa et al. 1986a, Kawai et al. 1987a) and *Coriolus versicolor* (Kamaya and Higuchi 1984c, d, Kawai et al. 1985a, b, 1987b) have been used to elucidate how lignin and dimeric lignin-related substances are degraded by white-rot fungi. Degradation of lignin and guaiacylglycerol-β-guaiacyl ether by *P. chrysosporium*, *C. versicolor*, *Lentinus edodes*, and *Coprinus aff. ellisii* were studied by Oki et al. (1981, 1982, 1986).

Although *Fusarium solani* is not a white-rot fungus it can degrade lignin-related dimers and has therefore been used in many studies by Higuchi and co-workers (Ohta et al. 1979, Iwahara et al. 1980, Iwahara 1983, Katayama et al. 1980, 1981, 1986a, b, 1987, Kamaya et al. 1981, 1983, Namba et al. 1983).

Before 1983 important work on degradation of lignin-related dimeric compounds also was done with *P. chrysosporium* by Gold and co-workers (Enoki et al. 1980, Goldsby et al. 1980, Weinstein et al. 1980, Enoki et al. 1981a, b, c,

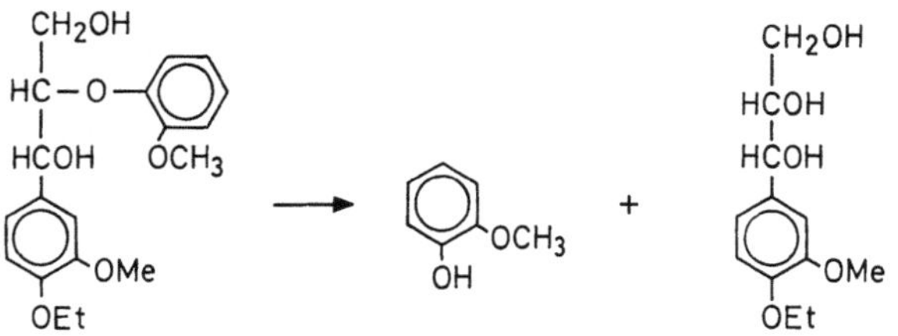

Fig. 4.43. Degradation of a β-aryl ether dimer by *Phanerochaete chrysosporium* to guaiacol and 4-ethoxy-3-methoxy-phenylglycerol (Enoki et al. 1981 a)

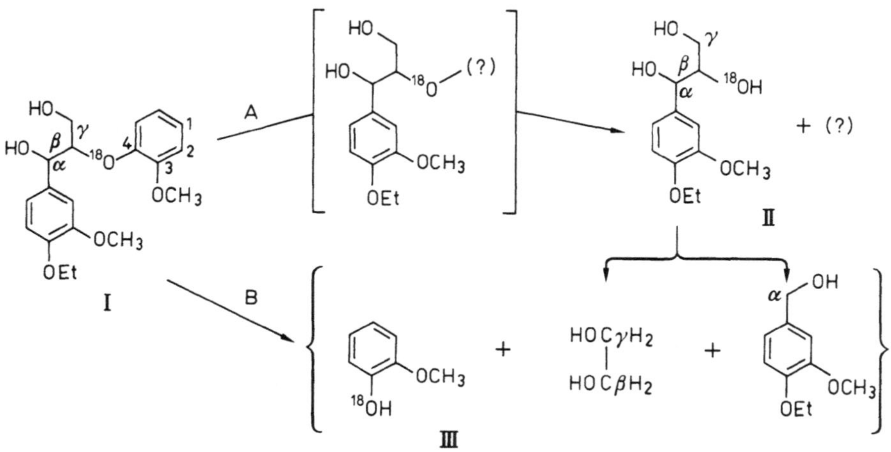

Fig. 4.44. Two different pathways (*A* and *B*) for degradation of the β-*O*-4 dimer (*I*) by *Phanerochaete chrysosporium*. Pathway *B* gives guaiacol (*III*) containing 98% of C_β-oxygen (Umezawa and Higuchi 1985 a)

Enoki and Gold 1982) and by Kirk and co-workers (Nakatsubo et al. 1982, Kirk and Nakatsubo 1983). In these early investigations the existence of an enzyme cleaving C_α-C_β bonds was suggested (Enoki et al. 1980, Nakatsubo et al. 1982). Furthermore, alkyl-phenyl cleavage was shown to occur in phenolic dimers, whereas in nonphenolic dimers, both β-ether cleavage and C_α-C_β cleavage were reported (Enoki et al. 1980, Goldsby et al. 1980). Enoki et al. (1981 a) used 4-ethoxy-3-methoxyphenylglycerol-β-guaiacyl ether (Fig. 4.43) and found guaiacol and 4-ethoxy-3-methoxyphenylglycerol as degradation products by *P. chrysosporium*. It was proposed that guaiacol and the phenylglycerol were counterparts in the β-*O*-4 cleavage. However, Umezawa et al. (1983 a, b) and Umezawa and Higuchi (1984), using deuterated substrates and ^{18}O incorporation from $H_2{}^{18}O$ and ^{18}O, found that this was not possible. The finding of Umezawa and Higuchi was later confirmed and in Fig. 4.44 two different pathways for degradation of the β-*O*-4 dimer (I) are shown (Umezawa and Higuchi 1985 a). Pathway A yields the arylglycerol (II) which contains 96% of the oxygen at C_β of the initial dimer (I). In pathway B, C_α-C_β cleavage gives 4-ethoxy-3-methoxy benzyl alcohol plus guaiacol and glycol. The guaiacol contains 98% C_β-oxygen of dimer (I). Thus, guaiacol is not a counterpart for the arylglycerol (II) in that cleavage (Umezawa and Higuchi 1985a, b).

4.5.4 Biomimetic Oxidation of Lignin Models

Lignin peroxidase is a heme protein containing protoporphyrin IX. Such porphyrins have therefore been tested for their ability to replace lignin peroxidase catalyzed oxidations. Different one-electron oxidizing compounds have also been used in order to help elucidate the mechanisms of lignin peroxidases. In such studies Shimada et al. (1984) found that tetraphenylporphyrinato-iron(III)chloride, in the presence of tert-butylhydroperoxide (t-BHP) at room temperature, oxidized the lignin model compound 1,2-diarylpropane-1,3-diol (Fig. 4.45, S) to compounds (I–V) with 4-*O*-ethylvanillin (I) being the major product. Habe et al. (1985a, b) found that hemin (Fig. 4.46), which is a natural iron porphyrin, catalyzed C_α-C_β cleavage of 1,2-bis(4-methoxyphenyl)propane-1,3 diol in the presence of hydrogen peroxide or t-BHP also at room temperature. Addition of 1 mM imidazol stimulated the cleavage reaction threefold. The reason for this activation is probably that histidine, structurally similar to imidazol, is part of the active site in lignin peroxidase. The optimum pH for C_α-C_β cleavage in 80% aqueous dimethylsulfoxide was pH 3, the same as for lignin peroxidase (Habe et al. 1985a, b). Wood et al. (1986) found that porphyrins oxidize arylglycerol-β-aryl ethers to give the same final reaction products as those found with lignin peroxidase. However, the porphyrins are unstable to two-electron oxidants such as hydrogen peroxide, iodosylbenzene, and t-BHP and can only catalyze a finite number of reactions before the catalyst is inactivated through tetrapyrrole ring oxidation, although stabilization can be accomplished by addition of electron-withdrawing substituents.

Dolphin et al. (1987) prepared water-soluble octachlorotetrasulfonato-*meso*-tetraphenyl hemin by adding four sulfonato groups to the porphyrin. This prevents the active iron centers of two porphyrins from coming too close to each other, which makes the porphyrin structure more stable against self-destruction. The sulfonatoporphyrin was reported to perform more than 1000 catalytic cycles without being destroyed (Dolphin et al. 1987).

Recently, Shimada et al. (1987a) used natural iron-porphyrin which was found to degrade β-1 and β-*O*-4 lignin substructures in a way which mimicked the lignin peroxidase-catalyzed reactions. A β-5 lignin substructure was also degraded but has not been studied with lignin peroxidase. Both natural hemin (Fig. 4.46) and tetraphenylporphyrin oxidized various lignin model dimers, but hemin was the more active compound. The mechanism for oxygenative ring-cleavage of veratryl alcohol by hemin was elucidated by Hattori et al. (1988). Paszczynski et al. (1988) used hemin, hematin, and other metalloporphyrins to study delignification of small birch chips and softwood kraft pulps. After 48-h reflux in water in the presence of t-BHP there was up to 40% lignin loss,

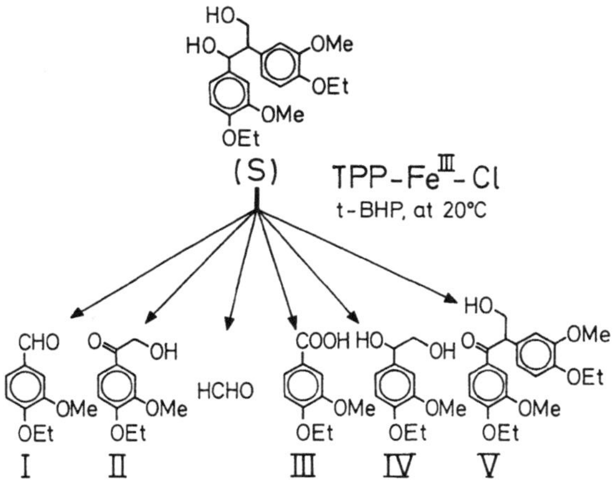

Fig. 4.45. Oxidation of the *β*-1 dimer (*S*) by tetraphenyl-porphyrinatoiron(*III*)chloride plus tert-butylhydroperoxide (Shimada et al. 1984)

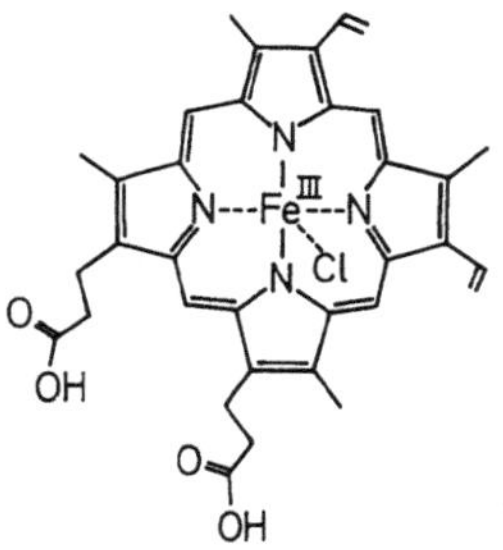

Fig. 4.46. Hemin (protoporphyrin IX) which can oxidize *β*-1, *β*-*O*-4, and *β*-5 lignin substructures (Shimada et al. 1987a)

a nearly 70% loss of hemicellulose, and a 10% loss of cellulose. The treated softwood pulps showed a decrease in kappa number (a measure of lignin content) from about 36 to less than 2 with a concomitant increase in brightness. The removal of lignin, as observed by transmission electron microscopy, was indistinguishable from that seen after decay by white-rot fungi (Paszczynski et al. 1988).

Not only porphyrins but also simpler inorganic complexes can mimic lignin peroxidase activity by transferring single electrons and bring about C_α-C_β cleavage. This is the case with cesium (IV)ammonium nitrate, cobalt(III)acetate, manganese(III)acetate, and tris(phenanthroline)iron(III) acting on non-phenolic diaryl propane structures or on veratryl alcohol (Harvey et al. 1985a, Schoemaker et al. 1985, Haemmerli et al. 1987, DiCosimo and Szabo 1988). Similarly, Huynh (1986) showed that copper peroxydisulfate mimics lignin peroxidase in its oxidative degradation of veratrylglycerol-*β*-guaiacyl ether and veratryl alcohol. However, the effect of these inorganic complexes on lignin has not been reported. In the future it seems possible that more stable porphyrins or other complexes (one-electron transferring agents) can be obtained for use in biotechnical applications to degrade lignin-related compounds or even lignin. To decrease the use of chlorine in bleaching of different pulps such compounds may be used in the future.

4.5.5 H_2O_2-Producing Enzymes

Since the discovery of lignin peroxidase and Mn-peroxidase, the source of H_2O_2 for these enzymes has become increasingly interesting. Earlier, Koenigs (1972) detected H_2O_2 production in a number of wood-rotting fungi and several oxidative enzymes oxidizing glucose, xylose, sorbose, galactose, galactosides, and other carbohydrates as well as oxalic acid were mentioned which could give rise to H_2O_2. Highley and Murmanis (1985) detected H_2O_2 production in five of six white-rot fungi tested. These authors also found that *Coriolus versicolor* generated H_2O_2 predominantly in the periplasmic space as well as in the cytoplasm and the hyphal sheath. Extracts of ligninolytic mycelium from *Phanerochaete chrysosporium* produced H_2O_2 as measured by the H_2O_2 catalase-aminotriazole assay (Forney et al. 1982a), while in whole cells H_2O_2 production was found to be localized in the periplasmic space of ligninolytic mycelium grown for 14 days in nitrogen-limited cultures (Forney et al. 1982b). This long incubation time indicates that some culture conditions were not optimal since lignin degradation in nitrogen-limited cultures usually starts at day 4. Intracellular glucose-1-oxidase activity appeared to be the main source of H_2O_2 production (Reddy et al. 1983) in *P. chrysosporium*. Later, Ramasamy et al. (1985) and Kelly et al. (1986) produced glucose oxidase-negative mutants from *P. chrysosporium* which did not degrade $^{14}C_{\beta}$-labeled DHP. In addition, the mutants did not decolorize poly R, did not produce ethylene from 2-keto-4-thiomethylbutyric acid, and did not produce lignin peroxidase or Mn-peroxidase. However, significant, although reduced, amounts of veratryl alcohol were produced by the mutants. All these lost properties were regained and often increased in five revertants (Kelley et al. 1986). This suggests that the lack of lignin degradation by the mutants may not only be due to the lack of glucose oxidase but also to the loss of lignin peroxidase and Mn-peroxidase. Possibly these enzymes, including glucose oxidase, are regulated by a common regulatory gene (Kelley et al. 1986). The above described glucose oxidase has now been purified and found to be a glucose-1-oxidase producing δ-D-gluconolactone which is in pH-equilibrium with D-gluconic acid:

$$\beta\text{-D-glucose} + O_2 \rightarrow \delta\text{-D-gluconolactone} + H_2O_2$$

(Kelley and Reddy 1986a, b). Simultaneously, Eriksson et al. (1986b) purified and partially characterized an intracellular glucose-2-oxidase from the same fungus. This enzyme, however, oxidized glucose at the C-2 carbon:

$$\beta\text{-D-glucose} + O_2 \rightarrow \text{D-arabino-2-hexulose} + H_2O_2.$$

Volc et al. (1985) detected intracellular glucose-2-oxidase in nine species of *Aphyllophorales* and in *Fomes fomentarius*, *Pycnoporus sanguineus*, *Coriolus versicolor*, *Kuehneromyces mutabilis*, and especially in *Oudemansiella mucida*. D-arabino-2-hexulose was released into the growth medium, suggesting that H_2O_2 also may be available for different peroxidases.

Kelley and Reddy (1986a) tested cell extracts of *Phanerochaete chrysosporium* for H_2O_2 production and found that D-glucose was the best substrate, while L-sorbose, D-cellobiose, D-mannose, and D-maltose gave decreasingly less H_2O_2. Oxygen was necessary for the oxidation. Several of the results indicated that glucose-1-oxidase seems to be the primary source of H_2O_2 production. Also, addition of ammonium sulfate or L-glutamate repressed both glucose-1-oxidase production and lignin degradation (Kelley and Reddy 1986a).

The glucose-1-oxidase purified and characterized by Kelley and Reddy (1986b) is a flavoprotein with a pH optimum between pH 4 and 5 and has a K_m for glucose of 38 mM. On the other hand, the glucose-2-oxidase studied by Eriksson et al. (1986b) has its pH-optimum at 7.5 and a K_m for glucose of 1.03 mM, suggesting that the two oxidases are different enzymes. In addition to glucose, δ-D-gluconolactone and xylose also could be oxidized (rate relative to glucose, 60% and 37%, respectively) by the glucose-2-oxidase. The activity against xylose for glucose-1-oxidase was only 4.8% compared to the activity against glucose. This means that glucose-2-oxidase has the ability to produce more H_2O_2 than glucose-1-oxidase. Since Kelley and Reddy (1986) and Eriksson et al. (1986b) used approximately the same assay method for their glucose oxidases, namely, the o-dianisidine-peroxidase system, it may be questioned why both enzymes were not detected in the fungal extracts. The use of different strains of *Phanerochaete chrysosporium*, ATCC 34541 and K-3, most likely is not responsible. However, the cultivation conditions were very different. Kelley and Reddy used a stationary low-nitrogen high-glucose (1%) medium according to Kirk et al. (1978a), while Eriksson et al. used 300-ml shaking cultures containing 6.7% glucose and 2.3% cornsteep at pH 5.5.

It is possible to increase H_2O_2 production by ground mycelia by adding milled wheat straw or Poly B-411 (Greene and Gould 1983). The best pH for this H_2O_2 generation was around pH 6. These authors also suggested that peroxisomal fatty acyl-CoA oxidase activity in *P. chrysosporium* is a source of extracellular H_2O_2 (Greene and Gould 1984).

As mentioned before, Mn-peroxidase can oxidize glutathione, NADH, NADPH, and dihydroxymaleic acid (Glenn et al. 1986, Paszczynski et al. 1986) to produce H_2O_2. Extracellular NADH and NADPH have indeed been found in *P. chrysosporium* grown on 5% glucose, 0.5% peptone, and 0.3% yeast extract for 2 days. After this time, however, the pyridine coenzymes started to be degraded (Kuwahara et al. 1984). The enzyme oxidizing NADH has been purified from ligninolytic cultures and is probably identical with Mn-peroxidase (Asada et al. 1986, 1987).

Another oxidase producing H_2O_2 has been detected in *P. chrysosporium*, namely, methanol oxidase (Buswell and Eriksson 1986, Eriksson et al. 1986a, Nishida and Eriksson 1987, Eriksson 1987). This intracellular enzyme contains

FAD and has a molecular weight of about 310000 with four subunits, each with a weight of 75000 (Nishida and Eriksson 1987). The enzyme in the presence of O_2 produces formaldehyde from methanol, and has a pH optimum near 7 (Buswell and Eriksson 1986, Nishida and Eriksson 1987). In addition to methanol, ethanol also is oxidized. This oxidation probably yields acetaldehyde, which may be used by the enzyme glyoxal oxidase described below. In any case, methanol formed during demethoxylation of lignin or lignin-related substances (Ander and Eriksson 1985, Ander et al. 1985) can be used for further production of H_2O_2 to be used by ligninolytic peroxidases to yield still more methanol which can be oxidized. However, one important question is not clear concerning the intracellular glucose and methanol oxidases; namely, how is the H_2O_2 produced transferred to the extracellular lignin peroxidases?

Kuwahara et al. (1987) also have purified an alcohol oxidase from *P. chrysosporium*. This enzyme oxidizes methanol to formaldehyde and H_2O_2 and has a molecular weight of 220000 as determined by GPC and 73000 as determined by SDS-PAGE. Highest activity was observed in low-nitrogen oxygenated cultures. It seems similar to the methanol oxidase described by Nishida and Eriksson (1987).

Still another H_2O_2-producing enzyme is the extracellular glyoxal oxidase found in *P. chrysosporium* culture filtrates (Kersten and Kirk 1987). This enzyme is produced under secondary metabolism and has a broad pH range with a peak around pH 6. It uses many substrates found as secondary metabolites in the fungal culture filtrate, i.e., methylglyoxal, glycolaldehyde, acetaldehyde, formaldehyde, glyoxal, glyoxylic acid, dihydroxyacetone, and glyceraldehyde.

Both intra- and extracellular H_2O_2 production has been found in the white-rot fungus *Panus tigrinus* (Golovleva et al. 1986) but the conditions for this production were not described. Glucose oxidase activity and H_2O_2 appeared simultaneously in the culture liquid. Lignin peroxidase was also detected in this fungus.

H_2O_2 production by brown-rot and white-rot fungi was also investigated by Highley (1987b), who found that low concentrations of nitrogen and carbohydrates (cellulose, glucose, xylose, and mannose) in an agar medium had little effect on H_2O_2 production by white-rot fungi but stimulated H_2O_2 production by brown-rot fungi.

The regulation of both glucose concentration and H_2O_2 production in *P. chrysosporium* is suggested to take place through the production and degradation of an extracellular glucan which is formed in low-nitrogen, high-glucose cultures during secondary metabolism (Eriksson et al. 1986a, Bes et al. 1987, Eriksson 1987). When the glucose concentration becomes low enough, the extracellular glucan is again degraded to glucose by an induced β-1,3-glucanase. This enzyme has been purified from the culture liquid of *P. chrysosporium* ME-446 (Bes et al. 1987). The glucan was shown to consist of a backbone of (1,3)-linked residues. Approximately every second residue was substituted at *O*-6, together with internal (1,6)-linked D-glucose residues in a ratio of 4.4:1. The glucan-degrading enzyme has a molecular weight of 80000 and is most active at pH 5. Addition of the glucanase to cultures releasing $^{14}CO_2$ from ^{14}C-

ring-labeled DHP resulted in decreased $^{14}CO_2$ release (Bes et al. 1987). Buchala and Leisola (1987) also have isolated and studied the β-D-glucan from *P. chrysosporium* (ATCC 24725). This glucan contained β-1,3-linked D-glucose and single β-1,6-linked D-glucose residues. Some differences in sugar composition after hydrolysis were reported by Buchala and Leisola (1987) compared to the results of Bes et al. (1987). This is probably due to differences in techniques used for extraction and purification of the polysaccharide.

4.5.6 Quinone-Reducing Enzymes

In lignin degradation experiments using agar cultures it was found that when *Coriolus versicolor* was grown on lignin agar plates supplemented with cellobiose or cellulose, the formation of chromophoric structures, most likely quinones, was strongly decreased. This did not occur when glucose was the supplement (Westermark and Eriksson 1974a). The enzyme responsible for this quinone reduction was subsequently purified from *Phanerochaete chrysosporium* P 127-1 (*Chrysosporium lignorum*, *Sporotrichum pulverulentum*) grown in cellulose shake cultures and named cellobiose: quinone oxidoreductase, CBQ, or cellobiose dehydrogenase(quinone), EC 1.1.5.1 (Westermark and Eriksson 1974b, 1975, 1988). It was found to be an FAD enzyme with a molecular weight of 58000 and a pH optimum of 4.5 – 5. During quinone reduction to the corresponding phenol, cellobiose is oxidized to cellobiono-δ-lactone which seems to be further oxidized by a lactonase to give cellobionic acid (see Fig. 2.11, Sect. 2.5.1). Whereas the enzyme only oxidized cellobiose and to some extent lactose (44%), the specificity for quinones was less restricted and both ortho- and paraquinones (Fig. 4.47) were reduced by CBQ. Enzyme activity was measured using either 5-tert-butyl-benzoquinone-(1,2) (Westermark and Eriksson 1974b) or 3,5-di-tert-butyl-o-benzoquinone (Ander and Eriksson 1977) together with cellobiose in acetate buffer at pH 4.5. It was shown by Westermark and Eriksson (1974a) that CBQ inhibited laccase-catalyzed polymerization of guaiacol, although the oxygen consumption by laccase was unaffected. Odier et al. (1988) studied possible inhibition by CBQ of ligninase-catalyzed polymerization of guaiacol, a dimethoxylated β-*O*-4 dimer, and ring-labeled DHP. It was found that CBQ did not inhibit polymerization caused by ligninase. Formation of the acetosyringone phenoxy radical by lignin peroxidase was not inhibited by CBQ as measured by ESR (Odier et al. 1988). Therefore, CBQ does not function as a phenoxy radical reductant.

Other results show that both laccase- and Mn-peroxidase-catalyzed decarboxylation of vanillic acid is strongly inhibited by CBQ (Ander et al. 1990, J. Biotechnol., in press). Similarly, lignin peroxidase oxidation of veratryl alcohol was inhibited by CBQ. Research concerning enzymatic oxidation and reduction is important in order to obtain an enzyme mixture which can depolymerize lignin, make it water soluble, and prevent repolymerization. That lignin peroxidase can polymerize different lignins and phenols was shown by Haemmerli et al. (1986a) and Kirk (1986).

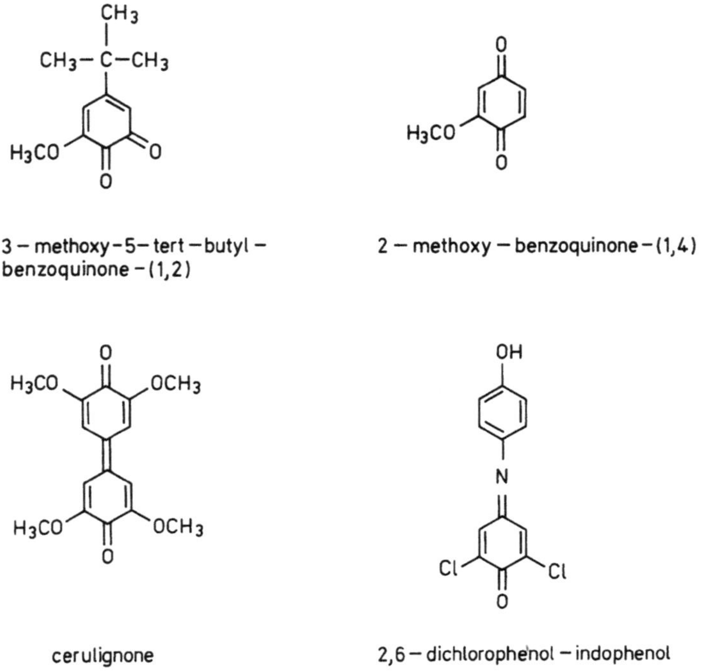

Fig. 4.47. Quinones which can be reduced by CBQ: 3-methoxy-5-tert-butyl-benzoquinone-(1,2), 2-methoxy-benzoquinone-(1,4), cerulignone, 2,6-dichlorophenol-indophenol (Westermark and Eriksson 1974b)

CBQ has now been found in many fungi, some of which are only cellulolytic and not ligninolytic (Dekker 1980, Coudray et al. 1982, Hütterman and Noelle 1982, Westermark and Eriksson 1988). As mentioned earlier (Sect. 4.3.3) the presence of cellulose increased lignin degradation by *Phanerochaete chrysosporium*, *Pleurotus ostreatus*, and other white-rot fungi (Ander and Eriksson 1975, 1977, Hiroi and Eriksson 1976, Hiroi et al. 1976). Ander and Eriksson (1977) detected CBQ in 25 different white-rot fungi; no evidence for the enzyme was obtained in six brown-rot fungi (Ander and Eriksson 1978).

The highest yield of CBQ is obtained in cultures with cellulose as a carbon source, although cellobiose cultures also give some activity. With glucose as a carbon source, no CBQ is obtained with *P. chrysosporium* P 127-1 (Westermark and Eriksson 1974a, b). The CBQ enzyme does not seem to be absolutely necessary for lignin degradation, since a cellulase-less mutant of *P. chrysosporium* could degrade kraft lignin and lignin in birch and pine without cellulose loss (Ander and Eriksson 1975). Furthermore, good lignin degradation is obtained on glucose media (Kirk et al. 1978a), in which CBQ is not produced.

Morpeth and Jones (1986) isolated and purified four forms of CBQ (or cellobiose quinone dehydrogenase as they called it) from *Phanerochaete chrysosporium* (CMI 172 727). The M_r was 57800–60000 as determined by sedimentation equilibrium and SDS/polyacrylamide gel electrophoresis. Activity was determined at pH 6 using cellobiose and 2,6-dichlorophenolindophenol. In the presence of cellobiose and after addition of catalase, all four isoenzymes evolved O_2, suggesting H_2O_2 production. Also, with the active isoenzymes, cytochrome c was reduced, indicating the production of superoxide radicals. The reduction of cytochrome c was decreased to less than 1% on addition of superoxide dismutase which scavenges $O_2^{\bar{\bullet}}$.

Using the same test methods as above, Morpeth (1985) found that the enzyme cellobiose oxidase, purified from *P. chrysosporium*, also produced $O_2^{\bar{\bullet}}$ and H_2O_2. This cellobiose oxidase has been previously purified from the same fungus (Ayers et al. 1978), but it is clearly different from CBQ in that the cellobiose oxidase is a flavocytochrome protein (Ayers et al. 1978, Morpeth 1985, see also Chap. 2.2). Some evidence for production of $O_2^{\bar{\bullet}}$ by pure cellobiose oxidase was obtained (Eriksson 1981b, Ayers and Eriksson 1982) using reduction of nitrotetrazolium blue to formazan as assay. Superoxide dismutase, which has been detected in many wood-rotting fungi (Eriksson 1981), dismutates $O_2^{\bar{\bullet}}$ to form H_2O_2 and O_2. Thus both CBQ and cellobiose oxidase may be additional possibilities for H_2O_2 production, although this could not be detected by Westermark and Eriksson (1974a, b, 1975) or by Ayers et al. (1978). The results mentioned above suggest that cellobiose oxidase incorporates one oxygen atom into cellobiose to give cellobionic acid, while the other oxygen atom is probably reduced to $O_2^{\bar{\bullet}}$ (Ayers and Eriksson 1982, Morpeth 1985).

Recently, Kelleher et al. (1987) studied the influence of temperature, oxygen concentration, agitation, nitrogen concentration, and carbon source on CBQ production in *Phanerochaete chrysosporium* Burds QM9145. The highest enzyme production was obtained at 38 °C in standing cultures and in the presence of 100% oxygen, and with 4% cellulose plus 0.15% acid-precipitated kraft black liquor in the growth medium. A low nitrogen medium, according to Kirk et al. (1978a), gave a lower yield than a high nitrogen medium as reported by Kelleher et al. (1987). These new culture conditions gave 30 times more CBQ compared with the earlier shaking conditions used by Westermark and Eriksson (1974a, b). Thus, CBQ production is not under nitrogen regulation in the same way as the lignin peroxidases, and the enzyme seems to be induced by degradation products from cellulose, as found by Westermark and Eriksson (1974a, b).

An intracellular quinone-reducing enzyme has also been found in *P. chrysosporium* P 127-1 (Buswell et al. 1979b). This enzyme reduces methoxyquinone to methoxyhydroquinone using NADH or NADPH as electron donors. In addition, toluquinone, 1,4-benzoquinone, and 4,5-dimethoxy-1,2-benzoquinone were all rapidly reduced to the corresponding hydroquinones. The enzyme is active over the pH range 4–8 with optimal activity near pH 5.6. It was proposed by Buswell et al. (1979b) that the enzyme, which was induced by vanillic acid, may be important for reduction of toxic quinones. Other quinones formed during veratryl alcohol degradation are also reduced by

NAD(P)H-dependent enzymes (Leisola et al. 1988, Schoemaker et al. 1989). Since quinones are common intermediates during lignin degradation, these authors suggested that rapid metabolism of quinone/hydroquinone-type intermediates is one way of shifting the polymerization-depolymerization equilibrium toward degradation.

Green (1977) proposed that glucose-1-oxidase from *Coriolus versicolor* could function as a glucose: quinone oxidoreductase and use quinones as co-substrate instead of oxygen at low oxygen tension. In nitrogen atmosphere, laccase-oxidized dichlorophenol-indophenol (DCIP) was reduced by glucose plus glucose-1-oxidase. Green (1977) suggested that glucose-1-oxidase, much in the same way as CBQ, may serve to protect the organism from accumulation of toxic quinones and to channel quinones back to degradative routes. Later Alberti and Klibanov (1982) as well as Szklarz and Leonowicz (1986) gave further support to these findings. The former authors immobilized D-glucose-1-oxidase onto alumina powder and in the presence of glucose and in an N_2-atmosphere obtained 99%–100% conversion of benzoquinone to hydroquinone on a preparative scale. Leonowicz et al. (1986) and Szklarz and Leonowicz (1986) found that laccase-generated quinoid intermediates from lignosulfonates were reduced by glucose plus glucose oxidase (probably glucose-1-oxidase). When both laccase and glucose oxidase were present, depolymerization of lignosulfonate was more extensive, as shown with GPC on a Sephadex G-50 column. Lobarzewski and Paszczynski (1985) and Lobarzewski (1987) immobilized cellulase, D-glucose oxidase, and fungal peroxidase on glass beads; when incubated together with wheat straw, phenolic compounds and D-glucose were formed.

Since lignin degradation is mainly an oxidative process, the putative involvement of glucose oxidase under reduced oxygen concentration conditions, however, is hard to understand. Except for the production of H_2O_2, real evidence for the importance of glucose oxidase in lignin degradation is still missing. Furthermore, the phenoxy radical of acetosyringone produced by ligninase was not reduced by glucose oxidase (Odier et al. 1988). Different aspects of laccase-glucose/cellobiose: oxidoreductase cycles have also been studied by Ishihara and Nishida (1983) and Iwahara (1983) using enzymes from *Fusarium* and *Pleurotus*. In addition, preliminary experiments indicate that a xylobiose: quinone oxidoreductase may be produced in *P. chrysosporium* P 127-1 (Westermark and Eriksson 1974b) and in *Pleurotus ostreatus* (Ishihara and Nishida 1983).

In Fig. 4.48, the results of Westermark and Eriksson (1974a, b), Eriksson et al. (1986a, b), and Leonowicz et al. (1986, 1988) together with the latest findings on aromatic ring cleavage by lignin peroxidase are combined. In this hypothetical scheme it is proposed that during fungal cellulose (xylan) degradation, CBQ and other quinone reductases, possibly glucose-1-oxidase and glucose-2-oxidase, are induced to reduce quinones, decrease polymerization, and produce H_2O_2. Presumably, enzymes reducing phenoxy radicals are present in the fungal culture. Lignin may be ring-cleaved in the polymer either di-

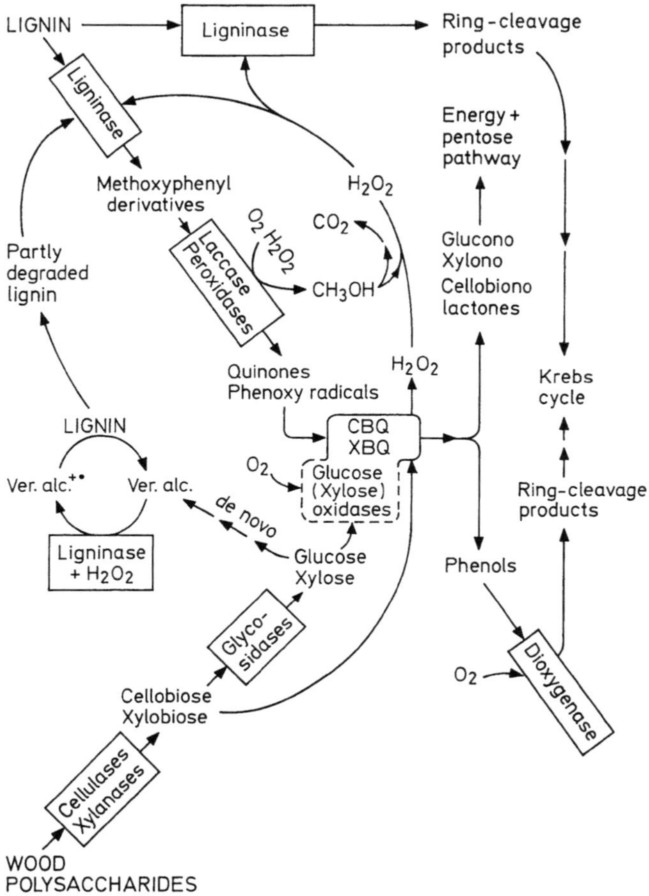

Fig. 4.48. Hypothetical scheme for degradation of lignin, cellulose, and xylan in wood. Adapted from results of Westermark and Eriksson (1974a, b), Leonowicz et al. (1986) as well as from results presented in Sect. 4.5.3

rectly by lignin peroxidase or may be attacked at the C_α-C_β bonds to produce methoxyphenyl derivatives. These are further oxidized by laccase or Mn-dependent peroxidase to produce phenoxy radicals, quinones, and methanol. The methanol can be oxidized by methanol oxidase (Nishida and Eriksson 1987) to produce H_2O_2 and formaldehyde, which is further oxidized to formic acid and ultimately to CO_2 and H_2O. Ring-cleavage products go to the Krebs cycle while the glucono- or cellobiono-δ-lactones are metabolized via the pentose pathway of the fungus. In Fig. 4.48 the veratryl alcohol cation radical, which acts as a mediator for charge transfer to lignin (Harvey et al. 1986), is also indicated.

4.5.7 Cleavage of Methoxyl Groups by Wood-Rotting Fungi

During decay of wood by wood-rotting fungi there is a significant decrease in the methoxyl content of the remaining lignin (Trojanowski and Leonowicz 1969, Trojanowski et al. 1966, Ander and Eriksson 1978, Crawford 1981 b, Oki et al. 1982, Tai et al. 1982, Chen and Chang 1985, Leonowicz et al. 1988). Thus, lignin from spruce wood rotted by the brown-rot fungus *Gloeophyllum trabeum* (*Lenzites trabea*) showed a 35% reduction in methoxyl content per C_9-unit compared with sound wood (Kirk 1975). After decay by the brown-rot fungus there was an increase in the content of aromatic hydroxyl groups due to both methoxyl cleavage and hydroxylation (Kirk and Adler 1970, Kirk et al. 1970, Kirk 1975). The methoxyl content of two lignin preparations obtained from spruce wood decayed by the white-rot fungi *Polyporus anceps* and *Coriolus versicolor* was 25% lower after decay to weight losses ranging between 45% and 50% (Kirk and Chang 1974, 1975).

In a wood meal culture, the brown-rot fungi *Gloeophyllum trabeum* and *Poria cocos* released 4%–8% $^{14}CO_2$ from methoxyl-labeled DHP in 25 days (Kirk et al. 1975). Somewhat lesser release of $^{14}CO_2$ was obtained by Ander et al. (1984,1988) with some brown-rot fungi. However, Haider and Trojanowski (1980) obtained 29% $^{14}CO_2$ from methoxyl-labeled DHP with *Gloeophyllum trabeum* in 15 days. Some other brown-rot fungi gave 0%–20% $^{14}CO_2$ from the same lignin.

Although Seifert (1966) found that the soft-rot fungus *Chaetomium globosum* strongly decreased the methoxyl content of beech wood lignin, more recent results indicate that conifer-type synthetic lignin (DHP) is very little degraded by soft-rot fungi. Thus, *Chaetomium globosum*, *Phialophora mutabilis*, and *Petriellidium boydii* released less than 2% $^{14}CO_2$ from ^{14}C-methoxyl- and

Table 4.10. Demethoxylation and methanol formation from vanillic acid, vanillyl alcohol, syringic acid, and vanillyl glycol by laccase and horseradish peroxidase (HRPO). Calculation is in percent of initial total amount of methoxyl groups (C.v. means *Coriolus versicolor*)

Treatment	Demethoxylation (%)	Reference
C.v. laccase + 1.2 mM vanillic acid, 3 h	2.2 – 2.6	Ishihara and Miyazaki (1974)
C.v. laccase + 1.3 mM vanillyl alcohol, 3 h	8.9 – 15.4	Ishihara and Miyazaki (1974)
C.v. laccase + 1.2 mM vanillic acid, 3 h	5 – 7	Ishihara and Ishihara (1975)
C.v. laccase + 1.0 mM syringic acid, 3 h	13 – 17	Ishihara and Ishihara (1975)
Laccase + 4 mM syringic acid, 3 h		
Pholiota mutabilis intracellular	37.4 – 45.6	Leonowicz et al. (1979)
Pholiota mutabilis extracellular	32.8 – 41.8	Leonowicz et al. (1979)
C.v. laccase + 2.4 mM vanillic acid, 2 h	1.4	Ander et al. (1983 a)
HRPO + 2.4 mM vanillic acid, 2 h	2.2	Ander et al. (1983 a)
C.v. laccase + ca. 1 μM vanillic acid, 2 h	12.5	Ander et al. (1983 a)
HRPO + ca. 1 μM vanillic acid, 2 h	10.2	Ander et al. (1983 a)
C.v. laccase + 22 mM vanillyl glycol, 100 h	15 – 20	Lundquist and Kristersson (1985)

ring-labeled DHP in 9–14 days (Ander et al. 1984, 1988). Using high nitrogen cultivation conditions, Haider and Trojanowski (1975) obtained 7.8% $^{14}CO_2$ in 10 days from ^{14}C-methoxyl-labeled DHP with one soft-rot fungus but much less with three others. Lignin degradation by brown- and soft-rot fungi is further discussed in Sect. 4.7.

Demethoxylation of low molecular weight phenols to produce methanol from vanillic acid, syringic acid, vanillyl alcohol, vanillyl glycol, acetosyringone, and 2,4,6-trimethoxyphenol has been found after incubation with laccase or peroxidase (Trojanowski et al. 1966, Caldwell and Steelink 1969, Harkin and Obst 1974, Ishihara and Miyazaki 1974, Ishihara and Ishihara 1975, 1976, Leonowicz et al. 1979, Ander et al. 1983a, Ishihara 1983, Lundquist and Kristersson 1985). Some of these results are shown in Table 4.10. It seems clear that both laccase and peroxidase demethoxylate these methoxylated phenols extensively, and that in the presence of a low concentration of the phenol (shown for vanillic acid) the enzymes give stronger demethoxylation compared to larger amounts (Ander et al. 1983a, 1985). Syringic acid appeared to be more easily demethoxylated than vanillic acid.

According to Frick and Crawford (1983), cleavage of methoxyl groups can be accomplished in three different ways (Fig. 4.49). The first mechanism is a direct hydrolysis of the ether bond to give methanol as in reaction (1). An enzyme reaction of this type, however, was considered doubtful by Frick and

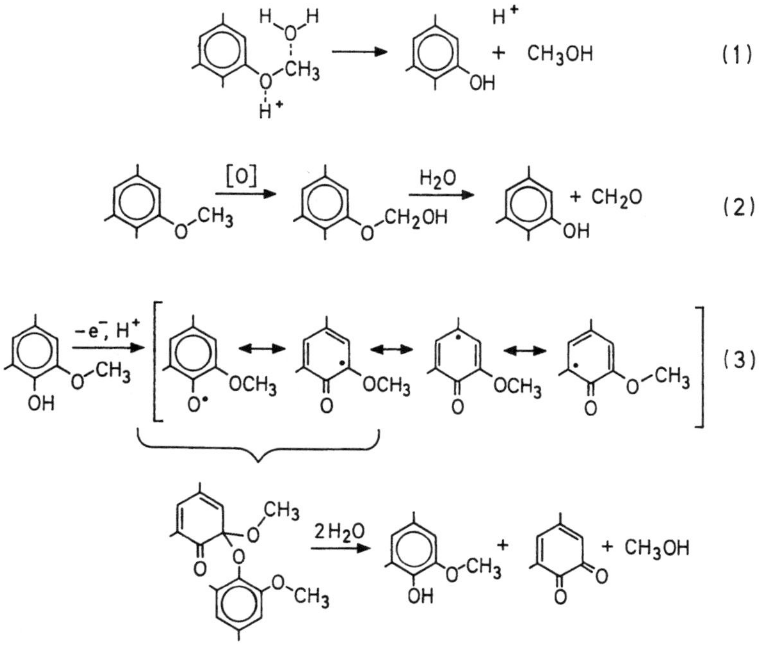

Fig. 4.49. Three mechanisms (*1-3*) for cleavage of methoxyl groups in lignin or in related phenols (Frick and Crawford 1983)

Crawford (1983). A reversed *O*-methyl transferase reaction would also be a possibility. Many bacteria and ascomycetes produce enzymes which, in the presence of NADH and oxygen, demethylate vanillic acid to give protocatechuic acid and formaldehyde (Ribbons 1970, Cain 1980). The mechanism in Eq. (2) can explain this mono-oxygenase catalyzed reaction. As discussed below, it seems as if soft-rot fungi use this mechanism for methoxyl cleavage of vanillic acid (Ander et al. 1988). In reaction (3) phenoloxidases (laccase, peroxidase) may form phenoxy radicals which can couple to form a hemiacetal, which is unstable and is hydrolyzed to the original phenol plus an o-quinone and methanol as in Figs. 4.20 and 4.49. Demethoxylation and release of methanol as mentioned in Table 4.10 may be explained in a similar way.

Recently, Ander et al. (1988) studied cleavage and metabolism of methoxyl groups from vanillic (VA) and ferulic acid (FA) by three brown-rot and three soft-rot fungi. After measuring methanol and/or formaldehyde formation from VA and FA and the release of $^{14}CO_2$ from ^{14}C-methoxyl labeled FA, $^{14}CH_3OH$ and $H^{14}CHO$, they found that brown-rot fungi remove methoxyl groups as methanol while soft-rot fungi appear to release formaldehyde as the initial product.

In vivo production of methanol from lignin-related substances by *Phanerochaete chrysosporium* K-3 was investigated by Ander et al. (1985). It was found (Table 4.11) that methanol formation, calculated as the percentage of theoretical yield, decreased in the order guaiacyl > syringyl > veratryl(3,4-dimethoxy) compounds. Methoxyhydroquinone and vanillic acid were most easily demethoxylated, whereas methanol production decreased with increasing molecular weight. That guaiacyl compounds gave more methanol than syringyl compounds is somewhat surprising, since it is generally accepted that syringyl/guaiacyl lignin is more easily degraded than guaiacyl lignin (Ander and Eriksson 1985, Faix et al. 1985, Blanchette et al. 1987). Also, in vitro a greater percentage of methanol is formed from syringic acid than from vanillic acid by laccase and peroxidase (Table 4.10; Ander, unpublished results). The use of as much as 2.4 mM lignin-related substances in the experiments by Ander et al. (1985) may result in toxic conditions for the fungus, which may be the reason for these unexpected results.

In order to produce a polymer useful for studies of methoxyl cleavage, Crawford and Frick (1982) synthesized $[O^{14}CH_3]$-polyguaiacol with a molecular weight of 5000–15000. Properties of a similar ring-labeled guaiacol polymer was also described by Crawford et al. (1981 b). In a concentrated culture filtrate from *Phanerochaete chrysosporium* an enzyme activity was found that in 48 h demethoxylated the methoxyl-labeled polyguaiacol to an extent of 2.3%. This figure increased to 3.6% if H_2O_2 was present (Frick and Crawford 1983). Some demethoxylation of DHP to methanol by horseradish peroxidase plus H_2O_2 was obtained by Frick and Crawford (1983) and Ander and Eriksson (1985). However, H_2O_2 alone (0.15–2.6 mM) also gave some demethoxylation. Both Ishihara and Miyazaki (1974) and Ander and Eriksson (1985) found very little (less than 2%) methoxyl cleavage of MWL or DHP by laccase B.

Table 4.11. Methanol production by *Phanerochaete chrysosporium* K-3 from lignin-related monomers on addition of these compounds on day 3 to a standing culture containing high glucose-low nitrogen medium. Methanol is calculated in percent of the theoretical yield (Ander et al. 1985)

Compounds (2.4 mM)	Methanol (%)		
	24 h	48 h	72 h
Methoxyhydroquinone	65.8	80.9	93.4
Vanillic acid	13.9	31.3	37.9
Acetovanillone	11.4	18.7	9.7
Vanillyl alcohol	4.7	5.97	12.06
Vanillin	1.17	2.40	9.22
Syringic acid	2.39	5.32	7.71
Syringaldehyde	0.98	2.50	1.87
Coniferyl alcohol	0	1.52	9.39
Ferulic acid	0	0.78	5.86
Veratric acid	1.31	1.77	0.68
Veratraldehyde	0.50	0.78	4.17
Veratryl alcohol	0.07	0.73	1.56
3,4-Dimethoxy cinnamyl alcohol	0.48	3.12	5.41
3,4-Dimethoxy cinnamaldehyde	0	0	1.94

Methoxyl metabolism in white-, brown-, and soft-rot fungi is repressed by 1% glucose and stimulated by nitrogen (Ander et al. 1984), as shown with methoxyl-labeled vanillic and syringic acid. Haider and Trojanowski (1980) compared release of $^{14}CO_2$ from differently labeled DHPs by white- and brown-rot fungi and observed that increasing concentrations of glucose, cellobiose, or xylose repressed release of $^{14}CO_2$ from methoxyl-labeled DHP. The reason was that methanol or formaldehyde from the methoxyl groups remained in culture solution and were not oxidized rapidly. Highest levels of methanol oxidase are obtained in a low-glucose high-nitrogen medium (Buswell and Eriksson 1986).

In a high-glucose low-nitrogen medium Ander and Eriksson (1985) later determined that *Phanerochaete chrysosporium* ME-446 released methanol from DHP as well as from spruce and birch MWL. Due to decreased oxidation, most methanol, 25.5% in 22 h, was accumulated in a large cultivation volume (15 – 20 ml) when DHP was added 6 days after inoculation (Fig. 4.50). Since methanol formation was strongly repressed by high nitrogen in the medium, by glutamate addition, or by culture agitation, it was concluded that demethoxylation of lignin takes place under secondary metabolism and conditions of lignin degradation. In percent of the total amount of methoxyl groups present in the lignins, birch MWL gave more methanol (11.6% in 28 h) than spruce MWL (4% in 28 h) (Ander and Eriksson 1985).

Laccase and horseradish peroxidase seem to have limited activity against lignin, although significant demethoxylation has been obtained with different lignin-related and other phenols. Whether Mn-peroxidase and lignin perox-

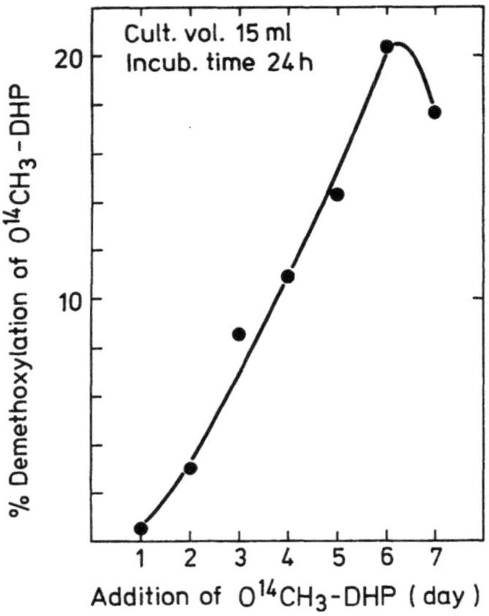

Fig. 4.50. Formation of methanol by *Phanerochaete chrysosporium* ME-446 from $O^{14}CH_3$-labeled lignin added at different days. Medium and calculation as in Table 4.11 (Ander and Eriksson 1985)

idase can cause more than limited methoxyl cleavage in lignin is not yet clear. In vivo Mn-peroxidase and lignin peroxidase from *Phanerochaete chrysosporium* may have some demethoxylating activity against lignin (Kirk et al. 1986d). As discussed by Chua et al. (1983) and Ander and Eriksson (1985), it is not clear whether methoxyl cleavage occurs in the lignin polymer or in the degradation products. Since Ander and Eriksson (1985) were able to detect 7.8% methanol from DHP only 8 h after addition to the culture, it may be concluded that methanol is released both from the polymer and from lignin degradation products during lignin degradation.

Paszczynski and Trojanowski (1977) purified intracellular veratrate-*O*-demethylases from the ascomycete *Chaetomium piluliferum* and from the basidiomycete *Xerocomus badius*. Both demethylases required NADH and O_2 but had to be purified and stored under anaerobic conditions and in the presence of dithioerythritol to avoid inactivation. In addition to veratric acid, 3,4-dimethoxycinnamic acid was also demethylated by these monooxygenases. Buswell et al. (1982b), however, did not find intracellular demethylase activity in some soft-rot fungi, possibly due to enzyme instability.

Huynh and Crawford (1985) detected an extracellular enzyme from *Phanerochaete chrysosporium* that could demethylate 2-methoxy-3-phenylbenzoic acid to 2-hydroxy-3-phenylbenzoic acid. The enzyme was a peroxidase, which was stimulated by Mn^{2+} and had a pH optimum at pH 5 in a Na-tartrate buffer. These properties suggest that the enzyme is very similar to the Mn-peroxidase discovered by Kuwahara et al. (1984). It was also found that a concentrated culture liquid of *P. chrysosporium* caused 10%–11% demethylation of ferulic acid in 17 h (Ander and Eriksson 1985, Ander et al. 1985). This activity is almost certainly due to Mn-peroxidase activity, and the question still remains whether, in addition to this enzyme, a special aromatic demethoxylase also operates in white-rot fungi.

Wood degradation and demethoxylation of lignin in wood can be detected by the use of osmium tetroxide (OsO_4), detectable in the transmission electron microscope as black particles (Messner et al. 1985, 1986). These "osmiophilic particles" reacted only with di- or trihydroxy phenols but not with monohydroxy phenols. Brown-rotted lignin, containing stable dihydroxy phenolic structures (Kirk 1975), reacted with OsO_4 and Messner et al. (1986) suggests the possible use of this oxide to localize demethoxylation in wood.

4.6 Metabolism of Monomeric Lignin-Related Compounds

Preparation of monomeric aromatic compounds for ring cleavage is often done via monooxygenases (Hayaishi 1974), which incorporate one atom of an oxygen molecule into a substrate molecule. The other oxygen atom is reduced to H_2O in the presence of an electron donor DH_2, such as NADH, NADPH, tetrahydrofolic acid, ascorbic acid, or GSH as follows:

$$X + O_2 + DH_2 \rightarrow XO + H_2O + D \ .$$

After the monooxygenase-catalyzed reaction, the substrate is more water-soluble and the ring can be split by dioxygenases. Monooxygenase activity has been found mainly in bacteria and Deuteromycetes but has also been found in a few basidiomycetes (Dagley 1971, Towers et al. 1974, Paszczynski and Trojanowski 1977, Cain 1980). To our knowledge, purification of a monooxygenase from a white- or brown-rot fungus has not been reported.

Dioxygenases are enzymes that incorporate both atoms of an oxygen molecule into the substrate:

$$DH_2 + O_2 \rightarrow D(OH)_2 \ .$$

Aromatic substrates, for example, catechol and protocatechuic acid, can be cleaved by dioxygenases either between the carbon atoms bearing the hydroxyl groups (intradiol; ortho- or 3,4-cleavage) or between one of the carbon atoms bearing the hydroxyl group and the next carbon atom (extradiol; meta- or 2,3/4,5-cleavage) (Figs. 4.51, 4.52). Ring-cleaving dioxygenases from bacteria have been extensively studied and also crystallized (Dagley 1971, Stanier and Ornston 1973, Hayaishi 1974), and Cain (1980) has studied partially purified dioxygenases also from Ascomycetes, Deuteromycetes, Basidiomycetes, and yeasts. Iyayi and Dart (1982) reported that cell-free intracellular extracts of *Schizophyllum commune* could cleave the aromatic ring of protocatechuic acid to form β-ketoadipic acid. *S. commune* is a rather weak lignin degrader but as reported by Iyayi and Dart (1982) it could even grow on sinapyl alcohol as the sole carbon source. Schmidt and Liese (1980) did not find extracellular catechol 1,2-dioxygenase in *S. commune*.

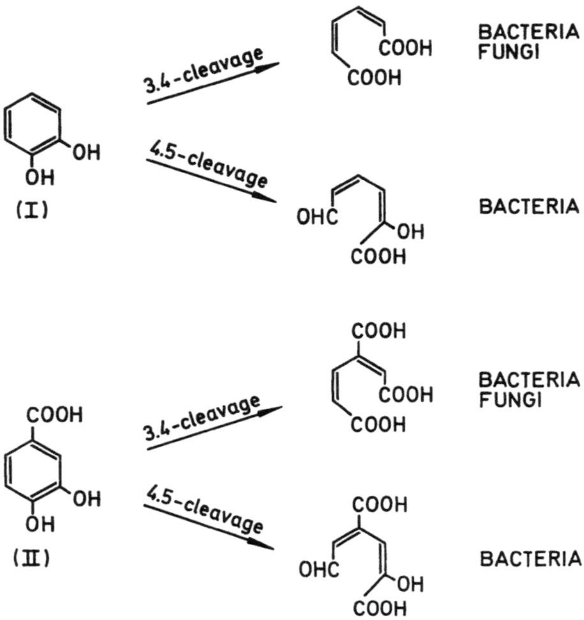

Fig. 4.51. *Ortho*-cleavage (3,4-cleavage) and *meta*-cleavage (4,5-cleavage) of catechol (*I*) and protocatechuic acid (*II*) by dioxygenases from bacteria and fungi

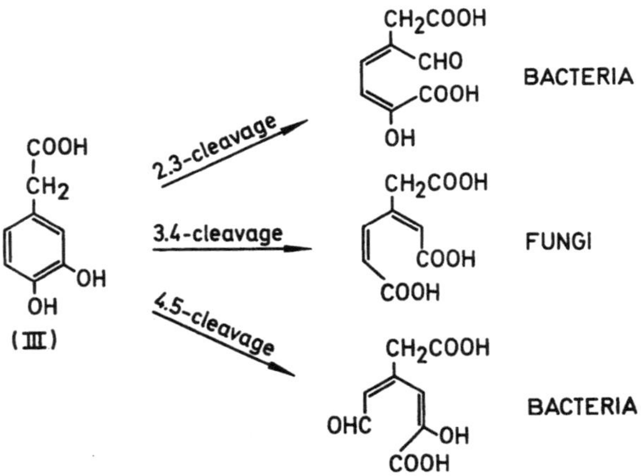

Fig. 4.52. Ring cleavage of homoprotocatechuic acid (*III*) by bacteria and fungi

Wojtas-Wasilewska and Trojanowski (1980) and Wojtas-Wasilewska et al. (1983) purified a protocatechuate 3,4-dioxygenase from the ascomycete *Chaetomium piluliferum* and from *Pleurotus ostreatus*, respectively. Both enzymes were intracellular and had pH optima at pH 8. The molecular weight of the *Chaetomium* enzyme was 76000, while that of *Pleurotus* was

Table 4.12. Substrate specificity of protocatechuate 3,4-dioxygenase from *Pleurotus ostreatus* mycelium induced with p-hydroxybenzoic acid (Wojtas-Wasilewska et al. 1983)

Substrate	Relative activity (%)
Protocatechuic acid	100
Homoprotocatechuic acid	83.3
4-Methylcatechol	65.5
Caffeic acid	37.9
3-Methylcatechol	34.5
Catechol	33.3
2,3-Dihydroxybenzoic acid	26.6
Protocatechuic aldehyde	20.6
DOPA	4.0
2,5-Dihydroxybenzoic acid	0
Vanillic acid	0
p-Hydroxybenzoic acid	0
4-Nitrocatechol	0

200000–205000. Both enzymes were induced by p-hydroxybenzoic acid. Storage of the *Pleurotus* enzyme in lyophilized form for 4 days at 4 °C gave rise to a 48% activity loss. A reducing agent such as mercaptoethanol added to the enzyme extracts decreased such losses. In Table 4.12, the substrate specificity of the *Pleurotus* protocatechuate 3,4-dioxygenase is recorded. The specificity is rather low and many aromatic compounds bearing o-dihydroxy groups are oxidized (Wojtas-Wasilewska et al. 1983). This intracellular dioxygenase, if incubated together with lignosulfonate, causes oxygen consumption in the same order of magnitude as obtained with protocatechuate and some decrease in absorbance of the lignosulfonate, suggesting that ring cleavage had occurred (Wojtas-Wasilewska and Luterek 1987). However, it is not known whether aromatic rings in wood lignin or in other water-insoluble lignins are cleaved by this dioxygenase.

The ring-cleavage products shown in Figs. 4.39 and 4.40 and obtained from β-aryl ether dimers and veratryl alcohol (Leisola et al. 1985b, Umezawa et al. 1986b) do not appear similar to the ring-cleavage products obtained from catechol and protocatechuic acid (Figs. 4.51, 4.52 and Cain 1980). Furthermore, the latter two phenols have not been found as main products from fungal lignin degradation (Kirk 1984, Chen and Chang 1985). Even if the reason for this may be rapid metabolism of these phenols, it also may indicate that the main pathway for fungal degradation of lignin in wood is *not via dioxygenases*, which are most active at pH 7–8, in contrast to the different lignin peroxidases which are active at pH 2–5. Nevertheless, intracellular enzymes such as the above-mentioned protocatechuate dioxygenase might be used for degradation of phenols in pulping waste waters after immobilization on glass beads, as described by Wojtas-Wasilewska et al. (1988).

Vanillic acid (vanillate) is a key intermediate in the degradation of lignin and lignin-related substances by white-rot fungi and other microorganisms (Chen et al. 1982, 1983a, b, Kirk 1984). Although earlier reports indicated that

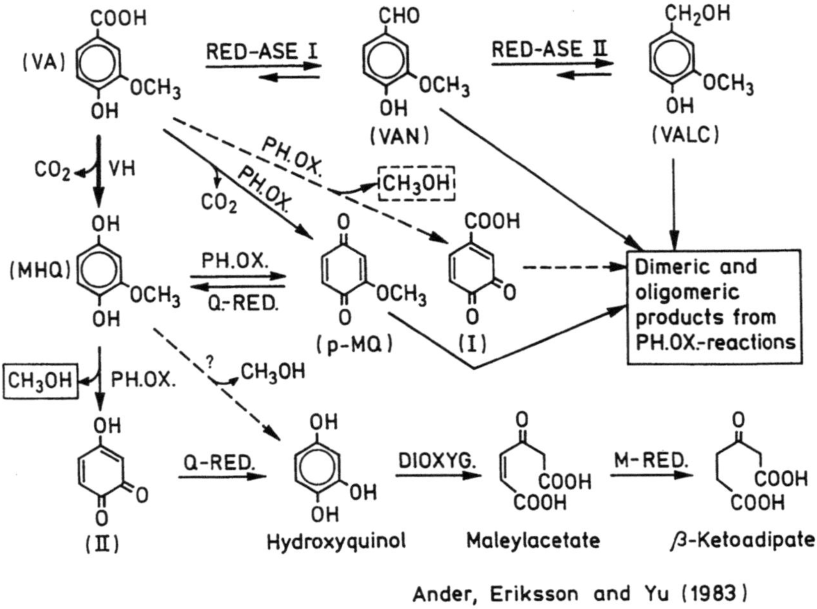

Fig. 4.53. Proposed scheme for the metabolism of vanillic acid to β-ketoadipate by *Phanerochaete chrysosporium*. The o-quinone (*II*) is a hypothetical structure (Ander et al. 1983a)

vanillate is metabolized through protocatechuate in *Coriolus versicolor* (Flaig and Haider 1961, Cain et al. 1968), it is now clear that such demethylation does not occur in white-rot fungi, and that the main degradative pathway for vanillate is through methoxyhydroquinone in white-rot fungi (Kirk and Lorenz 1973, Nishida and Fukuzumi 1978, Ander et al. 1980a, b, 1983a). After initial oxidative decarboxylation as shown in Fig. 4.53, methanol was detected as an early metabolic product appearing before $^{14}CO_2$ is released from ^{14}C-ring-labeled vanillate (Ander et al. 1983a). It is thus proposed that demethoxylation of vanillate occurs before ring cleavage, and that the formation of the trihydroxylated compound hydroxyquinol, which seems to be necessary for ring cleavage by *Phanerochaete chrysosporium* (*Sporotrichum pulverulentum*) (Buswell and Eriksson 1979), is achieved by oxidative decarboxylation, demethoxylation, and quinone reduction. The enzyme which decarboxylates and hydroxylates vanillic acid has been purified from *P. chrysosporium* (Buswell et al. 1979a, 1981, Yajima et al. 1979). Highest activity in the assay mixture of this intracellular vanillate hydroxylase was obtained in the presence of NADPH and FAD at pH 5.8–8 (Buswell et al. 1981).

In *Phanerochaete chrysosporium* cultures containing 1.0% glucose or cellobiose, vanillic acid was reduced to vanillin and vanillyl alcohol (Ander et al. 1980a, b). Krisnangkura and Gold (1979) found that vanillyl alcohol in vitro could be oxidized back to vanillin and vanillic acid and finally to methoxyhydroquinone (MHQ) by horseradish peroxidase (HRPO) in a yield of 0.3%.

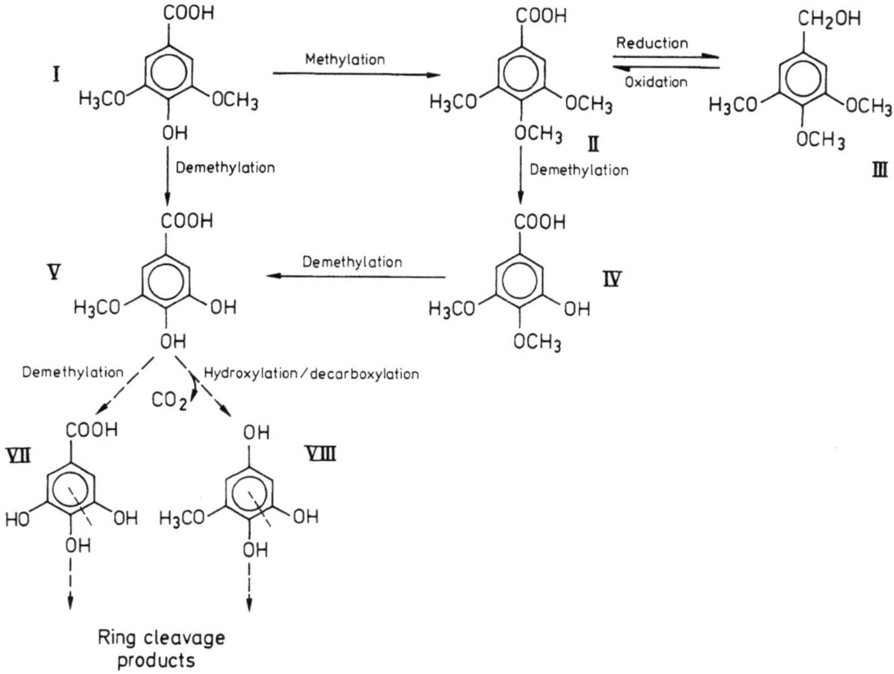

Fig. 4.54. Reactions proposed to take place during metabolism of syringic acid by *Phanerochaete chrysosporium*. *Dashed lines* indicate postulated metabolic pathways (Eriksson et al. 1984)

Ander et al. (1980b) also found reoxidation of vanillyl alcohol and vanillin by HRPO, but in 15–60 min no oxidation of vanillic acid to MHQ was detected. In any case, conversion of vanillic acid to MHQ by HRPO seems to be a minor reaction.

In low-glucose high-nitrogen cultures the vanillate hydroxylase-catalyzed decarboxylation of vanillate was rapid both during primary and secondary growth of *Phanerochaete chrysosporium* (Ander et al. 1983a, b, 1984). Ander et al. (1983b) suggested that monomers are degraded during both primary and secondary metabolism, while dimers and more complex lignin structures are attacked only during the secondary growth phase of the fungus.

Buswell et al. (1982b) studied the metabolism of vanillic acid by white-, brown-, and soft-rot fungi. All white- and brown-rot fungi decarboxylated vanillate to MHQ. Vanillate hydroxylase was found in white- and brown-rot fungi but not in soft-rot fungi, which instead contained small amounts of protocatechuate 3,4-dioxygenase. It was proposed by Buswell et al. (1982b) that the soft-rot fungi metabolize vanillate via demethylation to give formaldehyde, which is in accordance with the results of Ander et al. (1988).

Both syringic and vanillic acids are major intermediates of hardwood lignin degradation (Chen et al. 1983b, Tai et al. 1983a, b). Some metabolites formed from syringic acid during degradation are shown in Fig. 4.54. Eriksson et al.

(1984) suggested that demethoxylation to produce 3-*O*-methylgallate (V) and methanol is the first and main step in syringate metabolism by *Phanerochaete chrysosporium*. During studies of the metabolism of sinapyl alcohol by the white-rot fungus *Schizophyllum commune*, Iyayi and Dart (1982) found that syringate was metabolized directly to gallic acid (Fig. 4.54, VII) plus two methanol molecules. Since decarboxylation of 3-*O*-methylgallate by vanillate hydroxylase (Buswell et al. 1981) occurs at a high rate (75% of the rate observed for vanillate), it is thought that 3-*O*-methylgallate is either decarboxylated to 1,3,4-trihydroxy-5-methoxybenzene (VIII) or demethoxylated/demethylated to gallic acid (VII). Because Eriksson et al. (1984) could not detect compounds (VII) and (VIII), it was not possible to evaluate which of these two compounds were ring-cleaved. Umezawa et al. (1982) and Higuchi (1985b) suggested that phenylcoumaran and biphenyl substructures of lignin are mainly degraded to 5-carboxyvanillic acid, which is decarboxylated and demethylated to gallic acid, which is the final aromatic monomer. In addition to the catabolic reactions, syringic acid, like vanillic acid, is 4-*O*-methylated by *P. chrysosporium* (Eriksson et al. 1984). Since all degradation products and the enzymes in syringate metabolism have not been isolated or characterized, this scheme (Fig. 4.54) is still speculative.

Ferulic acid is also a main product from degradation of guaiacyl lignin (Ishikawa et al. 1963). The metabolism of this compound by a *Trametes* sp. (*Coriolus*) was studied by Nishida and Fukuzumi (1978). In a 2% ethanol medium, ferulic acid was metabolized to vanillic acid and methoxyhydroquinone, while in a 2% glucose medium reduction to coniferyl alcohol and vanillyl alcohol also occurred. Saturation of the side chain to give dihydroconiferyl alcohol was also found.

Degradation of ferulic acid by *Phanerochaete chrysosporium* has also been investigated by Enoki et al. (1981a, b, c) and Gupta et al. (1981). Main products identified were coniferyl alcohol, coniferyl aldehyde, dihydroferulic acid, and dihydroconiferyl alcohol. Small amounts of vanillyl alcohol, vanillic acid, and methoxyhydroquinone were also formed. It was suggested that ferulic acid is degraded via olefin saturation and subsequent α-hydroxylation followed by α,β-cleavage (Enoki et al. 1981c). This hydroxylation and cleavage would probably be catalyzed by lignin peroxidase, as shown for compounds III and IV in Fig. 4.28 (Gold et al. 1984). It was also shown by Enoki et al. (1981a, b, c) that 3,4-dimethoxycinnamic acid was metabolized by *P. chrysosporium* in a similar way as ferulic acid.

4.7 Lignin Degradation by Brown-Rot, Soft-Rot, and Other Fungi

4.7.1 Brown-Rot Fungi

Brown-rot fungi are generally associated with decay of softwoods, although in laboratory tests hardwoods are also attacked (Table 1.3 and Highley 1978,

Table 4.13. Weight, lignin, and carbohydrate losses from grass, maple hardwood, and spruce softwood caused by *Coriolus versicolor* (C.v.), *Poria placenta* (P.pl.), *Streptomyces viridosporus* (S.v.), and *Streptomyces setonii* (S.s.) after 12 weeks (After Antai and Crawford 1981, 1982)

Microbial strain	Weight loss (%)	Lignin loss (%)	Carbohydrate loss (%)
Grass lignocellulose			
C.v.	64.0	61.9	71.0
P.pl.	12.1	25.6	29.2
S.v.	56.7	44.2	68.8
S.s.	49.2	39.0	58.7
Control	5.1	5.7	5.1
Hardwood lignocellulose			
C.v.	46.2	63.0	55.4
P.pl.	14.6	18.9	27.8
S.v.	23.0	32.0	45.9
S.s.	19.3	29.5	37.1
Control	1.9	4.1	5.9
Softwood lignocellulose			
C.v.	43.9	64.2	65.5
P.pl.	33.3	16.1	62.6
S.v.	18.8	30.9	65.1
S.s.	20.8	34.1	47.7
Control	5.2	5.6	12.4

1987a). It was suggested that in nature white-rotters are more competitive on hardwoods compared to brown-rotters. The brown-rot fungus *Poria placenta* degraded grass lignin better than hardwood or softwood lignin as shown in Table 4.13 (Antai and Crawford 1982), and in softwood the carbohydrate loss was substantial. The degradation of wood and lignin by the white-rot fungus *Coriolus versicolor*, however, was much stronger than that of the brown-rot fungus *P. placenta*. Strains of actinomycetes (*Streptomyces*) degrade lignin better than brown-rot fungi (Table 4.13, see also Sect. 4.8). Kirk and Highley (1973) reported that three brown-rot fungi removed polysaccharides but not lignin from five species of conifer woods. In this context it was mentioned that lignin determination with the sulfuric acid method (Klason lignin) is problematic after decay by brown-rot fungi, which leave a modified lignin (Kirk 1971, Kirk and Highley 1973). White-rotted wood contains only small amounts of degraded lignin, which makes the lignin analyses more reliable.

The main feature of brown-rot decay is the rapid depolymerization of the holocellulose (Sects. 1.3 and 2.5). This ability seems to be unique for brown-rot fungi and should be used in defining different decay types (Nilsson 1985, Sect. 1.2).

As mentioned in Sect. 4.5.7, brown-rot fungi strongly decrease the methoxyl content of lignin in wood (Kirk and Adler 1970, Kirk et al. 1970, Kirk 1975, Fukuda and Haraguchi 1985). During this demethylation, aromatic hydroxyl groups are formed and new hydroxyl groups may also be introduced by direct hydroxylation of aromatic rings ortho to propanoid side chains (Kirk and Adler 1969, 1970, Kirk et al. 1970, Kirk 1975). In addition to an increased

Table 4.14. Analytical properties of lignin isolated and purified from spruce wood before and after decay by brown- and white-rot fungi (Kirk 1984)

Lignin	Formula for average C_9-unit	Functional groups (mol/C_9-unit)				
		Conjugated carbonyl	Total carboxyl	Hydroxyl		
				Phenolic	Aliphatic	Total
Sound	$C_9H_{8.66}O_{2.75}$ $(OCH_3)_{0.92}$	0.07	0.10	0.24	0.92	1.16
Brown-rotted[a]	$C_9H_{8.44}O_{3.75}$ $(OCH_3)_{0.61}$	0.14	0.23	0.58	0.78	1.36
White-rotted[b]	$C_9H_{7.70}O_{3.80}$ $(OCH_3)_{0.72}$	0.16	0.58	0.10	0.77	0.87

[a] Wood decayed by *Gloeophyllum trabeum* (= *Lenzites trabea*).
[b] Wood decayed by *Polyporus anceps*.

amount of phenolic hydroxyl groups (Table 4.14, Kirk 1984), there is also an increase in the content of oxygen due to formation of conjugated carbonyl and carboxyl groups. It may be observed from Table 4.14 that persisting phenolic hydroxyl groups are not obtained after degradation by white-rot fungi. The strong increase in total carboxyl for white-rotted lignin (Table 4.14) is certainly due to aromatic ring cleavage which is probably limited during brown-rot decay.

The above results, which were obtained mainly with the brown-rot fungus *Gloeophyllum trabeum*, are supported by many other investigations reviewed by Crawford (1981 b). More recently, Fukuda and Haraguchi (1985) found that MWL, isolated from spruce wood and decayed by the brown-rot fungus *Tyromyces palustris*, had similar changes in functional groups as reported by Kirk (1975, 1984).

Generally, $^{14}CO_2$ release from ^{14}C-labeled DHPs by brown-rot fungi is small, although Haider and Trojanowski (1980) obtained as much as 29% $^{14}CO_2$ in 15 days from methoxyl-labeled DHP with *Gloeophyllum trabeum* and Kirk et al. (1975) obtained 8% $^{14}CO_2$ with the same fungus and the same type of lignin during incubation for 25 days. This may reflect the demethylating ability of *G. trabeum* mentioned above. In these experiments, ring-labeled DHP gave only 10% and 2% $^{14}CO_2$ in 15 and 25 days (Haider and Trojanowski 1980 and Kirk et al. 1975), respectively.

More recently, Ander et al. (1984, 1988) found less than 2% $^{14}CO_2$ from both methoxyl- and ring-labeled DHP in 9–15 days with brown-rot fungi such as *Gloeophyllum trabeum*, *Fomitopsis pinicola*, and *Poria placenta*. The results of Haider and Trojanowski (1980) indicate that there is a wide variation among species, since eight brown-rot fungi examined by them released 0.4%–19.8% $^{14}CO_2$ from methoxyl-labeled DHP and 0.4%–8.7% $^{14}CO_2$ from ring-labeled DHP. Differently labeled corn stalk lignins were much more resistant to degradation by the brown-rot fungi studied by Haider and Trojanowski (1980). Kaplan and Hartenstein (1980) did not find any ability of brown-rot fungi to degrade DHP, probably reflecting unfavorable culture conditions. Indeed, the best white-rot fungus only yielded a maximum of 7.8% $^{14}CO_2$. In Table 1.3 (Chap. 1) substantial loss of lignin in birch and pine wood for eight brown-rot fungi is recorded. After 12 weeks rotting, 4.5%–27.3% lignin loss was obtained, as determined by the acid-insoluble lignin method of Effland (1977). Using 12 brown-rot fungi with Japanese cedar and Japanese beech as substrates, Enoki et al. (1988) found lignin losses ranging from 0.5%–54.0% in 12 weeks for these wood species. Thus, there is again a considerable variation among brown-rot fungal species in their lignin-degrading capacity.

Degradation of three lignin-related dimers and beech wood by a great number of white- and brown-rot fungi in low-nitrogen media was investigated by Enoki and Takahashi (1983) and Enoki et al. (1985). Generally, the brown-rot fungi had poor activity against the dimers under stationary-liquid or under agar-plate culture conditions. However, *Gloeophyllum trabeum* was again rather active against the dimers and also removed lignin in beech wood, especially on agar media. Agar-culture conditions were better than liquid conditions for all fungi for degradation of dimers and beech wood. From the results of dimer product analyses it was concluded that demethylation is a major initial reaction by brown-rot fungi, whereas for white-rot fungi cleavage of C_α-C_β bonds is a predominant reaction (Enoki et al. 1985).

Ethylene production from 2-keto-4-methiolbutyric acid (KTBA) and degradation of three dimers by brown- and white-rot fungi and ascomycetes were investigated by Tanaka et al. (1986). Ethylene formation from KTBA has been

Table 4.15. Degradation of lignin dimeric model compounds and the production of ethylene from KTBA by brown-rot fungi in agar cultures (Tanaka et al. 1986)

	Ethylene[a] production (16 days)	Proportion of remaining substrate[b] (30 days)		
		I	II	III[c]
Daedalea dickinsii	0.18	–	–	–
Fistulina hepatica	0.15	–	–	–
Fomitopsis pinicola	0.75	–	–	–
Gloeophyllum saepiarium	0.40	–	–	–
Gloeophyllum trabeum	10.47	+ + + +	+ + + +	+ + + +
Laetiporus sulphureus	0.49	–	+	+
Lentinus lepideus	4.17	+ + +	+	+
Phaeolus schweinitzii	0.03	–	–	–
Pholiota adiposa	4.43	+ + +	+ + +	+ + +
Spongiporus sinuosus	0.45	–	–	–
Tyromyces palustris	0.17	–	–	–

[a] Ethylene production: 10^{-8} mol/5 h.
[b] + + + +: 0–20%, + + +: 21–40%, + +: 41–60%, +: 61–80%, –: 81–100%.
[c] I: 4-ethoxy-3-methoxyphenylglycerol-β-guaiacyl ether; II: 1-(3′,4′-dimethoxyphenyl)-1,3-dihydroxy-2-(4″-methoxyphenyl)propane; III: dihydroanisoin.

used as a measurement of ligninolytic activity (Kelley and Reddy 1982, Sect. 4.5.1). From the results in Table 4.15 it is clear that most brown-rot fungi that generated ethylene from KTBA also degraded the three dimers. It appears from all the results presented in this Section that *Gloeophyllum trabeum* is a strong wood degrader that shares many properties with white-rot fungi. It must be emphasized, however, that in general brown-rot fungi are not as active as *G. trabeum* in lignin- and dimer-degrading ability.

Metabolism of lignin-related monomers like vanillic and syringic acid by brown-, white-, and soft-rot fungi was found to differ in some respects (Buswell et al. 1982b, Ander et al. 1984, 1988 and Eriksson et al. 1984). The brown-rot fungi grew more slowly on liquid media than the white- and soft-rot fungi tested (Ander et al. 1984, Eriksson et al. 1984) and metabolized vanillic acid and syringic acid at a lower rate. In all three types of wood-rotting fungi, metabolism of the methoxyl groups of vanillic and syringic acid was most rapid in low-glucose high-nitrogen media (Ander et al. 1984).

The enzyme cellobiose: quinone oxidoreductase (Sect. 4.5.6) does not appear to be produced by brown-rot fungi (Ander and Eriksson 1978). This was evidenced by cultivating the brown-rot fungi *Coniophora puteana*, *Daedalea quercina*, *Laetiporus sulphureus*, *Poria oleracea*, and *Tyromyces sericeomollis* on cellulose, wood meal, and kraft lignin-cellulose agar plates, as described by Ander and Eriksson (1977).

4.7.2 Soft-Rot Fungi

The term *soft rot* was first used by Savory (1954) to characterize the soft appearance of wood decayed by certain types of fungi. These fungi belong to the Ascomycetes and Fungi Imperfecti, are more common on hardwoods than on softwoods, and can grow at a high moisture content. The wood attack by soft-rot fungi is divided into Type 1 and Type 2 attack. The properties of these are discussed in detail in Chap. 1.3.

Degradation of wood by soft-rot fungi has mainly been studied with *Chaetomium globosum* and beech wood. A substantial degradation of lignin in beech wood by *C. globosum* was reported by Savory and Pinion (1958), Levi and Preston (1965), and Seifert (1966). At high weight losses, the lignin loss correlated with the loss of methoxyl groups. As shown in Table 1.5 (Chap. 1), it was found (Eslyn et al. 1975) that many soft-rot fungi removed some lignin, although more of the carbohydrates were degraded. In general, soft-rot fungi degrade or modify lignin to a greater extent than brown-rot fungi but more slowly than white-rot fungi (Seifert 1966). The soft-rot fungus *Daldinia concentrica* preferentially attacked carbohydrates but also degraded some lignin, the syringylpropane units being removed before guaiacylpropane units (Nilsson et al. 1989).

Some degradation of differently labeled DHPs by four soft-rot fungi (*Preussia fleischaki*, two strains of *Chaetomium piluliferum*, and *Stachybotrys chartarum*) was reported by Haider and Trojanowski (1975), but the total re-

Table 4.16. Mycelium growth and release of $^{14}CO_2$ from ring-labeled DHP at different pHs for two brown-rot (14 days) and three soft-rot (10 days) fungi[a]

	pH	Mycelium (mg) day			Accumulation of $^{14}CO_2$ after 14 or 10 days (%)
		3	6	9	
Brown-rot:					
Gloeophyllum trabeum	3.5	10	20	18	0.52
	4.0	11	19	17	0.63
	4.5	15	20	14	0.66
Poria placenta	3.5	2	19	18	0.29
	4.0	2	19	19	0.34
	4.5	3	15	16	0.39
Soft-rot:					
Chaetomium globosum	3.5	4	14	16	0.23
	4.3	21	52	54	0.25
	5.0	32	62	59	0.21
Pialophora mutabilis	3.5	7	24	30	0.38
	4.3	10	25	27	0.44
	5.0	15	28	31	0.27
Petriellidium boydii	3.5	2	2	8	0.21
	4.3	15	31	35	0.30
	5.0	17	33	40	0.22

[a] The pH in the cultures was kept constant with 30 mM dimethylsuccinate and flushing with oxygen was carried out every third day. Mycelium weights are from duplicate samples while $^{14}CO_2$ data are from four replicate cultures.

lease of $^{14}CO_2$ in 10 or 15 days was only about 2%–4% of the total added radioactivity. Later, it was shown that *C. piluliferum* released 20%–30% $^{14}CO_2$ from differently labeled corn stalk lignins in 7 weeks. In contrast to white-rot fungi, the soft-rot fungi degraded DHP and corn stalk lignin best in a shaking, high-nitrogen medium (Haider and Trojanowski 1975, 1980). Using different concentrations of glucose and nitrogen in liquid cultures, Ander et al. (1984) found little degradation of DHP by *C. globosum*, *Phialophora mutabilis*, and *Petriellidium boydii*. Table 4.16 shows the results obtained with two brown-rot and three soft-rot fungi cultivated on 1% glucose and a low-nitrogen medium according to Ander et al. (1984). Obviously these fungi could not degrade ring-labeled DHP under the conditions used. The agar technique used for degradation of dimeric model compounds by some brown-rot fungi (Enoki et al. 1985; discussed above) was not successful in obtaining DHP degradation by either brown- or soft-rot fungi (Ander et al. 1988, and unpublished results).

Lignin-related compounds such as vanillic, syringic, and ferulic acids and other phenols, however, are rapidly degraded by soft-rot fungi (Haider and Trojanowski 1975, 1980, Buswell et al. 1982b, Ander et al. 1984, 1988, Eriksson et al. 1984, Seigle-Murandi et al. 1986, Bugos et al. 1988).

In conclusion, many soft-rot fungi can degrade lignin to some extent, especially in hardwoods. They cause demethylation of lignin in wood but, to date, this has not been reported for isolated lignins such as DHP or kraft lignin. The mechanisms or the enzymes involved in methoxyl splitting in lignin are not known, although perhaps formaldehyde is the immediate cleavage product. Certain evidence (Ander et al. 1988) indicates that this is the case for monomers such as vanillic acid. Soft-rot fungi, in contrast to brown- or white-rot fungi, were found to contain low levels of intracellular protocatechuate 3,4-dioxygenase but not vanillate hydroxylase or hydroxyquinol 1,2-dioxygenase (Buswell et al. 1982b). It was thus proposed that vanillate undergoes initial demethylation by soft-rot fungi to protocatechuic acid which is ring-cleaved. However, as discussed in Sect. 4.5.7, *O*-demethylases are unstable and could not be detected by Buswell et al. (1982b). Bugos et al. (1988) found intracellular protocatechuate 3,4-dioxygenase and catechol 1,2-dioxygenase in the soft-rot fungus *Lecythophora hoffmannii*.

4.7.3 Other Fungi

Several Ascomycetes, Deuteromycetes and Phycomycetes, other than soft-rot fungi, have been reported to degrade lignin, but in these reports it has not been shown unequivocally that lignin is degraded (Kirk and Farrell 1987). Insignificant lignin degradation was reported for marine fungi (Sutherland et al. 1982) and for *Trichoderma harzianum* (Kern 1983a). Degradation of ^{14}C-labeled kraft lignin by *Aspergillus fumigatus* (Kadam and Drew 1986) may partly be due to degradation of the low molecular weight compounds. In both solid state and submerged cultivations with *Aspergillus japonicus*, *Coriolus versicolor*, *Trichoderma harzianum*, and *Lenzites trabea* on wheat straw, Milstein et al. (1986) found that *A. japonicus* degraded lignin equally well as *C. versicolor*. This was especially the case in submerged cultures containing 4 g l^{-1} NH_4NO_3, which probably inhibited lignin degradation by *C. versicolor* but not by *A. japonicus*.

Aspergillus flavus has been shown to grow on 2,6-dimethoxyphenol (Betts et al. 1987a) and on several lignin-related dimers, trimers, and tetramers (Betts et al. 1987b, Dart et al. 1987) as sole carbon sources. Formation of the dimer tetramethoxy-p-dibenz oquinone from 2,6-dimethoxyphenol was inhibited in the presence of glucose (Betts et al. 1987a). The reason for this is probably glucose repression of phenoloxidase production. Scanning electron microscopy of the fungal mycelium revealed formation of substrate crystals that were in intimate contact with or embedded in the mycelium.

Extensive research on degradation of lignin-related dimers and trimers by different *Fusarium* strains has been conducted by Higuchi, Iwahara, and coworkers, as cited in Sect. 4.5.3. However, little degradation of lignin by these species was found. Norris (1980) found that *Fusarium solani* released only 4%–5% of ^{14}C-labeled DHP to $^{14}CO_2$ in 30 days, while Sutherland (1983b) noted a maximum of 5% $^{14}CO_2$ from lignin-labeled spruce wood using 18 different *Fusarium* strains.

The gasteromycete, *Cyathus stercoreus*, which is associated with litter decomposition, can degrade lignin in wheat straw (Wicklow et al. 1980a, b,

1984). Other *Cyathus* species can also degrade wheat straw lignin (Abott and Wicklow 1984). Some of the strains are rather specific for lignin, both in straw and in hardwood.

4.8 Lignin Degradation by Bacteria

4.8.1 Actinomycetes

Actinomycetes are a diverse group of Gram-positive bacteria which participate in primary degradation and humification of organic matter in neutral or slightly alkaline soils or composts (Crawford and Sutherland 1979, Küster 1979, Goodfellow and Williams 1983, McCarthy 1987). Due to the growth of actinomycetes as branching hyphae in similarity with the filamentous fungi, they are well adapted to penetrate insoluble substrates such as lignocellulose. Some examples of common genera of lignocellulose-degrading actinomycetes are *Streptomyces*, *Micromonospora*, *Microbispora*, *Thermomonospora*, *Nocardia*, *Rhodococcus*, and *Arthrobacter* (Crawford and Sutherland 1979, Küster 1979, McCarthy and Broda 1984, McCarthy 1987).

The main body of research on lignocellulose degradation with *Streptomyces* has been done by Crawford and co-workers. This research, which was facilitated by the preparation of specifically labeled ^{14}C-lignin and ^{14}C-cellulose lignocelluloses (Crawford et al. 1977, Crawford and Crawford 1978), has been reported in several papers (Crawford 1978, 1981 a, Crawford and Sutherland 1979, 1980, Phelan et al. 1979, Antai and Crawford 1981, 1982, Barder and Crawford 1981). In Table 4.13 (Sect. 4.7) two *Streptomyces* strains are included which degraded substantial amounts of lignin; somewhat more of the lignin in grass was degraded compared to the lignin in hardwood or softwood (Antai and Crawford 1981, Crawford 1981 a). As reported by Barder and Crawford (1981), lignin degradation by *S. badius* is greatest in the presence of high levels of organic nitrogen. Furthermore, some degradation of ^{14}C-MWL (3.2% $^{14}CO_2$ in 30 days) occurred in a minimal medium, although degradation was greatly enhanced (to more than 11%) when 0.7% yeast extract plus cellulose was present in the medium. Thus, lignin degradation by *Streptomyces* is a primary metabolic activity in contrast to *Phanerochaete*, in which it is a secondary metabolic event. A similar result was obtained by McCarthy and Broda (1984) using strains of the genera *Thermomonospora* and *Streptomyces*, which degraded the lignin in wheat lignocellulose mainly during the first 72 h of growth.

The chemistry of degradation of lignin in different lignocelluloses has been investigated for several *Streptomyces* strains. For example, lignin degradation by *Streptomyces viridosporus* is oxidative and involves demethylations, aromatic ring cleavage, and oxidative attack on side chains (Crawford et al. 1982 a, 1983 b). Identified lignin fragments were p-hydroxybenzoic, vanillic, protocatechuic, p-coumaric, syringic, and ferulic acids (Crawford 1981 a). *S. virido-*

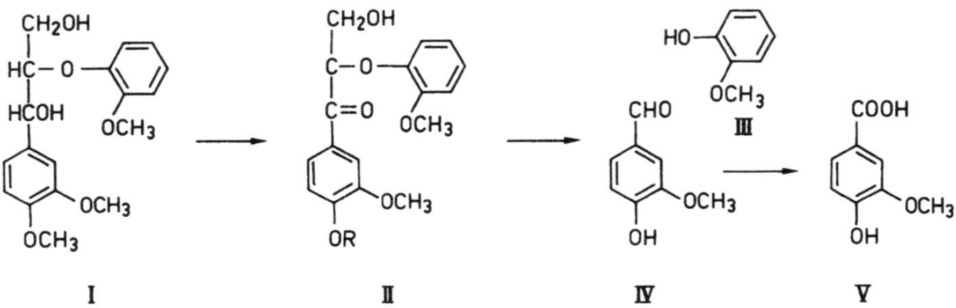

Fig. 4.55. Metabolism of veratrylglycerol-β-guaiacyl ether (*I*) by *Streptomyces viridosporus* T7A to guaiacol (*III*) and vanillin (*IV*) which was further oxidized to vanillic acid (*V*) (Crawford et al. 1984a)

sporus T7A and *S. badius* 252 metabolized veratrylglycerol-β-guaiacyl ether (I) by demethylation and oxidation of the α-carbon atom to an α-carbonyl group, followed by β-ether cleavage, resulting in the release of guaiacol as in Fig. 4.55 (Crawford et al. 1983b, 1984a). The β-ether cleaving activity was cell-associated and was induced by either lignin or APPL (see below; Crawford et al. 1983b, Deobald and Crawford 1987).

The main activity of actinomycetes on lignin is solubilization, whereas mineralization to CO_2 is much less than for white-rot fungi. Both *S. viridosporus* and *Thermomonospora mesophila* produce an Acid-Precipitable Polymeric Lignin (APPL) from corn and straw lignocellulose (Crawford et al. 1983a, Borgmeyer and Crawford 1985, McCarthy et al. 1986). The yield of APPL for these two actinomycetes was 22–72 mg g^{-1} corn or straw, while spruce or maple gave smaller amounts of APPL. The APPL is water-soluble, contains 7%–20% pentose-rich carbohydrates in addition to some protein, and has a molecular weight of more than 20000. If produced in enough quantities, APPL may be used as an antioxidant, a surfactant, and potentially as a component of adhesives and resins (Crawford et al. 1984b). It was therefore of interest to increase the yield of APPL. Improved *Streptomyces* strains with enhanced β-ether cleaving enzyme activity were obtained by UV-irradiation and protoplast fusion. These strains produced up to 90% more APPL from corn stover than the wild-type strains (Crawford et al. 1982b, 1984a, Pettey and Crawford 1984). Further improvements in APPL production by *S. viridosporus* were obtained using shake flasks or stirred reactors containing a slurry of 3% or 5% corn stalks (Adhi et al. 1988). After only a 2-week incubation time, 98 mg APPL g^{-1} corn was obtained compared to similar or lesser amounts obtained after 6–8 weeks in earlier solid state fermentations.

A comparison of APPL production in *S. viridosporus* and *S. badius* was made by Borgmeyer and Crawford (1985). It was suggested that APPL was released mainly by β-ether cleavage and other oxidative reactions, and that *S. badius* due to the presence of extracellular phenol oxidase activity repolymerized lignin degradation products together with APPL. *S. badius* produced APPL over a period of days, while *S. viridosporus* had to be cultivated for weeks to obtain the same amount of APPL (Borgmeyer and Crawford 1985). When APPL isolated from *S. viridosporus* was added to fresh cultures of some *Streptomyces* strains and to *Phanerochaete chrysosporium*, the APPL was degraded and several lignin-related aromatic acids could be isolated (Pometto and Crawford 1986a). Lignin degradation by *S. viridosporus* (Pometto and Crawford 1986b) was found to be extensive at pH 8, which is in contrast to white-rot fungi, which degrade lignin best at pH 4–6. Generally, filamentous fungi prefer a soil pH of 5.5 or less while actinomycetes prefer a more alkaline soil of pH 8. Thus, lignin-solubilizing actinomycetes are more competitive in alkaline soils than ligninolytic fungi (Pometto and Crawford 1986b).

Different extracellular enzymes have been isolated from cultures of *Streptomyces viridosporus*, for example, peroxidase, esterase (active against p-nitrophenyl acetate, as a measure of coumarate ester esterase in corn), endoglucanase, and xylanase (Deobald and Crawford 1987, Ramachandra et al. 1987, 1988). The peroxidase was assayed by using oxidation of L-3,4-dihydroxyphenylalanine (L-dopa) in the presence of 4.0 mM H_2O_2 at pH 7, or with 2,4-dichlorophenol at pH 5.5. The extracellular peroxidase appeared during the early logarithmic growth phase and reached maximum activity after 28 h. In the late stationary phase, the activity decreased rapidly (Ramachandra et al. 1987). The partially purified peroxidase, inducible with xylan from larch wood, catalyzed C_α-C_β cleavage in 1,2-diarylpropane and arylglycerol-β-aryl ether compounds. The enzyme had a Soret band at 408 nm, indicating that it is a heme-containing new type of lignin peroxidase (Ramachandra et al. 1987).

Deobald and Crawford (1987) concluded that xylanase, aromatic aldehyde oxidase, or coumarate ester esterase were not directly responsible for lignin solubilization and APPL production but found some involvement of endoglucanase (cellulase) in the lignin solubilization process. A crude extracellular enzyme preparation from *Streptomyces cyaneus* was found to solubilize 20%–30% of a ^{14}C-lignin labeled straw (Mason et al. 1988). This enzyme, however, was regarded as unlikely to be cellulase or xylanase. Although *S. viridosporus* T7A in the presence of lignocellulose oxidized veratryl alcohol to veratric acid, this strain could not degrade β-1 or β-*O*-4 dimers (Rüttiman et al. 1987).

Strains of the genus *Streptomyces* can metabolize a number of phenols and lignin-related aromatic acids. Thus, *Streptomyces setonii* metabolized phenol to catechol, which was cleaved by catechol 1,2-dioxygenase (Antai and Crawford 1983), whereas cinnamic acid, p-coumaric acid, ferulic acid, and vanillin were metabolized to catechol or protocatechuic acid as shown in Fig. 4.56 (Sutherland et al. 1983a). It may be observed that added vanillin was metabolized via vanillic acid and guaiacol to catechol. When vanillin was formed from ferulic acid, however, it was metabolized via vanillic acid to protocatechuic acid.

Dehydrodivanillin was rapidly metabolized by *Streptomyces viridosporus* to dehydrodivanillic acid, which was more slowly degraded to unidentified compounds (Crawford et al. 1981a). Four *Streptomyces* mutants, which could not oxidize dehydrodivanillin, however, degraded Douglas-fir lignin. Thus, it

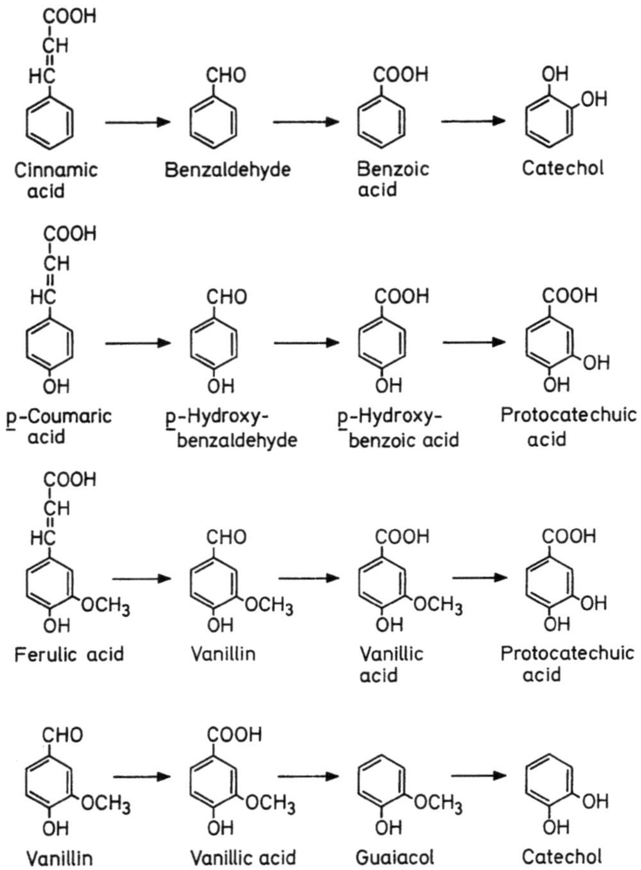

Fig. 4.56. Metabolism of lignin-related compounds by *Streptomyces setonii* (Sutherland et al. 1983a)

was concluded by Crawford et al. (1981a) that dehydrodivanillin is not a relevant model compound for the study of lignin polymer degradation by *S. viridosporus*.

Another actinomycete, which has been much studied in connection with degradation of lignin and lignin-related compounds, is *Nocardia*. It was reported that different *Nocardia* strains released 4.0% – 13.8% $^{14}CO_2$ from ^{14}C-methoxyl labeled DHP in 15 days, with lower yields of $^{14}CO_2$ from side-chain or ring-labeled DHP (Trojanowski et al. 1977, Gradziel et al. 1978, Haider et al. 1978). These authors also found that 13% – 15% $^{14}CO_2$ was released from side-chain and from methoxyl-labeled maize lignin by *Nocardia*. As expected, the *Nocardia* sp. investigated by Trojanowski et al. (1977) and Gradziel et al. (1978) rapidly metabolized differently labeled phenols including veratric, syringic, vanillic, and ferulic acids to $^{14}CO_2$. A remarkably strong $^{14}CO_2$ evolution (88% in 24 h) from ^{14}C-4-methoxyl labeled veratric acid was obtained, and Malarczyk (1984) accordingly isolated an inducible veratrate-*O*-demethylase from a *Nocardia* sp.

Eggeling and Sahm (1980, 1981) and Kern et al. (1984) studied the metabolism of coniferyl alcohol, veratric acid, 3-methoxy benzoic acid, and 4-methoxy benzoic acid by *Nocardia* sp. DSM 1069 and a *Xanthomonas* sp. These two bacteria converted coniferyl alcohol to coniferaldehyde, ferulic acid, vanillic acid, and finally to protocatechuic acid which was ring-cleaved to afford β-carboxy-cis-cis-muconic acid. Veratrate was metabolized via vanillic or isovanillic acid to protocatechuic acid. Although the *Nocardia* sp. of Trojanowski et al. (1977) only released 1.36% $^{14}CO_2$ from methoxyl-labeled coniferyl alcohol in 24 h, Eggeling and Sahm (1980) found good growth of their *Nocardia* strain on coniferyl alcohol.

4.8.2 Other Bacteria

Lignified cell walls are not attacked by most species of bacteria (Chap. 1.4, Schmidt 1980, Schmidt and Bauch 1980, Schmidt and Liese 1982, Schmidt et al. 1987), and degradation of labeled DHP and lignin in poplar wood to $^{14}CO_2$ by bacteria such as *Pseudomonas* spp. is slow (Haider et al. 1978, Odier et al. 1981, Janshekar and Fiechter 1982, Kerr et al. 1983). Lignin in thin wood sections (50–60 μm), however, is released more easily from the wood by *Pseudomonas, Acinetobacter*, and *Xanthomonas* (Monties et al. 1981, Odier et al. 1981) and by *Bacillus* (Schmidt and Bauch 1980), although the *Bacillus* strains did not respire the lignin. Substantial degradation (20%–40% in 7 days) of poplar dioxane lignin and MWL by *Pseudomonas*, *Acinetobacter*, and *Xanthomonas* as estimated on the basis of A_{280} measurement of residual lignin was obtained by Odier et al. (1981). However, the molecular weights of these lignins were not given and determination of residual lignin after microbial attack is a difficult problem (Odier et al. 1981). Extracted lignins such as kraft lignin and pine MWL were partially degraded by alkalophilic bacteria (Kawakami and Shumiya 1983).

Low molecular weight parts of ammonium lignosulfonate and kraft lignin may be metabolized by bacteria (Kerr et al. 1983, Ammar et al. 1986, Gonzales et al. 1986). Natural lignin, however, is easier to degrade by microorganisms than lignosulfonates (Haider and Trojanowski 1981, Kern 1981).

Small changes in ^{13}C-labeled synthetic lignin were found with ^{13}C-NMR spectroscopy after incubation with a *Xanthomonas* sp. (Haider et al. 1985), whereas Kern (1984) found that a *Xanthomonas* strain released more than 30% $^{14}CO_2$ from ring-labeled DHP in 20 days. However, in his comprehensive studies on fractionation of DHP, Kern also found that a substantial dissociation of the DHP polymer could occur, giving rise to smaller molecules, which could be metabolized by this bacterium. Later, Kern and Kirk (1987) extended this investigation, using five differently labeled synthetic lignins and one ^{14}C-methylated spruce lignin. The influence of fungal lignin peroxidase pretreatment was also examined. The *Xanthomonas* sp. evolved most $^{14}CO_2$ from the lignin samples containing the most low molecular weight material which was preferentially degraded up to a molecular weight of 600 to 1000. Pretreatment of the five DHPs with crude lignin peroxidase increased their molecular size and decreased their degradability by the bacterium. The methylated spruce lignin was reported by Tien and Kirk (1983) to be partly degraded by lignin peroxidase (Sect. 4.5.3). In the investigation of Kern and Kirk (1987), this lignin was both polymerized and depolymerized by ligninase, which resulted in a net decrease in degradability by the *Xanthomonas*. This strongly indicates that the *Xanthomonas* degrades lignin only up to a molecular weight of 600 to 1000.

Pellinen et al. (1984) isolated a mixed bacterial culture from industrial waste waters polluted by pulp and paper factories. In subsequent investigations (Jokela et al. 1985, 1987, Pellinen et al. 1987), it was found that these bacteria could degrade two tetrameric lignin model compounds and that 6% $^{14}CO_2$ was released from side chain-labeled DHP in 17 days. However, the tetramer degrading culture did not degrade kraft lignin, one of two chloroethanol-acetic acid lignins, or lignin isolated from an oxygen-bleaching liquor (Pellinen et al. 1987). A phenolic tetramer (MW 638) was decomposed readily, whereas a nonphenolic tetramer (MW 666) was degraded much more slowly (Jokela et al. 1985). Metabolism of both phenolic and nonphenolic tetramers started with oxidation of the α-hydroxyl groups in the propane side chain in two successive steps to give monoketones and diketones (Jokela et al. 1987). These ketone metabolites were then degraded via C_α-C_β cleavage, forming trimeric carboxylic acids, which were further metabolized via a second C_α-C_β cleavage. In conclusion, the results of Jokela et al. (1985, 1987) and Pellinen et al. (1984, 1987) show that the enzyme system of the lignin tetramer degrading bacteria resembles the lignin oxidation system of *Phanerochaete chrysosporium* with respect to C_α-oxidation, degradation of nonphenolic lignin model compounds, and cleavage of propane side chains at the C_α-C_β linkage. In *P. chrysosporium*, however, C_α-oxidation does not precede C_α-C_β cleavage and the bacteria studied by Jokela and Pellinen do not produce lignin peroxidase or Mn-peroxidase similar to white-rot fungi.

As discussed in detail in Sect. 1.4, ultrastructural studies have shown that certain tunneling bacteria (probably belonging to the *Myxobacteriales* or the *Cytophagales*) are able to degrade lignified cell walls. In addition, it has been shown that these new types of bacteria can also release up to 10%–11% of $^{14}CO_2$ from ring- and side chain-labeled DHP in 16 days (Nilsson and Daniel 1986, Daniel et al. 1987), as well as similar amounts of $^{14}CO_2$ from lignin-labeled spruce wood (Ander and Nilsson, unpublished results).

In conclusion, certain strains of actinomycetes can metabolize intact lignin to CO_2, although at a much lower rate than white-rot fungi. Except for tunneling bacteria, other kinds of bacteria do not metabolize intact wood lignin, although some degradation of extracted lignins or synthetic lignins up to a molecular weight of 600–1000 occurs.

Regarding lignin-related dimeric compounds, many reports on degradation of such compounds by bacteria have been published (Kawakami and Shumiya

1983, Pellinen et al. 1984, 1987, Jokela et al. 1985, 1987, Odier and Rolando 1985, Pelmont et al. 1985, Samejima et al. 1985, 1988, Gonzales et al. 1986, Bourbonnais and Paice 1987, Gurujeyalakshmi and Mahadevan 1987, Vicuña et al. 1987, Katayama et al. 1988). In general, these compounds are metabolized via β-ether cleavage and demethylation to give catechol or protocatechuic acid which are ring-cleaved (Crawford 1981 a, b, Vicuña et al. 1987). Dehydrodivanillic acid was metabolized by *Pseudomonas paucimobilis* to 3-*O*-methyl gallic acid, which was ring-cleaved by a protocatechuate 4,5-dioxygenase (Katayama et al. 1988).

4.9 Potential Applications of White-Rot Fungi

There are several areas in which technical application of white-rot fungi, mutants thereof, and isolated ligninolytic enzymes have industrial potential. The following applications are discussed briefly in the following:

1. Delignification of wood chips to reduce refining energy in (thermo)mechanical pulping and consumption of chemicals in kraft cooking.
2. Delignification of wood, straw, and bagasse to increase digestibility by ruminants.
3. Delignification, bleaching and modification of pulp fibers.
4. Modifications of different isolated lignins to produce more useful lignins or chemicals.
5. Treatment of waste bleach liquors to reduce color, toxicity, and mutagenicity.
6. Treatment of soils or waste waters to remove environmental pollutants like PCB, DDT, and dioxins.

Some reviews discussing the areas 1–6 have been published (Eriksson 1981 a–d, 1985, 1987, Kirk and Chang 1981, Bowman 1983, Kirk 1983 a, Kirk et al. 1983 a, Eriksson and Kirk 1985, Buswell and Odier 1987, Farrell 1987 a, Boman et al. 1988).

4.9.1 Biopulping and Biomechanical Pulping

In the pulp and paper industry, pulping of wood is done with the input of much energy (mechanical pulp) and chemicals (chemical pulp). The processes involve high temperatures and pressures and extreme pH values. Fungal pretreatment of wood chips or pulp, causing delignification, saves energy and reduces chemical consumption in mechanical or chemical pulping (Eriksson 1985, Eriksson and Kirk 1985).

It is possible to use both wild-type white-rot fungi and cellulase-less mutants of white-rot fungi for delignification purposes. However, there is always a possibility that sooner or later the wood polysaccharides are also degraded by wild-type fungi, although it has been shown (Chap. 1) that many naturally occurring fungi can selectively remove lignin from wood. Addition of malt extract (Ander and Eriksson 1977) or glucose (Ruel et al. 1984) to wood blocks, however, represses cellulose degradation by wild-type white-rot fungi. Poplar wood chips impregnated with a 1% glucose solution and subsequently treated with wild-type white-rot fungi could be refined with the input of less energy, giving a paper with good breaking length and tear strength (Akamatsu et al. 1984). Myers et al. (1988) treated glucose-amended aspen wood chips with *Phanerochaete chrysosporium* and *Dichomitus squalens*. Both fungal pretreatments gave better strength properties of paper prepared by refiner mechanical pulping compared to appropriate controls. However, brightness of paper sheets decreased. Pretreatment with *D. squalens* greatly altered the fiber size distribution of the pulp, resulting in a decrease of the content of fines (fibers passing a 0.149-mm screen) from 53% to 34%. However, it is doubtful whether the above-mentioned approaches with sugar addition would be technically feasible.

In order to avoid cellulose degradation, Eriksson and co-workers produced cellulase-less mutants of white-rot fungi by UV-irradiation of asexual conidiospores (Eriksson and Goodell 1974, Ander and Eriksson 1975, Eriksson et al. 1980a, b). Using *Phlebia radiata* and its Cel 26 mutant (Samuelsson et al. 1980), they found that treatment of pine wood chips for up to 14 days gave small savings in refining energy and also losses in strength properties. In other experiments with Cel 44 of *Sporotrichum pulverulentum* (*Phanerochaete chrysosporium*) (Eriksson and Vallander 1982), the refining energy in mechanical pulping decreased by 23% when glucose-impregnated spruce chips were used in the fungal pretreatment. However, there was a reduction in brightness. Without glucose addition little or no energy saving was obtained.

Not only wood chips but also unbleached kraft pulps and mechanical pulps have been used to study the effects of fungal treatment. It has been found that it is possible partly to delignify different pulps with wild-type *Phanerochaete chrysosporium*, but cellulose degradation is avoided only if glucose is present (Kirk and Yang 1979, Yang et al. 1980, Bar-Lev et al. 1982). The conditions which favor lignin degradation in pulps are the same as those reported by Kirk et al. (1978a) for synthetic lignins. These findings were later confirmed by Tran and Chambers (1987), using unbleached hardwood kraft pulp with a lignin content of 2.4%. In addition, they found that optimal delignification of the pulp with *P. chrysosporium* occurred at pH 3.5, whereas most H_2O_2 was found in the medium at pH 6.5. Delignification of kraft pulp and chemical thermomechanical pulp (CTMP) in the presence of glucose and Tween 80 by *P. chrysosporium* has recently been obtained under shaking conditions (Pellinen J, Abuhasan J, Joyce T W, Chang H-m 1989). Increased water retention values and better strength properties of mechanical pulp was obtained by Pilon et al. (1982a, b) after treatment with white-rot fungi.

The problem of the slow lignin degradation obtained with Cel^- mutants was addressed by Eriksson et al. (1983) who produced new cellulase-less mu-

Table 4.17. Weight and lignin loss of birch (*Betula verrucosa*) wood wafers determined after 2, 3, or 4 weeks of degradation by wild-type *Phanerochaete chrysosporium* K-3 and four Cel$^-$ strains (Johnsrud and Eriksson 1985)

Strain	Time (weeks)	Weight loss (%)	Lignin loss (%[a])
K-3	2	18.5	17.7
	3	19.9	17.2
	4	24.4	19.9
3113	2	6.0	5.9
	3	14.0	18.5
	4	19.0	29.5
13132-85	2	11.3	16.0
	3	18.2	28.5
	4	19.6	25.6
13132-118	2	6.9	13.3
	3	11.7	13.9
	4	10.5	18.4
85118	2	10.6	21.4
	3	14.5	22.7
	4	16.6	26.0

[a] In % of the initial amount of lignin.

tants of *Phanerochaete chrysosporium*. One of those mutants had a higher production of both phenoloxidase (probably Mn-peroxidase and/or ligninase) and xylanase and was also shown to degrade lignin better than earlier strains. Since xylan is associated with lignin and since xylose produced from the xylan can be used as a carbon source by the mutants, xylan degradation may stimulate lignin degradation. Selection, mutation, and intercrossing of homokaryotic strains of *P. chrysosporium* K-3 later gave rise to Cel$^-$ mutants with greater lignin-degrading capacity than all earlier Cel$^-$ mutants (Johnsrud and Eriksson 1985). Some of the new mutants could degrade lignin in birch as well as the wild-type strain (Table 4.17) and did not evolve $^{14}CO_2$ from ^{14}C-labeled cellulose, although $^{14}CO_2$ evolution from ^{14}C-ring-labeled DHP was strong (Fig. 4.57). The new mutants were used for treatment of birch chips, which were then pulped by the kraft process, O_2-alkali cooking, or chemimechanical pulping (Johnsrud and Eriksson 1986, Eriksson 1988). The results obtained from kraft cooking of treated birch chips show that for the rotted chips 1% less alkali could be used compared to the reference (percent effective alkali/percent sulfidity decreased from 17/40 to 16/40). The rotted chips were also delignified faster than the nontreated ones. The yield at a pre-set kappa number was the same for both the fungal-treated wood and the reference, whereas the viscosity of pulp from the rotted wood was higher than that from the reference wood (Fig. 4.58). Kraft pulps produced from pretreated and reference birch wood chips had the same properties at a density of 850 $kg\,m^{-3}$ except for the brightness which was 1.5 units higher in pulps from fungal-pretreated chips. However, the obtained savings of 1% alkali, higher pulp viscosity, and higher brightness are not considered to be enough to cover the extra costs for the fungal pretreatment.

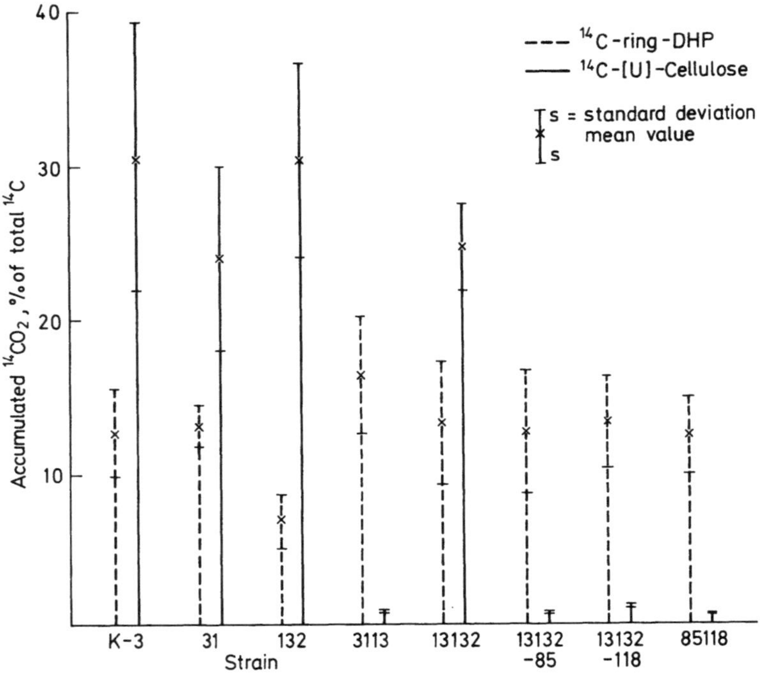

Fig. 4.57. Release of $^{14}CO_2$ from ^{14}C-ring-labeled DHP and from ^{14}C-[U]-cellulose by *Phanerochaete chrysosporium* K-3 and isolates of F_1, F_2, and F_3 generations. Cel^+ strains; *K-3, 31, 132, 13132*; Cel^- strains: *3113, 13132-85, 13132-118, 85118* (Johnsrud and Eriksson 1985)

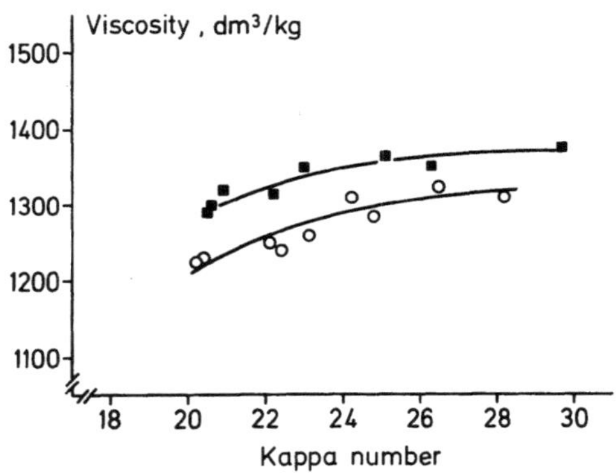

Fig. 4.58. Viscosity of kraft pulp produced from fungal pretreated birch chips (■) and reference chips (○) (Eriksson 1988, and Johnsrud, unpublished results)

Finally, it may be mentioned that fungal pretreatment of sugarcane bagasse, combined with the modified cold soda process for bagasse pulping developed in Cuba, gave a good quality pulp with less energy needed in the refining (Johnsrud et al. 1987, Eriksson 1988). The results obtained indicate that the Cel^- mutants are better suited for the rotting of grass materials (wheat straw and sugarcane bagasse) than for rotting of wood.

4.9.2 Treatment of Wood and Straw to Produce Feed and Food

Delignification of aspen, birch, and poplar wood by white-rot fungi has been undertaken to increase digestibility of these wood species for use as ruminant feed (Kirk and Moore 1972, Reid 1982, Reade and McQueen 1983, Reid 1985). The feed value of wood is limited since anaerobic bacteria in the rumen do not decompose the lignin that protects the polysaccharides from being utilized. In saccharification of cellulose by hydrolytic enzymes, not only the lignin but also crystalline cellulose is an obstacle (Grethlein 1984, Vallander and Eriksson 1985, Mes-Hartree et al. 1987). In order to improve the accessibility of the polysaccharides to enzymatic hydrolysis, steam explosion (Saddler et al. 1982, Deshpande and Eriksson 1984, Vallander and Eriksson 1985, 1987, Clark and Mackie 1986, Mes-Hartree et al. 1987) and alkaline delignifications in the presence of hydrogen peroxide (Gould 1984, 1985, Takagi 1987) or without hydrogen peroxide (MacDonald et al. 1983) have been performed.

Much of the research on conversion of wheat straw into feed has been reviewed by Kirk and Chang (1981). Usually the fungal treatment is used to increase digestibility, to facilitate saccharification, and to produce protein. Wicklow et al. (1980a, b, 1984) tested different strains of *Cyathus*, while Zadrazil (1985) investigated 235 different fungal strains for wheat straw delignification and effect on digestibility. Other recent papers on straw delignification are those of Hatakka and Pirhonen (1985), Agosin et al. (1986), and Milstein et al. (1986).

However, many problems in the scaling up of fungal treatment of wood, wood chips, pulps, bagasse, or straw are still unsolved. These include control of temperature, aeration, and humidity. The rate of lignin degradation, especially in softwoods, is still low and problems with infections by unwanted microorganisms may occur.

In many instances the economy of straw fermentation may be increased by simultaneous production of edible fungi (Kirk 1983a, Buswell and Odier 1987), but good digestibility and good yield of mushrooms may not occur simultaneously (Tsang et al. 1987). In Japan, the mushroom "shiitake" *Lentinula* (*Lentinus*) *edodes* is very popular and is cultivated on oak logs 1 m long and 5–20 cm thick (Leatham 1982, 1985). Harvest can be done for 3–7 years after fruitbody formation has begun. Methods for cultivation of *Agaricus*, *Lentinula*, and *Pleurotus* were summarized by Tautorus (1985).

4.9.3 Treatment of Spent Kraft Bleach Liquors

In kraft pulping 90%–95% of the lignin is removed and the 5%–10% lignin remaining in the pulp is responsible for the brown color of kraft pulp and papers (Eriksson and Kirk 1985). Almost half of the kraft pulp produced is bleached with chlorine and its oxides (C-stage) and extracted with alkali (E-stage). With the increased use of O_2-bleaching, especially in Sweden (Kringstad and McKague 1988), the use of chlorine, however, will decrease. The effluents produced during chlorine bleaching are major contributors to water pollution from the pulp industry and contain highly modified lignins and chlorolignins of different molecular weights together with phenols and chlorinated phenols (Kringstad and Lindström 1984, Österberg 1984, Eriksson and Kolar 1985, Eriksson et al. 1985).

In a chlorine bleach plant, most of the color, due to quinones and conjugated lignin structures, in the effluent comes from the first alkali extraction-stage (E_1)-effluent and is associated with the high molecular mass lignin. Therefore, efficient methods for decolorization and removal of these high molecular mass compounds are necessary. Bacteria-based aerated lagoons, activated sludge systems, or anaerobic systems used in many countries (Salkinoja-Salonen et al. 1984, Eriksson and Kirk 1985, Boman et al. 1988) do not degrade the high molecular mass material, which means that fungal systems have more promise because they can degrade these chlorolignins (Ek and Eriksson 1986).

In addition to unlabeled kraft lignin (Hiroi and Eriksson 1976), ^{14}C-labeled kraft lignin, bleached kraft lignin, and lignin sulfonates (Lundquist et al. 1977) are all degraded by *Phanerochaete chrysosporium*. Fukuzumi et al. (1977) and Fukuzumi (1980) found that the E_1-stage liquor was strongly bleached by *Tinctoporia borbonica* and other white-rot fungi. The presence of ethanol, glucose, or cellulose stimulated kraft lignin degradation in these investigations. Fungal decolorization of kraft bleach plant effluents by *P. chrysosporium* (Eaton et al. 1980, 1982, Sundman et al. 1981), and *Coriolus versicolor*, either free or immobilized (Livernoche et al. 1983, Royer et al. 1983) in the presence of glucose or sucrose, has been reported. Immobilization of the fungus in calcium alginate gels resulted in 80% decolorization of the bleach plant effluent in 3 days, which was better than without immobilization (Livernoche et al. 1983). Paice and Jurasek (1984b) found that treatment of a bleach plant effluent with peroxidase ($20\ mg\,l^{-1}$) plus $1-100$ mM H_2O_2 during the first 4 h was a faster process to remove color than to use only free *C. versicolor*. However, after 48 h, the fungus removed color better than peroxidase treatment alone.

Chang and co-workers have developed a continuous process for color removal from bleach plant effluents with *P. chrysosporium*. This MyCoR process (Mycelial Color Reduction) utilizes a rotating biological contactor in which the fungus is immobilized on the surface (Chang et al. 1983, 1986, 1987, Vasudevan et al. 1986). According to Eaton et al. (1982), residual cellulose in the primary sludge from a pulp mill can be used as co-substrate to obtain lignin degradation. In the MyCoR process chlorolignins are dechlorinated and degraded. Chlorinated derivatives of guaiacol and vanillin also disappear (Huynh et al. 1985). After 2 days, 70% of the organically bound chlorine (TOCl) is removed (Chang et al. 1987). The process has not yet been evaluated on a pilot scale.

Recently, Pellinen et al. (1988) found that *P. chrysosporium* (in the MyCoR process) could reduce the total organic chlorine content of chlorolignin almost by 50% during 1 day of treatment. A corresponding amount of inorganic chloride was liberated into the solution. Pellinen et al. (1988) speculated that lignin peroxidase could catalyze this dechlorination. Hammel and Tardone (1988) have shown that lignin peroxidase can catalyze oxidative 4-dechlorination of phenols with two to five chlorine atoms, which indicates the possibility that lignin peroxidase can also dechlorinate polymeric chlorolignins. Three bacterial species tested by Bourbonnais and Paice (1987) removed the color of a bleach kraft effluent, primarily by adsorption, and did not release significant amounts of $^{14}CO_2$ from the ^{14}C-labeled solids (mainly lignin and chlorolignin) obtained from the bleach effluent. The two white-rot fungi *Coriolus versicolor* and *Pleurotus sajor-caju* degraded/decolorized the labeled (chloro)lignins, thereby releasing $^{14}CO_2$. These results are in accordance with those obtained by Eriksson and Kolar (1985), who showed that bacteria from aerated lagoons slowly degraded ^{14}C-labeled chlorolignin, whereas *P. chrysosporium* more rapidly degraded the chlorolignin. Both the bacteria and the fungus methylated chlorinated catechols and guaiacols to the corresponding and more persistent veratrol compounds. Therefore, microbial methylation of chlorinated phenolic compounds may occur in nature and in the receiving water systems near the pulp industries (Eriksson et al. 1985).

Importantly, the ligninolytic system of *P. chrysosporium*, as discussed by Tien (1987), is also capable of (at least partly) degrading several persistent environmental pollutants such as benzo(a)pyrene (Haemmerli et al. 1986b, Sanglard et al. 1986), DDT (Bumpus et al. 1985, Bumpus and Aust 1986, 1987), PCB (Eaton 1985), dioxine (Hammel et al. 1986b, 1987), and chloroaniline/lignin conjugates (Arjmand and Sanderman 1985, 1986). Crystal violet, a triphenylmethane dye, found in soil and river sediments as a result of improper waste treatment was degraded by *P. chrysosporium*, and partly by lignin peroxidase (Bumpus and Brock 1988).

Since it has been reported that bleach pulp effluents are toxic, mutagenic, and maybe even carcinogenic (Ander et al. 1977, Eriksson et al. 1979, Kringstad et al. 1981, Kringstad and Lindström 1984, Strömberg 1984), all the different industrial and laboratory treatments described above have been investigated and considered for use in existing pulp factories. In the future, however, biological purification systems other than that with *P. chrysosporium* may be used. Anaerobic fluidized bed may be one system for purification of bleach effluents (Salkinoja-Salonen et al. 1984).

At present ultrafiltration of the E-stage effluent to remove high molecular weight (chloro)lignins (Fig. 4.59), mixing the permeate and the C-stage liquor,

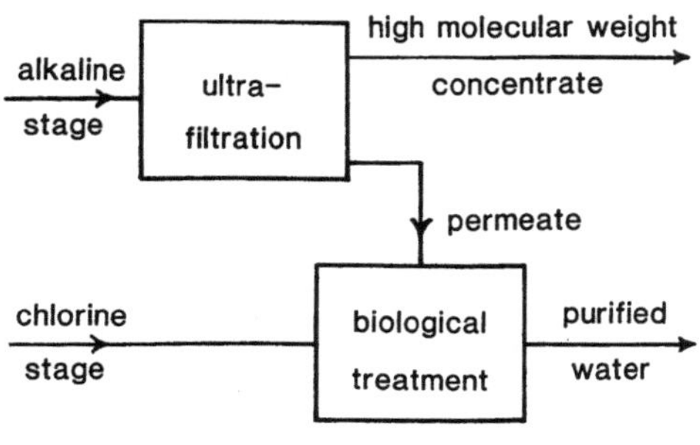

Fig. 4.59. Combined process utilizing ultrafiltration plus biological treatment for purification of waste bleach liquors (Boman et al. 1988)

followed by biological anaerobic plus aerobic treatment seems promising (Ek and Eriksson 1986, 1987, Boman et al. 1988). It is anticipated that such a combined process will give more than 95% BOD reduction, a complete chlorate reduction, and 70%–90% reduction of COD and AOX (adsorbable organic halogen) (Boman et al. 1988).

4.9.4 By-Products from Lignin

In the pulp and paper industry large amounts of kraft lignin and lignosulfonates are produced. The kraft lignin is mostly burned to generate energy for the process, while lignosulfonates are used as dispersing agents or as emulsifiers in the oil industry and as additives in rubber manufacture (Pearl 1982, Janshekar and Fiechter 1983). Vanillin is produced from lignosulfonates, while dimethylsulfide and dimethylsulfoxide are produced from kraft lignin. The market for these compounds, however, is limited. Since low molecular weight aromatic products are formed during lignin degradation, the possibility exists that controlled degradation of the very complex lignin structure, particularly by white-rot fungi or actinomycetes, could give valuable aromatic compounds (Crawford and Crawford 1980, Kirk and Chang 1981). The acid-precipitable polymeric lignin (APPL) produced by *Streptomyces* has antioxidant properties and could be used for production of adhesives and resins (Crawford et al. 1983 a, Crawford et al. 1984 a, b). Selective demethoxylation of kraft lignin by brown-rot fungi could possibly give a polymer useful in phenol-formaldehyde resins (Kirk 1983 a). Cultivation conditions to obtain substantial demethoxylation of isolated lignins by brown-rot fungi, however, are not known (Ander et al. 1988). Anaerobic fermentation is another potential way to produce chemicals from lignin wastes of lower molecular weight than 1000 (Kaiser and Hanselman 1982, Colberg and Young 1985 b). In addition to methane, a number of chemicals may be produced in this way – for example, vanillin and catechol and other phenols, and aliphatic compounds such as different adipic acids or acetic acid.

4.10 Some Examples of Future Research Possibilities

Applications of lignin peroxidases for pulp bleaching or for modifying kraft lignin or lignosulfonates depend partly on the production of these enzymes on a large scale. A step forward toward this goal is the sequencing and cloning of lignin peroxidase genes (or part thereof) from *Phanerochaete chrysosporium* into another organism (Rao and Reddy 1984, Zhang et al. 1986, Raeder et al. 1987, Tien 1987, Tien and Tu 1987, Smith et al. 1988). Complementary DNA (cDNA) sequences of lignin peroxidase from *P. chrysosporium* have now been published (de Boer et al. 1987, Tien and Tu 1987, Smith et al. 1988) and show that the corresponding polypeptide has 344 or 345 amino acids and a molecular weight of about 37000 without glycosylation. The cDNA sequence shows that the structural genes are preceded by a 28-residue leader which is consistent with the lignin peroxidase being an extracellular protein. In order to isolate genes related to ligninolytic activity, Trudel et al. (1988) cloned *Coriolus* (*Trametes*) *versicolor* genes induced by nitrogen starvation. A rapid method for preparation of DNA from fungi was reported by Raeder and Broda (1985) and may be useful in the above-mentioned research. Major hosts for expression of recombinant lignin peroxidase proteins are *Escherichia coli*, *Bacillus subtilis*, and *Saccharomyces cerevisiae* (Farrell et al. 1988). The best expression, that is, production of ligninase protein, was obtained with *E. coli* but this organism cannot glycosylate the protein, which is necessary to obtain better enzyme activity or stability. Another problem is to incorporate the heme group into the enzyme, which may be done either by the host organism itself or in vitro afterward.

Enzyme bleaching of kraft pulps has just begun to be studied, but 20%–30% reduction in the kappa number has been obtained with mixtures of ligninolytic enzymes (Farrell 1987b). However, lignin peroxidase does not only depolymerize lignin, polymerization also occurs (Haemmerli et al. 1986a, Kirk 1986). A major research aim is to find enzyme mixtures in which only depolymerization and degradation take place. Such a mixture will probably contain radical and/or quinone reducing enzymes. Probably enzymes other than those of the cellobiose: quinone oxidoreductase type are also involved in these reducing steps (Schoemaker et al. 1989). To satisfy the requirements of various types of peroxidases it is inevitable that H_2O_2-producing enzymes, including glucose oxidase or methanol oxidase, must also be present.

Degradation of lignin in mixtures of organic solvents and buffers using different peroxidases should be investigated thoroughly. Certain progress has been made by Klibanov's group (Dordick et al. 1986) using horseradish peroxidase for depolymerization of DHP, MWL, and kraft lignin. However, Lewis et al. (1987) could not repeat these promising results.

Since hemicelluloses are covalently bound to the residual lignin in pulp, hemicellulases have also been used with some success for bleaching of different pulps (Viikari et al. 1986, 1987). When hemicellulases were used for pretreatment of the pulp, a significant reduction of the chlorine dioxide consumption during bleaching was obtained, whereas lignin peroxidases had little effect.

Biomimetic degradation of lignin with hemes, porphyrins, or other inorganic complexes is an interesting field of research, which has attracted the interest of several research groups (see Sect. 4.5.4). Some encouraging results with bleaching of different pulps with hemes and similar compounds have already been obtained (Paszczynski et al. 1988, Pettersson et al. 1988). This type of bleaching, complemented with O_2-bleaching, may in the future produce satisfactorily bleached pulp and cause less environmental pollution than does chlorine bleaching. On a technical scale, however, the high concentration of H_2O_2 used by Paszczynski et al. (1988) must be decreased. Other problems in biobleaching to be attacked in future research are the stability of the bleaching agent, the long bleaching time, and reduction in pulp yield and viscosity.

References

Abbot TP, Wicklow DT 1984 Degradation of lignin by Cyathus species. Appl Environ Microbiol 47: 585–587

Abdurachim M 1965 Laboratory tests with Schizophyllum commune Fr. Rimba Indonesia 10: 34–46

Adaskaveg JE, Gilbertson RL 1986 In vitro decay studies of selective delignification and simultaneous decay by the white rot fungi Ganoderma lucidum and G. tsugae. Can J Bot 64: 1611–1619

Adhi TP, Korus RA, Pometto III, AL, Crawford DL 1988 Lignin degradation and production of microbially modified lignin polymers by Streptomyces viridosporus in slurry reactors. Appl Biochem Biotechnol 18: 291–301

Adler E 1977 Lignin chemistry – past, present and future. Wood Sci Technol 11: 169–218

Adolph P, Gerstetter E, Liese W 1972 Untersuchungen über einige Eigenschaften von Fichtenholz nach dreijähriger Wasserlagerung. Holzforschung 26: 18–25

Agosin E, Tollier MT, Brillouet JM, Thivend P, Odier E 1986 Fungal pretreatment of wheat straw: effects on the biodegradability of cell walls, structural polysaccharides, lignin and phenolic acids by rumen microorganisms. J Sci Food Agr 37: 97–106

Ahlgren E, Eriksson K–E 1967 Characterization of cellulases and related enzymes by isoelectric focusing, gel filtration and zone electrophoresis. II. Studies on Stereum sanguinolentum, Fomes anosus and Chrysosporium lignorum enzymes. Acta Chem Scand 21: 1193–1200

Aho PE, Seidler RJ, Evans HJ, Raju PN 1974 Distribution, enumeration and identification of nitrogen-fixing bacteria associated with decay in living fir trees. Phytopathology 64: 1413–1420

Ait N, Creuzet N, Cattaneo J 1979a Characterization and purification of thermostable β-glucosidase from Clostridium thermocellum. Biochem Biophys Res Commun 90: 537–546

Ait N, Creuzet N, Forget P 1979b Partial purification of cellulase from Clostridium thermocellum. J Gen Microbiol 113: 399–402

Akamatsu I, Yoshihara K, Kamishima H, Fujii T 1984 Influence of white-rot fungi on poplar chips and thermo-mechanical pulping of fungi-treated chips. Mokuzai Gakkaishi 30: 697–702

Alberti BN, Klibanov AM 1982 Preparative production of hydroquinone catalyzed by immobilized D-glucose oxidase. Enzyme Microb Technol 4: 47–49

Alexander JK 1968 Purification and specificity of cellobiose phosphorylase from Clostridium thermocellum. J Biol Chem 243: 2899–2904

Alic M, Gold MH 1985 Genetic recombination in the lignin-degrading basidiomycete Phanerochaete chrysosporium. Appl Environ Microbiol 50: 27–30

Alic M, Letzring C, Gold MH 1987 Mating system and basidiospore formation in the lignin-degrading basidiomycete Phanerochaete chrysosporium. Appl Environ Microbiol 53: 1464–1469

Allenca P, Scherl DS, Detroy RW, Leathers TD 1986 Hydrolysis of xylan by an immobilized xylanase from Aureobasidium pullulans. Biotechnol Bioeng Symp 17: 425–433

Almin KE, Eriksson K-E 1967 Enzymic degradation of polymers. I. Viscometric method for the determination of enzymic activity. Biochim Biophys Acta 139: 238–247

Almin KE, Eriksson K-E 1968 Influence of carboxymethyl cellulose properties on the determination of cellulase activity in absolute terms. Arch Biochem Biophys 124: 129–134

Almin KE, Eriksson K-E, Jansson C 1967 Enzymic degradation of polymers. II. Viscometric determination of cellulase activity in absolute terms. Biochim Biophys Acta 139: 248–253

Almin KE, Eriksson K-E, Pettersson B 1975 Extracellular enzyme system utilized by the fungus Sporotrichum pulverulentum (Chrysosporium lignorum) for the breakdown of cellulose. II Activities of five endo-1,4-β-glucanases towards carboxymethylcellulose. Eur J Biochem 51: 207–211

Aloni Y, Cohen R, Benziman M, Delmer DP 1983 Solubilization of the UDP-glucose: 1,4-β-D-glucan 4-β-D-glycosyltransferase (cellulose synthase) from Acetobacter xylinum. J Biol Chem 258: 4419–4423

Aloni Y, Delmer DP, Benziman M 1982 Achievement of high rates of in vitro synthesis of 1,4-β-D-glucan: activation of cooperative interaction of the Acetobacter xylinum enzyme system with GTP, polyethylene glycol and a protein factor. Proc Natl Acad Sci USA 79: 6448–6452

Amemura A, Ogawa T, Terui G 1967 Studies on fungal cellulases. III Fractionation and properties of the cellulase system components from Penicillium variabile. Hakko Kogaku Zasshi 45: 879–887

Ammar E. Deschamps AM, Lebeault JM 1986 Biodegradation of ammonium sulphite spent liquor by pure bacterial cultures. Appl Microbiol Biotechnol 24: 122–127

Ander P, Eriksson K-E 1975 Influence of carbohydrates on lignin degradation by the white-rot fungus Sporotrichum pulverulentum. Svensk Papperstidn 78: 643–652

Ander P, Eriksson K-E 1976 The importance of phenol oxidase activity in lignin degradation by the white-rot fungus Sporotrichum pulverulentum. Arch Microbiol 109: 1–8

Ander P, Eriksson K-E 1977 Selective degradation of wood components by white rot fungi. Physiol Plant 41: 239–248

Ander P, Eriksson K-E 1978 Lignin degradation and utilization by microorganisms. In: Bull MJ (ed) Progr Ind Microbiol, Vol 14. Elsevier, Amsterdam, 1–58

Ander P, Eriksson K-E 1985 Methanol formation during lignin degradation by Phanerochaete chrysosporium. Appl Microbiol Biotechnol 21: 96–102

Ander P, Eriksson K-E 1987 Determination of phenoloxidase activity using vanillic acid decarboxylation and syringaldazine oxidation. Biotechnol Appl Biochem 9: 160–169

Ander P, Eriksson MER, Eriksson K-E 1985 Methanol production from lignin-related substances by Phanerochaete chrysosporium. Physiol Plant 65: 317–321

Ander P, Eriksson K-E, Kolar M-C, Kringstad K, Rannug U, Ramel C 1977 Studies on the mutagenic properties of bleaching effluents. Svensk Papperstidn 80: 454–459

Ander P, Eriksson K-E, Månsson P, Pettersson B 1981 Lignin degradation by Sporotrichum pulverulentum: a new cultivation method to study fungal lignin degradation. The Ekman-Days 1981, Intern Symp Wood Pulp Chem. Stockholm, 3: 71–74

Ander P, Eriksson K-E, Yu H-s 1983a Vanillic acid metabolism by Sporotrichum pulverulentum: evidence for demethoxylation before ring-cleavage. Arch Microbiol 136: 1—6

Ander P, Eriksson K-E, Yu H-s 1983b Physiological requirements for degradation of lignin and lignin-related substances by Sporotrichum pulverulentum. Eur J Appl Microbiol Biotechnol 18: 374–380

Ander P, Eriksson K-E, Yu H-s 1984 Metabolism of lignin-derived aromatic acids by wood-rotting fungi. J Gen Microbiol 130: 63–68

Ander P, Hatakka A, Eriksson K-E 1980a Degradation of lignin and lignin-related substances by Sporotrichum pulverulentum (Phanerochaete chrysosporium). In: Kirk TK, Higuchi T, Chang H-m (eds) Lignin biodegradation: microbiology, chemistry and applications Vol II. CRC, Boca Raton, 1–15

Ander P, Hatakka A, Eriksson K-E 1980b Vanillic acid metabolism by the white-rot fungus Sporotrichum pulverulentum. Arch Microbiol 125: 189–202

Ander P, Stoytschev I, Eriksson K-E 1988 Cleavage and metabolism of methoxyl groups from vanillic and ferulic acids by brown-rot and soft-rot fungi. Cellul Chem Technol 22: 255–266

Andersson LA, Renganathan V, Chiu AA, Loehr TM, Gold MH 1985 Spectral characterization of diarylpropane oxygenase, a novel peroxide-dependent, lignin-degrading heme enzyme. J Biol Chem 260: 6080–6087

Andersson LA, Renganathan V, Loehr TM, Gold MH 1987 Lignin peroxidase: resonance Raman spectral evidence for compound II and for a temperature-dependent coordination-state equilibrium in the ferric enzyme. Biochemistry 26: 2258–2263

Antai SP, Crawford DL 1981 Degradation of softwood, hardwood, and grass lignocelluloses by two Streptomyces strains. Appl Environ Microbiol 42: 378–380

Antai SP, Crawford SL 1982 Degradation of extractive-free lignocelluloses by Coriolus versicolor and Poria placenta. Eur J Appl Microbiol Biotechnol 14: 165–168

Antai SP, Crawford DL 1983 Degradation of phenol by Streptomyces setonii. Can J Microbiol 29: 142–143

Araki T 1983 Purification and characterization of an endo-β-mannanase from Aeromonas sp. F-25. J Fac Agr, Kyushu Univ 27: 89–98

Araki T, Kitamikado M 1981 β-Mannanase of bacteria isolated from natural habitats. Bull Jap Soc Sci Fish 47: 753–760

Araki T, Kitamikado M 1982 Purification and characterization of a novel exo-β-mannanase from Aeromonas sp. F-25. J Biochem 91: 1181–1186

Araujo EF, Barros EG, Caldas RA, Silva DO 1983 Beta-glucosidase activity of a thermophilic cellulolytic fungus, Humicola sp. Biotechnol Lett 5: 781–784

Arjmand M, Sanderman Jr H 1985 Mineralization of chloroaniline/lignin conjugates and of free chloroanilines by the white rot fungus Phanerochaete chrysosporium. J Agr Food Chem 33: 1055–1060

Arjmand M, Sanderman Jr H 1986 Plant biochemistry of xenobiotics. Mineralization of chloroaniline/lignin metabolites from wheat by the white-rot fungus, Phanerochaete chrysosporium. Z Naturforsch 41c: 206–214

Arora DS, Sandhu DK 1985 Laccase production and wood degradation by a white-rot fungus Daedalea flavida. Enzyme Microb Technol 7: 405–408

Arora DS, Sandhu DK 1986 Degradation of lignocellulosic residues by Polyporus versicolor and the effect of moisture contents and phenolic compounds. Acta Biotechnol 6: 293–297

Arora DS, Sandhu DK 1987 Decomposition of angiospermic wood sawdust and laccase production by two Pleurotus species. J Basic Microbiol 27: 179–184

Asada Y, Miyabe M, Kikkawa M, Kuwahara M 1986 Oxidation of NADH by a peroxidase of a lignin-degrading basidiomycete, Phanerochaete chrysosporium, and its involvement in the degradation of a lignin model compound. Agr Biol Chem 50: 525–529

Asada Y, Miyabe M, Kikkawa M, Kuwahara M 1987 An extracellular NADH-oxidizing peroxidase produced by a lignin-degrading basidiomycete, Phanerochaete chrysosporium. J Ferment Technol 65: 483–487

Aspinall G 1973 Carbohydrate polymers of plant cell walls. In: Loewus F (ed) Biogenesis of plant cell wall polysaccharides. Academic Press, New York, 95–115

Aspinall GO 1980 Chemistry of cell-wall polysaccharides. In: Preiss J (ed) The biochemistry of plants, Vol 3. Academic Press, New York, 473–500

Asther M, Corrieu G, Drapron R, Odier E 1987 Effect of Tween 80 and oleic acid on ligninase production by Phanerochaete chrysosporium INA-12. Enzyme Microb Technol 9: 245–249

Asther M, Lesage L, Drapron R, Corrieu G, Odier E 1988 Phospholipid and fatty acid enrichment of Phanerochaete chrysosporium INA-12 in relation to ligninase production. Appl Microbiol Biotechnol 27: 393–398

Atalla RH 1983 The structure of cellulose: recent developments. In: Soltes EJ (ed) Wood and agricultural residues. Academic Press, New York, 59–77

Atalla RH, van der Hart DL 1984 Native cellulose: a composite of two distinct crystalline forms. Science 223: 283–285

Ayers AR, Ayers SB, Eriksson K-E 1978 Cellobiose oxidase, purification and partial characterization of a hemoprotein from Sporotrichum pulverulentum. Eur J Biochem 90: 171–181

Ayers AR, Eriksson K-E 1982 Cellobiose oxidase from Sporotrichum pulverulentum. Methods Enzymol 89: 129–135

Baecker AAW, King B 1980a Decay of wood by Actinomycetes. In: Oxley IA, Becker G, Allsopp D (eds) Biodeterioration, 4th Intern Symp, Berlin. Pitman, London, 53–58

Baecker AAW, King B 1980b A technique for the quantitative isolation of Actinomycetes from decayed wood. Intern Res Group on Wood Preserv Document No IRG/WP/116

Bahl OP, Agrawal KML 1969 Glycosidases of Aspergillus niger I. Purification and characterization of α- and β-galactosidases and β-N-acetylglucosaminidase. J Biol Chem 244: 2970–2978

Bailey MJ 1981 The effect of β-glucosidase on some assays for cellulolytic enzymes. Biotechnol Lett 3: 695–700

Bailey PJ, Liese W, Rosch R 1968 Some aspects of cellulose degradation in lignified cell walls. In: Walters AH, Elphick JJ (eds) Biodeterioration of materials. Elsevier, London, 456–557

Bailey IW, Vestal MR 1937 The significance of certain wood-destroying fungi in the study of the enzymatic hydrolysis of cellulose. J Arnold Arbor 18: 196–205
Baker CJ, Whalen CH, Bateman DF 1977 Xylanase from Trichoderma pseudokoningii: purification, characterization and effects on isolated plant cell walls. Physiol Biochem 67: 1250–1258
Barber GA, Elbein AD, Hassid WZ 1964 The synthesis of cellulose by enzyme systems from higher plants. J Biol Chem 239: 4056–4061
Barber GA, Hassid WZ 1964 Formation of guanosine diphosphate D-glucose by enzymes of higher plants. Biochim Biophys Acta 86: 397–399
Barder MJ, Crawford DL 1981 Effects of carbon and nitrogen supplement on lignin and cellulose decomposition by a Streptomyces. Can J Microbiol 27: 859–863
Barghoorn ES, Linder DH 1944 Marine fungi: their taxonomy and biology. Farlowia 1: 395–467
Bar-Lev SS, Kirk TK 1981 Effects of molecular oxygen on lignin degradation by Phanerochaete chrysosporium. Biochem Biophys Res Commun 99: 373–378
Bar-Lev SS, Kirk TK, Chang H-m 1982 Fungal treatment can reduce energy requirements for secondary refining of TMP. Tappi J 65(10): 111–113
Bartley T, Waldron C, Eveleigh D 1984 A cellobiohydrolase from a thermophilic Actinomycete, Microbispora bispora. Appl Biochem Biotechnol 9: 334–337
Bassham JT 1975 Cellulose as a chemical and energy resource. Biotech Bioeng Symp 5: 9–19
Bauch J, Seehann G, Fitzwer H 1976 Microspectrophotometrical investigations on lignin of decayed wood. Mater Org 3: 141–152
Bauchop T 1979 Rumen anaerobic fungi of cattle and sheep. Appl Environ Microbiol 38: 148–158
Bauchop T 1981 The anaerobic fungi in rumen fiber digestion. Agr Environ 6: 339–348
Bauchop T, Mountfort DO 1981 Cellulose fermentation by a rumen anaerobic fungus in both the absence and the presence of rumen methanogenes. Appl Environ Microbiol 42: 1103–1110
Bavendamm W 1928 Über das Vorkommen und den Nachweis von Oxydasen bei holzzerstörenden Pilzen. Z Pflanzenkr Pflanzenschutz 38: 257–276
Bayer EA, Kenig R, Lamed R 1983 Adherence of Clostridium thermocellum to cellulose. J Bacteriol 156: 818–827
Bayer EA, Lamed R 1986 Ultrastructure of the cell surface cellulosome of Clostridium thermocellum and its interaction with cellulose. J Bacteriol 167: 828–836
Bayer EA, Setter E, Lamed R 1985 Organization and distribution of the cellulosome in Clostridium thermocellum. J Bacteriol 163: 552–559
Béguin P 1983 Detection of cellulase activity in polyacrylamide gels using Congo red-stained agar replicas. Anal Biochem 131: 333–336
Béguin P, Eisen H 1978 Purification and partial characterization of three extracellular cellulases from Cellulomonas sp. Eur J Biochem 87: 525–531
Béguin P. Eisen H, Roupas A 1977 Free and cellulose-bound cellulases in a Cellulomonas sp. J Gen Microbiol 101: 191–196
Béguin B, Gilkes NR, Kilburn DG, Miller RC Jr, O'Neill GP, Warren RAJ 1987 Cloning of cellulase genes. CRC Crit Rev Biotechnol 6: 129–162
Beldman G, Searle-Van Leeuwen MF, Rombouts FM, Voragen FG 1985 The cellulase of Trichoderma viride. Eur J Biochem 146: 301–308
Benner R, Hodson R 1985 Thermophilic anaerobic biodegradation of ^{14}C lignin, ^{14}C cellulose, and ^{14}C lignocellulose preparations. Appl Environ Microbiol 50: 971–976
Benner R, Maccubbin AE, Hodson RE 1984 Anaerobic biodegradation of the lignin and polysaccharide components of lignocellulose and synthetic lignin by sediment microflora. Appl Environ Microbiol 47: 998–1004
Berg B 1975 Cellulase in Cellvibrio fulvus. Can J Microbiol 21: 51–57
Berg B 1978 Cellulose degradation and cellulase formation by Phialophora malorum. Arch Microbiol 118: 61–65
Berg B, Pettersson G 1977 Location and formation of cellulases in Trichoderma viride. J Appl Bacteriol 42: 65–75
Berg B, von Hofsten B, Pettersson LG 1972 Growth and cellulase formation by Cellvibrio fulvus. J Appl Bacteriol 35: 201–214

Berg B, von Hofsten B, Pettersson G 1972 Electron microscopic observations on the degradation of cellulose fibres by Cellvibrio fulvus and Sporocytophaga myxococcoides. J Appl Bacteriol 35: 215–219

Berghem LER, Pettersson LG 1973 The mechanism of enzymatic cellulose degradation. Purification of a cellulolytic enzyme from Trichoderma viride active on highly ordered cellulose. Eur J Biochem 37: 21–30

Berghem LER, Pettersson LG 1974 The mechanism of enzymatic cellulose degradation. Isolation and some properties of a β-glucosidase from Trichoderma viride. Eur J Biochem 46: 295–305

Berghem LER, Pettersson LG, Axiö-Fredriksson U-B 1975 The mechanism of enzymatic cellulose degradation. Characterization and enzymatic properties of a β-1,4-glucan cellobiohydrolase from Trichoderma viride. Eur J Biochem 53: 55–62

Berghem LER, Pettersson LG, Axiö-Fredriksson U-B 1976 The mechanism of enzymatic cellulose degradation. Purification and some properties of two different 1,4-β-glucan glucanohydrolases from Trichoderma viride. Eur J Biochem 61: 621–630

Bergman Ö, Nilsson T 1966 On the outside storage of pine chips at Lövholmen's paper mill. Res Notes R53 Dep Forest Prod Royal Coll Forest, Stockholm, 56 pp

Bergmeyer H-U, Bernt E 1963 Determination with glucose oxidase and peroxidase. In: Bergmeyer H (ed) Methods of enzymatic analysis. Academic Press, New York, 123–130

Berner KE, Chapman ES 1977 The cellulolytic activity of six Oomycetes. Mycologia 69: 1232–1236

Bernier R Jr, Driguez H, Desrochers M 1983 Molecular cloning of a Bacillus subtilis xylanase gene in Escherichia coli. Gene 26: 59–65

Bes B, Pettersson B, Lennholm H, Iversen T, Eriksson K-E 1987 Synthesis, structure and enzymatic degradation of an extracellular glucan produced in nitrogen-starved cultures of the white-rot fungus Phanerochaete chrysosporium. Biotechnol Appl Biochem 9: 310–318

Bes B, Ranjeva R, Boudet AM 1983 Evidence for the involvement of activated oxygen in fungal degradation of lignocellulose. Biochimie 65: 283–289

Betts WB, Dart RK, Ball MC 1987a Dimerization of 2,6-dimethoxy-phenol by Aspergillus flavus: evidence for the reaction occurring close to the mycelia. Microbios 49: 123–129

Betts WB, Dart RK, Ball MC 1987b Scanning electron microscope observations of fungal associations with lignin and insoluble dimeric lignin model compounds. Holzforschung 41: 337–341

Bhikhabhai R, Pettersson LG 1984 The cellulolytic enzymes of Trichoderma reesei as a system of homologous proteins. Cyanogen bromide peptides and partial sequence of endoglucanase II. FEBS Letters 167: 301–308

Bhikhabhai R,. Johansson G, Pettersson G 1984 Isolation of cellulolytic enzymes from Trichoderma reesei QM9414. J Appl Biochem 6: 336–345

Bhikhabhai R, Johansson G, Pettersson G 1985 Cellobiohydrolase from Trichoderma reesei. Internal homology and prediction of secondary structure. Int J Pept Protein Res 25: 368–374

Biely P 1985 Microbial xylanolytic systems. Trends Biotechnol 3: 286–290

Biely P, Kratky Z, Vrsanska M, Urmanicova D 1980 Induction and inducers of endo-1,4-β-xylanase in the yeast Cryptococcus albidus. Eur J Biochem 108: 323–329

Biely P, Markovic O 1988 Resolution of glucanases of Trichoderma reesei with respect to cellulose and xylan degradation. Biotechnol Appl Biochem 10: 99–106

Biely P, Markovic O, Mislovicoka D 1985a Sensitive detection of endo-1,4-β-glucanases and endo-1,4-β-xylanases in gels. Anal Biochem 144: 147–151

Biely P, McKenzie CR, Puls J, Schneider H 1986 Cooperativity of esterases and xylanases in the enzymatic degradation of acetyl xylan. Bio/Technol 4: 731–733

Biely P, Puls J, Schneider H 1985b Acetyl xylan esterases in fungal cellulolytic systems. FEBS Lett 186: 80–84

Bisaria VS, Ghose TK 1981 Biodegradation of cellulosic materials: substrates, microorganisms, enzymes and products. Enyzme Microb Technol 3: 90–104

Biswas-Hawkes D, Dodson APJ, Harvey P, Palmer JM 1987 Ligninases from white-rot fungi. In: Odier E (ed) Lignin enzymic and microbial degradation. Symp Intern INRA Paris, 125–130

Björkman A 1956 Studies on finely divided wood. Part 1. Extraction of lignin with neutral solvents. Svensk Papperstidn 59: 477–485

Björkman E, Samuelson O, Ringström E, Bergek T, Malm E 1949 Decay injuries in spruce forests and their importance for the production of chemical paper pulp and rayon pulp. Kgl Skogshögskolan Skr No 4, Stockholm, 73 pp

Björndahl H, Eriksson K-E 1968 Extracellular enyzme system utilized by the rot fungus Stereum sanguinolentum for the breakdown of cellulose. III. Characterization of two purified cellulase fractions. Arch Biochem Biophys 124: 149–153

Blackall LL, Hayward AC, Sly LI 1985 Cellulolytic and dextranololytic gram-negative bacteria: revival of the genus Cellvibrio. J Appl Bacteriol 59: 81–98

Blackwell J 1982 The macromolecular organization of cellulose and chitin. In: Brown RM Jr (ed) Cellulose and other natural polymer systems: biogenesis, structure and degradation. Plenum, New York, 403–428

Blanchette RA 1979 A study of progressive stages of discoloration and decay in Malus using scanning electron microscopy. Can J For Res 9: 464–469

Blanchette RA 1980a Wood decomposition by Phellinus (Fomes) pini: a scanning electron microscopy study. Can J Bot 58: 1496–1503

Blanchette RA 1980b Wood decay: a submicroscopic view. J For 78: 734–737

Blanchette RA 1982a Phellinus (Fomes) pini decay associated with sweetfern rust in sapwood of jack pine. Can J For Res 12: 304–310

Blanchette RA 1982b Progressive stages of discoloration and decay associated with the canker-rot fungus Inonotus obliquus, in birch. Phytopathology 72: 1272–1277

Blanchette RA 1983 An unusual decay pattern in brown-rotted wood. Mycologia 75: 552–556

Blanchette RA 1984a Selective delignification of eastern hemlock by Ganoderma tsugae. Phytopathology 74: 153–160

Blanchette RA 1984b Manganese accumulation in wood decayed by white rot fungi. Phytopathology 74: 725–730

Blanchette RA 1984c Screening wood decayed by white-rot fungi for preferential lignin degradation. Appl Environ Microbiol 48: 647–653

Blanchette RA, Obst JR, Hedges JI, Weliky K 1988 Resistance of hardwood vessels to degradation by white rot Basidiomycetes. Can J Bot 66: 1841–1847

Blanchette RA, Otjen L, Carlson MC 1987 Lignin distribution in cell walls of birch wood decayed by white rot basidiomycetes. Phythopathology 77: 684–690

Blanchette RA, Otjen L, Effland MJ, Eslyn WE 1985 Changes in structural and chemical components of wood delignified by fungi. Wood Sci Technol 19: 35–46

Blanchette RA, Reid ID 1986 Ultrastructural aspects of wood delignification by Phlebia (Merulius) tremellosus. Appl Environ Microbiol 52: 239–245

Blanchette RA, Shaw CG 1978 Associations among bacteria, yeasts and basidiomycetes during wood decay. Phytopathology 68: 631–637

Blanchette RA, Sutherland JB, Crawford DL 1981 Actinomycetes in discolored wood of living silver maple. Can J Bot 59: 1–7

Bland DE, Foster RC, Logan AF 1971 The mechanism of permanganate and osmium tetroxide fixation and the distribution of lignin in the cell wall of Pinus radiata. Holzforschung 25: 137–142

Boman B, Ek M, Eriksson K-E, Frostell B 1988 Some aspects on biological treatment of bleached pulp effluents. Nord Pulp Pap Res J 3: 13–18

Bone DH, Levonen-Munoz E 1984 Solid state fermentation of oat straw by Polyporus sp. A-336. Biotechnol Lett 6: 657–662

Bono JJ, Gas G, Boudet M, Fayret J, Delatour C 1984 Étude comparée de la dégradation de lignocelluloses par différentes souches de Fomes annosus. Can J Microbiol 29: 1683–1688

Bono JJ, Goulas P, Portet N, Séris JL 1987 Effect of Mn^{++} on the activity of two ligninolytic enzymes from Phanerochaete chrysosporium Burds. In: Odier E (ed) Lignin enzymic and microbial degradation. Symp Intern INRA, Paris, 87–91

Borgmeyer JR, Crawford DL 1985 Production and characterization of polymeric lignin degradation intermediates from two different Streptomyces spp. Appl Environ Microbiol 49: 273–278

Bourbonnais R, Paice MG 1987 The fate of ^{14}C-labelled high molecular weight chlorinated lignin and chromophoric material during microbial treatment of bleached kraft effluent. J Wood Chem Technol 7: 51–64

Boutelje JB, Bravery AF 1968 Observations on the bacterial attack of piles supporting a Stockholm building. J Inst Wood Sci 20: 47–57

Boutelje JB, Kiessling H 1964 On water-stored oak timber and its decay by fungi and bacteria. Arch Microbiol 49: 305–314

Bouveng H 1961 Phenylisocyanate derivatives of carbohydrates. II. Location of the O-acetyl groups in birch xylan. Acta Chem Scand 15: 96–100

Bowman JM 1983 Biodelignification. Pap Technol Ind (May), 89–94

Boyce JS 1961 Forest Pathology, 3rd edn, McGraw Hill, New York, 572 pp

Bracker CW, Ruiz-Herrera J, Bartnicki-Garcia S 1976 Structure and transformation of chitin synthetase particles (chitosomes) during microfibril synthesis in vitro. Proc Natl Acad Sci USA 73: 4570–4574

Braghoorn ES, Linder DH 1944 Marine fungi: their taxonomy and biology. Farlowia 1: 395–467

Bravery AF, Parameswaran N, Liese W 1974 Electron microscopic investigations on the effect of tri-n-butyl tin oxide on the decay of beech wood by Polystictus versicolor. Part I. Observations on solvent-treated control material. Mater Org 9: 133–144

Brice RE, Morrison IM 1982 The degradation of isolated hemicelluloses and lignin-hemicellulose complexes by cell-free, rumen hemicellulases. Carbohyd Res 101: 93–100

Brillouet J-M 1987 Mode of action of xylanase II from Polyporus tulipiferae on oat-spelt arbinoglucuronoxylan and larchwood 4-O-methylglucurono xylan. Carbohyd Res 159: 165–170

Brown JA, Ogawa K, Wood TM 1986 Studies on the preparation and regeneration of protoplasts from the cellulolytic fungus Penicillium pinophilum. Enzyme Microb Technol 9: 527–532

Brown RD, Gritzali M 1984 Microbial enzymes and lignocellulose utilization. In: Omenn GS, Hollaender A (eds) Genetic control of environmental pollutants. Plenum, New York, 239–265

Brown RM Jr 1979 Biogenesis of natural polymer systems with special reference to cellulose assembly and deposition. In: Proc 3rd Philip Morris Sci Symp, Philip Morris, USA Operation Center, Richmond, VA, USA, 52–123

Brown RM Jr 1981 Integration of biochemical and visual approaches to the study of cellulose biosynthesis and degradation. Proc Ekman Days Int Symp Wood Pulp Chem, Stockholm 3: 3–15

Brown RM Jr 1985 Cellulose microfibril assembly and orientation: recent developments. J Cell Sci Suppl 2: 13–32

Brown RM Jr, Montezinos D 1976 Cellulose microfibrils: visualization of biosynthetic and orienting complexes in association with the plasma membrane. Proc Natl Acad Sci USA 73: 143–147

Brown RM Jr, Willison JHM, Richardson CL 1976 Cellulose biosynthesis in Acetobacter xylinum: visualization of the site of synthesis and direct measurement of the in vivo process. Proc Natl Acad Sci USA 73: 4565–4569

Brownell HH, Saddler JN 1987 Steam pretreatment of lignocellulosic material for enhanced enzymic hydrolysis. Biotechnol Bioeng 29: 228–235

Bruchmann E-E 1978 Lactones, reductones, and enzymatic saccharification of cellulose. Chem-Ztg 102: 387–389

Bruchmann E-E, Graf H, Saad AA, Schrenk D 1978 Preparation of highly active cellulase preparations and optimization of enzymic cellulose hydrolysis. Chem-Ztg 102: 154–155

Bruchmann E-E, Schach H, Graf H 1987 Role and properties of lactonase in a cellulase system. Biotechnol Appl Biochem 9: 146–159

Bruinenberg PM, de Bot PHM, van Dijken JP, Scheffers WA 1984 NADH-linked aldose reductase: the key to anaerobic alcoholic fermentation of xylose by yeasts. Appl Microbiol Biotechnol 19: 256–260

Brunow G, Wallin H 1981 Studies concerning the preparation of synthetic lignin. The Ekman-Days 1981, Intern Symp Wood Pulp Chem, Stockholm, 4: 125–127

Buchala AJ, Leisola M 1987 Structure of the β-D-glucan secreted by Phanerochaete chrysosporium in continuous culture. Carbohyd Res 165: 146–149

Bucht B, Eriksson K-E 1968 Extracellular enzyme system utilized by the rot fungus Stereum sanguinolentum for the breakdown of cellulose. I. Studies of the enzyme production. Arch Biochem Biophys 124: 135–141

Bucht B, Eriksson K-E 1969 Extracellular enzyme system utilized by the rot fungus Stereum sanguinolentum for the breakdown of cellulose. IV Separation of cellobiase and aryl-β-glucosidase activities. Arch Biochem Biophys 129: 416–420

Bugos RC, Sutherland JB, Adler JH 1988 Phenolic compound utilization by the soft rot fungus Lecythophora hoffmannii. Appl Environ Microbiol 54: 1882–1885

Bumpus JA, Aust SD 1986 Biodegradation of environmental pollutants by the white rot fungus Phanerochaete chrysosporium: involvement of the lignin degrading system. Bio Essays 6: 166–170

Bumpus JA, Aust SD 1987 Biodegradation of DDT [1,1,1-Trichloro-2,2-bis(4-chlorophenyl)-ethane] by the white-rot fungus Phanerochaete chrysosporium. Appl Environ Microbiol 53: 2001–2008

Bumpus JA, Brock BJ 1988 Biodegradation of crystal violet by the white-rot fungus Phanerochaete chrysosporium. Appl Environ Microbiol 54: 1143–1150

Bumpus JA, Tien M, Wright D, Aust SD 1985 Oxidation of persistent environmental pollutants by a white-rot fungus. Science 228: 1434–1436

Burdsall HH 1981 The taxonomy of Sporotrichum pruinosum and Sporotrichum pulverulentum/Phanerochaete chrysosporium. Mycologia 73: 675–680

Burdsall HH, Eslyn WE 1974 A new Phanerochaete with a chrysosporium imperfect state. Mycotaxon 1: 123–133

Buswell JA, Ander P, Eriksson K-E 1982a Ligninolytic activity and levels of ammonia assimilating enzymes in Sporotrichum pulverulentum. Arch Microbiol 133: 165–171

Buswell JA, Ander P, Pettersson B, Eriksson K-E 1979a Oxidative decarboxylation of vanillic acid by Sporotrichum pulverulentum. FEBS Lett 103: 98–101

Buswell JA, Eriksson K-E 1979 Aromatic ring cleavage by the white-rot fungus Sporotrichum pulverulentum. FEBS Lett 104: 258–260

Buswell JA, Eriksson K-E 1986 Methanol metabolism in Phanerochaete chrysosporium. Proc 3rd Intern Conf Biotechnol in the Pulp and Paper Ind, Stockholm, 20–23

Buswell JA, Eriksson K-E, Gupta JK, Hamp SG, Nordh I 1982b Vanillic acid metabolism by selected soft-rot, brown-rot and white-rot fungi. Arch Microbiol 131: 366– 374

Buswell JA, Eriksson K-E, Pettersson B 1981 Purification and partial characterization of vanillate hydroxylase (decarboxylating) from Sporotrichum pulverulentum. J Chromatogr 215: 99–108

Buswell JA, Hamp S, Eriksson K-E 1979b Intracellular quinone reduction in Sporotrichum pulverulentum by a NAD(P)H: quinone oxidoreductase. FEBS Lett 108: 229–232

Buswell JA, Mollet B, Odier E 1984 Ligninolytic enzyme production by Phanerochaete chrysosporium under conditions of nitrogen sufficiency. FEMS Microbiol Lett 25: 295–299

Buswell JA, Odier E 1987 Lignin biodegradation. CRC Crit Rev Biotechnol 6: 1–60

Butcher JA, Howard M 1968 Outside storage of Pinus radiata wood chips in New Zealand. Tappi 51 (4): 117A–122A

Butcher JA, Nilsson T 1982 Influence of variable lignin content amongst hardwoods on soft-rot susceptibility and performance of CCA preservatives. Intern Res Group on Wood Preserv Document IRG/WP/1151

Cain RB 1980 The uptake and catabolism of lignin-related aromatic compounds and their regulation in microorganisms. In: Kirk TK, Higuchi T, Chang H-m (eds) Lignin biodegradation: microbiology, chemistry, and potential applications, Vol 1. CRC, Boca Raton, 21–60

Cain RB, Bilton RF, Darrah JA 1968 The metabolism of aromatic acids by micro-organisms. Biochem J 108: 797–828

Caldwell ES, Steelink C 1969 Phenoxy radical intermediates in the enzymatic degradation of lignin model compounds. Biochim Biophys Acta 184: 420–431

Calza RE, Irwin DC, Wilson DB 1985 Purification and characterization of two β-1,4-endoglucanases from Thermomonospora fusca. Biochemistry 24: 7797–7804

Campbell WA, Davidson RW 1940 Ustulina vulgaris decay in sugar maple and other hardwood. J For 38: 474–477

Campbell WA, Wiertelak J 1935 The chemistry of white rots of wood. IV. The effect on wood substance of Ustulina vulgaris Tul. Biochem J 35: 1318–1321

Campbell WG 1932 The chemistry of wood. III. The effect on wood substance of Ganoderma applanatum (Pers.) Pat., Fomes fomentarius (Linn.) Fr., Polyporus adustus (Willd.) Fr., Pleurotus ostreatus (Jacq.) Fr., Armillaria mellea (Vahl.) Fr., Trametes pini (Brot.) Fr., and Polystictus abietinus (Dicks.) Fr. Biochem J 26: 1829–1838

Canevascini G 1985 A cellulase assay coupled to cellobiose dehydrogenase. Anal Biochem 147: 419–427

Canevascini G, Meyer H-P 1979 β-Glucosidase in the cellulolytic fungus Sporotrichum thermophile Apinis. Exp Mycol 3: 203–214

Canevascini G, Coudray MR, Rey JP, Southgate RJG, Meier H 1979 Induction and catabolite repression of cellulase synthesis in the thermophilic fungus Sporotrichum thermophile. J Gen Microbiol 110: 291–303

Carpentier M, Robic D 1974 Dégradation de la cellulose par un micro-organisme du sol: Sporocytophaga myxococcoides: charactérisation d'une exoglucanase. CR Acad Sci Paris 279: 863–866

Carpita NC, Delmer DP 1980 Protection of cellulose synthesis in detached cotton fibers by polyethylene glycol. Plant Physiol 66: 911–916

Carpita NC, Delmer DP 1981 Concentration and metabolic turnover of UDP-glucose in developing cotton fibers. J Biol Chem 256: 308–315

Casagrande F, Ouellette GB 1971 A technique to study the development in wood of soft rot fungi and its application to Ceratocystis ulmi. Can J Bot 49: 155–159

Castanon M, Wilke CR 1980 Adsorption and recovery of cellulases during hydrolysis of newspaper. Biotechnol Bioeng 22: 1037–1053

Caulfield DF, Moore WE 1974 Effect of varying crystallinity of cellulose on enzymatic hydrolysis. Wood Sci 6: 375–379

Cavalcante MS, Eaton RA 1981 Inhibition of wood-inhibating fungi by actinomycetes. Intern Res Group on Wood Preserv Document IRG/WP/113

Chahal DS, Hawksworth DL 1976 Chaetomium cellulolyticum, a new thermotolerant and cellulolytic Chaetomium, I: Isolation, description and growth rate. Mycologia 68: 600–610

Chang H-m, Cowling EB, Brown W, Adler E, Miksche G 1975 Comparative studies on cellulolytic enzyme lignin and milled wood lignin of sweetgum and spruce. Holzforschung 29: 153–159

Chang H-m. Joyce TW, Campbell AG, Gerrard ED, Huynh V-B, Kirk TK 1983 Fungal decolorization of bleach plant effluents. In: Higuchi T, Chang H-m, Kirk TK (eds) Recent advances in lignin biodegradation research. Uni Publ, Tokyo, 257–268

Chang H-m, Joyce TW, Matsumoto Y, Yin C, Vasudevan B, Boechat C-A 1986 Decolorization and dechlorination of bleach plant effluents by a white-rot fungus. Proc 3rd Intern Conf Biotechnol in the Pulp and Paper Ind, Stockholm, 120–123

Chang H-m, Joyce TW, Vasudevan B, Kirk TK 1987 Optimizing culture parameters in the MyCoR process. Fourth Intern Symp Wood Pulp Chem, Paris, 1: 159–165

Chang SC, Steinkraus KH 1982 Lignocellulolytic enzymes produced by Volvariella volvacea, the edible straw mushroom. Appl Environ Microbiol 43: 440–446

Chanzy H, Henrissat B, Vuong R 1984 Colloidal gold labeling of 1,4-β-D-glucan cellobiohydrolase adsorbed on cellulose substrates. FEBS Lett 172: 193–197

Chanzy H, Henrissat B, Vuong R, Schülein M 1983 The action of 1,4-β-D-glucan cellobiohydrolase on Valonia cellulose microcrystals. An electron microscopic study. FEBS Lett 153: 113–117

Chapman ES, Evans E, Jacobelli MC, Logan AA 1975 The cellulolytic and amylolytic activity of Papulaspora thermophila. Mycologia 67: 608–615

Chauvet J-M, Comtat J, Noe P 1987 Assistance in bleaching of never-dried pulps by the use of xylanases, consequences on pulp properties. Proc 4th Int Congr on Wood and Pulping Chemistry, Paris, April 27–30, II: 325–327

Chen C-L, Chang H-m 1985 Chemistry of lignin biodegradation. In: Higuchi T (ed) Biosynthesis and biodegradation of wood components. Academic Press, Orlando, 535–556

Chen C-L, Chang H-m, Kirk TK 1982 Aromatic acids produced during degradation of lignin in spruce wood by Phanerochaete chrysosporium. Holzforschung 36: 3–9

Chen C-L, Chang H-m, Kirk TK 1983a Carboxylic acids produced through oxidative cleavage of aromatic rings during degradation of lignin in spruce wood by Phanerochaete chrysosporium. J Wood Chem Technol 3: 35–53

Chen C-L, Tai D, Terazawa M, Chang H-m, Kirk TK 1983b Chemistry of birch lignin biodegradation by Phanerochaete chrysosporium. Proc 1983 Intern Symp Wood Pulp Chem Tsukuba, 3: 13–18

Child JJ, Eveleigh DE, Sieben AS 1973 Determination of cellulose activity using hydroxyethyl-cellulose as substrate. Can J Biochem 51: 39–43

Christensen PJ 1977 The history, biology and taxonomy of the Cytophaga group. Can J Microbiol 23: 1599–1653

Chua GS, Chen C-L, Chang H-m, Kirk TK 1982 13C NMR spectroscopic study of spruce lignin degraded by Phanerochaete chrysosporium. Holzforschung 36: 165–172

Chua GS, Choi S, Kirk TK 1983 Mycelium binding and depolymerization of synthetic ^{14}C-labeled lignin during decomposition by Phanerochaete chrysosporium. Holzforschung 37: 55–61

Christman RF, Oglesby RT 1971 Microbiological degradation and the formation of humus. In: Sarkanen KV, Ludwig CH (eds) Lignins. Occurrence, formation, structure, and reactions. Wiley Interscience, New York, 769–795

Chum HL, Douglas LJ, Feinberg DA, Schroeder HA 1985 Evaluation of pretreatments of biomass for enzymic hydrolysis of cellulose. Solar Energy Research Institute, Golden, Colorado, USA

Civas A, Eberhard R, Le Dizet P, Petek F 1984a Glycosidases induced in Aspergillus tamarii: Mycelial α-D-galactosidases. Biochem J 219: 849–855

Civas A, Eberhard R, Le Dizet P, Petek F 1984b Glycosidases induced in Aspergillus tamarii: secreted α-D-galactosidase and β-D-mannanase. Biochem J 219: 857–863

Clark TA, Mackie KL 1986 Steam pretreatment of softwoods with sulphur dioxide addition; chemistry, characterization and digestibility. Proc 3rd Intern Conf Biotechnol in the Pulp Paper Ind, Stockholm, 101–104

Claeyssens M, van Leemputten E, Loontiens FG, De Bruyne CK 1966 Transfer reactions catalyzed by a fungal β-D-xylosidase: enzymic synthesis of phenyl β-D-xylobioside. Carbohyd Res 3: 32–37

Claeyssens M, Brown RD Jr, Deleyn F, De Bruyne CK 1980 Self transfer reactions catalyzed by a β-D-xylosidase from Penicillium wortmanni. Isolation and identification of some transfer products. J Carbohyd Nucleosides, Nucleotides 7: 203–217

Clermont S, Carpentier M, Percheron F 1970 Polysaccharidase de Sporocytophaga myxococcoides, β-mannanase, cellulase et xylanase. Bull Soc Chim Biol 52: 1481–1495

Cobb CE 1982 The non-enzymatic decomposition of cellulose by the brown-rot fungus Gloeophyllum trabeum Master's Thesis, Michigan Tech University, 125 pp

Cochet N, Tyagi RD, Ghose TK, Lebeault JM 1984 ATP measurement for cellulase production control. Biotechnol Lett 6: 155–160

Colberg PJ, Young LY 1982 Biodegradation of lignin-derived molecules under anaerobic conditions. Can J Microbiol 28: 886–889

Colberg PJ, Young LY 1985a Anaerobic degradation of soluble fractions of [^{14}C-lignin]lignocellulose. Appl Environ Microbiol 49: 345–349

Colberg PJ, Young LY 1985b Aromatic and volatile acid intermediates observed during anaerobic metabolism of lignin-derived oligomers. Appl Environ Microbiol 49: 350–358

Collmer A, Wilson DB 1983 Cloning and expression of a Thermomonospora YX endocellulase gene in E. coli. Biotechnology 1: 594–601

Commandy F, Macy JM 1985 Effect of substrate nitrogen on lignin degradation by Pleurotus ostreatus. Arch Microbiol 142: 61–65

Comtat J 1983 Isolation, properties and postulated role of some of the xylanases from the basidiomycete Sporotrichum dimorphosphorum. Carbohyd Res 118: 215–231

Comtat J, Joseleau J-P 1981 Mode of action of a xylanase and its significance for the structural investigation of the branched L-arabino-D-glucurono-D-xylan from red-wood (Sequoia sempervirens). Carbohyd Res 95: 101–112

Conrad D, Nöthen W 1984 Hydrolysis of xylodextrines (DP 2–6) by xylanolytic enzymes from Aspergillus niger. Proc 3rd Eur Congr Biotechnol, München, Verlag Chemie 2: 169–177

Corbett NH 1965 Micro-morphological studies on the degradation of lignified cell walls by Ascomycetes and Fungi Imperfecti. J Inst Wood Sci 4: 18–29

Core HA, Côté WA, Day AC 1979 Wood structure and identification. Syracuse University Press, Syracuse, 182 pp

Cornet P, Millet J, Béguin P, Aubert J-P 1983 Characterization of two cel (cellulose degradation) genes of Clostridium thermocellum coding for endoglucanases. Bio/Technology 1: 589–594

Côté WA 1974 Wood ultrastructure in relation to chemical composition. In: Loewus F, Runeckles WC (eds) The structure, biosynthesis and degradation of wood. Vol 11, chap 1. Plenum, New York, 1–44

Côté WA Jr, Day AC, Timell TE 1968 Studies on compression wood. VII. Distribution of lignin in normal and compression wood of tamarack (Larix laricina (Du Roi) K Koch). Wood Sci Technol 2: 13–37

Côté WA, Timell TE, Zabel RA 1966 Studies on compression wood. I. Distribution of lignin in compression wood of red spruce, Picea rubens Sarg. Holz Roh-Werkst 24: 432–438

Coudray M-R, Canevascini G, Meier H 1982 Characterization of a cellobiose dehydrogenase in the cellulolytic fungus Sporotrichum (Chrysosporium) thermophile. Biochem J 203: 277–284

Coughlan MP 1985 The properties of fungal and bacterial cellulases with comment on their production and application. Biotechnol Gen Eng Rev 3: 39–109

Coughlan MP, Hon-Nami K, Hon-Nami H, Ljungdahl LG, Paulin JJ, Rigsby WE 1985 The cellulolytic enzyme complex of Clostridium thermocellum is very large. Biochem Biophys Res Commun 130: 904–909

Coughlan MP, Ljungdahl LG 1988 Comparative biochemistry of fungal and bacterial cellulolytic enzyme systems. In: Aubert J-P, Béguin P, Millet J (eds) Biochemistry and genetics of cellulose degradation, FEMS symposium no 43. Academic Press, London, 11–30

Coulombel C, Clairmont S, Foglietti M-J, Percheron F 1981 Transglycosylation reactions catalyzed by two β-mannanases. Biochem J 195: 333–335

Courtois H 1963 Mikromorphologische Befallsymptome beim Holzabbau durch Moderfäulepilze. Holzforsch Holzverwert 15: 88–101

Courtois H 1966 Über den Zellwandabbau durch Bakterien im Nadelholz. Holzforschung 20: 148–154

Coutts AD, Smith RE 1976 Factors influencing the production of cellulases by Sporotrichum thermophile. Appl Environ Microbiol 31: 819–825

Cowling EB 1961 Comparative biochemistry of the decay of sweetgum by white-rot fungi. US Dep Agr Tech Bull 1258, 79 pp

Cowling EB, Merrill W 1966 Nitrogen in wood and its role in wood deterioration. Can J Bot 44: 1539–1554

Crawford DL 1978 Lignocellulose decomposition by selected Streptomyces strains. Appl Environ Microbiol 35: 1041–1045

Crawford DL 1981a Microbial conversion of lignin to useful chemicals using a lignin-degrading Streptomyces. Biotechnol Bioeng Symp 11: 275–291

Crawford RL 1981b Lignin biodegradation and transformation. Wiley, New York, 137 pp

Crawford DL, Barder MJ, Pometto III AL, Crawford R 1982a Chemistry of softwood lignin degradation by Streptomyces viridosporus. Arch Microbiol 131: 140–145

Crawford DL, Crawford RL 1980 Microbial degradation of lignin. Enzyme Microb Technol 2: 11–21

Crawford DL, Crawford RL, Pometto III AL 1977 Preparation of specifically labeled ^{14}C-(lignin)- and ^{14}C-(cellulose)-lignocelluloses and their decomposition by the microflora of soil. Appl Environ Microbiol 33: 1247–1251

Crawford DL, Pettey TM, Antai SP 1982b Genetic manipulation of lignin-degrading Streptomyces by protoplast fusion and transformation. 1982 TAPPI Res Developm Div Conf, 249–257

Crawford DL, Pettey TM, Thede BM, Deobald LA 1984a Genetic manipulation of ligninolytic Streptomyces and generation of improved lignin-to-chemical bioconversion strains. Biotech Bioeng Symp 14: 241–256

Crawford DL, Pometto III AL, Crawford RL 1983a Lignin degradation by Streptomyces viridosporus: isolation and characterization of a new polymeric lignin degradation intermediate. Appl Environ Microbiol 45: 898–904

Crawford DL, Pometto III AL, Crawford RL 1984b Production of useful modified lignin polymers by bioconversion of lignocellulose with Streptomyces. Biotech Adv 2: 217–232

Crawford DL, Pometto III AL, Deobald LA 1983b The pathway of lignin degradation by Streptomyces: chemistry and enzymology. In: Higuchi T, Chang H-m, Kirk TK (eds) Recent advances in lignin biodegradation research. Uni Publ Tokyo, 78–95

Crawford DL, Sutherland JB 1979 The role of Actinomycetes in the decomposition of lignocellulose. Dev Ind Microbiol 20: 143–151

Crawford DL, Sutherland JB 1980 Isolation and characterization of lignocellulose-decomposing actinomycetes. In: Kirk TK, Higuchi T, Chang H-m (eds) Lignin biodegradation: microbiology, chemistry and applications, CRC, Boca Raton, II: 95–101

Crawford RL, Crawford DL 1978 Radioisotopic methods for the study of lignin biodegradation. Dev Ind Microbiol 19: 35–49

Crawford RL, Crawford DL 1984 Recent advances in studies of the mechanisms of microbial degradation of lignins. Enzyme Microb Technol 6: 434–442

Crawford RL, Crawford DL, Dizikes GJ 1981a Catabolism of the lignin substructure model compound dehydrodivanillin by a lignin-degrading Streptomyces. Arch Microbiol 129: 204–209

Crawford RL, Frick TD 1982 New assays for ligninolytic activities. 1982 TAPPI Res Developm Div Conf, 259–261

Crawford RL, Robinson LE, Foster RD 1981b Polyguaiacol: a useful model polymer for lignin biodegradation research. Appl Environ Microbiol 41: 1112–1116

Creuzet N, Berenger J-F, Frixon C 1983 Characterization of exoglucanase and synergistic hydrolysis of cellulose in Clostridium stercorarium. FEMS Microbiol Lett 20: 347–350

Cromack K Jr, Sollins P, Todd RL, Fogel R, Todd AW, Fender WM, Crossley ME, Crossley DA Jr 1977 The role of oxalic acid and bicarbonate in calcium cycling by fungi and bacteria: some possible implications for soil animals. Ecol Bull 25: 246–252

Cromack K Jr, Todd RL, Monk CD 1975 Patterns of basidomycete nutrient accumulation in conifer and deciduous forest litter. Soil Biol Biochem 7: 265–268

Crook EM, Stone BA 1957 The enzymatic hydrolysis of β-glucosides. Biochemistry 65: 1–12

Crossley A 1980 The use of electron microscopy to compare wood decay mechanisms. PhD Thesis, Univ London, 317 pp

Dagley S 1971 Catabolism of aromatic compounds by micro-organisms. Adv Microbial Physiol 6: 1–46

Daniel G, Nilsson T 1984 Studies on the S_2 layer of Pinus silvestris. Sver Lantbruksuniv, Inst Virkeslära Rapport 154, 34 pp

Daniel G, Nilsson T 1985 Ultrastructural and TEM-EDXA studies on the degradation of CCA treated radiata pine by tunnelling bacteria. Intern Res Group on Wood Preserv Document IRG/WP/1260

Daniel G, Nilsson T 1986 Ultrastructural observations on wood degrading erosion bacteria. Intern Res Group on Wood Preserv Document IRG/WP/1283

Daniel GF, Nilsson T, Singh AP 1987 Degradation of lignocellulosics by unique tunnel-forming bacteria. Can J Microbiol 33: 943–948

Dart RK, Betts WB, Ball MC 1987 Physical aspects of the degradation of insoluble tri- and tetrameric lignin model compounds by Aspergillus flavus. Microbios 49: 151–159

Darvill A, McNeil M, Albersheim P, Delmer DP 1979 The primary cell walls of flowering plants. In: Tolbert NE (ed) The biochemistry of plants. Vol 1, chap. 3. Academic Press, New York, 91–162

Davidson RW, Campbell WA, Blaisdell DJ 1938 Differentiation of wood-decaying fungi by their reactions on gallic or tannic acid medium. J Agr Res 57: 683–695

de Bary A 1886 Ueber einige Sclerotinien und Sclerotienkrankheiten. Bot Z 44(22): 377–388

de Boer HA, Zhang YZ, Collins C, Reddy CA 1987 Analysis of nucleotide sequences of two ligninase cDNAs from a white-rot filamentous fungus, Phanerochaete chrysosporium. Gene 60: 93–102

Defaye J, Driguez H, John M, Schmidt J, Ohleyer E 1985 Induction of D-xylan-degrading enzymes in Trichoderma lignorum by nonmetabolizable inducers. A synthesis of 4-thioxylobiose. Carbohyd Res 139: 123–132

Dekker RFH 1980 Induction and characterization of a cellobiose dehydrogenase produced by a species of Monilia. J Gen Microbiol 120: 309–316

Dekker RFH 1983 Bioconversion of hemicellulose: aspects of hemicellulase production by Trichoderma reesei QM9414 and enzymatic saccharification of hemicellulose. Biotechnol Bioeng 25: 1127–1146

Dekker RFH 1985 Biodegradation of the hemicelluloses. In: Higuchi T (ed) Biosynthesis and biodegradation of wood components. Academic Press, Tokyo, 505–533

Dekker RFH 1986 Lipid-enhanced ethanol production from xylose by Pachysolen tannophilus. Biotechnol Bioeng 28: 605–608

Dekker RFH, Candy GP 1979 The β-mannanases elaborated by the phytopathogen Xanthomonas campestris. Arch Microbiol 122: 297–299

Dekker RFH, Richards GN 1974 Production of hemicellulases by three fungi pathogenic to sugar cane. Carbohyd Res 38: 257–265

Dekker RFH, Richards GN 1975a Purification, properties and mode of action of hemicellulase I produced by Ceratocystis paradoxa. Carbohyd Res 39: 97–114

Dekker RFH, Richards GN 1975b Purification, properties and mode of action of hemicellulase II produced by Ceratocystis paradoxa. Carbohyd Res 42: 107–123

Dekker RFH, Richards GN 1975c Structures of the oligosaccharides from the enzymic hydrolysis of hemicellulose by a hemicellulase of Ceratocystis paradoxa. Carbohyd Res 43: 335–344

Dekker RFH, Richards GN 1976 Hemicellulases: their occurrence, purification, properties, and mode of action. Adv Carbohyd Chem Biochem 32: 277–352

Dekker RFH, Richards GN, Shambe T 1975 Comparative properties and action patterns of hemicellulases from the phytopathogens Ceratocystis paradoxa and Cephalosporium sacchari. Biochem Soc Trans 3: 1081–1082

Deleyn F, Claeyssens M, De Bruyne CK 1982 β-D-Xylosidase from Penicillium wortmanni. Methods Enzymol 83: 639–644

Deleyn F, Claeyssens M, Van Beeumen J, De Bruyne CK 1978 Purification and properties of β-xylosidase from Penicillium wortmanni. Can J Biochem 56: 43–50

Delmer DP 1974 The biosynthesis of cellulose and other plant cell wall polysaccharides. In: Loewus F, Runeckles WC (eds) The structures, biosynthesis and degradation of wood. Vol 11. Plenum, New York, 45–77

Delmer DP 1982a Cellulose synthesis. In: Zaborsky O (ed) CRC Handbook of biosolar resources, 1(1): 351–355

Delmer DP 1982b Hemicellulose synthesis. In: Zaborsky O (ed) CRC Handbook of biosolar resources, 1(1): 357–361

Delmer DP 1983 Biosynthesis of cellulose. Adv Carbohyd Chem Biochem 41: 105–153

Delmer DP 1987 Cellulose biosynthesis. Ann Rev Plant Physiol 38: 259–290

Demeester J, Cooreman W, Bracke M, Lauwers A 1977 Determination of endocellulase activity by viscosimetry. Biochem Soc Trans 5: 115–117

Deobald LA, Crawford DL 1987 Activities of cellulase and other extracellular enzymes during lignin solubilization by Streptomyces viridosporus. Appl Microbiol Biotechnol 26: 158–163

Desai JD, Desai AJ, Patel NP 1982 Production of cellulases and β-glucosidases by shake culture of Scytalidium lignicola. J Ferment Technol 60: 117–124

Deshpande MV, Eriksson K-E 1984 Reutilization of enzymes for saccharification of lignocellulosic materials. Enzyme Microb Technol 6: 338–340

Deshpande MV, Eriksson K-E, Pettersson LG 1984 An assay for selective determination of exo-1,4-β-glucanases in a mixture of cellulolytic enzymes. Anal Biochem 138: 481–487

Deshpande V, Eriksson K-E, Pettersson B 1978 Production, purification and partial characterization of 1,4-β-glucosidase enzymes from Sporotrichum pulverulentum. Eur J Biochem 90: 191–198

Deshpande V, Mishra C, Ghadge GD, Seeta R, Rao M 1983 Separation and recovery of Penicillium funiculosum cellulase and glucose from cellulosic hydrolysates using polyacrylate gel. Biotechnol Lett 5: 391–394

Desrochers M, Jurasek L, Paice MG 1981 High production of β-glucosidase in Schizophyllum commune: isolation of the enzyme and effect of the culture filtrate on cellulose hydrolysis. Appl Environ Microbiol 41: 222–228

Dey PM 1978 Biochemistry of plant galactomannans. Adv Carbohyd Chem Biochem 35: 341–376

Dey PM, Pridham JB 1972 Biochemistry of α-galactosidases. Adv Enzymol 36: 91–130

DiCosimo R, Szabo H-c 1988 Oxidation of lignin model compounds using single-electron-transfer catalysts. J Org Chem 53: 1673–1679

Dill I, Kraepelin G 1986 Palo Podrido: Model for extensive delignification of wood by Ganoderma applanatum. Appl Environ Microbiol 52: 1305–1312

Dirol D 1976a Etude in vitro de la colonisation et de la dégradation structurale du bois de Hêtre par Coriolus versicolor (L.) Q Rev Mycol 40: 295–317

Dirol D 1978b Etude in vitro de la colonisation et de la dégradation structurale du bois d'aubier de Pin sylvestre par la Merule: Serpula lacrymans Schum ex Fr. SF Gray. Rev Mycol 42: 277–292

Dirol D 1980 Etude in vitro des phases initiales de dégradation du dusamen de pin sylvestre par le tramete du pin: Trametes pini (Thore) Fr. Crytog Mycol 1: 187–199

Dodson PJ, Evans CS, Harvey PJ, Palmer JM 1987 Production and properties of an extracellular peroxidase from Coriolus versicolor which catalyses C_{α}-C_{β} cleavage in a lignin model compound. FEMS Microbiol Lett 42: 17–22

Dolphin D, Nakano T, Maione TE, Kirk TK, Farrell R 1987 Synthetic model ligninases. In: Odier E (ed) Lignin enzymic and microbial degradation. Symp Intern Paris, INRA, 157–162

Dordick JS, Marletta MA, Klibanov A 1986 Peroxidases depolymerize lignin in organic media but not in water. Proc Natl Acad Sci USA 83: 6255–6257

Doudoroff M, Palleroni MJ 1974 Pseudomonas. In: Buchanan RE, Ribbons NE (eds) Bergey's manual of determinative bacteriology, 8th edn. Williams and Wilkins, Baltimore, 217–249

Dubeau H, Chahal DS, Ishaque M 1987 Xylanase of Chaetomium cellulolyticum: its nature of production and hydrolytic potential. Biotechnol Lett 9: 275–280

Duncan CG 1960 Wood attacking capacities and physiology of soft rot fungi. US For Serv Rep FPL-2173, 70 pp

du Preez JC, Bosch M, Prior BA 1986 The fermentation of hexose and pentose sugars by Candida shehatae and Pichia stipitis. Appl Microbiol Biotechnol 23: 228–233

Eaton DC 1985 Mineralization of chlorinated biphenyls by Phanerochaete chrysosporium: a ligninolytic fungus. Enzyme Microb Technol 7: 194–196

Eaton DC, Chang H-m, Joyce TW, Jeffries TW, Kirk TK 1982 Method obtains fungal reduction of the color of extraction-stage kraft bleach effluents. Tappi J 65(6): 89–92

Eaton DC, Chang H-m, Kirk TK 1980 Fungal decolorization of kraft bleach effluents. Tappi J 63(10): 103–106

Eaton RA, Dickinson DJ 1976 The performance of copper chrome arsenate treated wood in the marine environment. Mater Org 11: 521–529

Eberhart BM, Beck RS, Goolsby KM 1977 Cellulase of Neurospora crassa. J Bacteriol 130: 181–186

Effland MJ 1977 Modified procedure to determine acid-insoluble lignin in wood and pulp. Tappi J 60(10): 143–144

Eggeling L, Sahm H 1980 Degradation of coniferyl alcohol and other lignin-related aromatic compounds by Nocardia sp. DSM 1069. Arch Microbiol 126: 141–148

Eggeling L, Sahm H 1981 Degradation of lignin-related aromatic compounds by Nocardia spec. DSM 1069 and specificity of demethylation. In: Schaal/Pulverer (eds) Actinomycetes. Zbl Bakt Suppl 11, Gustav Fischer, Stuttgart, 362–365

Ehhalt DH 1976 The atmospheric cycle of methane. In: Schlegel HG, Gottschalk G, Pfennig N (eds) Microbial production and utilization of gases. Goltze, Göttingen, 13–22

Ek M, Eriksson K-E 1980 Utilization of the white-rot fungus Sporotrichum pulverulentum for water purification and protein production on mixed lignocellulosic waste waters. Biotechnol Bioeng 22: 2273–2284

Ek M, Eriksson K-E 1986 Treatment of bleach plant effluents with ultrafiltration and biological methods. Proc 3rd Intern Conf Biotechnol in the Pulp and Paper Ind, Stockholm, 128–129

Ek M, Eriksson K-E 1987 External treatment of bleach plant effluent. Proc 4th Intern Symp Wood Pulp Chem Vol 1, Paris, 175–177

Elbein AD, Adya S, Lee YC 1977 Purification and properties of a β-mannosidase from Aspergillus niger. J Biol Chem 252: 2026–2031

Elbein AD, Barber GA, Hassid WZ 1964 Synthesis of cellulose by an enzyme system from a higher plant. J Am Chem Soc 86: 309–310

Ellwood EL, Ecklund RA 1959 Bacterial attack of pine logs in pond storage. For Prod J 9: 283–292

Emert GH, Gum EK Jr, Lang JA, Liu TH, Brown RD Jr 1974 Cellulases. Adv Chem Ser 136: 79–100

Emi S, Fukumoto J, Yamamoto T 1972 Crystallization and some properties of mannanase. Agr Biol Chem 36: 991–1001

Enari T-M 1983 Microbial cellulases. In: Fogarty WM (ed) Microbial enzymes and biotechnology. Appl Sci, London, 183–223

Enari T-M, Niku-Paavola M-L 1987 Enzymatic hydrolysis of cellulose: is the current theory of the mechanisms of hydrolysis valid? CRC Crit Rev Biotechnol 5(3): 67–87

Enari T-M, Niku-Paavola M-L, Harju L, Lappalainen A, Nummi M 1981 Purification of Trichoderma reesei and Aspergillus niger β-glucosidase. J Appl Biochem 3: 157–163

Enoki A, Gold MH 1982 Degradation of the diarylpropane lignin model compound 1-(3′,4′-diethoxyphenyl)-1,3-dihydroxy-2-(4″-methoxyphenyl)-propane and derivatives by the basidiomycete Phanerochaete chrysosporium. Arch Microbiol 132: 123–130

Enoki A, Goldsby GP, Gold MH 1980 Metabolism of the lignin model compounds veratrylglycerol-β-guaiacyl ether and 4-ethoxy-3-methoxyphenylglycerol-β-guaiacyl ether by Phanerochaete chrysosporium. Arch Microbiol 125: 227–232

Enoki A, Goldsby GP, Gold MH 1981a β-Ether cleavage of the lignin model compound 4-ethoxy-3-methoxyphenylglycerol-β-guaiacyl ether and derivatives by Phanerochaete chrysosporium. Arch Microbiol 129: 141–145

Enoki A, Goldsby GP, Krisnangkura K, Gold MH 1981b Degradation of the lignin model compounds 4-ethoxy-3-methoxyphenylglycol β-guaiacyl and vanillic acid ethers by Phanerochaete chrysosporium. FEMS Microbiol Lett 10: 373–377

Enoki A, Takahashi M 1983 Metabolism of lignin-related compounds by various wood-decomposing fungi. In: Higuchi T, Chang H-m, Kirk TK (eds) Recent advances in lignin biodegradation research. Uni Publ, Tokyo, 119–133

Enoki A, Takahashi M, Tanaka H, Fuse G 1985 Degradation of lignin-related compounds and wood components by white-rot and brown-rot fungi. Mokuzai Gakkaishi 31: 397–408

Enoki A, Tanaka H, Fuse G 1988 Degradation of lignin-related compounds, pure cellulose, and wood components by white-rot and brown-rot fungi. Holzforschung 42: 85–93

Enoki A, Yajima Y, Gold MH 1981c Olefin saturation and acid reduction of 3,4-dimethoxycinnamic acid and derivatives by Phanerochaete chrysosporium. Phytochemistry 20: 1543–1546

Eriksen J, Goksöyr J 1976 The effect of temperature on growth and cellulase (β-1,4-endoglucanase) production in the compost fungus Chaetomium thermophile var. dissitum. Arch Microbiol 110: 233–238

Eriksson K-E 1967 Studies on cellulolytic and related enzymes. Svensk Kem Tidskr (Eng) 79: 660–680

Eriksson K-E 1969 In discussion of the paper by Leatherwood JM. Cellulose complex of Ruminicoccus and a new mechanism for cellulose degradation. In: Hajny GJ, Reese ET (eds) Cellulases and their applications. Adv Chem Ser 95: 58–59

Eriksson K-E 1978 Enzyme mechanisms involved in cellulose hydrolysis by the rot fungus Sporotrichum pulverulentum. Biotechnol Bioeng 70: 317–332

Eriksson K-E 1981a Cellulases of fungi. In: Hollaender A (ed) Trends in the biology of fermentation for fuels and chemicals. Plenum, New York, 19–32

Eriksson K-E 1981b Microbial degradation of cellulose and lignin. The Ekman-Days 1981, Intern Symp Wood Pulp Chem, Stockholm, 3: 60–65

Eriksson K-E 1981c Fungal degradation of wood components. Pure Appl Chem 53: 33–43

Eriksson K-E 1981d Swedish developments in biotechnology based on lignocellulosic materials. Adv Biochem Eng 20: 193–204

Eriksson K-E 1985 Swedish developments in biotechnology related to the pulp and paper industry. TAPPI J 68(7): 46–55

Eriksson K-E 1987 Production of H_2O_2 in Phanerochaete chrysosporium during lignin degradation. Philos Trans R Soc Lond A 321: 455–459

Eriksson K-E 1988 Microbial delignification – basics, potentials and applications. In: Aubert J-P, Beguin P, Millet J (eds) Biochemistry and genetics of cellulose degradation. Academic Press, London, 285–302

Eriksson K-E, Ander P, Pettersson B 1986a Regulation of lignin degradation in Phanerochaete chrysosporium. In: Proc 3rd Intern Conf Biotechnol in the Pulp and Paper Ind, Stockholm, 24–27

Eriksson K-E, Goodell EW 1974 Pleiotropic mutants of the wood-rotting fungus Polyporus adustus lacking cellulase, mannanase and xylanase. Can J Microbiol 20: 371–378

Eriksson K-E, Grünewald A, Nilsson T, Vallander L 1980a A scanning electron microscopy study of the growth and attack on wood by three white-rot fungi and their cellulase-less mutants. Holzforschung 34: 207–213

Eriksson K-E, Grünewald A, Vallander L 1980b Studies of growth conditions in wood for three white-rot fungi and their cellulase-less mutants. Biotechnol Bioeng 22: 363–776

Eriksson K-E, Gupta JK, Nishida A, Rao M 1984 Syringic acid metabolism by some white-rot, soft-rot and brown-rot fungi. J Gen Microbiol 130: 2457–2464

Eriksson K-E, Hamp SG 1978 Regulation of endo-1,4-β-glucanase production in Sporotrichum pulverulentum. Eur J Biochem 90: 183–190

Eriksson K-E, Johnsrud SC 1982 Mineralization of carbon. In: Burns RG, Slater JH (eds) Experimental microbial ecology. Blackwell, London, 134–153

Eriksson K-E, Johnsrud SC, Vallander L 1983 Degradation of lignin and lignin model compounds by various mutants of the white-rot fungus Sporotrichum pulverulentum. Arch Microbiol 135: 161–168

Eriksson K-E, Kirk TK 1985 Biopulping, biobleaching and treatment of kraft bleaching effluents with white-rot fungi. Comprehensive Biotechnology, Vol 3:15. Pergamon, New York, 271–294

Eriksson K-E, Kolar M-C 1985 Microbial degradation of chlorolignins. Environ Sci Technol 19: 1086–1089

Eriksson K-E, Kolar M-C, Kringstad K 1979 Studies on the mutagenic properties of bleaching effluents. Part 2. Svensk Papperstidn 82: 95–104

Eriksson K-E, Kolar M-C, Ljungquist P, Kringstad KP 1985 Studies on microbial and chemical conversion of chlorolignins. Environ Sci Technol 19: 1219–1224

Eriksson K-E, Pettersson B 1968a Extracellular enzyme system utilized by the rot fungus Stereum sanguinolentum for the breakdown of cellulose. II. Purification of the cellulase. Arch Biochem Biophys 124: 142–148

Eriksson K-E, Pettersson B 1972 Extracellular enzyme system utilized by the fungus Chrysosporium lignorum for the breakdown of cellulose. In: Walters AH, Hueck-van der Plas (eds) Biodeterioration of materials Vol 2. Applied Science, London, 116–120

Eriksson KE, Pettersson B 1973 A zymogram technique for the detection of carbohydrases. Anal Biochem 56: 618–620

Eriksson K-E, Pettersson B 1975a Extracellular enzyme system utilized by the fungus Sporotrichum pulverulentum (Chrysosporium lignorum) for the breakdown of cellulose. I Separation, purification and physico-chemical characterization of five endo-1,4-β-glucanases. Eur J Biochem 51: 193–206

Eriksson K-E, Pettersson B 1975b Extracellular enzyme system utilized by the fungus Sporotrichum pulverulentum (Chrysosporium lignorum) for the breakdown of cellulose. III, Purification and physico-chemical characterization of an exo-1,4-β-glucanase. Eur J Biochem 51: 213–218

Eriksson K-E, Pettersson B 1982 Purification and partial characterization of two acidic proteases from the white-rot fungus Sporotrichum pulverulentum. Eur J Biochem 124: 635–642

Eriksson K-E, Pettersson G 1968b Studies on cellulolytic enzymes. V Some structural properties of the cellulase from Penicillium notatum. Arch Biochem Biophys 124: 160–166

Eriksson K-E, Pettersson B, Westermark U 1974 Oxidation: an important enzyme reaction in fungal degradation of cellulose. FEBS Lett 49: 282–285

Eriksson K-E, Pettersson B, Volc J, Musilek V 1986b Formation and partial characterization of glucose-2-oxidase, a H_2O_2 producing enzyme in Phanerochaete chrysosporium. Appl Microbiol Biotechnol 23: 257–262

Eriksson K-E, Rzedowski W 1969a Extracellular enzyme system utilized by the fungus Chrysosporium lignorum for the breakdown of cellulose. I. Studies of the enzyme production. Arch Biochem Biophys 129: 683–688

Eriksson K-E, Rzedowski W 1969b Extracellular enzyme system utilized by the fungus Chrysosporium lignorum for the breakdown of cellulose. II. Separation and characterization of three cellulase peaks. Arch Biochem Biophys 129: 689–695

Eriksson K-E, Vallander L 1982 Properties of pulps from thermomechanical pulping of chips pretreated with fungi. Svensk Papperstidn 85: R33–R38

Eriksson K-E, Winell M 1968 Purification and characterization of a fungal β-mannanase. Acta Chem Scand 22: 1924–1934

Eriksson K-E, Wood TM 1985 Biodegradation of cellulose. In: Higuchi T (ed) Biosynthesis and biodegradation of wood components. Academic Press, New York, 469–503

Eslyn WE 1967 Outside storage of hardwood chips in the northeast. II Microbiological effects. Tappi 50: 297–303

Eslyn WE, Highley TL 1976 Decay resistance and susceptibility of sapwood of fifteen tree species. Phytopathology 66: 1010–1017

Eslyn WE, Kirk TK, Effland MJ 1975 Changes in the chemical composition of wood caused by six soft-rot fungi. Phytopathology 65: 473–476

Esteban R, Chordi A, Villa TG 1983 Some aspects of a 1,4-β-D-xylanase and a β-xylosidase secreted by Bacillus coagulans strain 26. FEMS Microbiol Lett 17: 163–166

Esteban R, Villanueva JR, Villa TG 1982 β-D-Xylanases of Bacillus circulans WL-12. Can J Microbiol 28: 733–739

Evans CS 1985 Laccase activity in lignin degradation by Coriolus versicolor in vivo and in vitro studies. FEMS Microbiol Lett 27: 339–343

Evans CS 1987 Lignin degradation. Process Biochem 22/4: 102–105

Evans CS, Farmer JY, Palmer JM 1984 An extracellular haem-protein from Coriolus versicolor. Phytochemistry 23: 1247–1250

Evans CS, Palmer JM 1983 Ligninolytic activity of Coriolus versicolor. J Gen Microbiol 129: 2103–2108

Fägerstam LG 1981 Cellulases from Trichoderma reesei QM9414. Enzymatical and structural properties. Abstracts of Uppsala dissertations from the faculty of science 610: 1–21

Fägerstam LG, Pettersson G 1979 The cellulolytic complex of Trichoderma reesei QM9414. An immunochemical approach. FEBS Lett 98: 363–367

Fägerstam LG, Pettersson LG 1980 1,4-β-Glucan cellobiohydrolases of Trichoderma reesei QM9414. A new type of cellulolytic synergism. FEBS Lett 119: 97–100

Fägerstam LG, Pettersson LG, Engström JÅ 1984 The primary structure of a 1,4-β-glucan cellobiohydrolase from the fungus Trichoderma reesei QM9414. FEBS Lett 167: 309–315

Fähnrich P, Irrgang K 1982 Conversion of cellulose to sugars and cellobionic acid by the extracellular enzyme system of Chaetomium cellulyticum. Biotechnol Lett 4: 775–780

Fåhraeus G. Reinhammar B 1967 Large scale production and purification of laccase from cultures of the fungus Polyporus versicolor and some properties of laccase A. Acta Chem Scand 21: 2367–2378

Faison BD, Kirk TK 1983 Relationship between lignin degradation and production of reduced oxygen species by Phanerochaete chrysosporium. Appl Environ Microbiol 46: 1140–1145

Faison BD, Kirk TK 1985 Factors involved in the regulation of a ligninase activity in Phanerochaete chrysosporium. Appl Environ Microbiol 49: 299–304

Faison BD, Kirk TK, Farrell RL 1986 Role of veratryl alcohol in regulating ligninase activity in Phanerochaete chrysosporium. Appl Environ Microbiol 52: 251–254

Faix O, Lange W, Besold G 1981 Molecular weight determinations of DHP's from mixtures of precursors by steric exclusion chromatography (HPLC). Holzforschung 35: 137–140

Faix O, Mozuch MD, Kirk TK 1985 Degradation of gymnosperm (guaiacyl) vs angiosperm (syringyl/guaiacyl) lignins by Phanerochaete chrysosporium. Holzforschung 39: 203–208

Falck R 1926 Über korrosive and destruktive Holzzersetzung und ihre biologische Bedeutung. Ber Deutsch Bot Ges 44: 652–664

Fan LT, Gharpuray MM, Lee Y-H 1987 Cellulose hydrolysis. Springer, Berlin Heidelberg New York, 198 pp

Farkas V, Sestak S, Gresik M, Kolarova N, Labudova I, Bauer S 1987 Induction of cellulase in Trichoderma reesei grown on lactose. Acta Biotechnol 7(5): 425–429

Farrell RL 1987a Industrial applications of lignin-transforming enzymes. Philos Trans R Soc Lond A 321: 549–553

Farrell RL 1987b A new key enzyme for lignin degradation. Ann NY Acad Sci 501: 150–158

Farrell RL, Belew ME, Maione TE, Sadownick BA 1988 Cloning ligninolytic enzymes. In: Aubert J-P, Beguin P, Millet J (eds) Biochemistry and genetics of cellulose degradation. Academic Press, London, 349–364

Fee JA 1982 On the question of superoxide toxicity and the biological function of superoxide dismutase. In: King TE, Mason HS, Morrison M (eds) Oxidases and related redox systems. Pergamon, Oxford, 101–149

Fengel D, Wegener G 1983 Wood: chemistry, ultrastructure, reactions. Walter de Gruyter, Berlin, 613 pp

Fenn P, Choi S, Kirk TK 1981 Ligninolytic activity of Phanerochaete chrysosporium: physiology of suppression by NH_4^+ and L-glutamate. Arch Microbiol 130: 66–71

Fenn P, Kirk TK 1979 Ligninolytic system of Phanerochaete chrysosporium: inhibition by o-phthalate. Arch Microbiol 123: 307–309

Fenn P, Kirk TK 1981 Relationship of nitrogen to the onset and suppression of ligninolytic activity and secondary metabolism in Phanerochaete chrysosporium. Arch Microbiol 130: 59–65

Fenn P, Kirk TK 1984 Effects of C_α-oxidation in the fungal metabolism of lignin. J Wood Chem Technol 4: 131–148

Fergus CL 1969 The cellulolytic activity of thermophilic fungi and actinomycetes. Mycologia 61: 120–129

Fergus BJ, Goring DAI 1970 The distribution of lignin in birch wood as determined by ultraviolet microscopy. Holzforschung 24: 118–124

Fergus BJ, Procter AR, Scott JAN, Goring DAI 1969 The distribution of lignin in sprucewood as determined by ultraviolet microscopy. Wood Sci Technol 3: 117–138

Fielding AH, Byrde RJW 1969 The partial purification and properties of an endopolygalacturonosidase and α-L-arabinofuranosidase from Scelerotina fructigena. J Gen Microbiol 58: 73–84

Finch P, Roberts JC 1985 Enzymatic degradation of cellulose. In: Nevell TP et al. (eds) Cellul Chem. Its Appl. Horwood, Chichester, UK, 312–343

Findlay WPK, Savory JG 1954 Moderfäule. Die Zersetzung von Holz durch niedere Pilze. Holz Roh Werks 12: 293–296

Flaig W, Haider K 1961 Die Verwertung phenolischer Verbindungen durch Weißfäulepilze. Arch Mikrobiol 40: 212–223

Folan MA, Coughlan MP 1981 Cellulase activity of colour "variants" of Talaromyces emersonii. Int J Biochem 13: 243–245

Forney LJ, Reddy CA, Pankratz HS 1982a Ultrastructural localization of hydrogen peroxide production in ligninolytic Phanerochaete chrysosporium. Appl Environ Microbiol 44: 732–736

Forney LJ, Reddy A, Tien M, Aust SD 1982b The involvement of hydroxyl radical derived from hydrogen peroxide in lignin degradation by the white-rot fungus Phanerochaete chrysosporium. J Biol Chem 257: 11455–11462

Forrester IT, Grabski AC, Burgess RR, Leatham GF 1988 Manganese, Mn-dependent peroxidases and the biodegradation of lignin. Biochem Biophys Res Commun 157: 992–999

Forss KG, Gadd GO, Lundell RO, Williamson HW 1974 Process for manufacturing of protein containing substances for fooder, food stuff and technical application. US patent 3,809,614

Fournier R, Frederick MM, Frederick JR, Reilley PJ 1985 Purification and characterization of endo-xylanases from Aspergillus niger III. An enzyme of pI 3.65. Biotechnol Bioeng 27: 539–546

Francis DM, Leightley LE 1983 Extracellular layers of wood decay fungi and copper tolerance. Intern Res Group on Wood Preserv. Document IRG/WP/1180

Francis DM, Leightley LE 1984 An unusual soft-rot decay pattern caused by the Ascomycete Hypoxylon mediterraneum (de Not.) J Miller. Intern Res Group on Wood Preserv Document IRG/WP/1222

Frazer AC. Bossert I, Young LY 1986 Enzymatic aryl-O-methyl-^{14}C labeling of model lignin monomers. Appl Environ Microbiol 51: 80–83

Frazer AC, Young LY 1986 Anaerobic C_1 metabolism of the O-methyl-^{14}C-labeled substituent of vanillate. Appl Environ Microbiol 51: 84–87

Frederick MM, Frederick JR, Fratzke AR, Reilly PJ 1981 Purification and characterization of xylobiose and xylose-producing endo-xylanase from Aspergillus niger. Carbohyd Res 97: 87–103

Frederick MM, Kiang C-H, Frederick JR, Reilly PJ 1985 Purification and characterization of endo-xylanases from Aspergillus niger I. Two isozymes active on xylan backbones near branch points. Biotechnol Bioeng 27: 525–532

Freer SN, Detroy RW 1982 Biological delignification of ^{14}C-labeled lignocelluloses by basidiomycetes: degradation and solubilization of the lignin and cellulose components. Mycologia 74: 943–951

Freudenberg K 1965 Lignin: its constitution and formation from p-hydroxycinnamyl alcohols. Science 148: 595–600

Freudenberg K, Friedrich K, Baumann I, Soff K 1932 Lignin and cellulose XVIII. Cellulose and Starch. Ann Chemie (Justus Liebigs) 494: 41–62

Freudenberg K, Neish AC 1968 Constitution and biosynthesis of lignin. Springer, Berlin Heidelberg New York, 129 pp

Frey-Wyssling A 1976 The plant cell wall. Encyclopedia of plant anatomy Vol. III. Part 4. Gebrüder Borntraeger, Berlin, 294 pp

Frick TD, Crawford RL 1983 Mechanisms of microbial demethylation of lignin model polymers. In: Higuchi T, Chang H-m, Kirk TK (eds) Recent advances in lignin biodegradation research. Uni Publ Tokyo, 143–152

Frohwein YZ, Zori U, Leibowitz J 1963 Enzyme action on fully and partially acetylated sugar derivatives IV. Enzymologia 26: 193–200

Fukazawa K, Imagawa H, Doi S 1975 Histochemical observation of decayed cell walls using ultraviolet and fluorescence microscopy. Res Bull Coll Exp For Hokkaido Univ 33: 104–114

Fukuda K, Haraguchi T 1985 Successive change of lignin through all the stage of wood decay by brown rot fungus Tyromyces palustris. Tokyo Noko Daigaku Nogakubu Enshurin Hokoku 21: 115–121

Fukuda M, Muramatsu T, Egami F 1969 β-Xylosidase from the liver of Charonia lampas I. Purification, properties and application in carbohydrate research. J Biochem 65: 191–199

Fukuzumi T 1980 Microbial metabolism of lignin-related aromatics. In: Kirk TK, Higuchi T, Chang H-m (eds) Lignin biodegradation: microbiology, chemistry, and potential applications. CRC, Boca Raton, 2: 73–94

Fukuzumi T 1987 Ligninolytic enzymes of Pleurotus sajor-caju. In: Odier (ed) Lignin enzymic and microbial degradation. Symp Intern INRA Paris, 137–142

Fukuzumi T, Nishida A, Aoshima K, Minami K 1977 Decolorization of kraft waste liquor with white-rot fungi. I. Screening of the fungi and culturing condition for decolorization of kraft waste liquor. Mokuzai Gakkaishi 23: 290–298

Garcia S, Latge JP 1987 A new colorimetric method for dosage of lignin. Biotechnol Tech 1: 63–68

Gardner KH, Blackwell J 1974 The structure of native cellulose. Biopolymers 13: 1975–2001

Garg SK, Neelakantan S 1982 Studies on the properties of cellulase enzyme from Aspergillus terreus GN1. Biotechnol Bioeng 24: 737–742

Gascoigne JA, Gascoigne MM 1960 Biological degradation of cellulose. Butterworths, London, 264 pp

Gaspar T, Penel C, Castillo FJ, Greppin H 1985 A two-step control of basic and acidic peroxidases and its significance for growth and development. Plant Physiol 64: 418–423

Geiger JP, Huguenin B, Nicole M, Nandris D 1986a Laccases of Rigidoporus lignosus and Phellinus noxius II. Effect of R. lignosus laccase L1 on thioglycolic lignin of hevea. Appl Biochem Biotechnol 13: 97–110

Geiger JP, Rio B, Nandris D, Nicole M 1986b Laccases of Rigidosporus lignosus and Phellinus noxius I. Purification and some physicochemical properties. Appl Biochem Biotechnol 12: 121–133

Ghose TK 1977 Cellulase biosynthesis and hydrolysis of cellulosic substances. Adv Biochem Eng 6: 39–76

Ghose TK 1987 Measurement of cellulase activities. Pure & Appl Chem 59: 257–268

Ghose TK, Pathak AN 1975 Cellulose – a source of future food. Indian Chem Eng 17(4): 3–16

Gibson TS, McCleary BV 1987 A simple procedure for the large-scale purification of β-D-xylanase from Trichoderma viride. Carbohyd Polym 7: 225–240

Giddings TH, Brower DL, Staehelin LA 1980 Visualization of particle complexes in the plasma membrane of Micrasterias denticulata associated with the formation of cellulose fibrils in primary and secondary walls. J Cell Biol 84: 327–329

Gilbert HJ, Jenkins G, Sullivan DA, Hall J 1987 Evidence for multiple carboxymethylcellulase genes in Pseudomonas fluorescence subsp. cellulosa. Mol Gen Genet 210: 551–556

Gilbertson RL 1980 Wood-rotting fungi of North America. Mycologia 72: 1–49

Gilbertson RL 1981 North American wood-rotting fungi that cause brown rots. Mycotaxon 12: 372–416

Gilkes NR, Langsford ML, Kilburn DG, Miller RC Jr, Warren RAJ 1984 Mode of action and substrate specificities of cellulases from cloned bacterial genes. J Biol Chem 259: 10455–10459

Gilligan W, Reese ET 1954 Evidence for multiple components in microbial cellulases. Can J Microbiol 1: 90–107

Glenn JK, Akileswaran L, Gold MH 1986 Mn(II) oxidation is the principal function of the extracellular Mn-peroxidase from Phanerochaete chrysosporium. Arch Biochem Biophys 251: 688–696

Glenn JK, Gold MH 1983 Decolorization of several polymeric dyes by the lignin-degrading basidiomycete Phanerochaete chrysosporium. Appl Environ Microbiol 45: 1741–1747

Glenn JK, Gold MH 1985 Purification and characterization of an extracellular Mn(II)-dependent peroxidase from the lignin-degrading basidiomycete, Phanerochaete chrysosporium. Arch Biochem Biophys 242: 329–341

Glenn JK, Morgan MA, Mayfield MB, Kuwahara M, Gold MH 1983 An extracellular H_2O_2-requiring enzyme preparation involved in lignin biodegradation by the white rot basidiomycete Phanerochaete chrysosporium. Biochem Biophys Res Commun 114: 1077–1083

Goksöyr J, Eriksen J 1980 Cellulases. In: Rose AH (ed) Economic microbiology: microbial enzymes and bioconversions. Vol 5. Academic Press, New York, 283–330

Gold MH, Akileswaran L, Wariishi H, Mino Y, Loehr TM 1987 Spectral characterization of Mn-peroxidase, an extracellular heme enzyme from Phanerochaete chrysosporium. In: Odier E (ed) Lignin enzymic and microbial degradation. Symp Intern INRA Paris, 113–118

Gold MH, Cheng TM 1978 Induction of colonial growth and replica plating of the white rot basidiomycete Phanerochaete chrysosporium. Appl Environ Microbiol 35: 1223–1225

Gold MH, Cheng TM, Alic M 1983a Formation, fusion, and regeneration of protoplasts from wild-type and auxotrophic strains of the white rot basidiomycete Phanerochaete chrysosporium. Appl Environ Microbiol 46: 260–263

Gold MH, Cheng TM, Mayfield MB 1982a Isolation and complementation studies of auxotrophic mutants of the lignin-degrading basidiomycete Phanerochaete chrysosporium. Appl Environ Microbiol 44: 996–1000

Gold MH, Glenn JK, Mayfield MB, Morgan MA, Kutsuki H 1983b Biochemical and genetic studies on lignin degradation by Phanerochaete chrysosporium. In: Higuchi T, Chang H-m, Kirk TK (eds) Recent advances in lignin biodegradation research. Uni Publ Co Ltd Tokyo, 219–232

Gold MH, Kutsuki H, Morgan MA 1983c Oxidative degradation of lignin by photochemical and chemical radical generating system. Photochem Photobiol 38: 647–651

Gold MH, Kuwahara M, Chiu AA, Glenn JK 1984 Purification and characterization of an extracellular H_2O_2-requiring diarylpropane oxygenase from the white rot basidiomycete, Phanerochaete chrysosporium. Arch Biochem Biophys 234: 353–362

Gold MH, Mayfield MB, Cheng TM, Krisnangkura K, Shimada M, Enoki A, Glenn JK 1982b A Phanerochaete chrysosporium mutant defective in lignin degradation as well as several other secondary metabolic functions. Arch Microbiol 132: 115–122

Goldsby GP, Enoki A, Gold MH 1980 Alkyl-phenyl cleavage of the lignin model compounds guaiacylglycol- and glycerol-β-guaiacyl ether by Phanerochaete chrysosporium. Arch Microbiol 128: 190–195

Golovleva LA, Maltseva OV, Myasoedova NM, Leontievsky AA 1986 Panus tigrinus 144 – degrading lignin. Proc 3rd Intern Conf Biotechnol in the Pulp and Paper Ind, Stockholm, 28–30
Gong C-S, Chey LF, Flickinger MC, Tsao GT 1981 Conversion of hemicellulosic carbohydrates. Adv Biochem Eng 20: 93–118
Gong C-S, Ladish MR, Tsao GT 1977 Cellobiase from Trichoderma viride: purification, properties, kinetics, and mechanism. Biotechnol Bioeng 19: 959–981
Gong C-S, Ladisch MR, Tsao GT 1978 Biosynthesis, purification, and mode of action of cellulases of Trichoderma reesei. Adv Chem Ser 181: 261–287
Gonzalez A, Grinbergs J, Griva E 1986 Biological transformation of wood into feed for cattle – "Palo podrido". Zbl Mikrobiol 141: 181–186
Gonzales B, Merino A, Almeida M, Vicuna R 1986 Comparative growth of natural bacterial isolates on various lignin-related compounds. Appl Environ Microbiol 52: 1428–1432
Goodfellow M, Williams ST 1983 Ecology of actinomycetes. Ann Rev Microbiol 37: 189–216
Gorbacheva IV, Rodionova NA 1977a Studies on xylan-degrading enzymes I. Purification and characterization of endo-1,4-β-xylanase from Aspergillus niger Str 14. Biochim Biophys Acta 484: 79–93
Gorbacheva IV, Rodionova NA 1977b Studies on xylan-degrading enzymes II. Action pattern of endo-1,4-β-xylanase from Aspergillus niger Str. 14 on xylan- and xylooligosaccharides. Biochim Biophys Acta 484: 94–102
Goring DAI, Timell TE 1962 Molecular weight of native celluloses. Tappi J 45: 454–460
Gould JM 1984 Alkaline peroxide delignification of agricultural residues to enhance enzymatic saccharification. Biotech Bioeng 26: 46–52
Gould JM 1985 Studies on the mechanism of alkaline peroxide delignification of agricultural residues. Biotech Bioeng 27: 225–231
Gow LA, Wood TM 1988 Breakdown of crystalline cellulose by synergistic action between cellulase components from Clostridium thermocellum and Trichoderma koningii. FEMS Microbiol Lett 50: 247–252
Gradziel K, Haider K, Kochmanska J, Malarczyk E, Trojanowski J 1978 Bacterial decomposition of synthetic ^{14}C-labeled lignin and lignin monomer derivatives. Acta Microbiol Polon 27: 103–109
Grajek W 1987 Comparative studies on the production of cellulases by thermophilic fungi in submerged and solid-state fermentation. Appl Microbiol Biotechnol 26: 126–129
Grbic-Galic D, Young LY 1985 Methane fermentation of ferulate and benzoate: anearobic degradation pathways. Appl Environ Microbiol 50: 292–297
Greaves H 1968 Occurrence of bacterial decay in copper-chrome-arsenic treated wood. Appl Microbiol 16: 1599–1601
Greaves H 1969 Micromorphology of the bacterial attack on wood. Wood Sci Technol 3: 150–166
Greaves H 1970a The effect of some wood-inhabiting bacteria on the permeability characteristics and microscopic features of Eucalyptus regnans sapwoods and Pinus radiata sapwood and heartwood. Holzforschung 24: 6–14
Greaves H 1970b The effect of selected bacteria and actinomycetes on the decay capacity of some wood-rotting fungi. Mater Org 5: 265–279
Greaves H 1971a The bacterial factor in wood decay. Wood Sci Technol 5: 6–16
Greaves H 1971b The effect of substrate availability on cellulolytic enzyme production by selected wood-rotting microorganisms. Aust J Biol Sci 24: 1169–1180
Greaves H, Levy JF 1965 Comparative degradation of the sapwood of scots pine, beech and birch by Lenzites trabea, Polystictus versicolor, Chaetomium globosum and Bacillus polymyxa. J Inst Wood Sci 15: 55–63
Green TR 1977 Significance of glucose oxidase in lignin degradation. Nature 268: 78–80
Greene RV, Gould JM 1983 Substrate-induced H_2O_2 production in mycelia from the lignin-degrading fungus Phanerochaete chrysosporium. Biochem Biophys Res Commun 117: 275–281
Greene RV, Gould JM 1984 Fatty acyl-coenzyme A oxidase activity and H_2O_2 production in Phanerochaete chrysosporium mycelia. Biochem Biophys Res Commun 118: 437–443
Grethlein HE 1984 Pretreatment for enhanced hydrolysis of cellulosic biomass. Biotech Adv 2: 43–62

Greve LC, Labawitch JM, Hungate RE 1984 α-L-Arabinofuranosidase from Ruminococcus albus 8: purification and possible role in hydrolysis of alfalfa cell wall. Appl Environ Microbiol 47: 1135–1140

Grisebach H 1981 Lignins. In: Conn EE (ed) The biochemistry of plants Vol 7. Academic Press, New York, 457–478

Gritzali M, Brown RD Jr 1979 The cellulase system of Trichoderma. Relationships between purified extracellular enzymes from induced or cellulose grown cells. Adv Chem Ser 181: 237–260

Groleau D, Forsberg CW 1983 Partial characterization of the extracellular carboxymethyl cellulase activity produced by the rumen bacterium Bacteroides succinogenes. Can J Microbiol 29: 504–517

Gross GG 1980 The biochemistry of lignification. In: Woolhouse HW (ed) Advances in botanical research Vol 8. Academic Press, New York, 26–63

Gross GG, Janse C, Elstner EF 1977 Involvement of malate, monophenols, and the superoxide radical in hydrogen peroxide formation by isolated cell walls from horseradish (Armoracia lapathifolia Gilib.). Planta 136: 271–276

Gum EK Jr, Brown RD Jr 1976 Structural characterization of a glycoprotein cellulase, 1,4-β-D-glucan cellobiohydrolase C from Trichoderma viride. Biochim Biophys Acta 446: 371–386

Gupta J, Hamp S, Buswell JA, Eriksson K-E 1981 Metabolism of trans-ferulic acid by the white-rot fungus Sporotrichum pulverulentum. Arch Microbiol 128: 349–354

Gupta DP, Heale JB 1971 Induction of cellulase (C_x) in Verticillium albo-atrum. J Gen Microbiol 63: 163–173

Gurujeyalakshmi G, Mahadevan A 1987 Degradation of guaiacol glyceryl ether (GGE) by Bacillus subtilis. Appl Microbiol Biotechnol 26: 289–293

Haars A, Hütterman A 1980 Function of laccase in the white-rot fungus Fomes annosus. Arch Microbiol 125: 233–237

Habe T, Shimada M, Higuchi T 1985a Biomimetic approach to lignin degradation I. H_2O_2-dependent C-C bond cleavage of the lignin model compounds with a natural iron porphyrin and imidazol complex. Mokuzai Gakkaishi 31: 54–55

Habe T, Shimada M, Okamoto T, Panijpan B, Higuchi T 1985b Incorporation of dioxygen into the hydroxylated product during the C-C single bond cleavage of 1,2-bis(p-methoxyphenyl)propane-1,3-diol catalysed by hemin. A novel model system for the hemoprotein ligninase. J Chem Soc Chem Commun 1985: 1323–1324

Habe T, Shimada M, Umezawa T, Higuchi T 1985c Evidence for deuterium retention in the products after enzymatic C-C and ether bond cleavages of deuterated lignin model compounds. Agr Biol Chem 49: 3505–3510

Hackett WF, Connors WJ, Kirk TK, Zeikus JG 1977 Microbial decomposition of synthetic ^{14}C-labeled lignins in nature: lignin biodegradation in a variety of natural materials. Appl Environ Microbiol 33: 43–51

Haemmerli SD, Leisola MSA, Fiechter A 1986a Polymerisation of lignins by ligninases from Phanerochaete chrysosporium. FEMS Microbiol Lett 35: 33–36

Haemmerli SD, Leisola MSA, Sanglard D, Fiechter A 1986b Oxidation of benzo(a)pyrene by extracellular ligninases of Phanerochaete chrysosporium. J Biol Chem 261: 6900–6903

Haemmerli SD, Schoemaker HE, Schmidt HWH, Leisola MSA 1987 Oxidation of veratryl alcohol by the lignin peroxidase of Phanerochaete chrysosporium. FEBS Lett 220: 149–154

Hägerdal BGR, Ferchak JD, Pye EK 1978 Cellulolytic enzyme system of Thermoactinomyces sp. grown on microcrystalline cellulose. Appl Environ Microbiol 36: 606–612

Hägerdal B, Harris H 1979 Association of β-glucosidase with intact cells of Thermoactinomyces. Biotechnol Bioeng 21: 345–355

Hägglund E, Lindberg B, McPehrson J 1956 Dimethylsulfoxide, a solvent for hemicellulose. Acta Chem Scand 10: 1160–1164

Haider K, Grabbe K 1967 Die Rolle der Phenoloxydase beim Ligninabbau durch Weißfäulepilze. Zentralbl Bakteriol Parasitenkd 205: 91–96

Haider K, Kern HW, Ernst L 1985 Intermediate steps of microbial lignin degradation as elucidated by ^{13}C NMR spectroscopy of specifically ^{13}C-enriched DHP-lignins. Holzforschung 39: 23–32

Haider K, Trojanowski J 1975 Decomposition of specifically ^{14}C-labelled phenols and dehydropolymers of coniferyl alcohol as models for lignin degradation by soft and white rot fungi. Arch Microbiol 105: 33–41

Haider K, Trojanowski J 1980 A comparison of the degradation of ^{14}C-labeled DHP and cornstalk lignins by micro- and macrofungi and by bacteria. In: Kirk TK, Higuchi T, Chang H-m (eds) Lignin biodegradation: microbiology, chemistry and potential applications Vol I, CRC, Boca Raton, 111–134

Haider K, Trojanowski J 1981 Mikrobieller Abbau von ^{14}C- und ^{35}S-markierten Ligninsulfonsäuren durch Pilze, Bakterien oder Mischkulturen. Holzforschung 35: 33–38

Haider K, Trojanowski J, Sundman V 1978 Screening for lignin degrading bacteria by means of ^{14}C-labelled lignins. Arch Microbiol 119: 103–106

Haigler CH, Brown RM Jr, Benziman M 1980 Calcofluor White ST alters the in vivo assembly of cellulose microfibrils. Science 210: 903–906

Håkansson U, Fägerstam L, Pettersson G, Andersson L 1978 Purification and characterization of a low molecular weight 1,4-β-glucan glucanohydrolase from the cellulolytic fungus Trichoderma viride QM9114. Biochim Biophys Acta 524: 385–392

Håkansson U, Fägerstam LG, Pettersson G, Andersson L 1979. A 1,4-β-glucanohydrolase from the cellulolytic fungus Trichoderma viride QM9414. Biochem J 179: 141–149

Hale MD, Eaton RA 1985a Oscillatory growth of fungal hyphae in wood cell walls. Trans Br Mycol Soc 84: 277–288

Hale MD, Eaton RA 1985b The ultrastructure of soft rot fungi. I. Fine hyphae in wood cell walls. Mycologia 77: 447–463

Hale MD, Eaton RA 1985c The ultrastructure of soft rot fungi. II. Cavity-forming hyphae in wood cell walls. Mycologia 77: 594–605

Hale MD, Eaton RA 1986a Soft rot cavity widening: a kinetic approach. Proc R Soc Lond 227: 217–226

Hale MD, Eaton RA 1986b Soft rot cavity formation in five preservative hardwood species. Trans Br Mycol Soc 86: 585–590

Hall PL 1980 Enzymatic transformations of lignin: 2. Enzyme Microb Technol 2: 170–176

Hall PL, Glasser WG, Drew SW 1980 Enzymatic transformation of lignin. In: Kirk TK, Higuchi T, Chang H-m (eds) Lignin biodegradation: microbiology, chemistry, and potential applications Vol 2. CRC, Boca Raton, 33–49

Hallaksela A-M 1984 Bacteria and their effect on the microflora in wounds of living Norway spruce (Picea abies). Commun Inst For Fenn 121: 1–25

Halliwell B 1981 The biological effects of the superoxide radical and its products. Clin Respir Physiol 17: 21–28

Halliwell G, Griffin M 1973 The nature and mode of action of the cellulolytic component C_1 of Trichoderma koningii on native cellulose. Biochem J 135: 587–594

Halliwell G, Riaz M 1971 Interactions between components of the cellulase complex of Trichoderma koningii in native substrates. Arch Microbiol 78: 295–309

Halliwell G, Vincent R 1981 The action on cellulose and its derivatives of a purified 1,4-β-glucanase from Trichoderma koningii. Biochem J 199: 409–417

Halliwell G, Griffin M, Vincent R 1972 Role of component C_1 in cellulolytic systems. Proc Biochem Soc 127: 43

Hammel KE, Kalyanaraman B, Kirk TK 1986a Substrate free radicals are intermediates in lignin catalysis. Proc Natl Acad Sci USA 83: 3708–3712

Hammel KE, Kalyanaraman B, Kirk TK 1986b Oxidation of polycyclic aromatic hydrocarbons and dibenzo(p)dioxins by Phanerochaete chrysosporium ligninase. J Biol Chem 261: 16498–16952

Hammel KE, Kalyanaraman B, Kirk TK 1987 Oxidation of aromatic pollutants by Phanerochaete chrysosporium ligninase. In: Odier E (ed) Lignin enzymic and microbial degradation. Symp Intern, INRA Paris, 45–49

Hammel KE, Tardone PJ 1988 The oxidative 4-dechlorination of polychlorinated phenols is catalyzed by extracellular fungal lignin peroxidase. Biochemistry 27: 6563–6567

Hammel KE, Tardone PJ, Moen MA, Price LA 1989 Biomimetic oxidation of nonphenolic lignin models by Mn(III): New observations on the oxidizability of guaiacyl and syringyl substructures. Arch Biochem Biophys 270: 404–409

Hammel KE, Tien M, Kalyanaraman B, Kirk TK 1985 Mechanisms of oxidative Cα-Cβ cleavage of a lignin model dimer by Phanerochaete chrysosporium ligninase. J Biol Chem 260: 8348–8353

Hanselman KW 1982 Lignochemicals. Experientia 38: 176–189

Harada H 1965 Ultrastructure of angiosperm vessels and ray parenchyma. In: Côté WA Jr (ed) Cellular ultrastructure of woody plants. Syracuse University Press, Syracuse, 235–249

Haraguchi T, Fukushima M, Fukuda S, Morohoshi N 1983 Degradation of milled-wood lignin by an enzyme prepared from a wood-destroying fungus, Stereum frustulatum. Biodeterioration 5: 75–83

Hardell H-L, Leary GJ, Stoll M, Westermark U 1980 Variations in lignin structure in defined morphological parts of birch. Svensk Papperstidn 83: 71–74

Harkin JM, Obst JR 1973 Lignification in trees: indication of exclusive peroxidase participation. Science 180: 296–298

Harkin JM, Obst JR 1974 Demethylation of 2,4,6-trimethoxyphenol by phenol oxidases. Tappi J 57(7): 118–121

Harmsen L, Nissen T 1965 Der Bakterienangriff auf Holz. Holz RohWerkst 23: 389–393

Harrer W, Kubicek CP, Rohr M, Wurth H, Marihart J 1983 The effect of carboxymethyl cellulose addition on extracellular enzyme formation in Trichoderma pseudokoningii. Eur J Appl Microbiol Biotechnol 17: 339–343

Hartig R 1878 Die Zersetzungserscheinungen des Holzes der Nadelholzbäume und der Eiche in forstlicher, botanischer und chemischer Richtung. Springer, Berlin Heidelberg New York, 151 pp

Harvey AE, Jurgensen MF, Larsen MJ 1976 Intensive fiber utilization and prescribed fire: effects on the microbial ecology of forests. US For Serv, Gen Tech Rep INT-28, 46 pp

Harvey PJ, Schoemaker HE, Bowen RM, Palmer JM 1985a Single-electron transfer processes and the reaction mechanism of enzymic degradation of lignin. FEBS Lett 183: 13–16

Harvey PJ, Schoemaker HE, Palmer JM 1985b Enzymic degradation of lignin and its potential to supply chemicals. Ann Proc Phytochem Soc Eur 26: 249–266

Harvey PJ, Schoemaker HE, Palmer JM 1986 Veratryl alcohol as a mediator and the role of radical cations in lignin biodegradation by Phanerochaete chrysosporium. FEBS Lett 195: 242–246

Harvey PJ, Schoemaker HE, Palmer JM 1987a Mechanism of lignin catalysis. In: Odier E (ed) Lignin enzymic and microbial degradation. Symp Intern, INRA Paris, 145–150

Harvey PJ, Schoemaker HE, Palmer JM 1987b Lignin degradation by white-rot fungi. Plant Cell Environ 10: 709–714

Hashimoto S, Muramatsu T, Funatsu M 1971 Studies on xylanase from Trichoderma viride. Part I. Isolation and some properties of crystalline xylanase. Agr Biol Chem 35: 501–508

Hassid WZ 1969 Biosynthesis of oligosaccharides and polysaccharides in plants. Science 165: 137–144

Hatakka A 1985 Degradation of veratric acid and other lignin-related aromatic compounds by the white-rot fungus Pycnoporus cinnabarinus. Arch Microbiol 141: 22–28

Hatakka A 1986 Degradation and conversion of lignin, lignin-related aromatic compounds and lignocellulose by selected white-rot fungi. Dissert, Univ of Helsinki, Helsinki, 97 pp

Hatakka A, Kantelinen A, Tervilä-Wilo A, Viikari L 1987 Production of ligninases by Phlebia radiata in agitated cultures. In: Odier E (ed) Lignin enzymic and microbial degradation. Symp Intern INRA Paris, 185–189

Hatakka A, Pirhonen TI 1985 Cultivation of white-rot fungi on agricultural lignocellulosic materials for the production of crude protein. Agr Wastes 12: 81–97

Hatakka A, Uusi-Rauva AK 1983 Degradation of ^{14}C-labeled poplar wood lignin by selected white-rot fungi. Arch Microbiol 17: 235–242

Hattori T, Schimada M, Umezawa T, Higuchi T, Leisola MSA, Fiechter A 1988 New mechanism for oxygenative ring cleavage of 3,4-dimethoxybenzyl alcohol catalyzed by the ligninase model. Agr Biol Chem 52: 879–880

Hayaishi O 1974 General properties and biological functions of oxygenases. In: Hayaishi O (ed) Molecular mechanisms of oxygen activation. Academic Press, New York, 1–28
Hayashida S, Yoshioka H 1980 Production and purification of thermostable cellulases from Humicola insolens YH-8. Agr Biol Chem 44: 1721–1728
Heale JB, Gupta DP 1971 The utilization of cellobiose by Verticillium albo-atrum. J Gen Microbiol 63: 175–181
Healy Jr JB, Young LY 1979 Anaerobic biodegradation of eleven aromatic compounds to methane. Appl Environ Microbiol 38: 84–89
Healy Jr JB, Young LA, Reinhard M 1980 Methanogenic decomposition of ferulic acid, a model lignin derivative. Appl Environ Microbiol 39: 436–444
Hedgecock GG, Long WH 1914 Heart rot of oaks and poplars caused by Polyporus dryophilus. J Agr Res 3: 65–77
Hejmanck M, Rypacek V 1954 O fungistatickem ucinku metabolikych produktu neckterych mikro-organismu. Ebenda 360: 225–237
Henningsson B 1967 The physiology, interrelationships and effect on the wood of fungi which attack birch and aspen pulpwood. Stud For Souec 53, 31 pp
Henningsson B, Henningsson M, Nilsson T 1972 Defibration of wood by the use of a white rot fungus. Res Notes 78, K Skogshögsk, Stockholm
Henrissat B, Driguez H, Viet C, Schülein M 1985 Synergism of cellulases from Trichoderma reesei in the degradation of cellulose. Bio/Technology 3: 722–726
Herr D, Baumer F, Dellweg H 1978a Purification and properties of an extracellular endo-1,4-β-glucanase from Lenzites trabea. Arch Microbiol 117: 287–292
Herr D, Luck G, Dellweg H 1978b Formation of cellulases and degradation of cellulose by several fungi. J Ferment Technol 56: 273–278
Hestrin S 1962 Synthesis of polymeric homosaccharides. In: Gunsalus IC and Stainer RI (eds) The bacteria Vol III. Academic Press, New York, 373–388
Highley TL 1973 Influence of carbon source on cellulase activity of white-rot and brown-rot fungi. Wood Fiber 5: 50–58
Highley TL 1975 Properties of cellulases of two brown-rot fungi and two white-rot fungi. Wood Fiber 6: 275–281
Highley TL 1977 Requirements for cellulose degradation by a brown-rot fungus. Mater Org 12: 161–174
Highley TL 1978 How moisture and pit aspiration affect decay of wood by white-rot and brown-rot fungi. Mater Org 13: 197–206
Highley TL 1980 Cellulose degradation by cellulose-clearing and non-cellulose clearing brown-rot fungi. Appl Environ Microbiol 40: 1145–1147
Highley TL 1982a Is extracellular hydrogen peroxide involved in cellulose degradation by brown-rot fungi? Mater Org 17: 205–214
Highley TL 1982b Influence of type and amount of lignin on decay by Coriolus versicolor. Can J For Res 12: 435–438
Highley TL 1987a Changes in chemical components of hardwood and softwood by brown-rot fungi. Mater Org 22: 39–45
Highley TL 1987b Effect of carbohydrate and nitrogen on hydrogen peroxide formation by wood decay fungi in solid medium. FEMS Lett 48: 373–377
Highley TL, Bar-Lev SS, Kirk TK, Larsen MJ 1983 Influence of O_2 and CO_2 on wood decay by heartrot and saprot fungi. Phytopathology 73: 630–633
Highley TL, Lutz JF 1970 Bacterial attack in water-stored bolts. For Prod J 20(4): 43–44
Highley TL, Murmanis LL 1985 Determination of hydrogen peroxide production in Coriolus versicolor and Poria placenta during wood degradation. Mater Org 20: 241–252
Highley TL, Murmanis LL 1987 Micromorphology of degradation in western hemlock and sweetgum by the white-rot fungus Coriolus versicolor. Holzforschung 41: 67–71
Highley TL, Murmanis L, Palmer JG 1983 Electron microscopy of cellulose decomposition by brown-rot fungi. Holzforschung 37: 272–279
Highley TL Murmanis L, Palmer JG 1985 Micromorphology of degradation in western hemlock and sweetgum by the brown-rot fungus Poria placenta. Holzforschung 39: 73–78

Highley TL, Wolter KE 1982 Properties of a carbohydrate-degrading enzyme complex from the brown-rot fungus Poria placenta. Mater Org 17: 127–134
Highley TL, Wolter KE, Evans FJ 1981 Polysaccharide-degrading complex produced in wood and liquid media by the brown-rot fungus Poria placenta. Wood Fiber 13: 265–274
Higuchi T 1971 Formation and biological degradation of lignins. Adv Enzymol 34: 207–283
Higuchi T 1980 Biochemistry of lignification. Wood Res (Kyoto) 66: 1–16
Higuchi T 1981 a Biosynthesis of lignin. In: Tanner W, Loewus FA (eds) Encyclopedia of plant physiology. New Series, Vol 13B. Springer, Berlin Heidelberg New York, 195–224
Higuchi T 1981 b Biodegradation of lignin. Wood Res (Kyoto) 67: 47–58
Higuchi T 1982 Biodegradation of lignin: biochemistry and potential applications. Experientia 38: 159–166
Higuchi T 1985 a Biosynthesis of lignin. In: Higuchi T (ed) Biosynthesis and biodegradation of wood components. Academic Press, San Diego, 141–160
Higuchi T 1985 b Degradative pathways of lignin model compounds. In: Higuchi T (ed) Biosynthesis and biodegradation of wood components. Academic Press, San Diego, 557–578
Higuchi T 1986 Catabolic pathways and role of ligninases for the degradation of lignin substructure models by white-rot fungi. Wood Res (Kyoto) 73: 58–81
Higuchi T, Shimada M, Nakatsubo F, Tanahashi M 1977 Differences in biosynthesis of guaiacyl and syringyl lignins in woods. Wood Sci Technol 11: 153–167
Hiley WE 1919 The fungal diseases of the common larch. Oxford University Press, London, 204 pp
Hillis WE 1968 Chemical aspects of heartwood formation. Wood Sci Technol 2: 241–259
Hillis WE 1987 Heartwood and tree exudates. Springer, Berlin Heidelberg New York, 268 pp
Hinds TE 1981 Cryptosphaeria canker and Libertella decay of aspen. Phytopathology 71: 1137–1145
Hiroi T, Eriksson K-E 1976 Microbial degradation of lignin. Part 1. Influence of cellulose on the degradation of lignins by the white-rot fungus Pleurotus ostreatus. Svensk Papperstidn 79: 157–161
Hiroi T, Eriksson K-E, Stenlund B 1976 Microbial degradation of lignin. Part 2. Influence of cellulose on the degradation of calcium lignosulfonate of various molecular sizes by the white-rot fungus Pleurotus ostreatus. Svensk Papperstidn 79: 162–166
Hiroi T, Tamai A 1983 Degradation of beech wood components by a white-rot fungus, Grifola frondosa. In: Higuchi T, Chang H-m, Kirk TK (eds) Res adv lignin biodegradation research, Uni Publ, Tokyo, 34–43
Hoebler C, Brillouet J-M 1984 Purification and properties of an endo-1,4-β-D-xylanase from Irpex lacteus (Polyporus tulipiferae). Carbohyd Res 128: 141–155
Hogness DS, Battley EH 1957 The α-galactosidase of Aerobacter aerogenes; its characteristics and induced synthesis. Am Soc Biol Chem 16: 197–198
Holt DM 1982 The effect of pre-conditioning wood with fungi on its subsequent degradation by cellulolytic prokaryotes. Mater Org 17: 1–20
Holt DM, Jones EBG 1978 Bacterial cavity formation in delignified wood. Mater Org 13: 13–30
Holt DM, Jones EBG 1983 Bacterial degradation of lignified wood cell walls in anaerobic aquatic habitats. Appl Environ Microbiol 46: 722–727
Hon-Nami K, Coughlan MP, Hon-Nami H, Carreira LH, Ljungdahl LG 1985 Properties of the cellulolytic enzyme system of Clostridium thermocellum. Biotechnol Bioeng Symp 15: 191–205
Hopp H 1938 The formation of colored zones by wood-destroying fungi in culture. Phytopathology 28: 601–620
Hoppe-Seyler F 1883 Gährung der Cellulose. Ber Deutsch Chem Ges 16: 122–123
Houston CW, Latimer SB, Mitchell ED 1974 Purification and properties of β-mannosidase from malted barley. Biochim Biophys Acta 370: 276–282
Hrmova M, Biely P, Vrsanska M 1986 Specificity of cellulase and β-xylanase induction in Trichoderma reesei QM9414. Arch Microbiol 144: 307–311
Hubert EE 1929 Sap stains of wood and their prevention. US Dep Commerce, Wood Utiliz 10th Rep, 77 pp
Hubert EE 1931 An outline of forest pathology. Wiley, New York, 543 pp

Hulme MA 1971 Viscometric determination of carboxymethylcellulase in standard international units. Arch Biochem Biophys 147: 49–54

Hulme MA, Stranks DW 1970 Induction and the regulation of production of cellulase by fungi. Nature 226: 469–470

Hurst HM, Burges NA 1967 Lignin and humic acids. In: McLaren AD, Peterson GH (eds) Soil biochemistry. Marcel Dekker, New York, 260–286

Hurst OL, Sullivan PA, Shepherd MG 1978 Substrate specificity and mode of action of a cellulase from Aspergillus niger. Biochem J 169: 389–395

Hütterman A, Noelle A 1982 Characterization and regulation of cellobiose dehydrogenase in Fomes annosus. Holzforschung 36: 283–286

Huynh V-B 1986 Biomimetic oxidation of lignin model compounds by simple inorganic complexes. Biochem Biophys Res Commun 139: 1104–1110

Huynh V-B, Chang H-m, Joyce TW, Kirk TK 1985 Dechlorination of chloro-organics by a white-rot fungus. Tappi J 68(7): 98–102

Huynh V-B, Crawford RL 1985 Novel extracellular enzymes (ligninases) of Phanerochaete chrysosporium. FEMS Microbiol Lett 28: 119–123

Huynh V-B, Paszczynski A, Olson P, Crawford R 1986a Transformations of arylpropane lignin model compounds by a lignin peroxidase of the white-rot fungus Phanerochaete chrysosporium. Arch Biochem Biophys 250: 186–196

Huynh V-B, Paszczynski A, Olson P, Crawford R 1986b Oxidation of lignin model compounds by a manganese-dependent enzyme from Phanerochaete chrysosporium as compared to chemically-generated Mn(III). Proc 3rd Intern Conf Biotechnol in the Pulp and Paper Ind, Stockholm, 42–45

Iiyama K, Wallis AFA 1988 An improved acetyl bromide procedure for determining lignin in woods and wood pulps. Wood Sci Technol 22: 271–280

Ikeda R, Yamamoto T, Funatsu M 1973 Chemical and enzymatic properties of acid-cellulase produced by Aspergillus niger. Agr Biol Chem 37: 1169–1175

Inglin M, Feinberg BA, Loewenberg JR 1980 Partial purification and characterization of a new intracellular β-glucosidase of Trichoderma reesei. Biochem J 185: 515–519

Inoue K, Inoue N 1957 The mannanase of Aerobacter mannanolyticus. J Ferment Technol 35: 354–355

Ishihara T 1983 Effect of pH in the oxidation of syringic acid by fungal laccase. Mokuzai Gakkaishi 29: 801–805

Ishihara T, Ishihara M 1975 Estimation of orthoquinone structure and methanol formed from lignin models by fungal laccase. Mokuzai Gakkaishi 21: 323–325

Ishihara T, Ishihara M 1976 Oxidation of syringic acid by fungal laccase. Mokuzai Gakkaishi 22: 371–375

Ishihara T, Miyazaki M 1972 Oxidation of milled wood lignin by fungal laccase. Mokuzai Gakkaishi 18: 415–419

Ishihara T, Miyazaki M 1974 Demethylation of lignin and lignin models by fungal laccase. Mokuzai Gakkaishi 20: 39–41

Ishihara T, Nishida A 1983 The role of oxidation and reduction enzymes in lignin biodegradation. In: Higuchi T, Chang H-m, Kirk TK (eds) Rec Adv Lignin Biodegradation Research. Uni Publishers, Tokyo, 134–142

Ishihara M, Shimizu K 1980 Hemicellulases of brown-rotting fungus, Tyromyces palustris IV. Purification and some properties of an extracellular mannanase. Mokuzai Gakkaishi 26: 811–818

Ishihara M, Shimizu K, Ishihara T 1975 Purification and enzymatic properties of intracellular xylanases from the mycelium of brown-rotting fungus, Tyromyces palustris. Mokuzai Gakkaishi 21: 680–685

Ishikawa H, Schubert W, Nord FF 1963 Investigations on lignins and lignification. XXVII. The enzymic degradation of softwood lignin by white-rot fungi. Arch Biochem Biophys 100: 131–139

(IUB) International Union of Biochemistry. Nomenclature Committee. Enzyme nomenclature 1984: Recommendations of the Nomenclature Committee of the International Union of Bio-

chemistry on the nomenclature and classification of enzyme catalyzed reactions. Academic Press, Orlando, 1984, 646 p.
Iwahara S 1983 Metabolism of lignin-related aromatic compounds by Fusarium species. In: Higuchi T, Chang H-m, Kirk TK (eds) Rec Adv Lignin Biodegradation Research. Uni Publishers, Tokyo, 96–111
Iwahara S, Nishihira T, Jomori T, Kuwahara M, Higuchi T 1980 Enzymic oxidation of α,β-unsaturated alcohols in the side chains of lignin-related aromatic compounds. J Ferment Technol 58: 183–188
Iwasaki T, Hayashi K, Funatsu M 1964 Purification and characterization of two types of cellulase from Trichoderma koningii. J Biochem 55: 209–212
Iyayi CB, Dart RK 1982 Degradation of sinapyl alcohol by the fungus Schizophyllum commune. Microbios 34: 167–176
Jäger A, Croan S, Kirk TK 1985 Production of ligninases and degradation of lignin in agitated submerged cultures of Phanerochaete chrysosporium. Appl Environ Microbiol 50: 1274–1278
Jain MK, Kapoor KK, Mishra MM 1979 Cellulase activity, degradation of cellulose and lignin, and humus formation by thermophilic fungi. Trans Br Mycol Soc 73: 85–89
Janshekar H, Fiechter A 1982 On the bacterial degradation of lignin. Eur J Appl Microbiol 14: 47–50
Janshekar H, Fiechter A 1983 Lignin: biosynthesis, application, and biodegradation. Adv Biochem Eng/Biotechnol 27: 117–178
Janshekar H, Fiechter A 1988 Cultivation of Phanerochaete chrysosporium and production of lignin peroxidases in submerged stirred tank reactors. J Biotechnol 8: 97–112
Jeffries TW, Choi S, Kirk TK 1981 Nutritional regulation of lignin degradation by Phanerochaete chrysosporium. Appl Environ Microbiol 42: 290–296
Jobe A, Bourgeois S 1972 Lac repressor-operator interaction VI. The natural inducer of the lac operon. J Mol Biol 69: 397–408
Joglekar AV, Karanth NG, Srinivasan MC 1983a Significance of β-D-glucosidase in the measurement of exo-β-D-glucanase activity of cellulolytic fungi. Enzyme Microb Technol 5: 25–29
Joglekar AV, Srinivasan MC, Manchanda AC, Jogdand VV, Karanth NG 1983b Studies on cellulase production by Penicillium funiculosum strain in an instrumented fermenter. Enzyme Microb Technol 5: 22–24
Johansson T, Nyman PO 1987 A manganese(II)-dependent extracellular peroxidase from the white-rot fungus Trametes versicolor. Acta Chem Scand B41: 762–765
John M, Schmidt B, Schmidt J 1979 Purification and some properties of five endo-1,4-β-D-xylanases and a β-xylosidase produced by a strain of Aspergillus niger. Can J Biochem 57: 125–134
John M, Schmidt J, Sahm H, Wandrey C 1981 Purification and characterization of two xylanases from Trichoderma lignorum. Biochem Soc Trans 9, 116 p
Johnson DB, Moore WE, Zank LC 1961 The spectrophotometric determination of lignin in small samples. Tappi J 44(11): 793–798
Johnson EA, Reese ET, Demain AL 1982 Inhibition of Clostridium thermocellum cellulase by end products of cellulolysis. J Appl Biochem 4: 64–71
Johnsrud SC, Eriksson K-E 1985 Cross-breeding of selected and mutated homokaryotic strains of Phanerochaete chrysosporium K-3: new cellulase deficient strains with increased ability to degrade lignin. Appl Microbiol Biotechnol 21: 320–327
Johnsrud SC, Eriksson K-E 1986 Pulp production from different fungal pretreated materials. Proc. 3rd Intern Conf Biotechnol in the Pulp and Paper Ind, Stockholm, 53
Johnsrud SC, Fernandez N, Lopez P, Gutierrez I, Saez A, Eriksson K-E 1987 Properties of fungal pretreated high yield bagasse pulps. Nord Pulp Pap J, Special Issue 2: 47–52
Jokela J, Pellinen J, Salkinoja-Salonen M 1987 Initial steps in the bacterial degradation of two tetrameric lignin model compounds. Appl Environ Microbiol 53: 2642–2649
Jokela J, Pellinen J, Salkinoja-Salonen M, Brunow G 1985 Biodegradation of two tetrameric lignin model compounds by a mixed bacterial culture. Appl Microbiol Biotechnol 23: 38–46
Joliff G, Béguin P, Aubert J-P 1986 Nucleotide sequence of the cellulase gene encoding endoglucanase D of Clostridium thermocellum. Nucl Acids Res 14: 8605–8613
Jones A 1974 World protein resources. Med & Techn Publ, Manchester, England, 381 pp

Jönsson L, Johansson T, Sjöström K, Nyman PO 1987 Purification of ligninase isozymes from the white-rot fungus Trametes versicolor. Acta Chem Scand B41: 766–769

Jurásek L 1955 Změny v mikrostrukture zdrevnatelé bunecné blány pri rozkladu drevokaznými houbami. (Changes in the microstructure of the lignified cell wall during attack by wood-destroying fungi). Biologia 10: 569–579

Jurásek L 1958 Mikrostruktura borové beli při rozkladu celulosovorními houbami. (Microstructure of Pinus sylvestris sapwood when attacked by brown rot fungi) Drev Vysk 3: 129–135

Jurásek L 1964 Změny v mikroskopické struktuře při rozkladu dřeva dřevokaznými houbami (Changes in the microstructure at the destruction of wood by wood destroying fungi). Drev Vysk 3: 127–144

Jutte SM, Wardrop AB 1970 Morphological factors relating to the degradation of wood fibres by cellulase preparations. Acta Bot Neerl 19: 906–917

Käärik A 1975 Successions of microorganisms during wood decay. In: Liese W (ed) Biological transformation of wood by microorganisms. Proc 2nd Intern Congr Plant Pathol, 1973, Minneapolis, 39–51

Kadam KL, Drew SW 1986 Study of lignin biotransformation by Aspergillus fumigatus and white-rot fungi using ^{14}C-labeled kraft lignins. Biotech Bioeng 28: 394–404

Kaiser J-P, Hanselman KW 1982 Aromatic chemicals through anaerobic microbial conversion of lignin monomers. Experientia 38: 167–176

Kaji A 1984 L-Arabinosidases. Adv Carbohyd Chem Biochem 42: 383–394

Kaji A, Sato M, Tsutsui Y 1981 An α-L-arabinofuranosidase produced by wild-type Streptomyces sp. No 17–1. Agr Biol Chem 45: 925–931

Kaji A, Tagawa K 1970 Purification, crystallization and amino acid composition of α-L-arabinofuranosidase from Aspergillus niger. Biochim Biophys Acta 207: 456–464

Kaji A, Tagawa K, Ishimi T 1969 Properties of purified α-L-arabinofuranosidase from Aspergillus niger. Biochim Biophys Acta 171: 186–188

Kaji A, Yoshihara O 1970 Production and properties of α-L-arabinofuranosidase from various strains of Corticium rolfsii. Agr Biol Chem 34: 1249–1253

Kaji A, Yoshihara O 1971 Properties of purified α-L-arabinofuranosidase from Corticium rolfsii. Biochim Biophys Acta 250: 367–371

Kamaya Y, Higuchi T 1983 Degradation of d,l-syringaresinol and its derivatives, β-β' linked lignin substructure models, by Phanerochaete chrysosporium. Mokuzai Gakkaishi 29: 789–794

Kamaya Y, Higuchi T 1984a Degradation of lignin substructure models with biphenyl linkage by Phanerochaete chrysosporium Burds. Wood Res (Kyoto) 70: 25–28

Kamaya Y, Higuchi T 1984b Metabolism of 1,2-disyringylpropane-1,3-diol by Phanerochaete chrysosporium. Mokuzai Gakkaishi 30: 237–239

Kamaya Y, Higuchi T 1984c Metabolism of non-phenolic diarylpropane lignin substructure model compound by Coriolus vesicolor. FEMS Microbiol Lett 22: 89–92

Kamaya Y, Higuchi T 1984d Metabolism of 3,4-dimethoxycinnamyl alcohol and derivatives by Coriolus versicolor. FEMS Microbiol Lett 24: 225–229

Kamaya Y, Nakatsubo F, Higuchi T 1983 Degradation of trimeric lignin model compounds, arylglycerol-β-syringaresinol ethers, by Fusarium solani M-13-1. Agr Biol Chem 47: 299–308

Kamaya Y, Nakatsubo F, Higuchi T, Iwahara S 1981 Degradation of d,l-syringaresinol, a β-β' linked model compound, by Fusarium solani M-13-1. Arch Microbiol 129: 305–309

Kamra DN, Zadrazil F 1985 Influence of oxygen and carbon dioxide on lignin degradation in solid state fermentation of wheat straw with Stropharia rugosoannulata. Biotechnol Lett 7: 335–340

Kanamoto J, Sakamoto R, Arai M, Murao S 1979 Enzymatic properties of two carboxymethyl cellulose hydrolyzing enzymes from Aspergillus aculeatus. J Ferment Technol 57: 163–168

Kanda T, Amano Y, Nisizawa K 1985 Purification and properties of two endo-1,4-β-xylanases from Irpex lacteus (Polyporus tulipiferae) J Biochem 98: 1545–1554

Kanda T, Noda I, Wakabayashi K, Nisizawa K 1983 Transglycosylation activities of exo- and endo-type cellulases from Irpex lacteus (Polyporus tulipiferae). J Biochem 93: 787–794

Kanda T, Wakabayashi K, Nisizawa K 1976 Synergistic action of two different types of endocellulase components from Irpex lacteus (Polyporus tulipiferae) in the hydrolysis of some insoluble celluloses. J Biochem 79: 997–1006

Kantelinen A, Waldner R, Niku-Paavola M-L, Leisola MSA 1988 Comparison of two lignin-degrading fungi: Phlebia radiata and Phanerochaete chrysosporium. Appl Microbiol Biotechnol 28: 193–198

Kaplan DL 1979 Reactivity of different oxidases with lignin and lignin model compounds. Phytochemistry 18: 1917–1919

Kaplan DL, Hartenstein R 1980 Decomposition of lignins by microorganisms. Soil Biol Biochem 12: 65–75

Katayama T, Kawai S, Sogo M, Higuchi T 1987 Degradation of guaiacylglycerol-α-vanillyl alcohol-β-vanillin diether, a lignin substructure model containing a non-cyclic benzyl aryl ether, by Fusarium solani M-13-1. Mokuzai Gakkaishi 33: 503–510

Katayama T, Nakatsubo F, Higuchi T 1980 Initial reactions in the fungal degradation of guaiacylglycerol-β-coniferyl ether, a lignin substructure model. Arch Microbiol 126: 127–132

Katayama T, Nakatsubo F, Higuchi T 1981 Degradation of arylglycerol-β-aryl ethers, lignin substructure models, by Fusarium solani. Arch Microbiol 130: 198–203

Katayama T, Nakatsubo F, Higuchi T 1986a Degradation of a phenylcoumaran, a lignin substructure model, by Fusarium solani M-13-1. Mokuzai Gakkaishi 32: 535–544

Katayama T, Sogo M, Higuchi T 1986b Degradation and stereoselective reduction of an α-ketone derivative of guaiacylglycerol-β-vanillin ether, a lignin substructure model, by Fusarium solani M-13-1. Holzforschung 40: 175–182

Katayama Y, Nishikawa S, Murayama A, Yamasaki M, Morohoshi N, Haraguchi T 1988 The metabolism of biphenyl structures in lignin by the soil bacterium (Pseudomonas paucimobilis SYK-6). FEBS Lett 233: 129–133

Kawai S, Umezawa T, Higuchi T 1985a Metabolism of a non-phenolic β-O-4 lignin substructure model compound by Coriolus versicolor. Mokuzai Gakkaishi 49: 2325–2330

Kawai S, Umezawa T, Higuchi T 1985b Arylglycerol-γ-formyl ester as an aromatic ring cleavage product of nonphenolic β-O-4 lignin substructure model compounds degraded by Coriolus versicolor. Appl Environ Microbiol 50: 1505–1508

Kawai S, Umezawa T, Higuchi T 1986a De novo synthesis of veratryl alcohol by Coriolus versicolor. Wood Res (Kyoto) 73: 18–21

Kawai S, Umezawa T, Higuchi T 1987a p-Benzoquinone monoketals, novel degradation products of β-O-4 lignin model compounds by Coriolus versicolor and lignin peroxidase of Phanerochaete chrysosporium. FEBS Lett 210: 61–65

Kawai S, Umezawa T, Higuchi T 1988 Degradation mechanisms of phenolic β-1 lignin substructure model compounds by laccase of Coriolus versicolor. Arch Biochem Biophys 262: 99–110

Kawai S, Umezawa T, Shimada M, Higuchi T 1986b C_α-C_β cleavage of β-1 lignin substructure models by the laccase of Coriolus versicolor. Proc 3rd Intern Conf Biotechnol in the Pulp Paper Ind, Stockholm, 39–41

Kawai S, Umezawa T, Shimada M, Higuchi T, Koide K, Nishida T, Morohoshi N, Haraguchi T 1987b C_α-C_β cleavage of phenolic β-1 lignin substructure model compound by laccase of Coriolus versicolor. Mokuzai Gakkaishi 33: 792–797

Kawakami H, Shumiya Y 1983 Degradation of lignin-related compounds and lignins by alkalophilic bacteria. In: Higuchi T, Chang H-m, Kirk TK (eds) Recent advances in lignin biodegradation research. Uni Publ, Tokyo, 64–77

Kawase K 1962 Chemical components of wood decayed under natural conditions and their properties. Hokkaido Diagaku Sapporo J Fac Agr 52: 186–245

Keilich G, Bailey PJ, Afting EG, Liese W 1969 Cellulase (β-1,4-glucan-4-glucan-hydrolase) from the wood rotting fungus Polyporus schweinitzii Fr II. Characterization. Biochim Biophys Acta 185: 392–401

Keilich G, Bailey P, Liese V 1970 Enzymatic degradation of cellulose, cellulose derivatives and hemicelluloses in relation to the fungal decay of wood. Wood Sci Technol 4: 273–283

Kelleher TJ, Montenecourt BS, Eveleigh DE 1987 Cellobiose-quinone oxidoreductase-application in monitoring cellobiohydrolase purification. Appl Microbiol Biotechnol 27: 299–305

Kelley RL, Ramasamy K, Reddy CA 1986 Characterization of glucose oxidase-negative mutants of a lignin degrading basidiomycete Phanerochaete chrysosporium. Arch Microbiol 144: 254–257

Kelley RL, Reddy CA 1982 Ethylene production from α-Oxo-γ-methylthiobutyric acid is a sensitive measure of ligninolytic activity by Phanerochaete chrysosporium. Biochem J 206: 423–425

Kelley RL, Reddy CA 1986a Identification of glucose oxidase activity as the primary source of hydrogen peroxide production in ligninolytic cultures of Phanerochaete chrysosporium. Arch Microbiol 144: 248–253

Kelley RL, Reddy CA 1986b Purification and characterization of glucose oxidase from ligninolytic cultures of Phanerochaete chrysosporium. J Bacteriol 166: 269–274

Kern HW 1981 Microbial degradation of lignosulfonates. FEMS Symp 12: 299–324

Kern HW 1983a Transformation of lignosulfonates by Trichoderma harzianum. Holzforschung 37: 109–115

Kern HW 1983b Action of the white-rot fungus Sporotrichum pulverulentum on lignosulfonates. Holzforschung 37: 287–292

Kern HW 1984 Bacterial degradation of dehydropolymers of coniferyl alcohol. Arch Microbiol 138: 18–25

Kern HW 1989 Improvement in the production of extracellular lignin peroxidases by Phanerochaete chrysosporium. Appl Microbiol Biotechnol 32: 223–234

Kern HW, Kirk TK 1987 Influence of molecular size and ligninase pretreatment on degradation of lignins by Xanthomonas sp. strain 99. Appl Environ Microbiol 53: 2242–2246

Kerr AJ, Goring DAI 1975 The ultrastructural arrangement of the wood cell wall. Cellul Chem Technol 9: 563–573

Kerr TJ, Kerr RD, Benner R 1983 Isolation of a bacterium capable of degrading peanut hull lignin. Appl Environ Microbiol 46: 1201–1206

Kersten PJ, Kirk TK 1987 Involvement of a new enzyme, glyoxal oxidase, in extracellular H_2O_2 production by Phanerochaete chrysosporium. J Bacteriol 169: 2195–2202

Kersten PJ, Kalyanaraman B, Hammel KE, Kirk TK 1987 Horseradish peroxidases oxidizes 1,2,4,5-tetramethoxybenzene by a cation radical mechanism. In: Odier (ed) Lignin enzymic and microbial degradation. Symp Intern INRA Paris, 75–79

Kersten PJ, Tien M, Kalyanaraman B, Kirk TK 1985 The ligninase of Phanerochaete chrysosporium generates cation radicals from methoxybenzenes. J Biol Chem 260: 2609–2612

Kersters-Hilderson H, Claeyssens M, Van Doorslaer E, De Bruyne CK 1976 Determination of the anomeric configuration of D-xylose with D-xylose isomerases. Carbohyd Res 47: 269–273

Kersters-Hilderson H, Claeyssens M, van Doorslaer E, Saman E, De Bruyne CK 1982 β-D-Xylosidase from Bacillus pumilus. Methods Enzymol 83: 631–639

Kersters-Hilderson H, Van Doorslaer E, De Bruyne CK 1978 β-D-Xylosidase from Bacillus pumilus PRL B12: Hydrolysis of aryl β-D-xylopyranosides. Carbohyd Res 65: 219–227

Kersters-Hilderson H, Van Doorslaer E, De Bruyne CK 1980 Binding of n-alkyl β-D-xylopyranosides and n-alkyl 1-thio-β-D-xylopyranosides to β-D-xylosidase from Bacillus pumilus PRL B12. Carbohyd Res 78: 163–172

Kersters-Hilderson H, Loontiens FG, Claeyssens M, De Bruyne CK 1969 Partial purification and properties of an induced β-D-xylosidase of Bacillus pumilus 12. Eur J Biochem 7: 434–441

Kertesz ZI 1963 Polyuronides. Compr Biochem 5: 233–245

Keyser P, Kirk TK, Zeikus JG 1978 Ligninolytic enzyme system of Phanerochaete chrysosporium: synthesized in the absence of lignin in response to nitrogen starvation. J Bacteriol 135: 790–797

Khan AW, Lamb KA, Forgie MA 1987 Formation of esterase, xylanase and xylosidase by cellulolytic anaerobes. Biomass 13: 135–146

Khan AW, Tremblay D, LeDuy A 1986 Assay of xylanase and xylosidase activities in bacterial and fungal cultures. Enzyme Microb Technol 8: 373–377

King B, Eaton RA, Baecker AAW 1978 A summary of current information on Actinomycetes and wood. Intern Res Group on Wood Preserv Document IRG/WP/177

King B, Eggins HOW 1977 Micromorphology of Streptomycete colonization of wood. J Inst Wood Sci 7: 24–29

King KW 1965 Enzymatic attack on highly crystalline hydrocellulose. J Ferment Technol (Japan) 43: 79–94

King KW, Vessal MI 1969 Enzymes of the cellulase complex. In: Hajny GJ, Reese ET (eds) Cellulases and their applications. Adv Chem Ser 95: 7–25

King NJ, Fuller DB 1968 The xylanase system of Coniophora cerebella. Biochem J 108: 571–576

Kirk TK 1971 Effect of microorganisms on lignin. Ann Rev Phytopathol 9: 185–210

Kirk TK 1973 Polysaccharide integrity as related to the degradation of lignin in wood by white-rot fungi. Phytopathology 63: 1504–1507

Kirk TK 1975 Effects of a brown-rot fungus, Lenzites trabea, on lignin in spruce wood. Holzforschung 29: 99–107

Kirk TK 1980 Physiology of lignin metabolism by white-rot fungi. In: Kirk TK, Higuchi T, Chang H-m (eds) Lignin biodegradation: microbiology, chemistry, and potential applications, Vol 2. CRC, Boca Raton, 51–63

Kirk TK 1981 Toward elucidating the mechanism of action of the ligninolytic system in basidiomycetes. In: Hollaender A (ed) Trends in the biology of fermentations for fuels and chemicals. Plenum, New York, 131–148

Kirk TK 1983a Degradation and conversion of lignocelluloses. In: Smith JE, Berry DR, Kristiansen B (eds) The filamentous fungi, Vol 4, Fungal technology. Edward Arnold, London, 266–295

Kirk TK 1983b What have studies of model compounds revealed about the fungal degradation of lignin. Proc 1983 Intern Symp Wood Pulp Chem, Vol 3, Tsukuba, 7–12

Kirk TK 1984 Degradation of lignin. In: Gibson DT (ed) Microbial degradation of organic compounds. Marcel Dekker, New York, 399–437

Kirk TK 1986 The action of ligninase on lignin model compounds and lignin. Proc TAPPI 1986 Res Develop Conf, 73–78

Kirk TK 1987 Lignin-degrading enzymes. Philos Trans R Soc London A 321: 461–474

Kirk TK 1988a Biochemistry of lignin degradation by Phanerochaete chrysosporium. In: Aubert JP, Beguin P, Millet J (eds) Biochemistry and genetics of cellulose degradation. Academic Press, London, 315–332

Kirk TK 1988b Lignin degradation by Phanerochaete chrysosporium. ISI Atlas of Science: Biochemistry 0894–3753: 71–76

Kirk TK, Adler E 1969 Catechol moieties in enzymatically liberated lignin. Acta Chem Scand 23: 705–707

Kirk TK, Adler E 1970 Methoxyl-deficient structural elements in lignin of sweetgum decayed by a brown-rot fungus. Acta Chem Scand 24: 3379–3390

Kirk TK, Chang H-m 1974 Decomposition of lignin by white-rot fungi I. Isolation of heavily degraded lignins from decayed spruce. Holzforschung 28: 217–222

Kirk TK, Chang H-m 1975 Decomposition of lignin by white-rot fungi II. Characterization of heavily degraded lignins from decayed spruce. Holzforschung 29: 56–64

Kirk TK, Chang H-m 1981 Potential applications of bio-ligninolytic systems. Enzyme Microb Technol 3: 189–196

Kirk TK, Chang H-m, Lorenz LF 1975 Topochemistry of the fungal degradation of lignin in birch wood as related to the distribution of guaiacyl and syringyl lignins. Wood Sci Technol 9: 81–86

Kirk TK, Connors WJ, Bleam RD, Hackett WF, Zeikus JG 1975 Preparation and microbial decomposition of synthetic (^{14}C)lignins. Proc Natl Acad Sci USA 72: 2515–2519

Kirk TK, Connors WJ, Zeikus JG 1976 Requirement for a growth substrate during lignin decomposition by two wood-rotting fungi. Appl Environ Microbiol 32: 192–194

Kirk TK, Connors WJ, Zeikus JG 1977 Advances in understanding the microbial degradation of lignin. In: Loewus FA, Runeckles VC (eds) Recent advances in phytochemistry Vol 2, The structure, biosynthesis, and degradation of wood. Plenum, New York, 369–394

Kirk TK, Cowling EB 1984 Biological decomposition of wood. In: Rowell RM (ed) The chemistry of solid wood. Adv Chem Ser 207, Am Chem Soc, Washington DC, 455–487

Kirk TK, Croan S, Tien M, Murtagh KE, Farrell RL 1986a Production of multiple ligninases by Phanerochaete chrysosporium: effect of selected growth conditions and use of a mutant strain. Enzyme Microb Technol 8: 27–32

Kirk TK, Farrell RL 1987 Enzymatic "combustion": the microbial degradation of lignin. Ann Rev Microbiol 41: 465–505

Kirk TK, Fenn P 1982 Formation and action of the ligninolytic system in basidiomycetes. In: Swift MJ, Frankland J, Hedger JN (eds) Decomposer basidiomycetes. Br Mycol Soc Symp 4, Cambridge University Press, Cambridge, 67–90

Kirk TK, Highley TL 1973 Quantitative changes in structural components of conifer woods during decay by white- and brown-rot fungi. Phytopathology 55: 739–745
Kirk TK, Higuchi T, Chang H-m 1980 Lignin biodegradation: microbiology, chemistry, and potential applications, Vol 1&2. CRC, Boca Raton, pp 241, 255
Kirk TK, Jeffries TW, Leatham GF 1983a Biotechnology: applications and implications for the pulp and paper industry. Tappi J 66(5): 45–51
Kirk TK, Kelman A 1965 Lignin degradation as related to the phenoloxidases of selected wood-decaying basidiomycetes. Phytopathology 55: 739–745
Kirk TK, Larsson S, Miksche GE 1970 Aromatic hydroxylation resulting from attack by a brown-rot fungus. Acta Chem Scand 24: 1470–1471
Kirk TK, Lorenz LF 1973 Methoxyhydroquinone, an intermediate of vanillate catabolism by Polyporus dichrous. Appl Microbiol 26: 173–175
Kirk TK, Moore WE 1972 Removing lignin from wood with white-rot fungi and digestibility of resulting wood. Wood Fiber 4: 72–79
Kirk TK, Mozuch MD, Tien M 1985 Free hydroxyl radical is not involved in an important reaction of lignin degradation by Phanerochaete chrysosporium Burds. Biochem J 226: 455–460
Kirk TK, Nakatsubo F 1983 Chemical mechanism of an important cleavage reaction in the fungal degradation of lignin. Biochim Biophys Acta 756: 376–384
Kirk TK, Nakatsubo F, Reid ID 1983b Further study discounts role for singlet oxygen in fungal degradation of lignin model compounds. Biochem Biophys Res Commun 111: 200–204
Kirk TK, Schultz E, Connors WJ, Lorenz LF, Zeikus JG 1978a Influence of culture parameters on lignin metabolism by Phanerochaete chrysosporium. Arch Microbiol 117: 277–285
Kirk TK, Shimada M 1985 Lignin biodegradation: the microorganisms involved and the physiology and biochemistry of degradation by white-rot fungi. In: Higuchi T (ed) Biosynthesis and biodegradation of wood components. Academic Press, Orlando, 579–605
Kirk TK, Tien M 1983 Biochemistry of lignin degradation by Phanerochaete chrysosporium: investigations with non-phenolic model compounds. In: Higuchi T, Chang H-m, Kirk TK (eds) Recent advances in lignin biodegradation research. Uni Publ, Tokyo, 233–245
Kirk TK, Tien M, Croan S, Mc Donagh T, Farrell RL 1986b Production of ligninases by Phanerochaete chrysosporium. Proc 3rd Intern Conf Biotechnol in the Pulp Paper Ind, Stockholm, 5–6
Kirk TK, Tien M, Faison BD 1984 Biochemistry of the oxidation of lignin by Phanerochaete chrysosporium. Biotech Adv 2: 183–199
Kirk TK, Tien M, Johnsrud SC, Eriksson K-E 1986c Lignin degrading activity of Phanerochaete chrysosporium Burds.: comparison of cellulase-negative and other strains. Enzyme Microb Technol 8: 75–80
Kirk TK, Tien M, Kersten J, Mozuch MD, Kalyanaraman B 1986d Ligninase of Phanerochaete chrysosporium. Mechanism of its degradation of the non-phenolic arylglycerol β-aryl ether substructure of lignin. Biochem J 236: 279–287
Kirk TK, Yang HH 1979 Partial delignification of unbleached kraft pulp with ligninolytic fungi. Biotech Lett 1: 347–352
Kirk TK, Yang HH, Keyser P 1978b The chemistry and physiology of the fungal degradation of lignin. Dev Ind Microbiol 19: 51–61
Kirkpatrick N, Palmer JM 1987 Semi-continuous ligninase production using foam-immobilised Phanerochaete chrysosporium. Appl Microbiol Biotechnol 27: 129–133
Kitpreechavanich V, Hayashi M, Nagai S 1986 Purification and characterization of extracellular β-xylosidase and β-glucosidase from Aspergillus fumigatus. Agr Biochem 50: 1703–1711
Knappert D, Grethlein HE, Converse AO 1980 Partial acid hydrolysis of cellulosic materials as a pretreatment for enzymatic hydrolysis. Biotechnol Bioeng 22: 1449–1463
Knösel D 1971 Continued investigation for pectolytic and cellulolytic activity of different Bacillus species. Zentralbl Bakteriol Parasitenkd Infektionskr Hyg Abt II 126: 604–609
Knowles J, Lehtovaara P, Teeri T 1987 Cellulase families and their genes. Tibtech 5: 255–261
Knuth DT 1964 Bacteria associated with wood products and their effects on certain chemical and physical properties of wood. PhD Thesis, Univ Wisconsin, Madison, 186 pp
Knuth DT, McCoy E 1962 Bacterial deterioration of pine logs in pond storage. For Prod J 12: 431–442

Koenigs JW 1972 Production of extracellular hydrogen peroxide and peroxidase by wood-rotting fungi. Phytopathology 62: 100–110

Koenigs JW 1974a Hydrogen peroxide and iron: a proposed system for decomposition of wood by brown-rot basidiomycetes. Wood Fiber 6: 66–80

Koenigs JW 1974b Production of hydrogen peroxide by wood rotting fungi in wood and its correlation with weight loss, depolymerization, and pH changes. Arch Microbiol 99: 129–145

Koevenig JL, Liu EH 1981 Carboxymethyl cellulase activity in the myxomycete Physarum polycephalum. Mycologia 73: 1085–1091

Kohlmeyer J 1980 Bacterial attack on wood and cellophane in the deep sea. In: Oxley TA, Becker G, Allsopp D (eds) Biodeterioration. Proc 4th Intern Symp, Berlin. Pitman, London, 187–192

Kolarova N, Farkas V 1983 Laminarinases, xylanases and amylases in the crude cellulolytic enzyme complex from Trichoderma reesei. Biologia (Bratislava) 38: 721–725

Komae K, Kaji A, Sato M 1982 An α-L-arabinofuranosidase from Streptomyces purpurascens IFO 3389. Agr Biol Chem 46: 1899–1905

Konishi K, Inoue Y, Higuchi T 1972 Decomposition of lignin by Coriolus versicolor. IV. Effect of laccase type enzyme on the interphenylpropane linkage of lignin. Mokuzai Gakkaishi 18: 571–576

Konishi K, Inoue Y, Higuchi T 1974 Decomposition of lignin by Coriolus versicolor. V. Formation of carboxyl groups in the lignin treated with fungal laccase. Mokuzai Gakkaishi 20: 26–30

Kozlik I, Schánĕl L 1974a The changes in the atmosphere during wood decay by fungi. Drev Vysk 19: 113–124

Kozlik I, Schánĕl L 1974b Changes of the atmosphere during wood decay by fungi under conditions of stopped gas diffusion. Drev Vysk 19: 169–179

Krahmer RL, DeGroot RC, Lowell EC 1982 Detecting incipient brown rot with fluorescence microscopy. Wood Sci 15: 78–80

Kringstad KP, Lindström K 1984 Spent liquors from pulp bleaching. Environ Sci Technol 18: 236A–248A

Kringstad KP, Ljungquist PO, de Sousa F, Strömberg LM 1981 Identification and mutagenic properties of some chlorinated aliphatic compounds in the spent liquor from kraft pulp chlorination. Environ Sci Technol 15: 562–566

Kringstad KP, McKague AB 1988 Bleaching and the environment. Tappi Proc 1988 Intern Pulp Bleaching Conf, Orlando Florida, 63–68

Krisnangkura K, Gold MH 1979 Peroxidase catalyzed oxidative decarboxylation of vanillic acid to methoxy-p-hydroquinone. Phytochemistry 18: 2019–2021

Kubačková M, Karácsonyi Š, Bilisics L, Toman R 1979 On the specificity and mode of action of a xylanase from Trametes hirsuta. Carbohyd Res 76: 177–188

Kubicek CP 1981 Release of carboxymethyl-cellulase and β-glucosidase from cell walls of Trichoderma reesei. Eur J Appl Microbiol Biotechnol 13: 226–231

Kuila D, Tien M, Fee JA, Ondrias MR 1985 Resonance Raman spectra of extracellular ligninase: evidence for a heme active site similar to those of peroxidases. Biochemistry 24: 3394–3397

Kundu AB, Ghosh BL, Bose RG 1980 An indirect viscometric method for assay of hemicellulase. Sci Cult 46: 115–116

Kusakabe I, Kawaguchi M, Yasui T, Kobayashi T 1977a Studies on the xylanase system of Streptomyces. Part VII. Purification and some properties of extracellular xylanase from Streptomyces sp. E-86. Nippon Nogei Kagaku Kaishi 51: 429–437

Kusakabe I, Park GG, Kumita N, Yasui T, Murakami K 1988 Specificity of β-Mannanase from Penicillum purpurogenum for konjac glucomannan. Agr Biol Chem 52: 519–524

Kusakabe I, Yasui T, Kobayashi T 1977b Studies on the xylanase system of Streptomyces. Part IX. Isolation and structures of the arabinoxylo-oligosaccharides from the hydrolytic products of corncobs arabinoxylan by a xylanase from Streptomyces. Nippon Nogei Kagaku Kaishi 51: 669–679

Kusakabe I, Yasui T, Kobayashi T 1977c Studies on the xylanase system of Streptomyces. Part VIII. The effects of Streptomyces xylanase on various xylans and xylooligosaccharides. Nippon Nogei Kagaku Kaishi 51: 439–448

Küster E 1979 Bedeutung der Aktinomyceten für den Abbau von Cellulose, Lignin und Huminstoffen im Boden. Z Pflanzenernähr Bodenkd 142: 365–374

Kutsuki H, Enoki A, Gold MH 1983 Riboflavin-photosensitized oxidative degradation of a variety of lignin model compounds. Photochem Photobiol 37: 1–7

Kutsuki H, Gold MH 1982 Generation of hydroxyl radical and its involvement in lignin degradation by Phanerochaete chrysosporium. Biochem Biophys Res Commun 109: 320–327

Kuwahara M, Asada Y, Kimura Y, Aokage M 1987 Isolation of mutants of a lignin-degrading basidiomycete, Phanerochaete chrysosporium, that produce lignin peroxidase in high nitrogen cultures. Mokuzai Gakkaishi 33: 821–823

Kuwahara M, Glenn JK, Morgan MA, Gold MH 1984 Separation and characterization of two extracellular H_2O_2-dependent oxidases from ligninolytic cultures of Phanerochaete chrysosporium. FEBS Lett 169: 247–250

Kyrklund B, Strandell G 1967 A modified chlorine number for evaluation of the cooking degree of high-yield pulps. Pap Puu 3: 99–106

Kyriacou A, MacKenzie CR, Neufeld RJ 1987 Detection and characterization of the specific and nonspecific endoglucanases of Trichoderma reesei: evidence demonstrating endoglucanase activity by cellobiohydrolase II. Enzyme Microb Technol 9: 25–32

Lacey J 1973 Actinomycetes in soils, composts and fodders. In: Sykes G, Skinner FA (eds) Actinomycetales: characteristics and practical importance. Academic Press, London, 231–251

Lachke AH, Deshpande MV, Srinivasan MC 1985 Extracellular β-D-xylosidase of Sclerotium rolfsii. Enzyme Microb Technol 7: 445–448

Lamed R, Bayer EA 1988 The cellulosome concept: exocellular/extracellular enzyme reactor centers for efficient binding and cellulolysis. In: Aubert J-P, Beguin P, Millet G (eds) Biochemistry and genetics of cellulose degradation, FEMS Symp no 43. Academic Press, London, 101–116

Lamed R, Kenig R, Setter E, Bayer EA 1985 Major characteristics of the cellulolytic system of Clostridium thermocellum coincide with those of the purified cellulosome. Enzyme Microb Technol 7: 37–41

Lamed R, Setter E, Bayer EA 1983b Characterization of a cellulose-binding, cellulase-containing complex in Clostridium thermocellum. J Bacteriol 156: 828–836

Lamed R, Setter E, Kenig R, Bayer EA 1983a The cellulosome – a discrete cell surface organelle of Clostridium thermocellum which exhibits separate antigenic, cellulose binding and various cellulolytic activities. Biotechnol Bioeng Symp 13: 163–181

Langsford ML, Gilkes NR, Wakarchuk WW, Kilburn DG, Miller RC Jr, Warren RAJ 1984 The cellulase of Cellulomonas fimi. J Gen Microbiol 130: 1367–1376

Lappalainen A 1986 Purification and characterization of xylanolytic enzymes from Trichoderma reesei. Biotechnol Appl Biochem 8: 437–448

Larsen MJ, Jurgensen MF, Harvey AE 1979 N_2 fixation associated with wood decayed by some common fungi in western Montana. Can J For Res 8: 341–345

Larsen MJ, Jurgensen MF, Harvey AE 1982 N_2 fixation in brown-rotted soil wood in an intermountain cedar-hemlock ecosystem. For Sci 28: 292–296

Larsen MJ, Lombard FF, Hodges CS Jr 1985 Hawaiian forest fungi V. A new species of Phellinus (Hymenochaetaceae) causing decay of Casuarina and Acacia. Mycologia 77: 345–352

Latham MJ, Brooker BE, Pettipher GL, Harris PJ 1978 Ruminococcus flavefaciens cell coat and adhesion to cotton cellulose and to cell walls in leaves of perennial ryegrass (Lolium perenne). Appl Environ Microbiol 35: 156–165

Leatham GF 1982 Cultivation of shiitake, the Japanes forest mushroom, on logs: a potential industry for the United States. For Prod J 32(8): 29–35

Leatham GF 1985 Extracellular enzymes produced by the cultivated mushroom Lentinus edodes during degradation of a lignocellulosic medium. Appl Environ Microbiol 50: 859–867

Leatham GF 1986 The ligninolytic activities of Lentinus edodes and Phanerochaete chrysosporium. Appl Microbiol Biotechnol 24: 51–58

Leatham GF, Kirk TK 1983 Regulation of ligninolytic activity by nutrient nitrogen in white-rot basidiomycetes. FEMS Microbiol Lett 16: 65–67

Leatham GF, Stahmann MA 1981 Studies on the laccase of Lentinus edodes: specificity, localization and association with the development of fruiting bodies. J Gen Microbiol 125: 147–157

Leathers TD 1986 Colour variants of Aureobacidium pullulans overproduce xylanase with extremely high specific activity. Appl Environ Microbiol 52: 1026–1030

Leathers TD, Detroy RW, Bothast RJ 1986 Induction and glucose repression of xylanase from a colour-variant strain of Aureobasidium pullulans. Biotechnol Lett 8: 867–872
Lee SB, Shin HS, Ryu DDY, Mandels M 1982 Adsorption of cellulase on cellulose. Effect of physicochemical properties of cellulose on adsorption and rate of hydrolysis. Biotechnol Bioeng 24: 2137–2153
Lee SF, Forsberg CW, Gibbins LN 1985 Xylanolytic activity of Clostridium acetobutylicum. Appl Environ Microbiol 50: 1068–1076
Lee SF, Forsberg CW, Rattray JB 1987 Purification and characterization of two endoxylanases from Clostridium acetobutylicum ATCC 824. Appl Environ Microbiol 53: 644–650
Lee YC, Wacek V 1970 Galactosidases from Aspergillus niger. Arch Biochem Biophys 138: 264–271
Lee Y-h, Fan LT 1980 Properties and mode of action of cellulase. In: Fiechter A (ed) Adv Biochem Eng, Vol 18, Springer, Berlin Heidelberg New York, 101–129
Leightley LE, Eaton RA 1977 Mechanisms of decay of timber by aquatic micro-organisms. Br Wood Preservers Assoc Convention Records, 1–26
Leisola M, Brown C, Laurila M, Ulmer D, Fiechter A 1982 Polysaccharide synthesis by Phanerochaete chrysosporium during degradation of kraft lignin. Eur J Appl Microbiol Biotechnol 15: 180–184
Leisola MSA, Fiechter A 1985a New trends in lignin biodegradation. Adv Biotechnol Processes 5: 59–89
Leisola MSA, Fiechter A 1985b Ligninase production in agitated conditions by Phanerochaete chrysosporium. FEMS Microbiol Lett 29: 33–36
Leisola MSA, Haemmerli SD, Waldner R, Schoemaker HE, Schmidt HWH, Fiechter A 1988 Metabolism of a lignin model compound, 3,4-dimethoxy benzyl alcohol by Phanerochaete chrysosporium. Cellul Chem Technol 22: 267–277
Leisola M, Kauppinen V 1978 Automatic assay of cellulase activity during fermentation. Biotechnol Bioeng 20: 837–846
Leisola M, Linko M 1977 Use of dyed avicel cellulose to determine the solubilizing activity of a cellulase complex. Finn Chem Lett 172–176
Leisola MSA, Kozulic B, Meussdoerffer F, Fiechter A 1987 Homology among multiple extracellular peroxidases from Phanerochaete chrysosporium. J Biol Chem 262: 419–424
Leisola MSA, Meussdoerffer F, Waldner R, Fiechter A 1985a Production and identification of extracellular oxidases of Phanerochaete chrysosporium. J Biotechnol 2: 379–382
Leisola MSA, Schmidt B, Thanei-Wyss U, Fiechter A 1985b Aromatic ring cleavage of veratryl alcohol by Phanerochaete chrysosporium. FEBS Lett 189: 267–270
Leisola MSA, Thanei-Wyss U, Fiechter A 1985c Strategies for production of high ligninase activities by Phanerochaete chrysosporium. J Biotechnol 3: 97–107
Leisola MSA, Ulmer D, Fiechter A 1983a Problem of oxygen transfer during degradation of lignin by Phanerochaete chrysosporium. Eur J Appl Microbiol Biotechnol 17: 113–116
Leisola MSA, Ulmer D, Haltmeier T, Fiechter A 1983b Rapid solubilization and depolymerization of purified kraft lignin by thin layers of Phanerochaete chrysosporium. Eur J Appl Microbiol Biotechnol 17: 117–120
Leisola MSA, Ulmer DC, Waldner R, Fiechter A 1984 Role of veratryl alcohol in lignin degradation by Phanerochaete chrysosporium. J Biotechnol 1: 331–339
Lejeune A, Eveleigh DE, Colson C 1988 Expression of an endoglucanase gene of Pseudomonas fluorescence var cellulosa in Zymomonas mobilis. FEMS Microbiol Lett 49: 363–366
Lemmel SA, Datta R, Frankiewicz JR 1986 Fermentation of xylan by Clostridium acetobutylicum. Enzyme Microb Technol 8: 217–221
Leonowicz A, Grzywnowicz K, Malinowska M 1979 Oxidative and demethylating activity of multiple forms of laccase from Pholiota mutabilis. Acta Biochim Polon 26: 431–434
Leonowicz A, Rogalski J, Szczodrak J, Fiedurek J 1986 The possible key role of glucose oxidase in transformation of lignocellulose. Proc 3rd Intern Conf Biotechnol in the Pulp and Paper Ind, Stockholm, 160–162
Leonowicz A, Szklarz G, Wojtas-Wasilewska M 1985 The effect of fungal laccase on fractionated lignosulphonates (Peritan Na). Phytochemistry 24: 393–396

Leonowicz A, Wojtas-Wasilewska M, Rogalski J, Luterek J 1988 Biological decomposition of lignocellulose. In: Blazej A, Zenek J (eds) Interbiotech 87: Enzyme technologies – Progress in Biotechnology Vol 4. Elsevier, Amsterdam, 415–451

Levi MP 1965 Decay patterns produced by Chaetomium globosum in beechwood fibers. Mater Org 1: 119–126

Levi MP, Merrill W, Cowling EB 1968 Role of nitrogen in wood deterioration. VI. Mycelial fractions and model nitrogen compounds as substrates for growth of Polyporus versicolor and other wood-destroying and wood-inhabiting fungi. Phytopathology 58: 626–634

Levi MP, Preston RD 1965 A chemical and microscopic examination of the action of the soft-rot fungus Chaetomium globosum on beechwood (Fagus sylv). Holzforschung 19: 183–190

Levonen-Munoz E, Bone DH 1985 Effect of different gas environments on bench-scale solid state fermentation of oat straw by white-rot fungi. Biotechnol Bioeng 27: 382–387

Levy JF 1975 Bacteria associated with wood in ground contact. In: Liese W (ed) Biological transformation of wood by microorganisms. Springer, Berlin Heidelberg New York, 64–73

Lewis NG, Razal RA, Yamamoto E 1987 Lignin degradation by peroxidase in organic media: a reassessment. Proc Natl Acad Sci USA 84: 7925–7927

Li LH, Flora RM, King KW 1965 Individual roles of cellulase components derived from Trichoderma viride. Arch Biochem Biophys 11: 439–447

Liese J 1950 Zerstörung des Holzes durch Pilze und Bakterien. In: Mahlke-Troschel (ed), Handbuch der Holzkonservierung. Springer, Berlin Göttingen Heidelberg

Liese W 1964 Über den Abbau verholzter Zellwände durch Moderfäulepilze. Holz Roh-Werkst 22: 289–295

Liese W 1970 Ultrastructural aspects of woody tissue disintegration. Ann Rev Phytopathol 8: 231–258

Liese W, Greaves H 1975 Micromorphology of bacterial attack. In: Liese W (ed) Biological transformation of wood by microorganisms. Springer, Berlin Heidelberg New York, 74–88

Liese W, Karnop G 1968 Über den Befall von Nadelholz durch Bakterien. Holz Roh-Werkst 26: 202–208

Liese W, Schmid R 1964 Über das Wachstum von Bläuepilzen durch verholzte Zellwände. Phytopathol Z 51: 385–393

Liese W, Schmid R 1966 Untersuchungen über den Zellwandabbau von Nadelholz durch Trametes pini. Holz Roh-Werkst 24: 454–460

Lin FZ, Brown RM Jr, Cooper JP, Delmer DP 1985 Synthesis of fibrils in vitro by a solubilized cellulose synthase from Acetobacter xylinum. Science 230: 822–825

Lindberg B, Rosell KG, Svensson S 1973a Positions of the *O*-acetyl groups in birch xylan. Svensk Papperstidn 76: 30–32

Lindberg B, Rosell KG, Svensson S 1973b Positions of the *O*-acetyl groups in pine glucomannan. Svensk Papperstidn 76: 383–384

Linko S 1988a Production and characterization of extracellular lignin peroxidase from immobilized Phanerochaete chrysosporium in a 10–1 bioreactor. Enzyme Microb Technol 10: 410–417

Linko S 1988b Continuous production of lignin peroxidase by immobilized Phanerochaete chrysosporium in a pilot scale bioreactor. J Biotechnol 8: 163–170

Linko Y-Y, Leisola M, Lindholm N, Troller J, Linko P, Fiechter A 1986 Continuous production of lignin peroxidase by Phanerochaete chrysosporium. J Biotechnol 4: 283–291

Livernoche D, Jurasek L, Desrochers M, Dorica J, Veliky IA 1983 Removal of color from kraft mill wastewaters with cultures of white-rot fungi and with immobilized mycelium of Coriolus versicolor. Biotechnol Bioeng 25: 2055–2065

Liwicki R, Paterson A, MacDonald MJ, Broda P 1985 Phenotypic classes of phenoloxidase-negative mutants of the lignin-degrading fungus Phanerochaete chrysosporium. J Bacteriol 2: 641–644

Ljungdahl LG, Eriksson K-E 1985 Ecology of microbial cellulose degradation. In: Marshall KC (ed) Adv Microb Ecol Vol 8. Plenum, New York, 237–299

Ljungdahl LG, Peterson B, Eriksson K-E, Wiegel J 1983 A yellow affinity substance involved in the cellulolytic system of Clostridium thermocellum. Curr Microbiol 9: 195–200

Lobarzewski J 1981 New affinity chromatography method for the separation of vanillic acid induced forms of fungal peroxidase from Trametes versicolor on porous glass beads activated with vanillin. Int J Biol Macromol 3: 77–81

Lobarzewski J 1984 A comparison of the effects of immobilized fungal mycelium fragments of Trametes versicolor and fungal peroxidase on Peritan-Na (Na-lignosulfonates). Holzforschung 38: 105–108

Lobarzewski J 1987 A lignin-biotransforming peroxidase from Trametes versicolor and its use in immobilized form. In: Odier E (ed) Lignin enzymic and microbial degradation. Symp Intern INRA Paris, 197–202

Lobarzewski J, Paszczynski A 1985 Lignocellulose biotransformation with immobilized cellulase, D-glucose oxidase and fungal peroxidases. Enzyme Microb Technol 7: 564–565

Lobarzewski J, Trojanowski J, Wojtas-Wasilewska M 1982 The effects of fungal peroxidase on Na-lignosulfonates. Holzforschung 36: 173–176

Lopez-Real JM, Swift MJ 1975 Formation of pseudosclerotia (zone lines) in wood decayed by Armillaria mellea and Stereum hirsutum. II. Formation in relation to the moisture content of wood. Trans Br Mycol Soc 64: 473–481

Lundquist K, Kirk TK 1978 De novo synthesis and decomposition of veratryl alcohol by a lignin-degrading basidiomycete. Phytochemistry 17: 1676

Lundquist K, Kirk TK, Connors WJ 1977 Fungal degradation of kraft lignin and lignin sulfonates prepared from synthetic ^{14}C-lignins. Arch Microbiol 112: 291–296

Lundquist K, Kristersson P 1985 Exhaustive laccase-catalysed oxidation of a lignin model compound (vanillyl glycol) produces methanol and polymeric quinoid products. Biochem J 229: 277–279

Lundström J 1972 Microscopic studies of cavity formation by soft rot fungi Allescheria terrestris Apinis, Margarinomyces luteo-virdis v. Beyma and Phialophora richardsiae (Nannf.) Conant. Stud For Suec 98: 18

Lynch JM, Slater JH, Bennett JA, Harper SHT 1981 Cellulase activities of some aerobic micro-organisms isolated from soil. J Gen Microbiol 127: 231–236

Lyr H 1963 Über das Vorkommen von Mannanase bei Pilzen. Z Allg Mikrobiol 3: 25–36

Lyr H 1972 Produktion von Xylanase durch Bacillus subtilis. Z Allg Mikrobiol 12: 135–242

Macdonald DG, Bakhshi NN, Mathews JF, Roychowdury A, Bajpai P, Moo-Young M 1983 Alkali treatment of corn stover to improve sugar production by enzymatic hydrolysis. Biotechnol Bioeng 25: 2067–2076

Macdonald MJ, Paterson A, Broda P 1984 Possible relationship between cyclic AMP and idiophasic metabolism in the white-rot fungus Phanerochaete chrysosporium. J Bacteriol 160: 470–472

MacKenzie CR 1986 Bacterial cellulases. In: Moo-Young M, Hasnai S, Lamptey J (eds) Biotechnology and renewable energy. Elsevier Appl Sci, London, 76–82

MacKenzie CR, Bilous D, Johnson KG 1984 Purification and characterization of an exo-glucanase from Streptomyces flavogriseus. Can J Microbiol 30: 1171–1178

MacKenzie CR, Patel GB, Bilous D 1987 Factors involved in hydrolysis of microcrystalline cellulose by Acetivibrio cellulolyticus. Appl Environ Microbiol 53: 304–308

Madan M, Bisaria R 1983 Cellulolytic enzymes from an edible mushroom, Pleurotus sajor-caju. Biotechnol Lett 5: 601–604

Malarczyk E 1984 Substrate-induction of veratric acid O-demethylase in Nocardia sp. Acta Biochim Polon 31: 383–395

Maloney MT, Chapman TW, Baker AJ 1985 Dilute acid hydrolysis of paper birch: kinetic studies of xylan and acetyl group hydrolysis. Biotechnol Bioeng 27: 355–361

Maltby D, Carpita N Montezimos D, Delmer DP 1979 β-1,3-Glucan of developing cotton fibers: structure, localization and relationship of synthesis to that of secondary wall cellulose. Plant Physiol 63: 1158–1164

Mandels M 1982 Cellulases. Ann Rep Ferment Proc 5: 35–78

Mandels M 1975 Microbial Sources of cellulase. Biotechnol Bioeng Symp 5: 81–105

Mandels M, Andreotti R, Roche C 1976 Measurement of saccharifying cellulase. Biotechnol Bioeng Symp 6: 21–33

Mandels M, Parrish FW, Reese ET 1962 Sophorose as an inducer of cellulase in Trichoderma viride. J Bacteriol 83: 400–408
Mandels M, Reese ET 1960 Induction of cellulase in fungi by cellobiose. J Bacteriol 79: 816–826
Mandels M, Reese ET 1964 Fungal cellulases and the microbial decomposition of cellulose fabric. Dev Ind Microbiol 5: 5–20
Mandels M, Weber J 1969 The production of cellulases. In: Hajny GJ, Reese ET (eds) Cellulases and their applications. Adv Chem Ser 95: 391–413
Manion PD 1981 Tree disease concepts. Prentice-Hall, Englewood Cliffs, 399 pp
Manning K, Wood DA 1983 Production and regulation of extracellular endocellulase by Agaricus bisporus. J Gen Microbiol 129: 1839–1847
Margaritis A, Merchant R 1983 Xylanase, CM-cellulase and avicelase production by the thermophilic fungus Sporotrichum thermophile. Biotechnol Lett 5: 265–270
Marshall KC 1980 Reactions of microorganisms, ions and macromolecules at interfaces. In: Elwood DC, Hedger JN, Latham MJ, Lynch JM, Slayter JH (eds) Contemporary microbial ecology. Academic Press, New York, 96–106
Marui M, Nakanishi K, Yasui T 1985 Purification and properties of three types of xylanases induced by methyl-β-xyloside from Streptomyces sp. Agr Biol Chem 49: 3399–3407
Marx-Figini M 1982 The control of molecular weight and molecular weight distribution in the biogenesis of cellulose. In: Brown, RM Jr (ed) Cellulose and other natural polymer systems: biogenesis, structure and degradation. Plenum, New York, 243–268
Marx-Figini M, Schultz GV 1966 Zur Biosynthese der Cellulose. Naturwissenschaften 53: 466–474
Mason JC, Richards M, Zimmerman W, Broda P 1988 Identification of extracellular proteins from actinomycetes responsible for the solubilization of lignocellulose. Appl Microbiol Biotechnol 28: 276–280
Matsumoto K, Endo Y, Tamiya N, Kano M, Miyaushi K, Abe J 1974 Studies on cellulase produced by the phytopathogens, purification and enzymatic properties of cellulase of Fusarium moniliformae. J Biochem 76: 563–572
Matsuo M, Fujie A, Win M, Yasui T 1987 Four types of β-xylosidases from Penicillum wortmanni IFO 7237. Agr Biol Chem 51: 2367–2379
Matsuo M, Yasui T 1984a Purification and some properties of β-xylosidase from Trichoderma viride. Agr Biol Chem 48: 1845–1852
Matsuo M, Yasui T 1984b Purification and some properties of β-xylosidase from Emericella nidulans. Agr Biol Chem 48: 1853–1860
Matsuo M, Yasui T, Kobayashi T 1977a Purification and some properties of β-xylosidase from Malbranchea pulchella var. sulphurae No 48. Agr Biol Chem 41: 1593–1599
Matsuo M, Yasui T, Kobayashi T 1977b Enzymatic properties of β-xylosidase from Malbranchea pulchella var. sulphurae No 48. Agr Biol Chem 41: 1601–1606
Mayer AM, Harel E 1979 Polyphenol oxidases in plants. Phytochemistry 18: 193–215
Mayer F, Coughlan MP, Mori Y, Ljungdahl LG 1987 Macromolecular organization of the cellulolytic enzyme complex of Clostridium thermocellum as revealed by electron microscopy. Appl Env Microbiol 53: 2785–2792
McBee RH 1948 The culture and physiology of a thermophilic cellulose-fermenting bacterium. J Bacteriol 56: 653–663
McCarthy AJ 1987 Lignocellulose-degrading actinomycetes. FEMS Microbiol Rev 46: 145–163
McCarthy AJ, Broda P 1984 Screening for lignin-degrading actinomycetes and characterization of their activity against [^{14}C]lignin-labelled wheat lignocellulose. J Gen Microbiol 130: 2905–2913
McCarthy AJ, Macdonald MJ, Paterson A, Broda P 1984 Degradation of [^{14}C]lignin-labelled wheat lignocellulose by white-rot fungi. J Gen Microbiol 130: 1023–1030
McCarthy AJ, Paterson A, Broda P 1986 Lignin solubilisation by Thermomonospora mesophila. Appl Microbiol Biotechnol 24: 347–352
McCleary BV 1979 Modes of action of β-mannanase enzymes of diverse origin on legume seed galactomannans. Phytochemistry 18: 757–763
McCleary BV 1988a β-D-Mannanase. Methods Enzymol 160: 596–614
McCleary B 1988b β-D-Mannosidase from Helix pomatia. Methods Enzymol 160: 614–619

McCleary BV 1988c α-D-Galactosidase from lucern and guar seed. Methods Enzymol 160: 627–632

McCleary BV, Gibson TS, Allen H, Gams TC 1986 Enzymatic hydrolysis and industrial importance of barley β-glucans and wheat flour pentosans. Stärke 38: 433–437

McCleary BV, Matheson NK 1983 Action patterns and substrate-binding requirements of β-D-mannanase with mannosaccharides and mannan-type polysaccharides. Carbohyd Res 119: 191–219

McCleary BV, Matheson NK, Small DM 1976 Galactomannans and the galactoglucomannan in legume seed endosperms: structural requirements for β-mannanase hydrolysis. Phytochemistry 15: 1111–1117

McCleary BV, Nurthen E, Taravel FR, Josleau J-P 1983 Characterization of the oligosaccharides produced on hydrolysis of galactomannan with β-D-mannanase. Carbohyd Res 118: 91–109

McFee WW, Stone EL 1966 The persistence of decaying wood in the humus layers of northern forests. Proc Soil Am 30: 513–516

McHale A, Coughlan MP 1980 Synergistic hydrolysis of cellulose by components of the extracellular cellulase system of Talaromyces emersonii. FEBS Lett 117: 319–322

McHale A, Coughlan MP 1981a The cellulolytic system of Talaromyces emersonii. Identification of the various components produced during growth on cellulosic media. Biochim Biophys Acta 661: 145–151

McHale A, Coughlan MP 1981b A convenient zymogram stain for cellulases. Biochem J 199: 267–268

Mega T, Matsushima Y 1979 Comparative studies of three exo-β-glycosidases of Aspergillus oryzae. J Biochem 85: 335–341

Meier H 1955 Über den Zellwandabbau durch Holzvermorschungspilze und die submikroskopische Struktur von Fichtentracheiden und Birkenholzfasern. Holz Roh-Werkst 13: 323–338

Meier H, Buchs L, Buchala AJ, Homewood T 1981 (1→3)-β-D-Glucan (callose) is a probable intermediate in biosynthesis of cellulose of cotton fibers. Nature 289: 821–822

Merrill W, French DW, Wood FA 1964 Decay of wood by species of the Xylariaceae. Phytopathology 54: 56–58

Mes-Hartree M, Yu EKC, Reid ID, Saddler JN 1987 Suitability of aspenwood delignified with Phlebia tremellosus for fermentation to ethanol or butanediol. Appl Microbiol Biotechnol 26: 120–125

Messner K, Foisner R, Stachelberger H, Röhr M 1985 Osmiophilic particles as a typical aspect of brown and white rot systems in transmission electron microscope studies. Trans Br Mycol Soc 84: 457–466

Messner K, Srebotnik E, Stachelberger H 1986 Electron microscopial indication of the demethylation of lignin in wood by osmium tetroxide. Proc 3rd Intern Conf Biotechnol in the Pulp and Paper Ind, Stockholm, 17–19

Messner K, Stachelberger H 1984a Transmission electron microscope observations of the micromorphology of brown rot caused by Fomitopsis pinicola with respect to osmiophilic particles. Trans Br Mycol Soc 83: 113–130

Messner K, Stachelberger H 1984b Transmission electron microscope observations of white rot caused by Trametes hirsuta with respect to osmiophilic particles. Trans Br Mycol Soc 83: 209–216

Meyer H-P, Canevascini G 1981 Separation and some properties of two intracellular β-glucosidases of Sporotrichum (Chrysosporium) thermophile. Appl Environ Microbiol 41: 924–931

Micales JA, Green F, Clausen CA, Highley TL 1987 Physical properties of β-1,4-xylanase produced by Postia (Poria) placenta: implications for the control of brown-rot. Intern Res Group Wood Preserv, 18th Ann Meeting, Honey Harbor, Ontario, May 17–22

Miki K, Renganathan V, Gold M 1986 Mechanisms of β-aryl ether dimeric lignin model compound oxidation by lignin peroxidase of Phanerochaete chrysosporium. Biochemistry 25: 4790–4796

Miki K, Renganathan V, Mayfield MB, Gold MH 1987 Aromatic ring cleavage of a β-biphenyl ether dimer catalyzed by lignin peroxidase of Phanerochaete chrysosporium. FEBS Lett 210: 199–203

Miller GL 1959 Use of dinitrosalicylic acid reagent for determination of reducing sugars. Anal Chem 31: 426–428
Millet MA, Baker AJ, Satter LD 1976 Physical and chemical pretreatments for enhancing cellulose saccharification. Biotechnol Bioeng Symp 6: 125–153
Milstein O, Vered Y, Sharma A, Gressel J, Flowers HM 1986 Heat and microbial treatments for nutritional upgrading of wheat straw. Biotechnol Bioeng 28: 381–386
Mino Y, Wariishi H, Blackburn NJ, Loehr TM, Gold MH 1988 Spectral characterization of manganese peroxidase, an extracellular heme enzyme from the lignin-degrading basidiomycete, Phanerochaete chrysosporium. J Biol Chem 263: 7029–7036
Mishra C, Keskar S, Rao M 1984 Production and properties of extracellular endoxylanase from Neurospora crassa. Appl Environ Microbiol 48: 224–228
Mishra MM, Yadav KS, Kapoor KK 1981 Degradation of lignocellulose by mixed cultures of cellulolytic fungi and their competitive ability. Zentralbl Bakteriol Parasitenkd Infektionskr Hyg Abt 136: 603–608
Moloney AP, Considine PJ, Coughlan MP 1983 Cellulose hydrolysis by the cellulases produced by Talaromyces emersonii when grown on different inducing substrates. Biotechnol Bioeng 25: 1169–1173
Montenecourt BS 1983 Tricoderma reesei cellulases. Trends Biotechnol 1: 156–161
Montenecourt BS, Eveleigh DE, Elmund GK, Parcells J 1978 Antibiotic disks – an improvement in the filter paper assay for cellulase. Biotechnol Bioeng 20: 297–300
Montenecourt BS, Kelleher TJ, Eveleigh DE, Pettersson LG 1980 Altered biochemical nature of cellulases from mutants of Trichoderma reesei. Biotechnol Bioeng Symp 10: 15–26
Montenecourt BS, Nhlapo SD, Trimino-Vazquez H, Cuskey S, Schamhart DHJ, Eveleigh DE 1981 Regulatory controls in relation to over-production of fungal cellulases. In: Hollaender A (ed) Trends in the biology of fermentations for fuels and chemicals, Plenum, New York, 33–35
Monties B, Odier E, Janin G, Czaninski 1981 Ultrastructural evidence of bacterial and chemical delignification of poplar wood. Holzforschung 35: 217–222
Moore WE Johnson DB 1967 Procedures for the chemical analysis of wood and wood products. US For Serv, For Prod Lab, Madison WI, Lab Manual
Moo-Young M, Chahal DS, Swan JE, Robinson CW 1977 SCP production by Chaetomium cellulolyticum, a new thermotolerant cellulolytic fungus. Biotechnol Bioeng 19: 527–538
Moo-Young M, Chahal DS, Vlach D 1978 Single cell protein from various chemically pretreated wood substrates using Chaetomium cellulolyticum. Biotechnol Bioeng 20: 107–118
Morohoshi N, Fujita K, Wariishi H, Haraguchi T 1987a Degradation of lignin by the extracellular enzymes of Coriolus versicolor V. Degradation of isolated lignin fractions by laccase I, II, and III. Mokuzai Gakkaishi 33: 310–315
Morohoshi N, Haraguchi T 1987a Degradation of lignin by the extracellular enzymes of Coriolus versicolor III. Degradation of various substrates by the phenol oxidase fraction. Mokuzai Gakkaishi 33: 143–150
Morohoshi N, Haraguchi T 1987b Degradation of lignin by the extracellular enzymes of Coriolus versicolor VI. Degradation of syringylglycerol-β-syringyl ether by laccase III-c purified from laccase III by isoelectric focusing. Mokuzai Gakkaishi 33: 495–502
Morohoshi N, Nakamura M, Yoshitake A, Katayama Y, Haraguchi T 1987b Isolation of a ligninolytic laccase and its messenger ribonucleic acid from Coriolus versicolor. Mokuzai Gakkaishi 33: 743–746
Morohoshi N, Wariishi H, Haraguchi T 1983 Degradation of lignin by the extracellular enzymes of Coriolus versicolor. Proc Intern Symp Wood Pulp Chem Vol 3. Tsukuba, 26–31
Morohoshi N, Wariishi N, Muraiso C, Nagai T, Haraguchi T 1987c Degradation of lignin by the extracellular enzymes of Coriolus versicolor IV. Properties of three laccase fractions fractionated from the extracellular enzymes. Mokuzai Gakkaishi 33: 218–225
Morosoli R, Bertrand J-L, Mondon F, Shareck F, Kluepfel D 1986a Purification and properties of a xylanase from Streptomyces lividans. Biochem J 239: 587–592
Morosoli R, Roy C, Yaguchi M 1986b Isolation and partial primary sequence of a xylanase from the yeast Cryptococcus albidus. Biochim Biophys Acta 870: 473–478
Morpeth FF 1985 Some properties of cellobiose oxidase from the white-rot fungus Sporotrichum pulverulentum. Biochem J 228: 557–564

Morpeth FF, Jones GD 1986 Resolution, purification and some properties of the multiple forms of cellobiose quinone dehydrogenase from the white-rot fungus Sporotrichum pulverulentum. Biochem J 236: 221–226

Morrison IM 1975 The rate and products of degradation of xylans by rumen hemicellulases. Biochem Soc Trans 3: 1992–1994

Mosbach R 1963 Purification and some properties of laccase from Polyporus versicolor. Biochim Biophys Acta 73: 204–212

Mountfort DO, Asher RA 1983 Role of catabolite regulatory mechanisms in control of carbohydrate utilization by the rumen anaerobic fungus Neocallimastix frontalis. Appl Environ Microbiol 46: 1331–1338

Mühlethaler K 1967 Ultrastructure and formation of plant cell walls. Ann Rev Plant Physiol 18: 1–24

Müller HW, Trösch W, Kulbe KD 1988 Effect of phenolic compounds on cellulose degradation by some white rot basidiomycetes. FEMS Microbiol Lett 49: 87–93

Mueller SC, Brown RM Jr 1980 Evidence for an intramembrane component associated with a cellulose microfibril-synthesizing complex in higher plants. J Cell Biol 84: 315–326

Mueller SC, Brown RM Jr 1982a The control of cellulose microfibril deposition in the cell wall of higher plants. I. Can directed membrane flow orient cellulose microfibrils? Indirect evidence from freeze-fractured plasma membranes of maize and pine seedlings. Planta 154: 489–500

Mueller SC, Brown RM Jr 1982b The control of cellulose microfibril deposition in the cell wall of higher plants. II. Freeze-fracture microfibril patterns in maize seedling tissues following experimental alteration with colchicine and ethylene. Planta 154: 501–515

Mukherjee M, Sengupta S 1985 An inducable xylanase of the mushrom Termitomyces clypeatus differing from the xylanase/amylase produced in dextrin medium. J Gen Microbiol 131: 1881–1885

Muramatsu T, Egami F 1967 α-Mannosidase and β-mannosidase from the liver of Turbo cortunus: purification, properties and application in carbohydrate research. J Biochem 62: 700–709

Murmanis L, Highley TL, Palmer JG 1984 An electron microscopy study of western hemlock degradation by the white rot fungus Ganoderma applanatum. Holzforschung 38: 11–18

Murphy N, McConnell DJ, Cantwell BA 1984 The DNA sequence of the gene and genetic control sites for the excreted B. subtilis enzyme β-glucanase. Nucleic Acids Res 12: 5355–5367

Musha Y, Goring DAI 1975 Distribution of syringyl and guaiacyl moieties in hardwoods as indicated by ultraviolet microscopy. Wood Sci Technol 9: 45–58

Mustranta A 1987 Production of peroxidase by Inonotus weirii. Appl Microbiol Biotechnol 27: 21–26

Myers GC, Leatham GF, Wegner TH, Blanchette RA 1988 Fungal pretreatment of aspen chips improves strength of refiner mechanical pulp. Tappi J 71(5): 105–108

Nagieb ZA, El-Meadaway SA, El-Gammal AA 1988 Chemical characterization of native and degraded lignins from both rice straw and cotton stalks decayed by white rot fungi. Holzforsch Holzverwert 40: 33–37

Nakajima T, Tsukamoto K, Watanabe T, Kainuma K, Matsuda K 1984 Purification and some properties of an endo-1,4-β-D-xylanase from Streptomyces sp. J Ferment Technol 62: 269–276

Nakatsubo F, Kirk TK, Shimada M, Higuchi T 1981a Metabolism of a phenylcoumaran substructure lignin model compound in ligninolytic cultures of Phanerochaete chrysosporium. Arch Microbiol 128: 416–420

Nakatsubo F, Reid ID, Kirk TK 1981b Involvement of singlet oxygen in the fungal degradation of lignin. Biochem Biophys Res Commun 102: 484–491

Nakatsubo F, Reid ID, Kirk TK 1982 Incorporation of $^{18}O_2$ and absence of stereospecificity in primary product formation during fungal metabolism of a lignin model compound. Biochim Biophys Acta 719: 284–291

Namba H, Nakatsubo F, Higuchi T 1983 Degradation of β-1 linked dilignols by Fusarium solani M-13-1. Wood Res 69: 52–60

Nelson N 1944 A photometric adaptation of the Somogyi method for the determination of glucose. J Biol Chem 153: 375–380

Neukom H, Markwalder HU, Schibli P 1977 The composition of the dietary fiber of wheat flour. Lebensm-Wiss Technol 10: 346–349
Nevalainen P 1987 Biomassojen hiötikäyttö. (English summary) Kemia-Kemi 14: 236–238
Ng TK, Weimer PJ, Zeikus JG 1977 Cellulolytic and physiological properties of Clostridium thermocellum. Arch Microbiol 114: 1–7
Ng TK, Zeikus JG 1981 Purification and characterization of an endoglucanase(1,4-β-D-glucan glucanohydrolase) from Clostridium thermocellum. Biochem J 199: 341–350
Niku-Paavola ML 1987 Ligninolytic enzymes of the white-rot fungus Phlebia radiata. In: Odier E (ed) Lignin enzymic and microbial degradation. Symp Intern INRA Paris, 119–123
Niku-Paavola ML, Lappalainen A, Enari T-M, Nummi M 1985 A new appraisal of the endoglucanases of the fungus Trichoderma reesei. Biochem J 231: 75–81
Niku-Paavola M-L, Lappalainen A, Enari T-M, Nummi M 1986 Trichoderma reesei cellobiohydrolase II. Purification by immunoadsorption and hydrolytic properties. Biotechnol Appl Biochem 8: 449–458
Nilsson T 1973 Studies on wood degradation and cellulolytic activity of microfungi. Stud For Suec 104: 1–40
Nilsson T 1974a Comparative study of the cellulolytic activity of white rot and brown rot fungi. Mater Org 9: 173–198
Nilsson T 1974b Formation of soft rot cavities in various cellulose fibres by Humicola alopallonella Meyers and Moore. Stud For Suec 112: 1–30
Nilsson T 1985 Defining fungal decay types – a proposal for discussion. Intern Res Group on Wood Preserv Document IRG/WP/1264
Nilsson T, Daniel G 1983a Formation of soft rot cavities in relation to concentric layers in wood fibre walls. Intern Res Group on Wood Preserv Document IRG/WP/1185
Nilsson T, Daniel G 1983b Micromorphology of Schizophyllum commune attack in pine (Pinus sylvestris) wood. Intern Res Group on Wood Preserv Document IRG/WP/1184
Nilsson T, Daniel G 1983c Tunnelling bacteria. Intern Res Group on Wood Preserv Document IRG/WP/1186
Nilsson T, Daniel G 1986 Ligninolytic activity of wood-degrading bacteria. Proc. 3rd Intern Conf Biotechnol in the Pulp and Paper Ind, Stockholm, 54–57
Nilsson T, Daniel G, Kirk TK, Obst JR 1989 Chemistry and microscopy of wood decay by some higher Ascomycetes. Holzforschung 43: 11–18
Nilsson T, Ginns J 1979 Cellulolytic activity and the taxonomic position of selected brown-rot fungi. Mycologia 71: 170–177
Nilsson T, Holt DM 1983 Bacterial attack occurring in the S_2 layer of wood fibers. Holzforschung 37: 107–108
Nilsson T, Singh AP 1984 Cavitation bacteria. Intern Res Group on Wood Preserv Document IRG/WP/1235
Nimz H 1974 Das Lignin der Buche – Entwurf eines Konstitutionsschemas. Angew Chem 86: 336–344
Nishida A, Eriksson K-E 1987 Formation, purification, and partial characterization of methanol oxidase, a H_2O_2-producing enzyme in Phanerochaete chrysosporium. Biotechnol Appl Biochem 9: 325–338
Nishida A, Fukuzumi T 1978 Formation of coniferyl alcohol from ferulic acid by the white rot fungus Trametes. Phytochemistry 17: 417–419
Nisizawa K, Kanda T, Shikata S, Wakabajashy K 1978 Mutarotation of hydrolysis products by different types of exo-cellulases from Trichoderma viride. J Biochem (Tokyo) 83: 1625–1630
Nisizawa T, Suzuki H, Nisizawa K 1971 Inductive formation of cellulase by sophorose in Trichoderma viride. J Biochem (Tokyo) 70: 375–385
Nisizawa T, Suzuki H, Nisizawa K 1972 Catabolite repression of cellulase formation in Trichoderma viride. J Biochem (Tokyo) 71: 999–1007
Nobles MK 1958 Cultural characteristics as a guide to the taxonomy and phylogeny of the Polyporaceae. Can J Bot 36: 893–926
Norris DM 1980 Degradation of ^{14}C-labeled lignins and ^{14}C-labeled aromatic acids by Fusarium solani. Appl Environ Microbiol 40: 376–380

Notario W, Villa TG, Villanueva JR 1979 Cell-wall associated 1,4-β-D-xylanase in Cryptococcus albidus var aerius: in situ characterization of the activity. J Gen Microbiol 114: 415–422
Novobranova TI 1972, Species novae fungorum imperfectorum e region Alma-Ataënsi. (New species of fungi imperfecti in the Alma-Ata region). Akad Nauk USSR, Bot Inst Nov Sist Nizh Rastenii 9: 180–187
Nummi M, Niku-Paavola M-L, Enari T-M, Raunio V 1980 Immunoelectrophoretic detection of cellulases. FEBS Lett 113: 164–166
Nummi M, Niku-Paavola M-L, Lappalainen A, Enari T-M. Raunio V 1983 Cellobiohydrolase from Trichoderma reesei. Biochem J 215: 677–683
Nummi M, Perrin J-M, Niku-Paavola M-L, Enari T-M 1985 Measurement of xylanase activity with insoluble xylan substrate. Biochem J 226: 617–620
Oberkotter LV, Rosenberg FA 1978 Extracellular endo-β-1,4-glucanase in Cellvibrio vulgaris. Appl Environ Microbiol 36: 205–209
Odegaard, BH, Andersson PC, Lowrien RE 1984 Resolution of the multienzyme cellulase complex of Trichoderma reesei QM 9414. J Appl Biochem 6: 156–183
Odier E, Janin G, Monties B 1981 Poplar lignin decomposition by gram-negative aerobic bacteria. Appl Environ Microbiol 41: 337–341
Odier E, Monties B 1983 Absence of microbial mineralization of lignin in anaerobic enrichment cultures. Appl Environ Microbiol 46: 661–665
Odier E, Mozuch MD, Kalyanaraman B, Kirk TK 1988 Ligninase-mediated phenoxy radical formation and polymerization uneffected by cellobiose: quinone oxidoreductase. Biochimie 70: 847–852
Odier E, Rolando C 1985 Catabolism of arylglycerol-β-aryl ethers lignin model compounds by Pseudomonas cepacia 122. Biochimie 67: 191–197
Ogawa K, Toyama N 1972 Resolution of the Trichoderma viride cellulolytic complex. J Ferment Technol 45: 671–680
Ohta M, Higuchi T, Iwahara S 1979 Microbial degradation of dehydrodiconiferyl alcohol, a lignin substructure model. Arch Microbiol 121: 23–28
Okazaki W, Akiba T, Horikoshi K, Akahoshi R 1984 Production and properties of two types of xylanases from alkalophilic, thermophilic Bacillus spp. Appl Microbiol Biotechnol 19: 335–340
Oki T, Senba Y, Ishikawa H 1982 The biodegradation of softwood lignin and guaiacylglycerol-β-guaiacyl ether by extra-cellular enzyme in Shiitake Lentinus edodes (Berk.) Sing. Mokuzai Gakkaishi 28: 800–808
Oki T, Shinmoto M, Ishikawa H 1986 Enzymatic degradation of guaiacylglycerol-β-guaiacyl ether. Mokuzai Gakkaishi 32: 448–456
Oki T, Watanabe H, Ishikawa H 1981 The biodegradation of lignin by Shiitake Lentinus edodes (Berk.) Sing. Mokuzai Gakkaishi 27: 696–702
Olah GM, Reisinger O 1974 Etude ultrastructurale et cytochimique de l'appareil sporifère chez Phialophora richardsiae. Can J Bot 52: 2473–2480
Omelianski W 1902 Über die Gärung der Cellulose. Zentralbl Bakteriol Parasitenkd Infektionskr Hyg Abt 2, 8: 193–201
O'Neill GP, Warren RAJ, Kilburn DG, Miller RC Jr 1986 Secretion of a Cellulomonas fimi exoglucanase by Escherichia coli. Gene 44: 331–336
Orpin CG 1975 Studies on the rumen flagellate Neocallimastix frontalis. J Gen Microbiol 91: 249–262
Orpin CG 1977 Invasion of plant tissues in the rumen by the flagellate Neocallimastix frontalis. J Gen Microbiol 99: 107–117
Osmundsvåg K, Goksöyr J 1975 Cellulases from Sporocytophaga myxococcoides: purification and properties. Eur J Biochem 57: 405–409
Österberg F 1984 Characterization of chlorinated organic substance in spent bleach liquors. Dissert Roy Inst Technol and STFI, Stockholm, 120 pp
Otjen L, Blanchette RA 1982 Patterns of decay caused by Inonotus dryophilus (Aphyllophorales: Hymenochaetaceae), a white-pocket rot fungus of oaks. Can J Bot 60: 2770–2779
Otjen L, Blanchette RA 1984 Xylobolus frustulatus decay of oak: patterns of selective delignification and subsequent cellulose removal. Appl Environ Microbiol 47: 670–676

Otjen L, Blanchette RA 1986a Selective delignification of birch wood (Betula paryrifera) by Hirschioporus pargamenus in the field and laboratory. Holzforschung 40: 183–189

Otjen L, Blanchette RA 1986b A discussion of microstructural changes in wood during decomposition by white rot basidiomycetes. Can J Bot 64: 905–911

Paice MG, Bourbonnais R, Desroches M, Jurasek L, Yaguchi M 1986 A xylanase gene from Bacillus subtilis: nucleotide sequence and comparison with B. pumilus gene. Arch Microbiol 144: 201–206

Paice MG, Desrochers M, Rho D, Jurasek L, Roy C, Rollin CF, De Miguel E, Yaguchi M 1984 Two forms of endoglucanase from the basidiomycete Schizophyllum commune and their relationship to other β-1,4-glycoside hydrolases. Bio/Technology 2: 535–539

Paice MG, Jurasek L 1984a Removing hemicellulose from pulps by specific enzymic hydrolysis. J Wood Chem Technol 4: 187–198

Paice MG, Jurasek L 1984b Peroxidase-catalyzed color removal from bleach plant effluent. Biotechnol Bioeng 26: 477–480

Palleroni NJ 1984 Pseudomonas. In: Krieg NR, Holt JG (eds) Bergey's manual of systematic bacteriology Vol 1. Williams and Wilkins, Baltimore, 141–199

Palma A, Buldt G, Jovanovic S 1976 Absolutes Molekulargewicht der nativen Cellulose des Alge Valonia. Makromol Chem 177: 1063–1072

Palmer JM, Evans CS 1983a The enzymic degradation of lignin by white-rot fungi. Philos Trans R Soc Lond B 300: 293–303

Palmer JM, Evans CS 1983b Extracellular enzymes produced by Coriolus versicolor in relationship to the degradation of lignin. Proc Intern Symp Wood Pulp Chem Vol 3. Tsukuba, 19–24

Palmer JM, Harvey PJ, Schoemaker HE 1987 The role of peroxidase, radical cations and oxygen in the degradation of lignin. Philos Trans R Soc Lond A 321: 495–505

Palmer JG, Murmanis L, Highley TL 1983a Visualization of hyphal sheath in wood-decay Hymenomycetes. I. Brown-rotters. Mycologia 75: 995–1004

Palmer JG, Murmanis L, Highley TL 1983b Visualization of hyphal sheath in wood-decay Hymenomycetes. II. White-rotters. Mycologia 75: 1005–1010

Panbangred W, Kawaguchi O, Tomita T, Shinmyo A, Okada H 1984 Isolation of two β-xylosidase genes of Bacillus pumilus and comparison of their gene products. Eur J Biochem 138: 267–273

Panbangred W, Shinmyo A, Kinoshita S, Okada H 1983 Purification and properties of endoxylanase produced by Bacillus pumilus. Agr Biol Chem 47: 957–963

Panshin AJ, de Zeeuw C 1980 Textbook of wood technology. Vol 1, 4th edn. McGraw-Hill, New York, 705 pp

Parameswaran N, Liese W 1982 Ultrastructural localization of wall components in wood cells. Holz Roh-Werkst 40: 145–155

Parham RA, Côté WA 1971 Distribution of lignin in normal and compression wood of Pinus taeda L. Wood Sci Technol 5: 49–62

Paszczynski A, Crawford RL, Blanchette RA 1988 Delignification of wood chips and pulps by using natural and synthetic porphyrins: models of fungal decay. Appl Environ Microbiol 54: 62–68

Paszczynski A, Huynh V-B, Crawford R 1985 Enzymatic activities of an extracellular, manganese-dependent peroxidase from Phanerochaete chrysosporium. FEMS Microbiol Lett 29: 37–41

Paszczynski A, Huynh V-B, Crawford R 1986 Comparison of ligninase-I and peroxidase-M2 from the white-rot fungus Phanerochaete chrysosporium. Arch Biochem Biophys 244: 750–765

Paszczynski A, Lobarzewski J 1984 Modifications of peroxidase polymorphism in Trametes versicolor fungus after short-time incubation under starvation conditions. Biochem Physiol Pflanz 179: 749–757

Paszczynski A, Trojanowski J 1977 An affinity-column procedure for the purification of veratrate O-demethylase from fungi. Microbios 18: 111–121

Paterson A, McCarthy AJ, Broda P 1984 The application of molecular biology to lignin degradation. In: Grainger JM, Lynch JM (eds) Microbiological methods for environmental biotechnology. Academic Press, London, 33–68

Patil RV, Sadana JC 1984 The purification and properties of 1,4-β-D-glucan cellobiohydrolase from Sclerotium rolfsii: substrate specificity and mode of action. Can J Biochem. Cell Biol 62: 920–926

Pavlova IN, Tinyanova NZ 1980 Isolation and characterization of microorganisms with mannanase activity. Prikl Biokhim Mikrobiol 16: 578–583

Payen M 1838 Mémoire sur la composition du tissue propre des plantes et du ligneux. Acad Sci 7: 1052–1056

Pearce PD, Bauchop T 1985 Glycosidases of the rumen anaerobic fungus Neocallimastix frontalis grown on cellulosic substrates. Appl Environ Microbial 49: 1265–1269

Peek R-D von, Liese W, Parameswaran N 1972 Infektion und Abbau des Wurzelholzes von Fichte durch Fomes annosus. Eur J For Path 2: 237–248

Pearl IA 1982 Utilization of by-products of the pulp and paper industry. Tappi Vol 65(5): 68–73

Pellinen J, Abuhasen J, Joyce TW, Chang H-m 1989 Biological delignification of pulp by Phanerochaete chrysosporium. J Biotechnol 10: 161–170

Pellinen J, Jokela J, Salkinoja-Salonen M 1987 Degradability of different lignins by bacteria. Holzforschung 41: 271–276

Pellinen J, Joyce TW, Chang H-m 1988 Dechlorination of high-molecular-weight chlorolignin by the white-rot fungus P. chrysosporium. Tappi J 71(9): 191–194

Pellinen J, Väisänen E, Salkinoja-Salonen M, Brunow G 1984 Utilization of dimeric lignin model compounds by mixed bacterial cultures. Appl Microbiol Biotechnol 20: 77–82

Pelmont J, Barrelle M, Hauteville M, Gamba D, Romdhane M, Dardas A, Beguin C 1985 A new bacterial dehydrogenase oxidizing the lignin model compound guaiacyglycerol β-O-4-guaiacyl ether. Biochimie 67: 973–986

Penttilä M, Lehtovaara P, Nevalainen H, Bhikhabhai R, Knowles J 1986 Homology between cellulase genes of Trichoderma reesei: complete nucleotide sequence of the endoglucanase I-gene. Gene 45: 253–263

Perlman D, Halvorson HO 1983 A putative signal peptidase recognition site and sequence in eukaryotic and prokaryotic signal peptides. J Mol Biol 167: 391–409

Petré J, Longin R, Millet J 1981 Purification and properties of an endo-β-1,4-glucanase from Clostridium thermocellum. Biochimie 63: 629–639

Pettersson B, Yang J-l, Eriksson K-E 1988 Biotechnical approaches to pulp bleaching. Nord Pulp Pap Res J 3: 198–202

Pettersson G 1968 Studies on cellulolytic enzymes. III Isolation of a cellulase from Penicillium notatum. Arch Biochem Biophys 123: 307–311

Pettersson G, Cowling EB, Porath J 1963 Studies on cellulolytic enzymes. I Isolation of a low-molecular-weight cellulase from Polyporus versicolor. Biochim Biophys Acta 67: 1–8

Pettersson G, Fägerstam L, Bhikhabhai R, Leandoer K 1981 The cellulase complex of Trichoderma reesei QM9414. The Ekman-Days 1981, Intern Symp Wood Pulp Chem, Stockholm 3: 39–42

Pettersson LG 1975 The mechanism of enzymatic hydrolysis of cellulose by Trichoderma viride. In: Bailey MJ, Enari T-M, Linko M (eds) Symposium on enzymatic hydrolysis of cellulose. Aluanko, Finland, 255–280

Pettersen RC, Schwandt VH, Effland MJ 1985 An analysis of the wood sugar assay using HPLC: a comparison with paper chromatography. J Chromatogr Sci 22: 478–484

Pettey TM, Crawford DL 1984 Enhancement of lignin degradation in Streptomyces spp. by protoplast fusion. Appl Environ Microbiol 47: 439–440

Pettipher GL, Latham ML 1979 Characterization of enzymes produced by Ruminococcus flavefaciens which degrade plant cell walls. J Gen Microbiol 110: 21–27

Phelan MB, Crawford DL, Pometto AL III 1979 Isolation of lignocellulose-decomposing actinomycetes and degradation of specifically ^{14}C-labeled lignocelluloses by six selected Streptomyces strains. Can J Microbiol 25: 1270–1276

Philippi F 1893 Die Pilze Chiles, soweit dieselben als Nahrungsmittel gebraucht werden. Hedwigia 32: 115–118

Pilon L, Barbe MC, Desrochers M, Jurasek L 1982a Fungal treatment of mechanical pulps – its effect on paper properties. Biotechnol Bioeng 24: 2063–2076

Pilon L, Desrochers M, Jurasek L, Neumann PJ 1982b Increasing water retention of mechanical pulp by biological treatments. Tappi 65(6): 93–96

Polakis ES, Barley W, Meek GA 1965 Changes in the activities of respiratory enzymes during the aerobic growth of yeast on different carbon sources. Biochem J 97: 298–302

Pometto AL III, Crawford DL 1986a Catabolic fate of Streptomyces viridosporus T7A-produced, acid-precipitable polymeric lignin upon incubation with ligninolytic Streptomyces species and Phanerochaete chrysosporium. Appl Environ Microbiol 51: 171–179

Pometto AL III, Crawford DL 1986b Effects of pH on lignin and cellulose degradation by Streptomyces viridosporus. Appl Environ Microbiol 52: 246–250

Poutanen K 1988a Characterization of xylanolytic enzymes for potential applications. Diss Techn Res Centre, Finland, Publications 47

Poutanen K 1988b An α-L-arabinofuranosidase of Trichoderma reesei. J Biotechnol 7: 271–282

Poutanen K, Puls J 1988 Characteristics of Trichoderma reesei β-xylosidase and its use in the hydrolysis of solubilized xylans. Appl Microbiol Biotechnol 28: 425–432

Poutanen K, Puls J, Linko M 1986 Hydrolysis of steamed birch wood hemicellulose by enzymes produced by Trichoderma reesei and Aspergillus awamori. Appl Microbiol Biotechnol 23: 487–490

Poutanen K, Rättö M, Puls J, Vikarii L 1987 Evaluation of different microbial xylanolytic systems. J Biotechnol 6: 49–60

Poutanen K, Sundberg M 1988 An acetyl esterase of Trichoderma reesei and its role in the hydrolysis of acetyl xylans. Appl Microbiol Biotechnol 28: 419–424

Preston RD 1974 Physical biology of plant cell walls. Chapman and Hall, London, 491 pp

Priest FG 1977 Extracellular enzyme synthesis in the genus Bacillus. Bacteriol Rev 41: 711–753

Proctor P Jr 1941 Penetration of the walls of wood cells by the hyphae of wood-destroying fungi. Yale Univ Sch For Bull 47: 31

Pryor WA, Tang RH 1978 Ethylene formation from methional. Biochem Biophys Res Commun 81: 498–503

Puls J, Schmidt O, Granzow C 1987 α-Glucuronidase in microbial xylanolytic systems. Enzyme Microb Technol 9: 83–88

Puls J, Sinner M, Dietrichs H-H 1978 Trägergebundene xylanolytische Enzyme I. Wirkungsweise xylanolytischer Enzyme. Stärke 30: 294–299

Rabinovich ML, Martyanov VA, Chumak GA and Klyosov AA 1982 [Substituted cellodextrin – a new substrate for the selective determination of exoglucosidase in cellulase complexes.] Bioorg Khim 8(3): 396–403 (in Russian)

Raeder U, Broda P 1985 Rapid preparation of DNA from filamentous fungi. Lett Appl Microbiol 1: 17–20

Raeder U, Thompson W, Broda P 1987 Establishing molecular genetics for Phanerochaete chrysosporium. Philos Trans R Soc Lond A 321: 475–483

Ramachandra M, Crawford DL, Hertel G 1988 Characterization of an extracellular lignin peroxidase of the lignocellulolytic Actinomycete Streptomyces viridosporus. Appl Environ Microbiol 54: 3057–3063

Ramachandra M, Crawford DL, Pometto AL III 1987 Extracellular enzyme activities during lignocellulose degradation by Streptomyces spp.: a comparative study of wild-type and genetically manipulated strains. Appl Environ Microbiol 53: 2754–2760

Ramasamy K, Kelley RL, Reddy CA 1985 Lack of lignin degradation by glucose oxidase-negative mutants of Phanerochaete chrysosporium. Biochem Biophys Res Commun 131: 436–441

Ramasamy K, Verachtert H 1980 Localization of cellulase components in Pseudomonas sp. isolated from activated sludge. J Gen Microbiol 117: 181–191

Rao M, Deshpande V, Keskar S, Srinivasan MC 1983a Cellulase and ethanol production from cellulose by Neurospora crassa. Enzyme Microb Technol 5: 133–136

Rao MNA, Mithal BM, Thakur RN, Sastry KSM 1983b Productions of cellulase from Pestalotiopsis versicolor. Biotechnol Bioeng 25: 2395–2398

Rao TR, Reddy CA 1984 DNA sequences from a ligninolytic filamentous fungus Phanerochaete chrysosporium capable of autonomous replication in yeast. Biochem Biophys Res Commun 118: 821–827

Rappi P, Wagner F 1986 Production and properties of xylan-degrading enzymes from Cellulomonas uda. Appl Environ Microbiol 51: 746–752

Rättö M, Poutanen K 1988 Production of mannan-degrading enzymes. Biotechnol Lett 10: 661–664

Rayner ADM, Todd NK 1977 Intraspecific antagonism in natural populations of wood-decaying Basidiomycetes. J Gen Microbiol 103: 85–90

Reade AE, McQueen RE 1983 Investigation of white-rot fungi for the conversion of poplar into a potential feedstuff for ruminants. Can J Microbiol 29: 457–463

Reddy CA, Forney LJ, Kelley RL 1983 Involvement of hydrogen peroxide-derived hydroxyl radical in lignin degradation by the white-rot fungus Phanerochaete chrysosporium. In: Higuchi T, Chang H-m, Kirk TK (eds) Recent advances in lignin biodegradation research, Uni Publ Tokyo, 153–163

Reese ET 1956 A microbial progress report. Enzymatic hydrolysis of cellulose. Appl Microbiol 4: 39–45

Reese ET 1982 Elution of cellulase from cellulose. Proc Biochem 17: 2–6

Reese ET, Lola JE, Parrish FW 1969 Modified substrates and modified products as inducers of carbohydrases. J Bacteriol 100: 1151–1154

Reese ET, Maguire A 1971 Increase in cellulase yields by addition of surfactants to cellobiose cultures of Trichoderma viride. Dev Ind Microbiol 12: 212–224

Reese ET, Maguire A, Parrish FW 1973 Production of β-D-xylopyranosidases by fungi. Can J Microbiol 19: 1065–1074

Reese ET, Mandels M, Weiss AH 1972 Cellulose as a novel energy source. In: Ghose TK, Fiechter A, Blackbreuogh N (eds) Adv Biochem Eng 2: 181–200

Reese ET, Shibata Y 1965 β-Mannanases of fungi. Can J Microbiol 11: 167–183

Reese ET, Siu RGH, Levinson HS 1950 The biological degradation of soluble cellulose derivatives and its relationship to the mechanism of cellulose hydrolysis. J Bacteriol 59: 485–497

Reid ID 1979 The influence of nutrient balance on lignin degradation by the white-rot fungus Phanerochaete chrysosporium. Can J Bot 57: 2050–2058

Reid ID 1982 Increasing digestibility of lignocellulosic materials by solid-state fermentation with white-rot fungi. Tappi Res Dev Div Conf 237–240

Reid ID 1983a Effects of nitrogen supplements on degradation of aspen wood lignin and carbohydrate components by Phanerochaete chrysosporium. Appl Environ Microbiol 45: 830–837

Reid ID 1983b Effects of nitrogen sources on cellulose and synthetic lignin degradation by Phanerochaete chrysosporium. Appl Environ Microbiol 45: 838–842

Reid ID 1985 Biological delignification of aspen wood by solid-state fermentation with the white-rot fungus Merulius tremellosus. Appl Environ Microbiol 50: 133–139

Reid ID, Chao EE, Dawson PSS 1985 Lignin degradation by Phanerochaete chrysosporium in agitated cultures. Can J Microbiol 31: 88–90

Reid ID, Seifert KA 1980 Lignin degradation by Phanerochaete chrysosporium in hyperbaric oxygen. Can J Microbiol 26: 1168–1171

Reid ID, Seifert KA 1982 Effect of an atmosphere of oxygen on growth, respiration and lignin degradation by white rot fungi. Can J Bot 60: 252–260

Renganathan V, Gold MH 1986 Spectral characterization of the oxidized states of lignin peroxidase, an extracellular heme enzyme from the white rot basidiomycete Phanerochaete chrysosporium. Biochemistry 25: 1626–1631

Renganathan V, Miki K, Gold MH 1985 Multiple molecular forms of diarylpropane oxygenase, an H_2O_2-requiring, lignin-degrading enzyme from Phanerochaete chrysosporium. Arch Biochem Biophys 241: 304–314

Renganathan V, Miki K, Gold MH 1986 Role of molecular oxygen in lignin peroxidase reactions. Arch Biochem Biophys 246: 155–161

Rho D, Desrochers M, Jurasek L, Driguez H, DeFaye J 1982 Induction of cellulase in Schizophyllum commune: thiocellobiose as a new inducer. J Bacteriol 149: 47–53

Ribbons DW 1970 Stoichiometry of O-demethylase activity in Pseudomonas aeruginosa, FEBS Lett 8: 101–104

Richards GN, Whistler RL 1973 Isolation of two pure polysaccharides from the hemicellulose of slash pine (Pinus elliottii). Carbohyd Res 31: 47–55

Ristproph DL, Humphrey AE 1985 Genetic characterization of the extracellular xylanases of Thermomonospora sp. Biotechnol Bioeng 27: 832–836

Robert D, Chen C-L 1987 Structural analysis of biodegraded lignins from spruce wood decayed by Phanerochaete chrysosporium by quantitative ^{13}C NMR spectroscopy. Proc 4th Intern Symp Wood Pulp Chem Vol 1, Paris, 139–145
Robert D, Chen C-L 1989 Biodegradation of lignin in sprucewood by Phanerochaete chrysosporium: Quantitative analyses of lignins by ^{13}C NMR spectroscopy. Holzforschung 43: 323–332
Robinson DG 1977 Plant cell wall synthesis. Adv Bot Res 5: 89–151
Robson LM, Chambliss GH 1984 Characterization of the cellulolytic activity of a Bacillus isolate. Appl Environ Microbiol 47: 1039–1046
Robson LM, Chambliss GH 1986 Cloning of the Bacillus subtilis DLG β-1,4-glucanase gene in Escherichia coli and B. subtilis. J Bacteriol 165: 612–619
Robson LM, Chambliss GH 1987 Endo-β-1,4-glucanase gene of Bacillus subtilis DLG. J Bacteriol 169: 2017–2025
Roch P, Odier E, Monties B, Eriksson K-E 1983 The influence of simple sugar supplementation on lignin degradation by different white-rot fungi. Proc Intern Conf Biotechnol in the Pulp and Paper Ind, London, 323–332
Rodionova NA, Tavobilov IM, Bezborodov AM 1983 β-Xylosidase from Aspergillus niger 15: purification and properties. J Appl Biochem 5: 300–312
Rogers CJ, Coleman E, Spino DF, Purcell TC 1972 Production of fungal protein from cellulose and waste cellulosics. Environ Sci Technol 6: 715–718
Rogers JD 1979 The Xylariaceae: systematic, biological and evolutionary aspects. Mycologia 71: 1–42
Romanelli RA, Houston CW, Barrett SM 1975 Studies on thermophilic cellulolytic fungi. Appl Microbiol 30: 276–281
Rosén C-G 1982 Continuous fermentation. In: Duckworse HE (ed) International symposium on ethanol from biomass. The Royal Society of Canada, Winnipeg, Canada, 85–117
Ross P, Aloni Y, Weihaus Z, Michaeli D, Weinburger-Ohana P, Meyer R, Benziman M 1985 An unusual guanyl oligonucleotide regulates cellulose synthesis in Acetobacter xylinum. FEBS Lett 186: 191–196
Ross P, Weinhaus H, Aloni Y, Michaeli D, Weinberger-Ohana P, Mayer R, Braun S, de Vroom E, van der Marel GA, van Boom JH, Benziman M 1987 Regulation of cellulose synthesis in Acetobacter xylinum by cyclic diguanylic acid. Nature 325: 279–281
Rossell SE, Abbot EGM, Levy JF 1973 Bacteria in wood – a review of the literature relating to the presence, action and interaction of bacteria in wood. J Inst Wood Sci 6: 28–35
Rouau X, Foglietti MJ 1985 Purification and partial characterization of three endoglucanases from Dicomitus squalens. Carbohyd Res 142: 299–314
Royer G, Livernoche D, Desrochers M, Jurasek L, Rouleau D, Mayer RC 1983 Decolorization of kraft mill effluent: kinetics of a continuous process using immobilized Coriolus versicolor. Biotechnol Lett 5: 321–326
Ruel K, Barnoud F, Eriksson K-E 1981 Micromorphological and ultrastructural aspects of spruce wood degradation by wild-type Sporotrichum pulverulentum and its cellulase-less mutant Cel 44. Holzforschung 25: 157–171
Ruel K, Barnoud F, Eriksson K-E 1984 Ultrastructural aspects of wood degradation by Sporotrichum pulverulentum. Observations on spruce wood impregnated with glucose. Holzforschung 38: 61–68
Ruel K, Barnoud F, Goring DAI 1978 Lamellation in the S_2 layer of softwood tracheids and demonstrated by scanning transmission electron microscopy. Wood Sci Technol 12: 287–291
Ruel K, Joseleau J-P, Johnsrud SC, Eriksson K-E 1986 Ultrastructural aspects of birch wood degradation by Phanerochaete chrysosporium and two of its cellulase deficient mutants. Holzforschung 40: 5–9
Russell JD, Henderson MEK, Farmer VC 1961 Metabolism of lignin model compounds by Polystictus versicolor. Biochim Biophys Acta 52: 565–570
Rüttimann C, Seelenfreund D, Vicuna R 1987 Metabolism of low molecular weight lignin-related compounds by Streptomyces viridosporus T7A. Enzyme Microb Technol 9: 526–530
Rypácek V 1975 Biological characteristics of decomposition of wood by fungi. Drev Vysk 20: 1–22
Ryu DDY, Mandels M 1980 Cellulases: biosynthesis and applications. Enzyme Microb Technol 2: 91–102

Sadana JC, Patil RV 1985 Synergism between enzymes of Sclerotium rolfsii involved in the solubilization of chrystalline cellulose. Carbohyd Res 140: 111–120
Sadana JC, Shewale JG, Desphande MV 1979 Enhanced cellulase production by a mutant of Sclerotium rolfsii. Appl Environ Microbiol 38: 730–733
Sadana JC, Shewale JG, Patil RV 1983 β-D-Glucosidases of Sclerotium rolfsii. Substrate specificity and mode of action. Carbohyd Res 118: 205–214
Saddler JN, Brownell HH, Clermont LP, Levitin N 1982 Enzymatic hydrolysis of cellulose and various pretreated wood fractions. Biotechnol Bioeng 24: 1389–1402
Saddler JN, Khan AW 1980 Cellulase production by Acetivibrio cellulolyticus. Can J Microbiol 26: 760–765
Saddler JN, Khan AW 1981 Cellulolytic enzyme system of Acetovibrio cellulolyticus. Can J Microbiol 27: 288–294
Saka S, Goring DAI 1985 Localization of lignins in wood cell walls. In: Higuchi T (ed) Biosynthesis and biodegradation of wood components. Academic Press, Orlando, 51–62
Saka S, Thomas RJ 1982a Evaluation of the quantitative assay of lignin distribution by SEM-EDXA technique. Wood Sci Technol 16: 1–18
Saka S. Thomas RJ 1982b A study of lignification in loblolly pine tracheids by the SEM-EDXA technique. Wood Sci Technol 16: 167–179
Sakakibara A 1980 A structural model of softwood lignin. Wood Sci Technol 14: 89–100
Sakakibara A 1983 Chemical structure of lignin related mainly to degradation products. In: Higuchi T, Chang H-m, Kirk TK (eds) Recent advances in lignin biodegradation research. Uni Publ, Tokyo, 12–33
Salkinoja-Salonen M, Apajalahti J, Silakoski L, Hakulinen R 1984 Anaerobic fluidised bed for the purification of effluents from chemical and mechanical pulping. Biotech Adv 2: 357–375
Saloheimo M, Lehtovaara P, Penttilä M, Teeri TT, Ståhlberg J, Johansson G, Pettersson G, Claeyssens M, Tomme P, Knowles JKC 1988 EG III, a new endoglucanase from Trichoderma reesei: the characterization of both gene and enzyme. Gene 63: 11–21
Samejima M, Abe S-I, Saburi Y, Yoshimoto T 1988 Metabolism of lignin model compounds with β-aryl ether linkage by Pseudomonas sp. TMY1009 and its mutants. Cellul Chem Technol 22: 279–286
Samejima M, Saburi Y, Yoshimoto T, Fukuzumi T, Nakazawa T 1985 Catabolic pathway of guaiacylglycerol-β-guaiacyl ether by Pseudomonas sp. TMY1009. Mokuzai Gakkaishi 31: 956–958
Samuelsson L, Mjöberg PJ, Hartler N, Vallander L, Eriksson K-E 1980 Influence of fungal treatment on the strength versus energy relationship in mechanical pulping. Svensk Papperstidn 83: 221–225
Sanglard D, Leisola MSA, Fiechter A 1986 Role of extracellular ligninases in biodegradation of benzo(a)pyrene by Phanerochaete chrysosporium. Enzyme Microb Technol 8: 209–212
Sannia G, Giardina P, Luna M, Rossi M, Buonocore V 1986 Laccase from Pleurotus ostreatus. Biotechnol Lett 8: 797–800
Sarkanen KV, Ludwig CH 1971 Definition and Nomenclature. In: Sarkanen KV, Ludwig CH (eds) Lignins. Occurrence, formation, structure, and reactions. Wiley-Interscience, New York, 1–18
Sasaki T, Tanaka T, Nakagawa S, Kainuma K 1983 Purification and properties of Cellvibrio gilvus cellobiose phosphorylase. Biochem J 209: 803–807
Sashihara N, Kudo T, Horikoshi K 1984 Molecular cloning and expression of cellulase genes of alkalophilic Bacillus sp. strain N-4 in Escherichia coli. J Bacteriol 158: 503–506
Savory JG 1954 Breakdown of timber by Ascomycetes and Fungi Imperfecti. Ann Appl Biol 41: 336–347
Savory JG, Pinion LC 1958 Chemical aspects of decay of beech wood by Chaetomium globosum. Holzforschung 12: 99–103
Saxena SK, Bisaria VS, Verma J, Gopalkrishnan KS 1985 Lignin degrading potential of a new isolate of Coriolus hirsutus. J Ferment Technol 63: 307–310
Schacht H 1863 Über die Veränderungen durch Pilze in abgestorbenen Pflanzenzellen. Jahrb Wiss Bot 3: 442–483
Schafer ML, King KW 1965 Utilization of cellulose oligosaccharides by Cellvibrio gilvus. J Bacteriol 89: 113–116

Schmid R, Liese W 1964 Über die mikromorphologischen Veränderungen der Zellwandstrukturen von Buchen- und Fichtenholz beim Abbau durch Polyporus versicolor (L.) Fr. Arch Mikrobiol 47: 260–276

Schmid R, Liese W 1966 Elektronenmikroskopische Beobachtungen über die Struktur der Hyphen einiger Holzpilze. Mater Org 1: 252–261

Schmidt O 1980 Über den bakteriellen Abbau der chemisch behandelten verholzten Zellwand. Mater Org 15: 207–224

Schmidt O, Bauch J 1980 Lignin in woody tissues after chemical pretreatment and bacterial attack. Wood Sci Tecnol 14: 229–239

Schmidt O, Liese W 1980 Variability of wood degrading enzymes of Schizophyllum commune. Holzforschung 34: 67–72

Schmidt O, Liese W 1982 Bacterial decomposition of woody cell walls. Int J Wood Preserv 2: 13–19

Schmidt O, Nagashima Y, Liese W, Schmitt U 1987 Bacterial wood degradation studies under laboratory conditions and in lakes. Holzforschung 41: 137–140

Schmidt O, Puls J, Sinner M, Dietrichs H-H 1979 Current yield of mycelium and xylanolytic enzymes from extracts of steamed birchwood, oat husks and wheat straw. Holzforschung 33: 192–196

Schmitz H 1919 Studies on the physiology of the fungi VI. The relation of bacteria to cellulose fermentation induced by fungi with special reference to the decay of wood. Ann Missouri Bot Gard 6: 93–136

Schmuck M, Pilz I, Hayn M, Esterbauer H 1986 Investigation of cellobiohydrolase from Trichoderma reesei by small angle X-ray scattering. Biotechnol Lett 8: 397–402

Shoemaker HE, Harvey PJ, Bowen RM, Palmer JM 1985 On the mechanism of enzymatic lignin breakdown. FEBS Lett 183: 7–12

Shoemaker HE, Meijer EM, Leisola MSA, Haemmerli SD, Waldner R, Sanglard D, Schmidt HWH 1989 Oxidation and reduction in lignin biodegradation. In: Lewis NG, Paice MG (eds) ACS Symp Ser No 399 Plant Cell Wall Polymers: Biogenesis and Biodegradation Ch 33, Amer Chem Soc, Washington, DC, 454–471

Schulze B, Theden G 1938 Polarisationsmikroskopische Untersuchungen über den Abbau des Werkstoffes Holz durch holzzerstörende Pilze. Holz Roh-Werkst 1: 548–554

Schulze E 1891 Zur Kenntnis der chemischen Zusammensetzung der pflanzlichen Zellmembranen. Chem Ber 24: 2277–2287

Seidler RJ, Aho PE, Raju PM, Evan HJ 1972 Nitrogen fixation by bacterial isolates from decay in living white fir trees [Abies concolor (Gord and Glend) Lindl]. J Gen Microbiol 73: 413–416

Seifert K 1964 Die Veränderung der chemischen Holzzusammensetzung durch den Bläuepilz Pullularia pullulans (de Bary) Beckhout [= Aureobasidium pullulans (de Bary) Arnaud.] Holz Roh-Werkst 22: 405–413

Seifert K 1966 Die chemische Veränderung der Buchenholz-Zellwand durch Moderfäule (Chaetomium globosum Kunze). Holz Roh-Werkst 24: 185–189

Seigle-Murandi F, Steiman R, Rahouti M 1986 Metabolization of phenolic compounds by soft-rot fungi. Proc 3rd Intern Conf Biotechnol in the pulp and Paper Ind, Stockholm, 143–146

Selby K 1969 The purification and properties of the C_1-component of the cellulase complex. In: Hajny GJ, Reese ET (eds) Cellulases and their applications. Adv Chem Ser 95: 34–50

Selby K, Maitland CC 1965 The fractionation of Myrotecium verrucaria cellulase by gel filtration. Biochem J 94: 578–583

Selby K, Maitland CC 1967 The cellulase of Trichoderma viride cellulolytic complex. Biochem J 104: 716–724

Sen S, Abraham TK, Chakrabarty SL 1981 Characteristics of the cellulase produced by Myceliophthora thermophila D-14. Can J Microbiol 28: 271–277

Setliff EC, Eudy WW 1980 Screening white-rot fungi for their capacity to delignify wood. In: Kirk TK, Higuchi T, Chang H-m (eds) Lignin biodegradation: microbiology, chemistry and potential applications, Vol I. CRC, Boca Raton, 135–149

Shawky B, Kasulke U, Phillip B, Schultz G, Hirte W 1984 Sorption of Penicillium-cellulase on cellulose and lignin. Acta Biotechnol 4: 267–274

Shei JC, Fratzke AR, Frederick MM, Frederick JR and Riley PJ 1985 Purification and characterization of endo-xylanases from Aspergillus niger II An enzyme of pI 4.5. Biotechnol Bioeng 27: 533–538

Sheir-Neiss G, Montenecourt BS 1984 Characterization of the secreted cellulases of Trichoderma reesei wild-type and mutants during controlled fermentations. Appl Microbiol Biotechnol 20: 46–53

Shepherd MG, Tong CC, Cole AL 1981 Substrate specificity and mode of action of the cellulases from the thermophilic fungus Thermoascus aurantiacus. J Biochem 193: 67–74

Sheth K, Alexander JK 1969 Purification and properties of β-1,4-oligoglucan: orthophosphate glycosyltransferase from Clostridium thermocellum. J Biol Chem 244: 457–464

Shigo AL 1965 Organism interaction in decay and discoloration in beech, birch and maple. US For Serv Res Pap NE-43: 23

Shigo AL 1967 Successions of microorganisms in discoloration and decay of wood. Int Rev For Res 2: 237–299

Shigo AL 1986 A new tree biology dictionary. Terms, topics and treatments for trees and their problems and proper tree care. Shigo and Trees Assoc. Durham, NH, 132 pp

Shigo AL, Hillis WE 1973 Heartwood, discolored wood and microorganisms in living trees. Ann Rev Phytopathol 11: 197–222

Shimada M 1972 Biochemical studies on bamboo lignin and methoxylation in hardwood and softwood lignins. Wood Res 53: 19–65

Shimada M, Habe T, Higuchi T, Okamoto T, Panijpan B 1987a Biomimetic approach to lignin degradation II. The mechanism of oxidative C-C bond cleavage reactions of lignin model compounds with natural iron(III) porphyrin chloride as a heme-enzyme model system. Holzforschung 41: 277–285

Shimada M, Habe T, Umezawa T, Higuchi T, Okamoto T 1984 The C-C bond cleavage of a lignin model compound, 1,2-diarylpropane-1,3-diol, with a heme-enzyme model catalyst tetraphenylporphyrinatoiron(III)chloride in the presence of tert-butylhydroperoxide. Biochem Biophys Res Commun 122: 1247–1252

Shimada M, Hattori T, Umezawa T, Higuchi T, Uzura K 1987b Regiospecific oxygenations during ring cleavage of a secondary metabolite, 3,4-dimethoxybenzyl alcohol catalyzed by lignin peroxidase. FEBS Lett 221: 327–331

Shimada M, Nakatsubo F, Kirk TK, Higuchi T 1981 Biosynthesis of the secondary metabolite veratryl alcohol in relation to lignin degradation in Phanerochaete chrysosporium. Arch Microbiol 129: 321–324

Shimahara H, Suzuki H, Sugiyama N, Nisizawa K 1975 Partial purification of β-mannanases from the konjac tubers and their substrate specificity in relation to the structure of konjac glucomannan. Agr Biol Chem 39: 301–312

Shimizu K, Ishihara M 1987 Immobilization of cellulolytic and hemicellulolytic enzymes on inorganic supports. Biotechnol Bioeng 29: 236–241

Shoemaker SP, Brown RD Jr 1978a Characterization of endo-1,4-β-D-glucanases purified from Trichoderma viride. Biochim Biophys Acta 523: 133–146

Shoemaker SP, Brown RD 1978b Enzymic activities of endo-1,4-β-D-glucanases purified from Trichoderma viride. Biochim Biophys Acta 523: 147–161

Shoemaker S, Schweickart V, Ladner M, Gelfand D, Kwok S, Myambo K, Innis M 1983a Molecular cloning of exocellobiohydrolase I derived from Trichoderma reesei strain L27. Bio/Technology 1: 691–696

Shoemaker S, Watt K, Tsitovsky G, Cox R 1983b Characterization and properties of cellulases purified from Trichoderma reesei strain L27 Bio/Technology 1: 687–690

Shortle WC 1979 Compartmentalization of decay in red maple and hybrid poplar trees. Phytopathology 69: 410–413

Singh AP, Butcher JA 1985 Degradation of CCA treated Pinus radiata posts by erosion bacteria. J Inst Wood Sci 10: 140–144

Singh AP, Nilsson T, Daniel G 1987 Ultrastructure of the attack of the wood of two high lignin tropical hardwood species, Alstonia scholaris and Homalium foetidum, by tunnelling bacteria. J Inst Wood Sci 11: 26–42

Sinha SN, Ghosh GL, Ghose SN 1981 Detection of cellulase inhibitor in the wheat bran culture of Aspergillus terreus. Can J Microbiol 27: 1334–1340
Sinitsyn AP, Bungay HR, Clesceri LS 1983 Enzyme management in the Iotech process. Biotechnol Bioeng 25: 1393–1399
Sinner M, Dietrichs HH 1975 Enzymatische Hydrolyse von Laubholzxylanen III. Kennzeichnung von fünf isolierten β-1,4-Xylanasen. Holzforschung 29: 207–214
Sinner M, Dietrichs HH 1976 Enzymatische Hydrolyse von Laubholz-Xylanen IV. Abbau von isolierten Xylanen. Holzforschung 30: 50–59
Siu RGH 1951 Microbial decomposition of cellulose. Reinhold, New York, 531 pp
Sjoblad RD, Bollag J-M 1981 Oxidative coupling of aromatic compounds by enzymes from soil microorganisms. In: Paul EA, Ladd JN (eds) Soil biochemistry Vol 5. Marcel Dekker, New York, 113–152
Sjöström E 1981 Wood chemistry. Fundamentals and applications. Academic Press, New York, 223 pp
Smiley CJ, Ray J, Huggins LM 1975 Preservation of Miocene fossils in unoxidized lake deposits, Clarkia, Idaho. J Paleontol 51: 885–901
Smith RS 1975 Economic aspects of bacteria in wood. In: Liese W (ed) Biological transformation of wood by microorganisms. Springer, Berlin Heidelberg New York, 89–102
Smith TL, Schalch H, Gaskell J, Covert S, Cullen D 1988 Nucleotide sequence of a ligninase gene from Phanerochaete chrysosporium. Nucleic Acids Res 16: 1219
Somogyi M 1952 Notes on sugar determination. J Biol Chem 195: 19–23
Sone Y, Misaki A 1978 Purification and characterization of β-D-mannosidase and β-N-acetyl-D-hexoseaminidase of Tremella fuciformis. J Biochem 83: 1135–1144
Sprey B 1986 Localization of β-glucosidase in Trichoderma reesei cell walls with immunoelectron microscopy. FEMS Microbiol Lett 36: 287–292
Sprey B, Lambert C 1983 Titration curves of cellulases from Trichoderma reesei: demonstration of a cellulase-xylanase-β-glucosidase containing complex. FEMS Microbiol Lett 18: 217–222
Stack RJ, Cotta MA 1986 Effect of 3-phenylpropanic acid on growth of and cellulose utilization by cellulolytic ruminal bacteria. Appl Environ Microbiol 52: 209–210
Stack RJ, Hungate RE 1984 Effect of 3-phenylpropanoic acid on capsule and cellulases of Ruminococcus albus 8. Appl Environ Microbiol 48: 218–223
Ståhlberg J, Johansson G, Pettersson G 1988 A binding-site-deficient, catalytically active, core protein of endoglucanase III from the culture filtrate of Trichoderma reesei. Eur J Biochem 173: 179–183
Stanier RY, Ornston LN 1973 The β-ketoadipate pathway. Adv Microbial Physiol 9: 89–151
Stark N 1972 Nutrient cycling pathways and litter fungi. Bioscience 22: 355–360
Stephens GR, Heichel GH 1975 Agricultural and forest products as sources of cellulose. Biotechnol Bioeng Symp 5: 27–42
Sternberg D 1976 β-Glucosidase of Trichoderma: its biosynthesis and role in saccharification of cellulose. Appl Environ Microbiol 31: 648–654
Sternberg D, Mandels GR 1979 Induction of cellulolytic enzymes in Trichoderma reesei by sophorose. J Bacteriol 139: 761–769
Sternberg D, Vijayakumar P, Reese ET 1977 β-Glucosidase: microbial production and effect of enzymatic hydrolysis of cellulose. Can J Microbiol 23: 139–147
Stewart JC, Lester A, Milburn B, Pany JB 1983 Xylanase and cellulase production by Aspergillus fumigatus fresenius. Biotechnol Lett 5: 543–548
Storvick WO, Cole FE, King KW 1963 Mode of action of a cellulase component from Cellvibrio gilvus. Biochemistry 2: 1106–1110
Streamer M, Eriksson K-E, Pettersson B 1975 Extracellular enzyme system utilized by the fungus Sporotrichum pulverulentum (Chrysosporium lignorum) for the breakdown of cellulose. IV. Functional characterization of five endo-1,4-β-glucanases and one exo-1,4-β-glucanase. Eur J Biochem 59: 607–613
Strömberg L 1984 Mutagenic compounds formed in the chlorination of softwood kraft pulp as determined by the Ames test. Dissert Royal Inst Technol and STFI, Stockholm, 87 pp
Stubblefield SP, Taylor TN 1986 Wood decay in silicified gymnosperms from Antarctica. Bot Gaz 147: 116–125

Stubblefield SP, Taylor TN, Beck CB 1985 Studies of Paleozoic fungi. IV. Wood decaying fungi in Callixylon newberryi from the Upper Devonian. Am J Bot 72: 1765–1774

Stüttgen E, Sahm H 1982 Purification and properties of endo-1,4-β-xylanase from Trichosporon cutaneum. Eur J Appl Microbiol Biotechnol 15: 93–99

Stutzenberger FJ 1972a Cellulolytic activity of Thermomonospora curvata: nutritional requirements for cellulase production. Appl Microbiol 24: 77–82

Stutzenberger FJ 1972b Cellulolytic activity of Thermomonospora curvata: optimal assay conditions, partial purification and product of the cellulase. Appl Microbiol 24: 83–90

Stutzenberger FJ, Kahler G 1986 Cellulase biosynthesis during degradation of cellulose derivatives by Thermomonospora curvata. J Appl Bacteriol 61: 225–233

Stutzenberger FJ, Lupo D 1986 pH-dependent thermal activation of endo-1,4-β-glucanase in Thermomonospora curvata. Enzyme Microb Technol 8: 205–208

Sudo T, Nagayama H, Tamari K 1977 Effect of NaCl on cellulase activities of various strains of Pyricularia oryzae. Agr Biol Chem 41: 597–598

Sugiyama N, Shimahara H, Andoh T, Takemoto M 1973 Konjac-mannanase from the tubers of Amorphophallus konjac C. Koch. Agr Biol Chem 37: 9–17

Sukeno T, Tarentino AL, Plummer TH Jr, Maley F 1972 Purification and properties of α-D- and β-D-mannosidases from hen oviduct. Biochemistry 11: 1493–1501

Sundman G, Kirk TK, Chang H-m 1981 Fungal decolorization of kraft bleach plant effluent. Fate of chromophoric material. Tappi J 64(9): 145–148

Sutherland JB, Blanchette RA, Crawford DL, Pometto AL 1979 Breakdown of Douglas-fir phloem by a lignocellulose-degrading Streptomyces. Curr Microbiol 2: 123–126

Sutherland JB, Crawford DL, Pometto III AL 1983a Metabolism of cinnamic, p-coumaric, and ferulic acids by Streptomyces setonii. Can J Microbiol 29: 1253–1257

Sutherland JB, Crawford DL, Speedie MK 1982 Decomposition of ^{14}C-labeled maple and spruce lignin by marine fungi. Mycologia 74: 511–513

Sutherland JB, Pometto III, AL, Crawford DL 1983b Lignocellulose degradation by Fusarium species. Can J Bot 61: 1194–1198

Suzuki H, Li S-C, Li Y-T 1970 α-Galactosidase from Mortierella vinacea. J Biol Chem 245: 781–786

Szakacs G, Réczey K, Hernadi P, Dobozi M 1981 Penicillium verruculosum WA30, a new source of cellulase. Eur J Appl Microbiol Biotechnol 11: 120–124

Szklarz G, Leonowicz A 1986 Cooperation between fungal laccase and glucose oxidase in the degradation of lignin derivatives. Phytochemistry 25: 2537–2539

Tai D, Terazawa M, Chen C-L, Chang H-m, Kirk TK 1983a Characterization of biodegraded lignins isolated from birch wood decayed by Phanerochaete chrysosporium. Proc Intern Symp Wood Pulp Chem Vol 4, Tsukuba, 144–149

Tai D, Terazawa M, Chen C-L, Chang H-m, Kirk TK 1983b Biodegradation of guaiacyl and guaiacyl-syringyl lignins in wood by Phanerocheate chrysosporium. In: Higuchi T, Chang H-m, Kirk TK (eds) Recent advances in lignin biodegradation research. Uni Publ, Tokyo, 44–63

Tai D, Terazawa M, Huynh V-B, Chen C-L, Chang H-m, Kirk TK 1982 Degradation of lignin in birch wood by the white-rot fungus, Phanerochaete chrysosporium. Proc 1982 TAPPI Res Developm Div Conf, 263–272

Takagi M 1987 Pretreatment of lignocellulosic materials with hydrogen peroxide in presence of manganese compounds. Biotechnol Bioeng 29: 165–170

Takahashi M, Kutsumi S 1979 Purification and properties of xylanase from Gliocladium virens. J Ferment Technol 57: 434–439

Takahashi M, Nishimoto K 1976 Action of soft rot- and white rot fungi on partially delignified softwoods. Wood Res (Kyoto) 59/60: 19–32

Takahashi R, Kusakabe I, Kobayashi H, Murakami K, Maekawa A, Suzuki T 1984 Purification and some properties of mannanase from Streptomyces sp. Agr Biol Chem 48: 2189–2195

Takenishi S, Tsujisaka Y 1973 Structures of the oligosaccharides from the enzymic hydrolysate of rice-straw arabinoxylan by a xylanase of Aspergillus niger. Agr Biol Chem 37: 1385–1391

Takenishi S, Tsujisaka Y, Fukumoto J 1973 Studies on hemicellulases. IV Purification and properties of the β-xylosidase produced by Aspergillus niger van Tieghem. J Biochem 73: 335–343

Tan LUL, Wong KKY, Yu EKC, Saddler JN 1985a Purification and characterization of two D-xylanases from Trichoderma harzianum. Enzyme Microb Technol 7: 425–430

Tan LUL, Wong KKY, Saddler JN 1985b Functional characteristics of two D-xylanases purified from Trichoderma harzianum. Enzyme Microb Technol 7: 431–436

Tanahashi M, Higuchi T 1981 Dehydrogenative polymerization of monolignols by peroxidase and H_2O_2 in a dialysis tube. I. Preparation of highly polymerized DHPs. Wood Res (Kyoto) 67: 29–42

Tanahashi M, Takada S, Aoki T, Goto T, Higuchi T, Hanai S 1982 Characterization of explosion wood. 1. Structure and physical properties. Wood Res (Kyoto) 69: 36–51

Tanaka H, Enoki A, Fuse G 1986 Correlation between ethylene production from α-oxo-γ-methylthiobutyric acid and degradation of lignin dimeric model compounds by wood-inhabiting fungi. Mokuzai Gakkaishi 32: 125–135

Tanaka M, Takenawa S, Matsuno R, Kamikubo T 1977 Purification and properties of cellulases from Pellicularia filamentosa. J Ferment Technol 55: 137–142

Tanaka M, Taniguchi M, Matsuno R, Kamikubo T 1981 Purification and properties of cellulases from Eupenicillium javanicum. J Ferment Technol 59: 177–183

Tanaka T, Yamamoto R, Oi S, Nevins DJ 1982 Purification and some properties of β-transglucosylases of Sclerotinia libertiana, having the ability to synthesize higher cello-oligosaccharides from cellotriose or cellotetraose. Carbohyd Res 106: 131–142

Tangnu SK, Blanch HW, Wilke CR 1981 Enhanced production of cellulase, hemicellulase and β-glucosidase by Trichoderma reesei RUT C-30. Biotechnol Bioeng 23: 1837–1849

Tassinari TH, Macy CF, Spano LA 1982 Technology advances for continuous compression milling pretreatment of lignocellulosics for enzymatic hydrolysis. Biotech Bioeng 24: 1495–1505

Tautorus TE 1985 Mushroom fermentation. Adv Biotechnol Proc 5: 227–273

Taylor BF 1983 Aerobic and anaerobic catabolism of vanillic acid and some other methoxy-aromatic compounds by Pseudomonas sp. strain PN-1. Appl Environ Microbiol 46: 1286–1292

Teeri T, Lehtovaara P, Kauppinen S, Salovuori I, Knowles J 1987 Homologous domains in Trichoderma reesei cellulolytic enzymes: gene sequence and expression of cellobiohydrolase II: Gene 51: 43–52

Teeri T, Salovuori I, Knowles J 1983 The molecular cloning of the major cellulase gene from Trichoderma reesei. Bio/Technology 1: 696–700

Terazawa M, Kayama T, Chen C-L, Tai D 1987 Further characterization of solvent soluble components in the pre-extracted birch chips degraded by the fungus. Phanerochaete chrysosporium. Proc Fourth Intern Symp Wood Pulp Chem Vol 1, Paris, 147–150

Terazawa M, Tai D, Chen C-L, Chang H-m, Kirk TK 1983 Identification of the constituents of low molecular weight fraction obtained from birch wood degraded by Phanerochaete chrysosporium. Proc Intern Symp Wood Pulp Chem Vol 4, 150–155

Thakur RN, Sastry KSM 1981 Leaf blight of Mucuna prurita. Ind Phytopathol 34: 394–395

Thayer DW 1978 Carboxymethylcellulase produced by facultative bacteria from the hind-gut of the termite Reticulitermes hesperus. J Gen Microbiol 106: 13–18

Thomke S, Rundgren M, Eriksson S 1980 Nutritional evaluation of the white-rot fungus Sporotrichum pulverulentum as a feedstuff to rats, pigs, and sheep. Biotechnol Bioeng 22: 2285–2303

Tien M 1987 Properties of ligninase from Phanerochaete chrysosporium and their possible applications. CRC Crit Rev Microbiol 15:2, 141–168

Tien M, Kersten PJ, Kirk TK 1987 Selection and improvement of lignin-degrading microorganisms: potential strategy based on lignin model-amino acid adducts. Appl Environ Microbiol 53: 242–245

Tien M, Kirk TK 1983 Lignin-degrading enzyme from the hymenomycete Phanerochaete chrysosporium Burds. Science 221: 661–663

Tien M, Kirk TK 1984 Lignin-degrading enzyme from Phanerochaete chrysosporium: purification, characterization, and catalytic properties of a unique H_2O_2-requiring oxygenase. Proc Natl Acad Sci USA 81: 2280–2284

Tien M, Kirk TK, Bull C, Fee JA 1986 Steady-state and transient-state kinetic studies on the oxidation of 3,4-dimethoxybenzyl alcohol catalyzed by the ligninase of Phanerochaete chrysosporium Burds. J Biol Chem 261: 1687–1693

Tien M, Tu C-P 1987 Cloning and sequencing of a cDNA for a ligninase from Phanerochaete chrysosporium. Nature 326: 520–523

Timell TE 1962 Enzymatic hydrolysis of a 4-*O*-methylglucuronoxylan from the wood of white birch. Svensk Papperstidn 65: 435–447

Timell TE 1964 Wood hemicelluloses. Part I. Adv Carbohyd Chem 19: 247–302

Timell TE 1965a Wood hemicelluloses. Part II. Adv Carbohyd Chem 20: 409–483

Timell TE 1965b Wood and bark polysaccharides In: Côté WA (ed) Cellular ultrastructure of woody plants, Syracuse University Press, Syracuse, New York, 127–156

Timell TE 1967 Recent progress in the chemistry of wood hemicelluloses. Wood Sci Technol 1: 45–70

Timell TE 1982 Recent progress in the chemistry and topochemistry of compression wood. Wood Sci Technol 16: 83–122

Timell TE 1986 Compression wood in gymnosperms. Springer, Berlin Heidelberg New York 2150 pp

Tomme P, Van Tillbeurgh H, Pettersson G, Van Damme J, Vandekerckhove J, Knowles J, Teeri T, Claeyssens M 1988 Studies of the cellulolytic system of Trichoderma reesei QM9414. Analysis of domain function in two cellobiohydrolases by limited proteolysis. Eur J Biochem 170: 575–581

Tong CC, Cole AL, Shepherd MG 1980 Purification and properties of the cellulases from the thermophilic fungus Thermoascus aurantiacus. Biochem J 191: 83–94

Tönnesen BA, Ellefsen Ö 1971 Sub-microscopical investigations. In: Bikales, N.M., Segal, L. (eds) Cellulose and cellulose derivatives. Wiley, New York, Vol. 5, Part IV, 265–304

Tonon F, Odier E 1988 Influence of veratryl alcohol and hydrogen peroxide on ligninase activity and ligninase production by Phanerochaete chrysosporium. Appl Environ Microbiol 54: 466–472

Towers GHN, Vance CP, Nambudiri AMD 1974 Photoregulation of phenylpropanoid and styrylpyrone biosynthesis. Rec Adv Phytochem 8: 82–94

Toyama N, Ogawa K, Toyama H 1983 Sugar and ethanol production from cellulosic resources with Trichoderma cellulase. Bull Fac Agr Miyazaki Univ 30: 57–78

Tran AV, Chambers RP 1987 Delignification of an unbleached hardwood kraft pulp by Phanerochaete chrysosporium. Appl Microbiol Biotechnol 25: 484–490

Trivedi LS, Rao KK 1980 Factors influencing cellulase induction in Fusarium sp. Curr Microbiol 3: 219–224

Trivedi LS, Rao KK 1981 Production of cellulolytic enzymes by Fusarium species. Biotechnol Lett 3: 281–284

Trojanowski J, Haider K, Sundman V 1977 Decomposition of ^{14}C-labelled lignin and phenols by a Nocardia sp. Arch Microbiol 114: 149–153

Trojanowski J, Hüttermann A 1987 Screening of wood inhabiting fungi for their capacity to degrade and to solubilize ^{14}C-labelled lignin. Microbios 50: 91–97

Trojanowski J, Hüttermann A, Haider K, Wessels JGH 1985 Degradation of lignin and lignin related compounds by protoplasts isolated from Fomes annosus. Arch Microbiol 140: 326–330

Trojanowski J, Leonowicz A 1969 The biodeterioration of lignin by fungi. Microbios 3: 247–251

Trojanowski J, Leonowicz A, Hampel B 1966 Exoenzymes in fungi degrading lignin. II. Demethoxylation of lignin and vanillic acid. Acta Microbiol Polon 15: 17–22

Troller J, Smit JDG, Leisola MSA, Kallen J, Winterhalter KH, Fiechter A 1988 Crystallization of a lignin peroxidase from the white-rot fungus Phanerochaete chrysosporium. Biotechnology 6: 571–573

Trudel P, Courchesne D, Roy C, Chartrand P 1988 Cloning of Trametes versicolor genes induced by nitrogen starvation. Appl Environ Microbiol 54: 1550–1556

Tsang LJ, Reid ID, Coxworth EC 1987 Delignification of wheat straw by Pleurotus spp. under mushroom-growing conditions. Appl Environ Microbiol 53: 1304–1306

Tsujisaka Y, Hiyama K, Takenishi S, Fukomoto J 1972 Studies of the hemicellulases Part III. Purification and some properties of mannanase from Aspergillus niger van Tieghem sp. J Agr Chem Soc Jap 46: 155–161

Uchino F, Nakane T 1981 A thermostable xylanase from a thermophilic acidophilic Bacillus sp. Agr Biol Chem 45: 1121–1227

Ueda K, Ishikawa S, Asai T 1952 Studies on the aerobic mesophilic cellulose-decomposing bacteria, part 5-2. Taxonomical study on genus Pseudomonas. J Agr Chem Soc Jap 26: 35–45
Ulmer DC, Leisola MSA, Fiechter A 1984 Possible induction of the ligninolytic system of Phanerochaete chrysosporium. J Biotechnol 1: 13–24
Umezawa T 1988 Mechanisms for chemical reactions involved in lignin biodegradation by Phanerochaete chrysosporium. Wood Res 75: 21–79
Umezawa T, Higuchi T 1984 Incorporation of $H_2{}^{18}O$ into the C_α but not the C_β position in degradation of a β-O-4 lignin substructure model by Phanerochaete chrysosporium. Agr Biol Chem 48: 1917–1921
Umezawa T, Higuchi T 1985a Role of guaiacol in the degradation of arylglycerol-β-guaiacyl ether by Phanerochaete chrysosporium. FEMS Microbiol Lett 26: 123–126
Umezawa T, Higuchi T 1985b Aromatic ring cleavage in degradation of β-O-4 lignin substructure by Phanerochaete chrysosporium. FEBS Lett 1982: 257–259
Umezawa T, Higuchi T 1985c A novel C_α-C_β cleavage of a β-O-4 lignin model dimer with rearrangement of the β-aryl group by Phanerochaete chrysosporium. FEBS Lett 192: 147–150
Umezawa T, Higuchi T 1985d Arylglycerol formation in degradation of O-4 lignin substructures with and without α-O-γ bonds by Phanerochaete chrysosporium. Wood Res (Kyoto) 71: 25–31
Umezawa T, Higuchi T 1986 Aromatic ring cleavage of β-O-4 lignin model dimers without prior demeth(ox)ylation by lignin peroxidase. FEBS Lett 205: 293–298
Umezawa T, Higuchi T 1987a Mechanism of aromatic ring cleavage of β-O-4 lignin substructure models by lignin peroxidase. FEBS Lett 218: 255–260
Umezawa T, Higuchi T 1987b Formation of a muconate in aromatic ring cleavage of a β-O-4 lignin substructure model by lignin peroxidase. Agr Biol Chem 51: 2281–2284
Umezawa T, Higuchi T 1990 Cleavages of aromatic ring and β-O-4 bond of synthetic lignin (DHP) by lignin peroxidase. FEBS Lett 242: 325–329
Umezawa T, Higuchi T, Nakatsubo F 1983a Difference in $^{18}O_2$ incorporation in oxygenative degradation of β-O-4 and β-1 lignin substructures by Phanerochaete chrysosporium. Agr Biol Chem 47: 2945–2948
Umezawa T, Kawai S, Yokota S, Higuchi T 1986a Aromatic ring cleavage of various β-O-4 lignin model dimers by Phanerochaete chrysosporium. Wood Res (Kyoto) 73: 8–17
Umezawa T, Nakatsubo F, Higuchi T 1982 Lignin degradation by Phanerochaete chrysosporium: metabolism of a phenolic phenylcoumaran substructure model compound. Arch Microbiol 131: 124–128
Umezawa T, Nakatsubo F, Higuchi T 1983b Degradation pathway of arylglycerol-β-aryl ethers by Phanerochaete chrysosporium. Agr Biol Chem 47: 2677–2681
Umezawa T, Shimada M, Higuchi T, Kusai K 1986b Aromatic ring cleavage of β-O-4 lignin substructure model dimers by lignin peroxidase of Phanerochaete chrysosporium. FEBS Lett 205: 287–292
Umezurike GM 1971 Kinetic properties of β-glucosidase from Botryodiplodia theobromae Pat. Biochim Biophys Acta 250: 182–191
Umezurike GM 1975 The subunit structure of β-glucosidase from Botryodiplodia theobromae Pat. Biochem J 145: 361–368
Umezurike GM 1979 The β-glucosidase from Botryodiplodia theobromae. Mechanism of enzyme-catalysed reactions. Biochem J 179: 503–507
Umezurike GM 1981 The β-glucosidase from Botryodiplodia theobromae. Mechanism of enzyme action. Biochem J 199: 203–209
Unligil HH 1972 Penetrability and strength of white spruce after ponding. For Prod J 22: 92–100
Urbanek H, Zalewska-Sobczak J, Borowinska A 1978 Isolation and properties of extracellular cellulase – hemicellulase complex of Phoma hibernica. Arch Microbiol 118: 265–269
US Patent 4275159 (1981) Process for the production of xylose by the enzymatic hydrolysis of xylan (Puls J, Sinner M, Dietrichs H-H) Projektierung chemische Verfahrenstechnik GmbH, Ratingen FRG, 30 pp
Uziie M, Matsuo M, Yasui T 1985a Purification and some properties of Chaetomium trilaterale β-xylosidase. Agr Biol Chem 49: 1159–1166
Uziie M, Matsuo M, Yasui T 1985b Possible identity of β-xylosidase and β-glucosidase of Chaetomium trilaterale. Agr Biol Chem 49: 1167–1173

Vaheri MP 1982a Acidic degradation products of cellulose during enzymatic hydrolysis by Trichoderma reesei. J Appl Biochem 4: 153–160

Vaheri MP 1982b Oxidation as a part of degradation of crystalline cellulose by Trichoderma reesei. J Appl Biochem 4: 356–363

Vaheri M, Leisola M, Kauppinen V 1979a Transglycosylation products of cellulase system of Trichoderma reesei. Biotechnol Lett 1: 41–46

Vaheri MP, Vaheri MEO, Kauppinen VS 1979b Formation and release of cellulolytic enzymes during growth of Trichoderma reesei on cellobiose and glycerol. Eur J Appl Microbiol Biotechnol 8: 73–80

Vallander L, Eriksson K-E 1985 Enzymic saccharification of pretreated wheat straw. Biotechnol Bioeng 27: 650–659

Vallander L, Eriksson K-E 1987 Enzyme recirculation in saccharification of lignocellulosic materials. Enzyme Microb Technol 9: 714–720

Van Arsdell JN, Kwok S, Schweikart VL, Ladner MB, Gelfand DH, Innis MA 1987 Cloning, characterization, and expression in Saccharomyces cervisiae of endoglucanase I from Trichoderma reesei. Biotechnology 5: 60–64

Vance CP, Kirk TK, Sherwood RT 1980 Lignification as a mechanism of disease resistance. Ann Rev Phytopathol 18: 259–288

Vance I, Topham CM, Blayden SL, Tampion J 1980 Extracellular cellulase production by Sporocytophaga myxococcoides NCIB 8639. J Gen Microbiol 117: 235–241

Van der Valk P, Marchant R, Wessels JGH 1977 Ultrastructural localization of polysaccharides in the wall and septum of the basidiomycete Schizophyllum commune. Exp Mycol 1: 69–82

Van Doorslaer E, Kersters-Hilderson H, De Bruyne CK 1980 Binding of alkyl β-D-xylopyranosides, containing branched-chain, cyclic, and substituted aglycon groups, to β-D-xylosidase from Bacillus pumilus PRL B12. Carbohyd Res 78: 317–326

Van Doorslaer E, Kersters-Hilderson H, De Bruyne CK 1985 Hydrolysis of β-D-xylose-oligosaccharides by β-D-xylosidase from Bacillus pumilus. Carbohyd Res 140: 342–346

Van Tilbeurgh H, Bhikhabhai R, Pettersson LG, Claeyssens M 1984 Separation of endo- and exo-type cellulases using a new affinity chromatography method. FEBS Lett 169: 215–218

Van Tilbeurgh H, Claeyssens M, De Bruyne CK 1982 The use of 4-methylumbelliferyl and other chromophoric glycosides in the study of cellulolytic enzymes. FEBS Lett 149: 152–156

Van Tilbeurgh H, Pettersson G, Bhikhabhai R, De Boeck H, Claeyssens M 1985 Studies of the cellulolytic system of Trichoderma reesei QM9414. Reaction specificity and thermodynamics of interactions of small substrates and ligands with the 1,4-β-glucan-cellobiohydrolase II. Eur J Biochem 148: 329–334

Van Tilbeurgh H, Tomme P, Claeyssens M, Bhikhabhai R, Pettersson G 1986 Limited proteolysis of the cellobiohydrolase I from Trichoderma reesei. FEBS Lett 204: 223–227

Varadi J 1972 The effect of aromatic compounds on cellulase and xylanase production of the fungi Schizophyllum commune and Chaetomium globosum. In: Walters AH, Hueck-van der Plas EH (eds) Biodeterioration of materials, Vol 2. Applied Science, London, 129–135

Vardanis A, Finkelman M 1981 A radiometric microassay for cellulase activity. Anal Biochem 115: 78–80

Vasseur E 1948 Spectrophotometric study on the orcinol reaction with carbohydrates. Acta Chem Scand 2: 693–701

Vasudevan B, Joyce TW, Chang H-m 1986 The synergistic effects of nitrogen and mineral in the presence of Tween 80 on decolorization of bleach plant effluents using the Mycor process. Proc TAPPI Res Develop Conf, 59–64

Vicuña R, González B, Mozuch MD, Kirk TK 1987 Metabolism of lignin model compounds of the arylglycerol-β-aryl ether type by Pseudomonas acidovorans D_3. Appl Environ Microbiol 53: 2605–2609

Viikari L, Ranua M, Kantelinen A, Linko M, Sundquist J 1987 Application of enzymes in bleaching. Proc 4th Intern Symp Wood Pulp Chem Vol 1, Paris, 151–154

Viikari L, Ranua M, Kantelinen A, Sundquist J, Linko M 1986 Bleaching with enzymes. Proc 3rd Intern Conf Biotechnol in the Pulp and Paper Ind, Stockholm, 67–69

Viljoen JA, Fred ED, Peterson WH 1926 The fermentation of cellulose by thermophilic bacteria. J Agr Sci 16: 1–17

Villarroya H, Petek F 1976 Purification and properties of a β-mannanase from alfalfa seeds. Biochim Biophys Acta 438: 200–211

Vogels GD 1979 The global cycle of methane. Antonie Leeuwenhoek. J Microbiol Serol 45: 347–352

Vogt KA, Edmonds RL 1980 Patterns of nutrient concentration in basidiocarps in western Washington. Can J Bot 58: 694–698

Vogt KA, Edmonds RL, Grier CC 1981 Biomass and nutrient concentrations of sporocarps produced by mycorrhizal and decomposer fungi in Abies amabilis. Oecologia 50: 170–175

Volc J, Denisova NP, Nerud F, Musilek V 1985 Glucose-2-oxidase activity in mycelial cultures of basidiomycetes. Folia Microbiol 30: 141–147

von Aufsess H, von Pechmann H, Graessle H 1968 Fluoreszenzmikroskopische Beobachtungen an pilzbefallenem Holz. Holz Roh-Werkst 26: 50–61

von Hofsten B, von Hofsten A 1974 Ultrastructure of a thermotolerant basidiomycete possibly suitable for production of food protein. Appl Microbiol 27: 1142–1148

von Peek R-D, Liese W, Parameswaran N 1972 Infektion und Abbau des Wurzelholzes von Fichte durch Fomes annosus. Eur J For Pathol 2: 237–248

Vrsanska M, Gorbacheva IV, Kratky Z, Bieley P 1982 Reaction pathways of substrate degradation by an acidic endo-1,4-β-xylanase of Aspergillus niger. Biochim Biophys Acta 704: 114–122

Wakabayashi K, Nisizawa K 1964 On the cellulase system of Irpex lacteus. J Ferment Technol 42: 347–355

Waksman SA 1967 The Actinomycetes, a summary of current knowledge. Ronald, New York, 280 pp

Waldner R 1987 Ligninolytic systems of white-rot fungi. Dissert, Swiss Federal Institute of Technology, Zürich, 91 pp

Waldner R, Leisola MSA, Fiechter A 1988 Comparison of ligninolytic activities of selected fungi. Appl Microbiol Biotechnol 29: 400–407

Walling C 1982 The nature of the primary oxidants in oxidations mediated by metal ions. In: King TE, Mason HS, Morisson M (eds) Oxidases and related redox systems. Pergamon, New York, 85–97

Walseth CS 1952 Occurrence of cellulases in enzyme preparations from microorganisms. Tappi 35: 228–233

Wan CC, Muldrey JE, Li S-C, L Y-T 1976 β-Mannosidase from the mushroom Polyporus sulfureus. J Biol Chem 251: 4384–4388

Wariishi H, Akileswaran L, Gold MH 1988 Manganese peroxidase from the basidiomycete Phanerochaete chrysosporium: spectral characterization of the oxidized states and the catalytic cycle. Biochemistry 27: 5365–5370

Wariishi H, Morohoshi N, Haraguchi T 1985 Degradation of β-O-4 lignin substructure model compound by an extracellular phenol oxidase of Coriolus versicolor. Abstr Lignin Biodegrad Workshop Vancouver, B.C, 28 pp

Wariishi H, Morohoshi N, Haraguchi T 1987 Degradation of lignin by the extracellular enzymes of Coriolus versicolor VII. Effective degradation of syringyl type β-aryl ether lignin model compound by laccase III. Mokuzai Gakkaishi 33: 892–898

Warren RAJ, Beck CF, Gilkes NR, Kilburn DG, Langsford ML, Miller RC Jr, O'Neill GP, Wong WKR 1987 Unusual structural organization of two cellulases from Cellulomonas fimi. In: Kennedy JF, Philips GO, Williams PA (eds) Proc Cellucon '86: Wood and cellulosics. Ellis Horwood, Chichester, 263–266

Webb EC 1984 Enzyme Nomenclature 1984. Academic Press, Orlando, 646 pp

Wegener G 1982 Die Rolle des Holzes als Chemierohstoff und Energieträger. Teil 2. Verwertungsmöglichkeiten für Cellulose, Polyosen und Lignin. Holz Roh-Werkst 40: 209–214

Weinstein L, Albersheim P 1979 Structure of plant cell walls. IX Purification and partial characterization of a wall-degrading endo-arabanase and an arabinosidase from Bacillus subtilis. Plant Physiol 63: 425–432

Weinstein DA, Krisnangkura K, Mayfield MB, Gold MH 1980 Metabolism of radiolabeled β-guaiacyl ether-linked lignin dimeric compounds by Phanerochaete chrysosporium. Appl Environ Microbiol 39: 535–540

Westermark U 1982 Calcium promoted phenolic coupling by superoxide radical – a possible lignification reaction in wood. Wood Sci Technol 16: 71–78

Westermark U 1985 Bromination of different morphological parts of spruce wood (Picea abies). Wood Sci Technol 19: 323–328

Westermark U, Eriksson K-E 1974a Carbohydrate-dependent enzymic quinone reduction during lignin degradation. Acta Chem Scand B28: 204–208

Westermark U, Eriksson K-E 1974b Cellobiose: quinone oxidoreductase, a new wood-degrading enzyme from white-rot fungi. Acta Chem Scand B28: 209–214

Westermark U, Eriksson K-E 1975 Purification and properties of cellobiose: quinone oxidoreductase from Sporotrichum pulverulentum. Acta Chem Scand B29: 419–424

Westermark U, Eriksson K-E 1988 Cellobiose dehydrogenase (quinone). Methods Enzymol 160: 463–468

Whitaker DR, Thomas R 1963 Improved procedures for preparation and characterization of Myrothecium cellulase. Part I. Production of enzyme. Can J Biochem Physiol 41: 667–670

White AR, Brown RM Jr 1981 Enzymatic hydrolysis of cellulose: visual characterization of the process. Proc Natl Acad Sci USA 78: 1047–1051

Wicklow DT, Detroy RW, Adams S 1980a Differential modification of the lignin and cellulose components in wheat straw by fungal colonists of ruminant dung: ecological implications. Mycologia 72: 1065–1076

Wicklow DT, Detroy RW, Jessee BA 1980b Decomposition of lignocellulose by Cyathus stercoreus (Schw) de Toni NRRL 6473, a "white rot" fungus from cattle dung. Appl Environ Microbiol 40: 169–170

Wicklow DT, Langie R, Crabtree S, Detroy RW 1984 Degradation of lignocellulose in wheat straw versus hardwood by Cyathus and related species (Nidulariaceae). Can J Microbiol 30: 632–636

Wiegel J, Carreira LH, Mothersed CP, Puls J 1983 Production of ethanol from biopolymers by anaerobic, thermophilic and extreme thermophilic bacteria II. Thermoanaerobacter ethanolicus JW 200 and its mutants in batch cultures and resting cell experiments. Biotechnol Bioeng Symp 13: 193–205

Wiegel J, Dykstra M 1984 Clostridium thermocellum: adhesion and sporulation while adhered to cellulose and hemicellulose. Appl Microbiol Biotechnol 20: 59–65

Wiegel J, Ljungdahl LG 1986 The importance of thermophilic bacteria in biotechnology. CRC Crit Rev Biotechnol 3: 39–107

Wiegel J, Mothersed CP, Puls J 1985 Differences in xylan degradation by various non-cellulolytic thermophilic anaerobs and Clostridium thermocellum. Appl Environ Microbiol 49: 656–659

Wilcox WW 1968 Changes in wood microstructure through progressive stages of decay. US For Serv Res Pap FPL-70, 46 pp

Wilcox WW, Parameswaran N, Liese W 1974 Ultrastructure of brown-rot in wood treated with pentachlorophenol. Holzforschung 28: 211–217

Wilke CR, Yang RD, Sciamanna AF, Freitas RP 1981 Raw materials evaluation and process development studies for conversion of biomass to sugars and ethanol. Biotechnol Bioeng 23: 163–183

Willershausen H, Jäger A, Graf H 1987 Ligninase production of Phanerochaete chrysosporium by immobilization in bioreactors. J Biotechnol 6: 239–243

Williams AG, Withers SE 1981 Hemicellulose-degrading enzymes synthesized by rumen bacteria. J Appl Bacteriol 51: 375–385

Willison JHM, Brown RM Jr 1977 An examination of the developing cotton fiber: wall and plasmalemma. Protoplasma 92: 21–41

Willison JHM, Rowe AJ 1980 Replica, shadowing and freeze-etching techniques. In: Glauert AM (ed) Practical methods in electron microscopy. North Holland, Amsterdam, 300 pp

Willoughby GA, Leightley LE 1984 Patterns of bacterial decay in preservative treated Eucalypt power transmission poles. Intern Res Group on Wood Preserv Document IRG/WP/1223

Win M, Matsuo M, Yasui T 1987 Immunological relationships of four types of β-xylosidase from Penicillium wortmanni IFO 7237. Agr Biol Chem 51: 3151–3152

Winterbourn CC 1987 The ability of scavengers to distinguish OH production in the iron-catalyzed Haber-Weiss reaction: comparison of four assays for $OH^{\bullet}$. Free Radical Biol Medic 3: 33–39

Winston GW, Cederbaum AI 1982 Oxidative decarboxylation of benzoate to carbon dioxide by rat liver microsomes: a probe for oxygen radical production during microsomal electron transfer. Biochemistry 21: 4265–4270

Withers SG, Dombroski D, Berven LA, Kilburn DG, Miller RC Jr, Warren RAJ, Gilkes NR 1986 Direct ^{1}H NMR determination of the stereochemical course of hydrolysis catalysed by glucanase components of the cellulase complex. Biochem Biophys Res Commun 139: 487–494

Wojtas-Wasilewska M, Luterek J 1987 The effect of fungal protocatechuate 3,4-dioxygenase on sodium lignosulphonate fractions. Phytochemistry 26: 2671–2674

Wojtas-Wasilewska M, Luterek J, Leonowicz A, Dawidowicz A 1988 Dearomatization of lignin derivatives by fungal protocatechuate 3,4-dioxygenase immobilized on porosity glass. Biotechnol Bioeng 32: 507–511

Wojtas-Wasilewska M, Trojanowski J 1980 Purification and properties of protocatechuate 3,4-dioxygenase from Chaetomium piluliferum induced with p-hydroxybenzoic acid. Acta Biochim Polon 27: 21–34

Wojtas-Wasilewska M, Trojanowski J, Luterek J 1980 Biotransformation of sodium lignosulfonates of different molecular weights by the fungus Pleurotus ostreatus. Acta Microbiol Polon 29: 353–364

Wojtas-Wasilewska M, Trojanowski J, Luterek J 1983 Aromatic ring cleavage of protocatechuic acid by the white-rot fungus Pleurotus ostreatus. Acta Biochim Polon 30: 291–301

Wolff BR, Mudry TA, Glick BR, Pasternak JJ 1986 Isolation of endoglucanase genes from Pseudomonas fluorescens subsp. cellulosa and a Pseudomonas sp. Appl Environ Microbiol 56: 1367–1369

Wolter KE, Highley TL, Evans FJ 1980 A unique polysaccharide- and glycoside-degrading enzyme complex from the wood decay fungus Poria placenta. Biochem Biophys Res Commun 97: 1499–1504

Wong KKY, Tan LUL, Saddler JN 1986 Purification of a third distinct xylanase from the xylanolytic system of Trichoderma harzianum. Can J Microbiol 32: 570–576

Wood DA 1980 Production, purification and properties of extracellular laccase of Agaricus bisporus. J Gen Microbiol 117: 327–338

Wood DA 1985 Useful biodegradation of lignocellulose. Ann Proc Phytochem Soc Eur 26: 295–309

Wood DA, Goodenough PW 1977 Fruiting of Agaricus bisporus. Changes in extracellular enzyme activities during growth and fruiting. Arch Microbiol 114: 161–165

Wood JM, Folsom BR, Eudy WW 1986 Biomimetic oxidation of lignin model compounds by substituted porphyrins. Proc 3rd Intern Conf Biotechnol in the Pulp and Paper Ind, Stockholm, 31–35

Wood TM 1969 The cellulase of Fusarium solani. Resolution of the enzyme complex. Biochem J 115: 457–464

Wood TM 1971 The cellulase of Fusarium solani. Purification and specificity of the β-1,4-glucanase and the β-D-glucosidase components. Biochem J 121: 353–362

Wood TM 1975 Properties and mode of action of cellulases. Biotechnol Bioeng Symp 5: 111–137

Wood TM 1980 Enzymes and mechanisms involved in fungal degradation of cellulosic materials. O.E.C.D. Workshop, Conversion of lignocellulosic materials to simple carbohydrates. Amersfoort, Holland, Oct 8–10, 247–269

Wood TM 1983 Biochemistry of cellulase complex. In: Proceedings from the Seminar, La biomasse, source d'intermédiaires industriels, ADEPRINA, Paris

Wood TM 1985a Observations and speculations on the complex interactions involved in the solubilization of native cellulose. 16th FEBS meeting Moscow, 251–256

Wood TM, McCrae SI 1972 The purification and properties of the C_1 component of Trichoderma koningii cellulase. Biochem J 128: 1183–1192

Wood TM, McCrae SI 1975 The cellulase complex of Trichoderma coningi. In: Bailey M, Enari T-M, Linko M (eds) Symposium on enzymatic hydrolysis of cellulose. SITRA, Helsinki, 231–254

Wood TM, McCrae SI 1977a Cellulase from Fusarium solani: purification and properties of the C_1 component. Carbohyd Res 57: 117–133

Wood TM, McCrae SI 1977b The mechanism of cellulase action with particular reference to the C_1-component. In: Ghose TK (ed) Bioconversion of cellulosic substances into energy chemicals and microbial protein. Symposium IIT, New Delhi, 111–141
Wood TM, McCrae SI 1978 The cellulase of Trichoderma koningii, purification and properties of some endoglucanase components with special reference to their action on cellulose when acting alone and in synergisms with the cellobiohydrolase. Biochem J 171: 61–72
Wood TM, McCrae SI 1979 Synergism between enzymes involved in the solubilization of native cellulose. Adv Chem Ser 181: 181–209
Wood TM, McCrae SI 1982 Purification and some properties of a (1,4)-β-D-glucan glucohydrolase associated with the cellulase from the fungus Penicillium funiculosum. Carbohyd Res 110: 291–303
Wood TM, McCrae SI 1986a The cellulase of Penicillium pinophilum: synergism between enzyme components in solubilizing cellulose with special reference to the involvement of two immunologically distinct cellobiohydrolases. Biochem J 234: 93–99
Wood TM, McCrae SI 1986b Studies of two low-molecular-weight endo(1,4)-β-D-xylanases constitutively synthesized by the cellulolytic fungus Trichoderma koningii. Carbohyd Res 148: 321–330
Wood TM, McCrae SI 1986c Purification and properties of a cellobiohydrolase from Penicillium pinophilum. Carbohyd Res 148: 331–344
Wood TM, McCrae SI, MacFarlane CC 1980 The isolation, purification, and properties of the cellobiohydrolase component of Penicillium funiculosum cellulase. Biochem J 189: 51–65
Wood TM, Phillips DR 1969 Another [fungal] source of cellulases. Nature 222: 986–987
Wood TM, Wilson CA 1984 Some properties of the endo-(1,4)-β-D-glucanase synthesized by the anaerobic cellulolytic rumen bacterium Ruminococcus albus. Can J Microbiol 30: 316–321
Wood TM, Wilson CA, McCrae SI, Joblin KN 1986 A highly active extracellular cellulase from the anaerobic rumen fungus Neocallimastix frontalis. FEMS Microbiol Lett 34: 37–40
Woodward J 1984 Xylanases: functions, properties and applications. Top Enzyme Ferment Biotechnol 8: 9–30
Wu JHD, Demain AL 1988 Proteins of the Clostridium thermocellum cellulase complex responsible for degradation of crystalline cellulose . In: Aubert J-P, Beguin P, Millet J (eds) Biochemistry and genetics of cellulose degradation, FEMS Symp no 43. Academic Press, London, 117–131
Wynne EC, Pemberton JM 1986 Cloning of a gene cluster from Cellvibrio mixtus which codes for cellulase, chitinase, amylase, and pectinase. Appl Environ Microbiol 52: 1362–1367
Yaguchi M, Roy C, Rollin CF, Paice MG, Jurasek L 1983 A fungal cellulase shows sequence homology with the active site of hen egg-white lysozyme. Biochem Biophys Res Commun 116: 408–411
Yajima Y, Enoki A, Mayfield MB, Gold MH 1979 Vanillate hydroxylase from the white rot Basidiomycete Phanerochaete chrysosporium. Arch Microbiol 123: 319–321
Yamane K, Suzuki H, Nisizawa K 1970 Purification and properties of extracellular and cell-bound cellulase components of Pseudomonas fluorescence var. cellulosa. J Biochem 67: 19–35
Yamazaki Von N, Sinner M, Dietrichs HH 1976. Isolierung und Eigenschaften einer β-1,4-Mannanase aus Aspergillus niger. Holzforschung 30: 101–109
Yang H-h, Effland MJ, Kirk TK 1980 Factors influencing fungal degradation of lignin in a representative lignocellulosic, thermomechanical pulp. Biotechnol Bioeng 22: 65–77
Yasui T, Nguyen F, Nakanishi K 1984 Inducers for xylanase production by Cryptococcus flavus. J Ferment Technol 62: 353–359
Yokota S, Umezawa T, Higuchi T 1988 Influence of alkoxyl ring substituents on the degradability of β-O-4 lignin substructure model dimers by Phanerochaete chrysosporium. Mokuzai Gakkaishi 34: 65–74
Yoshihara K, Kamijima H, Akamatsu I 1987 Effect of urea on lignin degradation by Coriolus hirsutus. Mokuzai Gakkaishi 33: 61–70
Yoshihara K, Kobayashi T, Fujii T, Akamatsu I 1984 A novel modification of Klason lignin quantitative method. J Jap Tappi 38: 466–475

Yoshikawa T, Suzuki H, Nisizawa K 1974 Biogenesis of multiple cellulase components of Pseudomonas fluorescence var. cellulosa I. Effects of culture conditions on the multiplicity of cellulase. J Biochem 75: 531–540

Yoshikawa T, Suzuki H, Nisizawa K 1975 Biogenesis of multiple cellulase components of Pseudomonas fluorescens var. cellulosa. II. In vitro changes in multiplicity patterns of cellulase components. Sci Rep Tokyo Kyoiku Daigaku 16(242): 87–104

Yoshioka H, Nagato N, Chavanich S, Nilubol N, Hayashida S 1981 Purification and properties of thermostable xylanase from Talaromyces byssochlamydoides YH-50. Agr Biol Chem 45: 2425–2432

Yu H-s, Eriksson K-E 1985 Influence of oxygen on the degradation of wood and straw by white-rot fungi. Svensk Papperstidn 88: R57–R60

Zaar K 1979 Visualization of pores (export sites) correlated with cellulose production in the envelope of the gram-negative bacterium Acetobacter xylinum. J Cell Biol 80: 773–777

Zabel R, Lombard FF, Wang CJK, Terracina F 1985 Fungi associated with decay in treated southern pine utility poles in the eastern United States. Wood Fiber Sci 17: 75–91

Zadrazil F 1985 Screening of fungi for lignin decomposition and conversion of straw into feed. Angew Botanik 59: 433–452

Zadrazil F, Brunnert H 1980 The influence of ammonium nitrate supplementation on degradation and in vitro digestibility of straw colonized by higher fungi. Eur J Microbiol Biotechnol 9: 37–44

Zadrazil F, Grinbergs J, Gonzales A 1982 "Palo Podrido" – decomposed wood used as feed. Eur J Appl Microbiol Biotechnol 15: 167–171

Zainal AA 1976 The soft rot fungi: The effect of lignin. Mater Org 3: 121–127

Zeikus JG 1981 Lignin metabolism and the carbon cycle: polymer biosynthesis, biodegradation, and environmental recalcitrance. Adv Microb Ecol 5: 211–243

Zeikus JG, Wellstein AL, Kirk TK 1982 Molecular basis for the biodegradative recalcitrance of lignin in anaerobic environments. FEMS Microbiol Lett 15: 193–197

Zhang YZ, Zylstra GJ, Olsen RH, Reddy CA 1986 Identification of cDNA clones for ligninase from Phanerochaete chrysosporium using synthetic oligonucleotide probes. Biochem Biophys Res Commun 13: 649–656

Zhu YS, Wu YQ, Chen W, Tan C, Gao JH, Fei JX, Shih CN 1982 Induction and regulation of cellulase synthesis in Trichoderma pseudokoningii mutants EA_3-867 and N_2-78. Enzyme Microb Technol 4: 3–12

Subject Index

MIX
Papier aus verantwortungsvollen Quellen
Paper from responsible sources
FSC® C105338
FSC
www.fsc.org

If you have any concerns about our products,
you can contact us on
ProductSafety@springernature.com

In case Publisher is established outside the EU,
the EU authorized representative is:
Springer Nature Customer Service Center GmbH
Europaplatz 3, 69115 Heidelberg, Germany

Printed by Libri Plureos GmbH
in Hamburg, Germany